Serum Free Light Chain Analysis
(plus Hevylite)
6th Edition

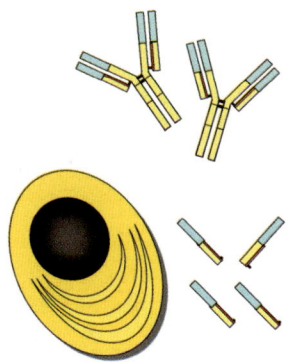

Distributors:
The Binding Site Group Ltd., P.O. Box 11712, Birmingham, B14 4ZB, UK.
The Binding Site Inc.: 5889 Oberlin Drive, Suite 101, San Diego, Ca 92121, USA.
The Binding Site GmbH, Robert-Bosch-Str. 2A, D-68723 Schwetzingen, Germany.
The Binding Site, Centre Atoll, 14 rue des Glairaux, BP 226, 38522 Saint Egrève, France.
The Binding Site, Balmes 243 4º 3ª, 08006 Barcelona, Spain.
The Binding Site s.r.o., Sinkulova 55, 140 00 Prague 4, Czech Republic.
The Binding Site SPRL/BVBA, Brusselsstraat 51, 2018 Antwerpen, Belgium.

Published by The Binding Site Group Ltd., PO Box 11712, Birmingham, B14 4ZB, UK
Printed in UK by HSW Ltd., Rhondda, Wales, UK
This book was produced using QuarkXpress 8 and Powerpoint 2003
A CIP record for this book is available from the British Library.

ISBN: 9780704427969

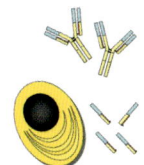

Serum Free Light Chain Analysis

(plus Hevylite)

6th Edition

AR Bradwell MB ChB, FRCP, FRCPath.

Professor of Immunology, The Medical School, University of Birmingham, B15 2TT, UK and The Binding Site Group Ltd., PO Box 11712, B14 4ZB, UK.

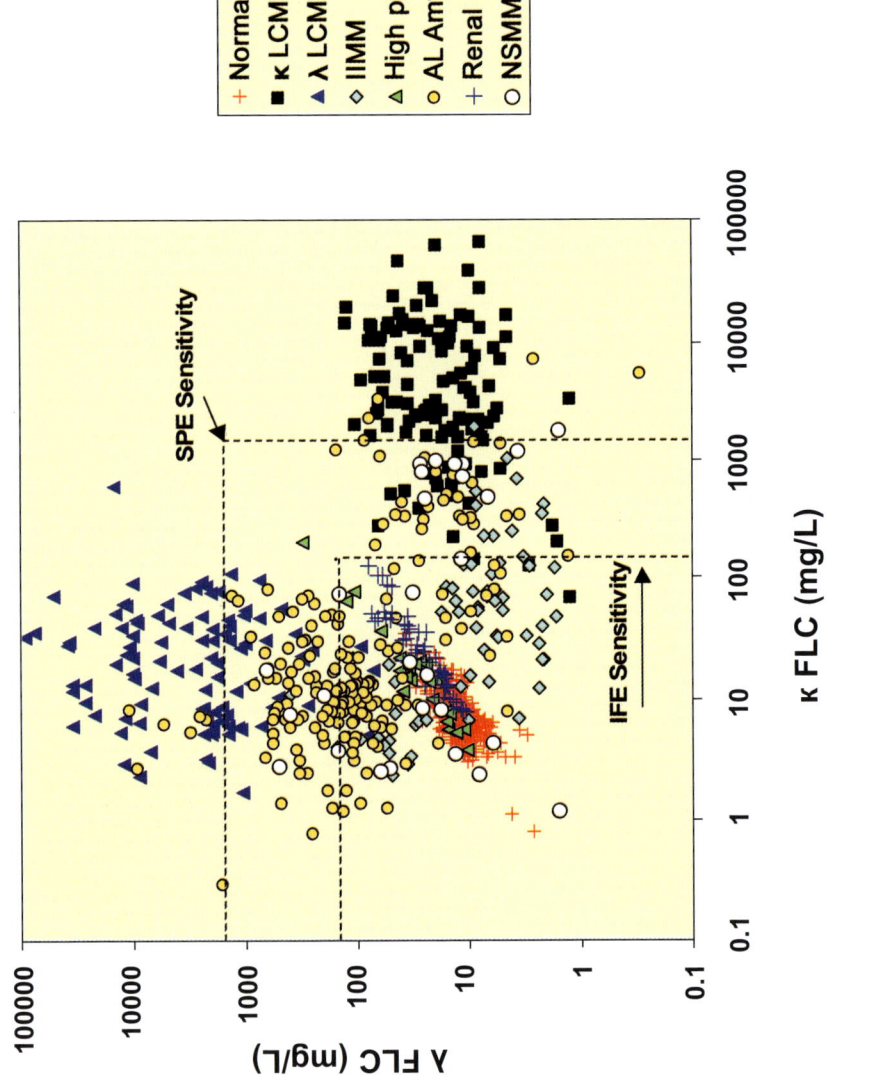

Composite figure of serum free light chain concentrations in various diseases, LCMM = light chain multiple myeloma; IIMM = intact immunoglobulin multiple myeloma; High pIgG = polyclonal hypergammaglobulinaemia; NSMM = nonsecretory multiple myeloma.

Contents

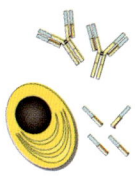

Contents	v
Preface to sixth edition 2010	xiii
Wikilite.com	xv
Dedication	xv
Acknowledgements	xv
Abbreviations	xvi
Overview - The clinical importance of serum free light chain analysis	xviii

SECTION 1

Immunoglobulin free light chains and their analysis	1
1 Introduction	1
2 Dr Bence Jones and the history of free light chains	6
3 Biology of immunoglobulin light chains	10
3.1 Structure	10
3.2 Synthesis	12
3.3 Production	12
3.4 Clearance and metabolism	14
4 Immunoassays for free light chain measurement	18
4.1. Introduction	18
4.2. Immunoassays for FLCs	20
A. Polyclonal antisera versus monoclonal antibodies	20
B. Antisera specificity	20
C. Accuracy and standardisation	23
D. Assay sensitivity	24
E. Assay ranges	25
F. Antigen excess	25
G. Precision	27
H. Linearity	27
I. FLC polymerisation	29
J. Stability	30
K. Serum and plasma comparisons for FLC assays	30
4.3. Comparison of FLC immunoassays on different instruments	31
4.4. Comparison of different serum FLC immunoassays	32
4.5. Quality control of FLC antisera and kits	32
4.6. Scale-up of antiserum production	34

5 Normal ranges and reference intervals   **36**

5.1.   Serum free light chain (sFLC) normal ranges   **36**

5.2.   Normal ranges in children   **39**

5.3.   Variations in normal/reference ranges - hospital ranges   **39**

5.4.   Borderline results   **41**

5.5.   Utility of sFLC assays for disease diagnosis   **41**

5.6.   Urine free light chain (uFLC) normal ranges   **42**

6 Sensitivity of serum free light chain assays   **45**

6.1.   Introduction   **45**

6.2.   Serum and urine protein electrophoresis (SPE and UPE)   **46**

6.3.   Serum immunofixation electrophoresis (sIFE)   **47**

6.4.   Capillary zone electrophoresis (CZE)   **49**

6.5.   Total κ and λ immunoassays   **50**

6.6.   Urine tests for free light chains (FLCs)   **50**

6.7.   Discordant serum and urine FLC results   **52**

SECTION 2

Free light chains and monoclonal gammopathies   **57**

SECTION 2A. *(Chapters 7-14)*

7 Multiple myeloma (MM) - Introduction   **59**

8 Light chain multiple myeloma (LCMM)   **61**

8.1.   Diagnosis of LCMM using sFLCs   **61**

8.2.   Monitoring LCMM using sFLCs   **64**

   Clinical case history No 1   **66**

   Clinical case history No 2   **68**

9 Nonsecretory multiple myeloma (NSMM)   **69**

9.1.   Introduction   **69**

9.2.   Diagnosis of nonsecretory multiple myeloma (NSMM)   **70**

9.3.   Monitoring NSMM   **72**

   Clinical case history No 3   **73**

   Clinical case history No 4   **75**

10 Intact immunoglobulin multiple myeloma (IIMM)-theoretical considerations of sFLC and Ig measurements   **77**

10.1.   Introduction   **77**

10.2.   Half-life of sFLCs   **78**

10.3.   Half-life of IgG, IgA and IgM and recyling by FcRn receptors   **79**

10.4.   Tumour killing rates and half-lives of immunoglobulins   **80**

10.5.   Blood volume changes in monoclonal gammopathies   **82**

10.6.   Changes in sFLCs and immunoglobulins during treatment   **83**

11 Intact immunoglobulin multiple myeloma (IIMM) - sFLCs at diagnosis   **86**
- 11.1. Introduction   **86**
- 11.2. Myeloma cell diversity: production of monoclonal Igs and FLCs   **86**
- 11.3. sFLC concentrations at disease presentation   **88**
- 11.4. Disease stage and sFLCs   **91**

12 Monitoring patients with IIMM using sFLCs   **95**
- 12.1. Introduction   **95**
- 12.2. Overview of sFLCs for monitoring patients with IIMM   **95**
- 12.3. Bone marrow responses and short sFLC half life   **97**
- 12.4. Speed of response using sFLCs   **98**
- 12.5. Complete response and residual disease   **100**
- 12.6. sFLCs during disease relapse   **102**
- 12.7. Free light chain escape (breakthough)   **104**
- 12.8. sFLCs after high dose melphalan and bone marrow transplant   **105**
- 12.9. Long term variations in sFLC levels after high dose therapy   **107**
- 12.10. Drug selection using rapid responses in sFLCs   **108**
- 12.11. Utility of sFLCs in IIMM   **109**

13 The kidney and monoclonal free light chains   **111**
- 13.1. Introduction   **111**
- 13.2. Normal FLC clearance and metabolism   **112**
- 13.3. Nephrotoxicity of monoclonal FLCs   **113**
- 13.4. Diagnosis of myeloma kidney using sFLC analysis   **116**
- 13.5. Removal of FLCs by plasma exchange   **117**
- 13.6. Model of FLC removal by plasma exchange and haemodialysis   **117**
- 13.7. Removal of FLCs by haemodialysis   **119**
- 13.8. Recovery from renal failure following FLC removal by haemodialysis   **121**

14 Asymptomatic (smouldering) multiple myeloma (ASMM)   **126**

SECTION 2B. *(Chapters 15-17)*

Diseases with monoclonal light chain deposition   **130**

15 AL Amyloidosis   **130**
- 15.1. Introduction   **130**
- 15.2. sFLCs at disease presentation   **132**
- 15.3. Initial sFLC measurements correlate with survival   **137**
-      Clinical case history No 5   **137**
- 15.4. Monitoring patients with AL amyloidosis   **138**
-      Clinical case history No 6   **142**
- 15.5. Stem cell transplantation and sFLCs   **143**
- 15.6. Solid organ transplantation and sFLCs   **144**

15.7. Brain naturetic peptide (NT-proBNP) and sFLCs in AL
amyloidosis   144
15.8. sFLCs in renal failure complicating AL amyloidosis   145
Clinical case history No 7   145

16 Localised amyloid disease   **148**

17 Light chain deposition disease (LCDD)   **149**
17.1. Introduction   149
17.2. Diagnosis of LCDD using sFLC assays   149
Clinical case history No 8   150
17.3. Monitoring LCDD using sFLC assays   151
Clinical case history No 9   152

SECTION 2C. *(Chapters 18-19)*
Other diseases with monoclonal free light chains   **154**

18 Other malignancies with monoclonal FLCs   **154**
18.1. Solitary plasmacytoma of bone   154
18.2. Extramedullary plasmacytoma   156
18.3. Multiple solitary plasmacytoma (+/- recurrent)   156
18.4. Plasma cell leukaemia   157
18.5. Waldenström's macroglobulinaemia   157
18.6. B-cell, non-Hodgkin lymphomas   159
B-cell, non-Hodgkin lymphoma complicated by AL amyloidosis 162
18.7. B-cell, chronic lymphocytic leukaemia   162
18.8. POEMS syndrome   164
18.9. Cryoglobulinaemia   165

**19 Monoclonal gammopathies of undetermined
significance (MGUS)**   **166**
19.1. MGUS: Definition and frequency   166
19.2. MGUS and monoclonal FLCs   167
19.3. Risk stratification of MGUS using serum FLC concentrations 168
19.4. Serum free light chain MGUS   170
19.5. MGUS as the precursor condition for MM   173

SECTION 3
Diseases with increased polyclonal free light chains   **175**

20 Renal diseases and free light chains   **175**
20.1. Introduction   175
20.2. Effect of renal impairment on sFLC concentrations   176

20.3.   Effect of renal impairment on serum κ/λ ratios   178
20.4.   Removal of FLCs by dialysis in chronic renal failure (CRF)   179
20.5.   Monoclonal FLCs in CRF   181
        Clinical case history No 10   182

21 Immune stimulation and elevated polyclonal free light chains   184
21.1.   Introduction   184
21.2.   Rheumatic diseases   184
            Systemic Lupus Erythematosus (SLE)   185
            Primary Sjögren's Syndrome (pSS)   187
            Rheumatoid Arthritis (RA)   188
            Dermatitis   188
21.3.   Diabetes mellitus   189
21.4.   Infectious diseases with elevated polyclonal sFLCs   190
21.5.   Lymphoma   192
21.6.   Free light chains as bioactive molecules in inflammatory
        diseases   192

22 Cerebrospinal fluid and free light chains   194

SECTION 4

General applications of free light chain assays   197

23 Screening studies using serum free light chain analysis   197
23.1.   Introduction   197
23.2.   Diagnostic protocols for monoclonal gammopathies   198
23.3.   Screening studies using sFLC analysis   199
23.4.   Audit of sFLC usage   207

24 Serum versus urine tests for free light chains   208
24.1.   Introduction   208
24.2.   Renal threshold for FLC excretion   208
24.3.   Problems collecting satisfactory urine samples   210
24.4.   Problems measuring urine samples   211
24.5.   Clinical benefits of sFLC analysis   212
24.6.   Elimination of urine studies when screening for
        monoclonal gammopathies   213
24.7.   Comparison of sFLCs and urinalysis for monitoring patients   217
24.8.   Organisational, cost and other benefits of sFLC analysis   218
        Clinical case history No 11   219

Contents

25 Guidelines for use of serum free light chain assays   **221**

25.1.   International guidelines for the classification of MM and MGUS   221

25.2.   International Myeloma Working Group guidelines for serum free light chain analysis in multiple myeloma and related disorders (2009)   222

25.3.   UK Myeloma Forum, Nordic Myeloma Study Group and the British Committee for Standards in Haematology (2006)   223

25.4.   UK Myeloma Forum and Nordic Myeloma Study Group: Guidelines for the investigation of newly detected M-proteins and management of MGUS (2009)   223

25.5.   International Staging System for Multiple Myeloma (2005)   224

25.6.   "Uniform Response Criteria for MM" incorporating FLCs (2009)   224

25.7.   National Comprehensive Cancer Network. Clinical Practice Guidelines in Oncology: Multiple myeloma 2007   226

25.8.   European Society of Medical Oncology (ESMO)   226

25.9.   USA National Academy of Clinical Biochemistry guidelines   226

25.10   Guidelines for AL amyloidosis (2005)   228

25.11   International Myeloma Working Group guidelines for MGUS and smoldering (asymptomatic) multiple myeloma (2010).   229

25.12   Consensus statement for the screening, evaluation and management of MGUS (2010)   230

SECTION 5

Practical aspects of serum free light chain testing   **231**

26 Implementation and interpretation of free light chain assays   **231**

26.1.   Why measure FLCs in serum?   231

26.2.   Getting started   232

26.3.   Use and interpretation of sFLC results   233

26.4.   Limitations of sFLC analysis   236

27 Instrumentation for free light chain immunoassays   **239**

Introduction   239

27.1.   Beckman Coulter AU® (400, 640, 2700 and 5400)   241

27.2.   Beckman Coulter IMMAGE® and IMMAGE 800®   242

27.3.   Binding Site MININEPHPLUS™   243

27.4.   Binding Site SPAPLUS bench-top analyser   244

27.5.   Radim Delta™   245

27.6.   Roche Cobas® c501 automated analyser   246

27.7.   Roche Cobas Integra® (400, 400plus and 800) automated analyser   246

27.8.   Roche Hitachi 911/912/917 and Modular P   247

27.9   Siemens Advia® 1650, 1800, 2400   248
27.10. Siemens BN™II   249
27.11. Siemens BN ProSpec®   250

28 Quality assurance for serum free light chain analysis   252
28.1.   Introduction   252
28.2.   The Binding Site QA scheme (QA003)   252
28.3.   College of American Pathologists (CAP) QA scheme   254
28.4.   AFSSAPS QA Scheme (Agencé Française de Securité
        Sanitaire des Produits de Santé)   254
28.5.   UK NEQAS Monoclonal Protein Identification Scheme   255
28.6.   Randox Laboratories Ltd. UK   255
28.7.   Instand e.V. Laboratories (Germany)   255
28.8.   Practical aspects of The Binding Site QA scheme QA003   255

SECTION 6
Appendices   259

29 Classification of diseases with increased immunoglobulins   259
29.1.   Monoclonal gammopathies   259
29.2.   Polyclonal gammopathies   260

30 Questions and answers about free light chains   263
30.1.   Questions about urine testing for FLCs   263
30.2.   Clinical questions about serum testing for FLCs   264
30.3.   Laboratory questions about serum testing for FLCs   267

31 Serum free light chain publications
31.1.   sFLC publications   272

SECTION 7
32 Analysis of immunoglobulin heavy chain/light chain pairs
       (Hevylite™)   301
32.1.   Introduction: limitations of immunoglobulin measurements   301
32.2.   Concept: immunoglobulin heavy chain/light chain assays   302
32.3.   Antibody specificity   302
32.4.   Normal ranges of Hevylite (HLC) assays   304
32.5.   Clinical sensitivity of HLC assays for monoclonal
        gammopathies   306
32.6.   HLC assays for monitoring monoclonal gammopathies   309
32.7.   Prognostic value of Hevylite assays in monoclonal
        gammopathies   312
32.8.   HLC assays in non-Hodgkin's lymphoma (NHL)   317
32.9.   HLC assays for immunohistochemistry   317

32.10. Publications on FcRn receptors and related issues **318**
32.11. Publications on HLC assays. **319**

33 References used in the book **321**

Index **337**

Acknowledgements for figures **350**

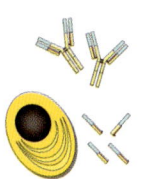

Preface to 6th edition 2010

Since those early days of serum free light chain (FLC) tests, eleven years ago, there has been huge progress in understanding their clinical importance. There are now nearly one thousand publications analysing their measurement; there have been six international conferences and now a sixth edition of 'Serum Free Light Chain Analysis.' I had imagined, maybe even wished, that the rate of new indications would slow down. This would permit a clear focus on implementing the current clinical uses on a wide-spread basis. But, it seems that this is not to be because there are a range of new and important applications.

First, there are novel observations regarding *monoclonal* free light chains. It has recently been shown that FLC MGUS is a precursor abnormality for light chain multiple myeloma. This completes the story of pre-malignant markers in the disease and presumably similar observations will apply to other monoclonal gammopathies. Additionally, the diagnostic importance of *monoclonal* serum FLCs has been validated in several large studies and its use is fully established, alongside SPE, for identifying monoclonal proteins. This reduces serum and urine IFE to the secondary role of typing monoclonal proteins, except under exceptional circumstances such as when AL amyloidosis is suspected. Also, serum FLC tests should replace urine tests in new cases of acute and chronic renal failure as soon as light chain diseases are suspected. Flowing from these observations has been the application of 'high cut-off' dialysers for treating light chain cast nephropathy. Patients with this type of acute renal failure have benefitted from rapid removal of toxic light chains by dialysis, when combined with novel chemotherapeutics such as Velcade. Reported renal recovery rates are now approaching 80%.

Second, there are novel observations regarding *polyclonal* FLCs. Since these molecules are produced by all B-cells, concentrations increase with any inflammatory process and, it seems, in a diagnostically sensitive manner. Publications attest to this in, systemic rheumatic diseases, chronic liver diseases, HIV infections, acute and chronic infections, to name but a few. Furthermore, increases resulting from intense B-cell stimulation such as in Sjogren's syndrome, HIV and hepatitis C infections, predict malignant transformation to lymphomas. In addition, patients with Hodgkin lymphoma have raised *polyclonal* FLCs in a prognostically dependant manner for, as yet, unexplained reasons. Serum *polyclonal* FLC concentrations, moreover, increase when their metabolism is impaired. Hence, serum FLCs are very sensitive markers of impaired glomerular filtration, in fact considerably more sensitive than creatinine clearance or urine albumin/creatinine ratios. Early observations also indicate that *polyclonal* FLCs are sensitive markers of liver damage. Impaired hepatic function, particularly by ongoing hepatic inflammation appears to cause marked increases in FLCs in relation to organ destruction. Even in cardiovascular disease, where C-reactive protein (CRP) tests have proven diagnostic utility, *polyclonal* FLCs appear to have additional clinical sensitivity. These many preliminary observations suggest that serum

FLC analysis may be helpful in a multitude of inflammatory conditions.

Third, there is the application of protein ratios to monoclonal intact immunoglobulin measurements – so called *Hevylite tests*. Historically, IgG, IgA and IgM tests have comprised either monoclonal measurements by SPE or total immunoglobulins by immunochemical tests. Neither of these measurements takes into account the suppression of the un-involved immunoglobulins – which underpins the clinical value of serum FLC analysis. Studies of *Hevylite* ratios have now been completed in a variety of clinical situations. In multiple myeloma, they are more accurate for the diagnosis of some patients with IgA disease, they are more sensitive than IFE when monitoring clinical changes and, in particular, are of prognostic significance. This latter observation relates largely to suppression of the un-involved immunoglobulin of the same isotype as the tumour. Similar observations have been made in MGUS. These and other results lead to a host of interesting questions. What is the underlying mechanism of the suppression? Does it apply to other monoclonal gammopathies? Are there immunoglobulin-specific niches in the bone marrow? Should *Hevylite* tests be used alongside IFE because of their prognostic value? Do *Hevylite* tests predict transformation of smouldering myeloma to multiple myeloma?

So that myself and others might keep abreast of these new developments, it seemed appropriate to write a sixth edition of *The Book*. But, to prevent a pleasure from becoming an obsession, many people have helped with the text, particularly Richard Hughes. My new role as 'editor in chief,' allows more reflection with less direct input. A big thank you to all those who have helped make the writing process easier.

AR Bradwell, August 2010

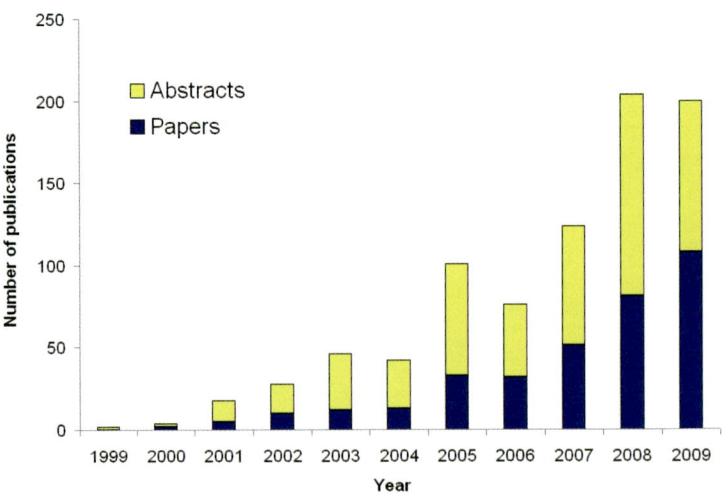

Serum free light chain publications per year.

Wikilite.com

In June 2009, a full electronic version of the 5th edition of *Serum Free Light Chain Analysis (plus Hevylite)* was published on-line, called **wikilite.com**. The format is similar to that of Wikipedia, with a search engine, hyperlinks with the literature and an escalating data base. **Wikilite.com** is available on a read-only basis but dedicated writers will regularly monitor the literature and add new data while a chief editor will ensure consistency of content and style. In this manner, my colleagues and I intend to maintain a fully up-to-date literature review for all those who are interested in *Serum Free Light Chain Analysis (and Hevylite)*.

Dedication

This 6th edition is dedicated to my wife Barbara and my children, Edward, Annie and Susie. They have given me the indulgence of time, discovery and innovation. This has allowed faster entry of serum free light chain assays into clinical practice. I much appreciate their love and patience.

Acknowledgements

Many people have been involved in the research, development and application of free light chain and Hevylite immunoassays. They include the following at The Binding Site Group Ltd:- Graham Mead, Hugh Carr-Smith, Paul Showell, Stephen Harding, Richard Hampton, Laura Smith, Nicolas Fourrier and Jenny Harris. The clinical data were obtained in collaboration with the following:- Mark Drayson, Paul Cockwell and Colin Hutchison, University of Birmingham, UK; Jerry Katzmann, Robert Kyle, Roshini Abraham, Vincent Rajkumar and Raynell Clark, The Mayo Clinic, USA; David Keren, University of Michigan Medical School, USA; Philip Hawkins and Helen Lachmann, UK Amyloidosis Centre, Royal Free Hospital, London; Peter Hill and Julia Forsyth, Derby NHS Trust, UK; G Pratt, Heartlands Hospital, Birmingham; M Chappell and N Evans, University of Warwick; S Abdulla, St Mary's Hospital, London; Mohammad Nowrousian, Essen University, Germany; and numerous other physicians who have sent samples to the laboratory. I am also indebted to Richard Hughes for his contributions towards writing and coordinating the editing of this 6th edition, Michelle Surmacz for the setting and presentation of this book, Josie Hobbs, Alex Legg, Lakhvir Assi and Roger Williams for proof reading. My grateful thanks go to them all.

The initial stages of the project were supported by a grant from the Department of Trade and Industry (No. WMR/26799/SP).
* Figure acknowledgements see page 350.

Abbreviations

ABCM	adriamycin (doxorubicin), busulphan, cyclophosphamide, melphalan
ACD	acid citrate dextrose
ACR	urine albumin/creatinine ratio
ARF	acute renal failure
ASCT	autologous stem-cell transplant
ASMM	asymptomatic multiple myeloma
β_2M	beta-2 microglobulin
B-CLL	B cell chronic lymphocytic leukaemia
BJP	Bence Jones protein
BM	bone marrow
BMPC	bone marrow plasma cells
BrdU	bromo deoxyuridine
CAP	College of American Pathologists
CCP	cyclic citrullinated peptides
CKD	chronic kidney disease
CR	complete response
CrCl	creatinine clearance
CRF	chronic renal failure
CRP	C-reactive protein
CZE	capillary zone electrophoresis
CSF	cerebrospinal fluid
CT	computerised tomography
C-VAMP	cyclophosphamide and VAMP
Dex	dexamethasone
DSS	disease specific survival
ECLAM	European Consensus Lupus Activity Measurement
ESR	erythrocyte sedimentation rate
FcRn	Brambell Fc receptor
FITC	fluorescein isothiocyanate
FLC(s)	free light chain(s)
GFR	glomerular filtration rate
GP	general practitioner
HDM	high dose melphalan
HLC	hevylite
(s)IFE	(serum) immunofixation electrophoresis
IgG	immunoglobulin G
Ig(s)	immunoglobulin(s)
IIMM	intact immunoglobulin multiple myeloma
IMIg(s)	intact monoclonal immunoglobulin(s)
κ	kappa
κ/λ	kappa/lambda (ratio)
λ	lambda
LCDD	light chain deposition disease
LCMM	light chain multiple myeloma
LPL	lymphoplasmacytic lymphoma
LM	lymphomatous meningitis
Mabs	monoclonal antibodies
MALToma	mucosa-associated lymphoid tissue lymphoma
MDRD	Modification of Diet in Renal Disease study
MGUS	monoclonal gammopathy of undetermined significance
M-Igs	monoclonal intact immunoglobulins
MM	multiple myeloma
M-protein	monoclonal protein
MRI	magnetic resonance imaging
MZL	marginal zone lymphoma
ND	not determined
NHL	non-Hodgkin lymphoma
NPV	negative predictive value
NR	normal range
NSC	new scientific company
NSMM	nonsecretory multiple myeloma
NT-proBNP	amino-terminal fragment of type B naturetic peptide
PBSCT	peripheral blood stem cell transplant
PEL	primary effusion lymphoma
PET	positron emission tomography
PIgs	polyclonal immunoglobulins
POEMS	polyneuropathy, organ omegaly, endocrinopathy, monoclonal protein, skin changes; osteosclerotic myeloma

PPV	positive predictive value
PR	partial response
pSS	primary Sjögren's syndrome
QA	quality assurance
RA	rheumatoid arthritis
RBCs	red blood cells
ROC	receiver operating characteristic
SAP	serum amyloid P (scans)
sCR	stringent complete response
SD	stable disease
SDS-AGE	sodium dodecyl sulfate-agarose gel electrophoresis
SDS-PAGE	sodium dodecyl sulfate-polyacrylamide gel electrophoresis
sFLC	serum free light chain
sIFE	serum IFE
SLE	systemic lupus erythematosus
SPE	serum protein electrophoresis
TRITC	tetramethylrhodamine isothiocyanate
TSP	total serum protein (in IFE tests)
uFLC(s)	urine free light chain(s)
uIFE	urine IFE
UPE	urine protein electrophoresis
VAD	vincristine, adriamycin (doxorubicin), dexamethasone
VAMP	vincristine, adriamycin (doxorubicin), melphalan, methyl-prednisolone
VED	vincristine, epirubicin, dexamethasone
WM	Waldenström's macroglobulinaemia

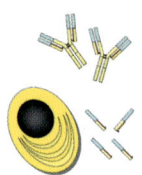

For more than 150 years, the presence of Bence Jones protein (immunoglobulin free light chains [FLCs]) in the urine has been an important diagnostic marker for multiple myeloma (MM). Indeed, it was the first in vitro cancer test to be used, a century before any others *(Jones 1)*. In the last few years, interest in FLCs has undergone a renaissance with the development of serum tests for free kappa (κ) and free lambda (λ) light chains. The clinical importance of these tests continues to grow as new applications for their use emerge *(Bradwell 3)*. By way of comparison, the management of diabetes mellitus was hugely improved when blood replaced urine for glucose analysis.

From a physiological viewpoint, blood tests for small molecular weight proteins have clear advantages over urine tests. Serum FLCs are rapidly cleared through the renal glomeruli with half-lives of between 2 and 6 hours before being metabolised in the proximal tubules of the nephrons. Under normal circumstances, little protein escapes to the urine so serum FLC concentrations have to increase many-fold before absorption mechanisms are overwhelmed *(Chapter 3)*. Hence, urinalysis is a fickle witness to changes in FLC production. Conversion to a serum test has provided clarity in assessing disease processes that were previously hidden from view.

Serum concentrations of FLCs are dependent upon the balance between production by plasma cells (and their progenitors) and renal clearance. When there is increased polyclonal immunoglobulin production and/or renal impairment, both κ and λ FLC concentrations can increase 30-40 fold. However, the relative concentrations of κ to λ (i.e. the κ/λ ratio) remain unchanged, or only slightly increase *(Chapter 20)*. In contrast, tumours produce a monoclonal excess of only one of the light chain types, often with bone marrow suppression of the alternate light chain, so that κ/λ ratios become highly abnormal. Accurate measurement of κ/λ ratios underpins the utility of serum FLC immunoassays and provides a numerical indicator of clonality *(Katzmann 6)*. Urine κ/λ ratios are not as dependable because the non-tumour light chain production is too low to guarantee consistent passage through the nephrons. In contrast, electrophoretic tests can only be used to quantify monoclonal light chain peaks because they are not sensitive enough to identify the non-tumour FLC concentrations.

Early clinical studies with serum FLC tests were performed in patients with Bence Jones (light chain) multiple myeloma. In two studies, on 252 sera taken at clinical presentation, highly abnormal serum FLC concentrations were found in every case *(Bradwell 1; Abraham 3)*. Furthermore, during chemotherapy, urine tests frequently normalised whereas serum tests remained abnormal, indicating their increased sensitivity for detection of residual disease. In this patient group, analysis of urine can now be replaced by serum FLC tests. This is particularly helpful for frail, elderly patients because 24-hour urine samples are difficult to collect and results may be unreliable *(Alyanakian)*.

Around 3-4% of patients with multiple myeloma have so-called nonsecretory disease *(Chapter 9)*. By definition these patients have no monoclonal proteins when assessed by serum and urine electrophoretic tests. Nevertheless, in a study by Drayson *(2)* et al., serum FLC tests identified monoclonal proteins in 70% of 28 such patients. A study by Katzmann *(3)* et al. found that all of 5 patients with untreated nonsecretory myeloma had abnormal FLC concentrations. It is apparent that these patients' tumour cells produced small amounts of monoclonal protein. Their serum FLC concentrations were below the sensitivity of serum electrophoretic tests and below the threshold for clearance into the urine. Importantly, these patients can now be closely monitored by serum FLC tests rather than repeated bone marrow biopsies or whole body scans.

Approximately 20% of all patients with myeloma have light chain or nonsecretory myeloma. Among the remaining patients, those who produce intact monoclonal immunoglobulins, FLCs are abnormal in 96% at disease presentation *(Mead 1) (Chapter 10)*. Interestingly, the serum concentrations of FLCs and intact monoclonal immunoglobulins are not correlated. Monoclonal serum FLCs are therefore independent markers of the disease process and have been shown to have prognostic value *(Snozek 2)*. This is of potential clinical use when the tumour produces large amounts of FLCs and small amounts of intact monoclonal immunoglobulins. Thus, patients who are in apparent remission, as assessed by measurement of intact monoclonal immunoglobulins, may still have residual disease as judged by elevated monoclonal FLCs. Using a similar argument, when these patients relapse, FLC concentrations may increase first *(Dawson)*. Free light chain "breakthrough" is estimated to occur in 2 - 5% of patients who relapse after modern, intensive treatment *(Kühnemund 2) (Chapter 12.7)*.

An additional feature of FLC molecules is that, in contrast to intact immunoglobulins, they are frequently nephrotoxic *(Chapter 13)*. Indeed, "myeloma kidney", which presents as acute renal failure, occurs in approximately 10% of patients *(Hutchison 2)*. The life expectancy of these patients is then often significantly reduced. Myeloma kidney is particularly common in light chain only disease, but in many patients with intact monoclonal immunoglobulins the serum FLC concentrations are >1,000mg/L (50-100 times normal) and there is associated renal damage. This is characteristic of patients with IgD multiple myeloma but is also apparent in 10-15% of IgG- and IgA-producing patients. The FLC assays allow assessment of the pre-renal load of monoclonal light chains in virtually all of these patients.

There is early evidence that treatment should be aimed to rapidly reduce serum FLC concentrations in order to prevent further renal damage *(Abdalla SH)*. Furthermore, rapid removal of nephrotoxic FLCs, using high cut-off dialysers, can lead to renal recovery *(Hutchison 2, 8) (Chapter 20.4)*. This important new development, when used in combination with aggressive chemotherapy, should lead to significant increases in survival in this serious disease.

One particularly interesting aspect of serum FLCs is their short half-life in the blood (κ 2-4 hours, λ 3-6 hours) *(Chapter 12.4)*. This is approximately 100-200 times shorter than the 21-day half-life of IgG molecules. Hence, responses to treatment are seen in "real time". This is apparent from the good correlation between bone marrow assessment of disease status and FLC concentrations but poor correlation with IgG concentrations *(Mead 1)*. Thus, FLC concentrations allow more rapid assessment of the effects of chemotherapy than do monoclonal IgG or IgA concentrations. The impact of this phenomenon is likely to be considerable: for instance, the resistance of patients to particular drugs or drug combinations can be observed quickly and alternative treatments considered. In addition, the short half-lives of FLCs often allow assessment of complete tumour responses after one or two cycles of chemotherapy and before stem cell transplantation *(Hassoun 3)*. Thus, IgG with its 21-day half life is a slow marker of treatment responses whereas FLC analysis allows more accurate assessments *(Pratt 2; Van Rhee; Orlowski)*.

Serum FLC tests are also of considerable importance in AL (primary) amyloidosis. *(Chapter 15)*. Characteristically, light chain fibrils are deposited in various organs and tissues and lead directly to disease. The origin of the fibrils is monoclonal FLCs produced by a slowly growing clone of plasma cells. While concentrations of these are often insufficient for measurement by serum electrophoretic tests, serum FLC assays provide quantification of the circulating fibril precursors in 88-98% of patients *(Lachmann; Abraham 4)*. Furthermore, the tests allow assessment of treatment responses and disease relapses that, in turn, correlate with survival. As stated by Dispenzieri *(1)* et al. from The Mayo Clinic:

"Introduction of the serum immunoglobulin free light chain assay has revolutionized our ability to assess hematological responses in patients with low tumor burden".

Further support for the role of serum FLC analysis in AL amyloidosis was provided by Katzmann *(3)* et al. The combination of serum FLC and serum immunofixation electrophoretic tests identified 109 of 110 patients at diagnosis. FLC analysis alone identified 91% of the patients while immunofixation electrophoresis identified only 69%, and urinalysis failed to identify the sole patient that was normal by both serum tests. A similarly high sensitivity of the FLC assays has been reported for light chain deposition disease *(Katzmann 3, 6)*.

Many national and international guidelines, including those from the UK and USA, and the Consensus Opinion in AL Amyloidosis support the use of serum FLC measurements *(Bird; Durie; Gertz 1) (Chapter 25)*. Reduction of the κ/λ ratio to normal, alongside the intact monoclonal immunoglobulins, is the benchmark for complete or stringent complete serological responses to therapy in these and other monoclonal diseases.

An emerging role for serum FLC analysis is in the assessment of risk of progression in individuals with monoclonal gammopathies of undetermined significance (MGUS) *(Chapter 19)*. These are premalignant states that progress to multiple myeloma, AL amyloidosis or other plasma cell dyscrasias at a rate of approximately 1% per year. Rajkumar *(5)* et al. have shown that the presence of an abnormal serum FLC κ/λ ratio is a major independent risk factor for progression. In particular, the 40% of MGUS patients with low levels of IgG M-spike (<15g/L) and normal κ/λ ratios had a 21-fold lower risk of progression than patients with an M-spike of >15g/L, abnormal κ/λ ratios and non-IgG immunoglobulin class.

The predictive value of baseline serum FLCs for progression or prognosis extends to most monoclonal gammopathies. This has been shown for asymptomatic myeloma *(Dispenzieri 3)*, solitary plasmacytoma *(Dingli 1)*, AL amyloidosis *(Dispenzieri 4)*, Waldenström's macroglobulinaemia *(Itzykson)* and, most recently, multiple myeloma *(Kyrtsonis 3)*. Indeed, initial serum FLC concentrations are independent of serum albumin and β_2-microglobulin as a risk factor for myeloma progression, suggesting that FLCs should be added into the current international staging system (ISS) for multiple myeloma *(Snozek 2)*.

The high sensitivity of serum FLC immunoassays for tumour detection indicates that they have a role in the initial screening for plasma cell dyscrasias. Historically, symptomatic patients were assessed using serum and urine protein electrophoretic tests. Since urine is frequently unavailable, it is logical to add serum FLC analysis to current test protocols *(Chapter 24.6)*. In a study of 1,003 consecutive unknown samples by Bakshi et al., serum FLC analysis identified an additional 16 patients with monoclonal proteins above the 39 detected by serum capillary zone electrophoresis. B-cell/plasma cell tumours were present in 9 of the 16, including 2 with light chain and 1 with nonsecretory multiple myeloma. Confirmation of the sensitivity of a serum panel was provided by Katzmann *(4)* et al., who showed that among 428 patients positive for urine monoclonal immunoglobulins, all except one were identified by serum tests. The one sample that was only urine positive was of no clinical consequence. Other studies have demonstrated similar benefits and, in each, little or no gain was achieved from urine tests *(Hill; Abadie 2; Katzmann 8)*.

All of these findings indicate that the combination of serum electrophoresis, serum immunofixation, and FLC analysis is a clinically sensitive strategy for identifying patients with MM that was recently recommended by the International Myeloma Working Group *(Dispenzieri 7)*. For clinically suspected amyloidosis, it was recommended that urine immunofixation be added to the panel *(Dispenzieri 7)*. As a consequence, serum free light chain analysis for assessing monoclonal gammopathies is being widely adopted. However, the question of whether serum is additional to or can replace urine tests for monitoring patients remains unresolved.

FLC concentrations have also been assessed in cerebrospinal fluid (CSF) samples. *(Chapter 22)*. In a study by Fischer et al. kappa concentrations provided information comparable with oligoclonal band measurements. They concluded that CSF kappa FLC measurements may be a useful diagnostic procedure for detecting, and potentially monitoring, intrathecal immunoglobulin synthesis.

In summary, serum FLC tests are assuming an important role in the detection and monitoring of monoclonal gammopathies, bringing benefits to the multitude of patients with plasma cell dyscrasias. The increasing clinical use of this new approach is clearly apparent from the number of recent reviews *(Rajkumar 6; Jagannath; Mayo; Pratt 3; Siegel 1)*.

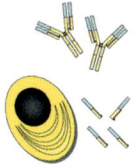

SECTION 1 - Immunoglobulin free light chains and their analysis

Chapter 1. *Introduction*   *1*
Chapter 2. *Dr Bence Jones and the history of free light chains*   *6*
Chapter 3. *Biology of immunoglobulin light chains*   *10*
Chapter 4. *Immunoassays for free light chain measurement*   *18*
Chapter 5. *Normal ranges and reference intervals*   *36*
Chapter 6. *Sensitivity of serum free light chain assays*   *45*

Figure 1.0. Three-dimensional structure of a κ light chain molecule.

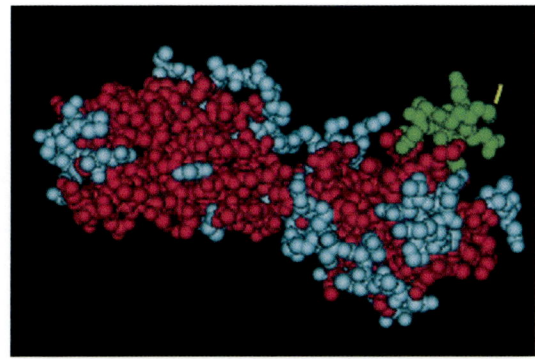

Chapter

1

SECTION 1 - Immunoglobulin free light chains and their analysis

Introduction

Multiple myeloma (MM) is a disease with many faces. It usually presents in old age but may occur in youth. Bone pain and fractures are characteristic yet soft tissue involvement by plasmacytoma occurs. Patients may die within weeks of presentation while others "smoulder" for years. Patients can develop renal failure, acute and chronic infections or AL amyloidosis, and many will require stem cell transplantation or intensive chemotherapy. Consequently, many specialists, including haematologists, nephrologists, immunologists, orthopaedic surgeons and chemical pathologists become involved in disease management. Furthermore, the prevalence of MM is increasing due to a slowly rising incidence and a longer life expectancy *(Katzel; Kumar 3; Jemal)*.

Despite the complexity of this disease, one feature has been a great lighthouse in the fog, alerting the unwary to the diagnosis and guiding the hand of management: the presence of monoclonal immunoglobulins. Produced in excess, and in a variety of shapes and sizes, these molecules have been linked to MM since they were first identified by Henry Bence Jones over 150 years ago *(Jones 1, 2)*. Notwithstanding their substantial history and great utility, measurement of these tumour markers, particularly free light chains (FLCs), remained imperfect for many years. Techniques used for their

Tumour Type	Cancer Deaths (USA) (*Jemal*)	Tumour Markers	Specificity	Sensitivity	Tumour Detection	Clinical Utility
Lung + bronchus	28%	Neuron specific enolase	Poor	Poor	Late	Poor
Colon + rectum	9%	Carcinoembryonic antigen (CEA)	Poor	Modest	Late	Modest
Breast	7%	CA 15-3; CEA	Poor	Modest	Late	Modest
Pancreas	6%	CA 19-9; CEA	Poor	Poor	Late	Poor
Prostate	5%	Prostate-specific antigen	Modest	Good	Good	Good
Stomach	2%	CEA; CA 19-9	Modest	Modest	Late	Poor
Ovary	2.5%	CA 125; PLAP	Modest	Modest	Intermediate	Good
Liver	3%	Alpha feto-protein (αFP)	Good	Good	Intermediate	Good
Myeloma	**1.9%**	**Monoclonal protein/FLC**	**Good**	**Good**	**Early**	**Very good**
AL amyloidosis	**0.3%**	**Monoclonal protein/FLC**	**Good**	**Good**	**Early**	**Very good**
Germ cell	~0.1%	αFP; human chorionic gonadotrophin	Good	Good	Early	Very good
Choriocarcinoma	<0.1%	Human chorionic gonadotrophin	Good	Good	Early	Very good
Neuroendocrine	<0.1%	Chromogranin A, gastrin	Modest	Good	Early	Very good

Table 1. Some common serum tumour markers and their clinical utility. All of these analytes are measured using highly sensitive immunoassays apart from monoclonal proteins. FLC = free light chains

measurement had failed to keep pace with analytical developments in other fields until the 21st century when serum FLC analysis was introduced.

Most serum cancer tests are now based on state-of-the-art immunoassays and are highly automated. *(A selection of the more common serum markers is shown in Table 1)*. In contrast, tests for MM and AL amyloidosis are primarily based on serum and urine electrophoretic techniques. These are relatively insensitive, require considerable experience for interpretation and are often labour-intensive. How ironic that the first tumour marker to be identified should be the last to benefit from modern technology.

It is, perhaps, not surprising that these techniques produce errors in FLC measurements *(Ward)*. And, why use urine? It is hard to imagine a less attractive fluid in which to evaluate these molecules. An important function of the kidneys is to **prevent** the loss of FLCs and other small protein molecules into the urine. Furthermore, urine samples are voluminous, difficult to obtain, awkward to transport and need to be concentrated prior to analysis.

An alternative strategy is to measure FLCs in serum. In 1981, it was shown that serum concentrations of FLCs were elevated when Bence Jones proteinuria occurred *(Chapter 2)*, and that measuring serum rather than urine was diagnostically more accurate in patients with renal failure *(Solling 4; Sinclair 1)*. Why, therefore, have serum immunoassays not been used before? It is now apparent that the overriding barrier was the difficulty in developing satisfactory antibodies for use in the assays. To function correctly, these antibodies must not only be of high affinity to allow measurement of low concentrations of serum FLCs, but must also be highly specific. Serum FLC concentrations are several orders of magnitude lower than serum light chains bound to intact immunoglobulins, so even minor antibody crossreactivity produces unacceptable results. Only recently have suitable antibodies been developed that bind exclusively to the hidden epitopes of FLC molecules *(Chapter 4)*. These antibodies have facilitated the development of serum FLC assays that are specific, sensitive and quantitative *(Bradwell 2; Katzmann 6)*.

Serum FLC immunoassays include the following benefits:
- Better sensitivity and precision than current electrophoretic assays for FLC molecules.
- Numerical results for disease monitoring.
- Convenience of using serum as a test medium.
- Sensitive measurement of FLCs in AL amyloidosis and nonsecretory multiple myeloma patients who have no detectable monoclonal proteins by conventional tests.
- Accurate marker of disease remission.
- Short half-life marker for rapid assessment of treatment responses.
- Marker of increased risk of progression in individuals with monoclonal gammopathy of undetermined significance and other monoclonal gammopathies.
- More sensitive than urine tests when screening symptomatic patients.
- Diagnosis and monitoring of patients with myeloma kidney.

General statements have been formulated to describe the clinical applications of cancer markers *(Chan)*. Most of these have now been applied to FLC immunoassays, as follows:

1. Differential diagnosis in symptomatic patients. Serum FLC analysis is most helpful in the differential diagnosis of patients with bone pain, fractures, unexplained renal impairment and other features of MM and AL amyloidosis *(Section 2 and 4)*.

2. Clinical staging of disease. Serum FLC concentrations show a relationship with the staging of monoclonal diseases and are helpful in assessing residual disease after treatment *(Chapter 10 and 12)*.

3. Estimating tumour burden. Serum FLC concentrations correlate poorly with other tumour markers at the time of diagnosis but changing concentrations correlate well with changing tumour burden during treatment *(Section 2)*.

4. Prognostic indicator for disease progression. Serum FLC concentrations are of prognostic value in MM at the time of clinical presentation. Also, patients who are in remission but have elevated FLC levels, in the absence of other abnormalities, are at risk of early relapse. Of particular interest is the use of the assays for predicting risk of progression of monoclonal gammopathies. Studies from the Mayo Clinic indicate that elevated monoclonal FLCs are a sensitive risk factor for progression to myeloma and other plasma cell dyscrasias *(Section 2)*.

5. Evaluating the success of treatment. FLC analysis is particularly helpful in assessing treatment responses in patients with AL amyloidosis, light chain multiple myeloma (LCMM) and nonsecretory multiple myeloma (NSMM). There are also indications that the short half-life of FLCs can be used to assess early treatment responses in nearly all patients with monoclonal proteins *(Section 2)*.

6. Detecting the recurrence of cancer. Serum FLCs are useful for detecting disease recurrence and are often more sensitive than other tests in patients with AL amyloidosis, LCMM and NSMM. They are useful in a proportion of patients with intact immunoglobulin myeloma and some other monoclonal gammopathies *(Section 2)*.

7. Screening symptomatic patients. Historically, patients with symptoms of MM or related disorders were screened for monoclonal proteins using serum and urine electrophoretic tests. Several recent studies have shown that serum FLC analysis in combination with serum electrophoresis and/or serum immunofixation identifies more patients and can replace urine tests in most instances *(Chapter 23 and 24)*.

8. Screening the general population. There is evidence that serum FLC measurements identify a new set of MGUS patients in otherwise healthy people *(Chapter 19)*. Indeed, progression to light chain and nonsecretory myeloma has been observed in patients whose only preceding abnormalities were abnormal sFLC ratios. Because the criteria for a successful screening test include a beneficial outcome for the population screened, it will be many years before it is clear whether serum FLCs are useful in this context.

In addition, serum FLC assays have allowed new guidelines to be written for the diagnosis and monitoring of patients with MM, AL amyloidosis and other diseases producing excess clonal FLCs *(Chapter 25)*.

In the following chapters, the discovery of Bence Jones protein, the structure and synthesis of light chain molecules, and assays for serum FLC quantification are described. A detailed account of the current use of the assays in clinical and laboratory practice is presented, together with potential applications in other settings. Appendices for guidance in their clinical and laboratory use are also provided. In addition, Chapter 32 provides an introduction to the analysis of heavy chain/light chain pairs (Hevylite™). These are new immunoassays that measure the concentrations of different light chain types of each immunoglobulin class, and provide a ratio that may assist in diagnosis, staging and monitoring treatment responses of MM and other plasma cell dyscrasias.

Figure 2.1. Henry Bence Jones. (A) Albumen print by Maull c. 1850s. (B) Charcoal and chalk on paper by Richmond 1865 (reproduced with permission from The Royal Institution, London and the Bridgeman Art Library).

Although immunoglobulin free light chains (FLCs) are synonymous with Bence Jones proteins, history might have been more generous to others involved in their discovery *(Clamp; Rosenfeld; Kyle 3,4; Hajdu)*.

On Friday, October 30th 1845, the 53 year-old Dr William MacIntyre, physician to the Western General Dispensary, St. Marylebone, London, left his rooms in Harley Street. He had been called to see Mr Thomas Alexander McBean, a 45 year-old, highly respectable grocer, who had severe bone pain and fractures and had been under the care of his general practitioner, Dr Thomas Watson, for several months. Upon examination of the patient, William MacIntyre noted the presence of oedema. Considering the possibility of nephrosis, he tested the urine for albumin. To his consternation, the albuminous protein precipitate found on warming the urine uncharacteristically re-dissolved when heated to 75°C.

Both Dr MacIntyre and Dr Watson then sent urine samples to the chemical pathologist at St.George's Hospital. A note accompanying the urine sent by Dr Watson read as follows:

"Dear Dr Bence Jones,
   The tube contains urine of very high specific gravity. When boiled it becomes highly opaque. On the addition of nitric acid, it effervesces, assumes a reddish hue, and becomes quite clear; but as it cools assumes the consistence and appearance which you see. Heat reliquifies it. What is it?"

Over the next two months the patient deteriorated, became emaciated, weak and racked with pain. He eventually died on January 1st 1846, in full possession of his mental faculties. Dr MacIntyre subsequently published the post-mortem examination and the description of the peculiar urine in 1850 *(MacIntyre)*. Unfortunately for him, Henry Bence Jones *(1,2)* had already described the patient's urinary findings in two single-author articles, one of which was published in *The Lancet*, in 1847. He considered the protein to be a *"hydrated deutoxide of albumen"*. He wisely commented:

"I need hardly remark on the importance of seeking for this oxide of albumen in other cases of mollities ossium".

Bence Jones's reputation was assured, while the contributions of his colleagues were consigned to the footnotes of history.

For all the apparent injustice to his colleague William MacIntyre, Henry Bence Jones achieved much else in his career. He published over 40 papers and became rich and famous based on his clinical practice, lecturing and original observations and was elected to fellowship of The Royal Society at the tender age of 33. Florence Nightingale once described him as *"the best chemical doctor in London."* Surprisingly, there was no mention of Bence Jones protein in his obituary and the eponym (and the hyphen in his name) was not used until after his death *(Rosenfeld)*.

By 1909, over 40 cases of Bence Jones proteinuria had been reported *(Weber FP)*, and the protein was thought to originate in bone marrow plasma cells, first identified by Waldeyer in 1875. In 1922, Bayne-Jones and Wilson characterised two types of Bence Jones protein by observing precipitation reactions using antisera from rabbits immunised with the urine of several patients. The proteins were classified as group I and group II types. However, it was not until 1956 that Korngold and Lapiri, using the Ouchterlony technique, showed that antisera raised against the different groups also reacted with myeloma proteins. As a tribute to their observations the two types of Bence Jones protein were designated kappa and lambda (κ and λ). Edelman and Gally, in 1962, subsequently showed that FLCs prepared from IgG monoclonal proteins were the same as Bence Jones protein. It had taken 117 years from the original observation for the function of Bence Jones protein to be finally determined. Remarkably, the format of the urine test had remained unchanged for a similar period. The following is a protocol from Levinson and MacFate's *"Clinical Laboratory Diagnosis"*, a standard textbook published in 1946.

Bence-Jones* Protein Test

Principle. Bence-Jones protein is soluble in urine at room and body temperature. When the urine is heated to 40°C, a white cloud appears and at 60°C, a distinct precipitate forms. The precipitate disappears on boiling but reappears on cooling. Excessive amounts of acid or salt will prevent the appearance of the precipitate.
Reagent. Acetic Acid, approximately 10% aqueous solution.
Procedure. Add urine to a 6-inch test tube until it is about two-thirds full. Place in a water bath and heat slowly. Do not allow the bottom of the test tube to touch

the bottom of the beaker, as it may become hotter than the rest of the water. Suspend a thermometer in the water bath. Note the temperature and the appearance of the urine every few minutes, especially between the temperatures of 40°C - 60°C. If a slight cloud appears when the urine is heated to the boiling point, 100°C, add a few drops of acetic acid. This will dissolve any phosphates that have separated out. If a precipitate forms at the boiling point, it is due to albumin. Boil the urine with a few drops of dilute acetic acid, filter rapidly, and repeat the test on the cooled filtrate.

Interpretation. Bence-Jones protein is found in urine in many cases of multiple myeloma, osteogenic sarcoma, osteomalacia, carcinomatosis.

(The hyphen was only added after his death)

In parallel with the clinical and scientific observations of the role of Bence Jones protein, electrophoretic techniques for protein separation were entering clinical laboratories. Longsworth et al. in 1939 recognised tall, narrow-based, *"church spire"* peaks in the sera of patients with multiple myeloma (MM), using moving boundary protein electrophoresis. Electrophoresis was subsequently improved by the use of paper as a substrate followed by cellulose acetate and then agarose in the 1950's and 1960's. Finally, immunofixation electrophoresis became established in the 1980's *(Whicher)*.

Clear identification of κ and λ molecules was possible with the use of antibodies specific for each type of protein. Immunodiffusion was initially used *(Ouchterlony)*, followed by immunoelectrophoresis in 1953 *(Grabar)*, radial immunodiffusion and ultimately nephelometry and turbidimetry. However, serum assays for Bence Jones protein (serum FLCs - sFLCs) remained unattainable because the antibodies could not distinguish between sFLCs and the overwhelming amounts of light chains bound in intact immunoglobulin molecules.

The first successful attempt to measure FLCs in serum was in 1975: size-separation column chromatography *(Sölling 1,2; Cole)* was used to isolate them from intact immunoglobulins prior to analysis. Although the results were accurate and showed the potential use of serum analyses, these assays were clearly impractical for routine use. Subsequent assays focussed on the use of antibodies directed against *"hidden"* epitopes on FLC molecules. These are located at the interface between the light and heavy chains of intact immunoglobulins and become detectable when the FLCs are unbound. Radio-immunoassays and enzyme immunoassays using polyclonal antisera against FLCs were employed to analyse urine samples but specificity remained inadequate for serum measurements *(Robinson; Brouwer)* and variations in FLC polymerisation caused measurement errors *(Sölling 4; Heino)*.

The use of monoclonal antibodies was an obvious approach to improving specificity but satisfactory reagents were difficult to develop *(Ling; Axiak; Nelson; Abe)* and their use was restricted to radioimmunoassays and enzyme immunoassays. Attempts were made to develop turbidimetric *(Hemmingsen 1; Tillyer)* and latex-enhanced nephelometric assays *(Wakasugi 2)*, using polyclonal antibodies but they could not detect normal sFLC concentrations, and cross-reactions with intact immunoglobulins were unacceptable. In 2001, polyclonal-antibody-based immunoassays that could measure FLCs at normal serum concentrations were developed *(Bradwell 2)*. Their utility was immediately apparent when monoclonal FLCs were detected in the sera of most patients classified as having *"nonsecretory myeloma" (Drayson 2)*. Furthermore, as described in *The Lancet*, all of 224 patients with light chain multiple myeloma who had Bence Jones proteinuria, also had elevated sFLC concentrations *(Bradwell 1)*. The serum tests were also better than urinalysis at detecting residual disease. Further studies have shown that sFLC assays can be used for screening symptomatic patients *(Chapter 23)*, are more sensitive than urine tests *(Chapter 24)* and are markers for progression in monoclonal gammopathy of undetermined significance (MGUS) *(Chapter 19)*.

These and many other results heralded the widespread use of sFLC immunoassays. The enduring story of Bence Jones protein and Bence Jones proteinuria may be entering its final chapter, 165 years after Dr MacIntyre's original observation.

Test questions
1. Who was the first person to observe "Bence Jones proteinuria?"
2. What is the origin of the names, kappa and lambda?

Answers
1. Dr William MacIntyre in 1845 (page 6).
2. The first letters of Korngold and Lapiri who showed that light chains were located on immunoglobulin molecules in 1956 (page 7).

3.1.	Structure	10
3.2.	Synthesis	12
3.3.	Production	12
3.4.	Clearance and metabolism	14

Summary: Serum free light chains:

1. Are produced in excess by plasma cells.
2. Have a half-life of a few hours because of rapid renal clearance.
3. Bind to Tamm-Horsfall protein in the distal tubules and may obstruct urine flow.
4. Are more frequently abnormal in serum than in urine because of renal metabolism.
5. Contain constant region epitopes that are hidden in intact immunoglobulins.

3.1. Structure

Antibody molecules have a two-fold symmetry and are composed of two identical heavy and light chains, each containing variable and constant domains. The variable domains of each light chain/heavy chain pair combine to form an antigen-binding site, so that both chains contribute to the antigen-binding specificity of the antibody molecule. Light chains are of two types, κ and λ, and in any given antibody molecule

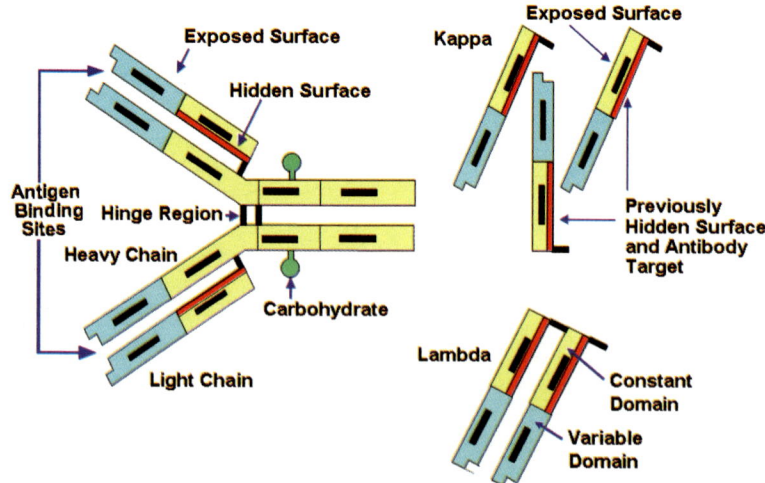

Figure 3.1. An antibody molecule showing the heavy and light chain structure, together with free κ and λ FLCs.

only one type occurs. Approximately twice as many κ as λ molecules are produced in humans but in other mammals this ratio can vary. Each free light chain (FLC) molecule contains approximately 220 amino acids in a single polypeptide chain that is folded to form the constant and variable region domains *(Figures 3.1 and 3.2)*.

Domains are constructed from two β sheets. These are elements of protein structure made up of strands of the polypeptide chain (β strands) packed together in a particular shape. The sheets are linked by a disulfide bridge and together form a roughly barrel-shaped structure known as a β barrel *(Figure 3.2)*.

Each FLC molecule is composed of an N-terminal domain that contains the variable (VL) region, and a C-terminal domain that contains the constant (CL) region. These domains are linked by a joining (J) gene segment. The most variability is seen in the hypervariable (CDR) regions of the V domain. Genes that encode antibodies are assembled in B lymphocytes by joining DNA segments that are far apart in the DNA of germline and other types of somatic cells. One of each V and J region genes in the genome is joined to a single C gene. Consequently, immunoglobulin genes differ depending on which V and J is used *(Singer)*.

The constant (C) domains of FLCs show little variation except for the amino acid substitutions found in the 3 KM allotypes on κ molecules *(Table 3.1)* and the Mcg, Oz and Kern isotypes on λ molecules *(Udey; Niewold)*. There is no evidence that these variants affect FLC measurements.

In contrast, the variable (V) domain has huge structural diversity, particularly in

Allotype	Amino Acids	
KM1	Val (153)	Leu (191)
KM1,2	Ala (153)	Leu (191)
KM3	Ala (153)	Val (191)

Table 3.1. Amino acid substitutions in κ constant domains.

Figure 3.2. A κ FLC molecule showing the constant region (left), and the variable region (right) with its alpha helix (red). (Courtesy of J Hobbs).

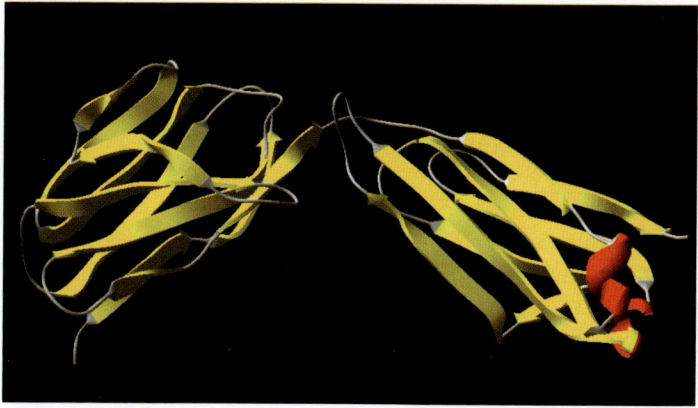

association with the antigen binding amino acids. In addition, the first 23 amino acids of the 1st variable domain framework region have a limited number of variations known as subgroups. Using monoclonal antibodies, 4 kappa (Vκ1 -Vκ4) and 6 lambda subgroups (Vλ 1 -Vλ 6) can be identified *(Soloman 1)*. The specific subgroup structures influence the potential of the FLCs to polymerise such that AL amyloidosis is associated with Vλ6 and light chain deposition disease (LCDD) with Vκ1 and Vκ4.

3.2. Synthesis

κ FLC molecules (chromosome 2) are constructed from approximately 40 functional Vκ gene segments, five Jκ gene segments and a single Cκ gene. λ FLC molecules (chromosome 22) are constructed from about 30 Vκ gene segments, four pairs of functional Jλ gene segments and a single Cλ gene *(Figure 3.3)*.

FLCs are incorporated into immunoglobulin molecules during B lymphocyte development and are expressed initially on the surface of pre-B-cells. Production of FLCs occurs throughout the rest of B-cell development and in plasma cells, where secretion is highest. Tumours associated with the different stages of B-cell maturation will secrete monoclonal FLCs into the serum where they may be detected by FLC immunoassays *(Figure 3.4 and Chapter 18)*.

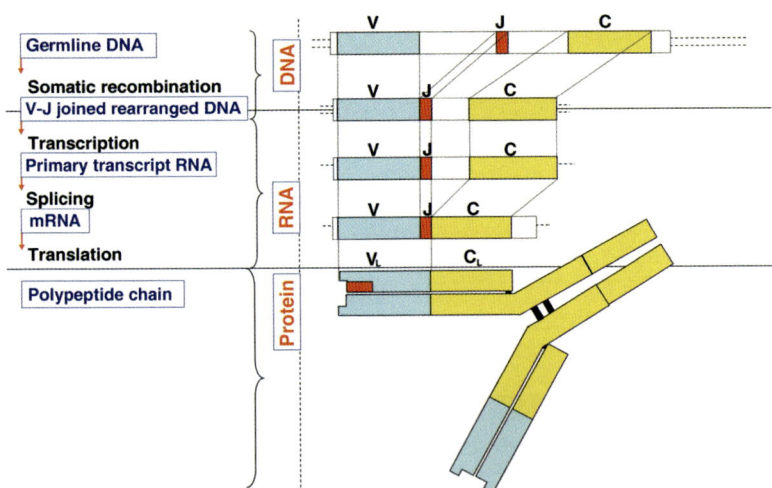

Figure 3.3. Construction of light chains.

3.3. Production

In normal individuals, approximately 500mg of FLCs are produced each day from bone marrow and lymph node cells *(Solomon 1; Waldman)*. The molecules enter the blood and are rapidly partitioned between the intravascular and extravascular compartments. The normal plasma cell content of the bone marrow is about 1%, whereas in multiple myeloma (MM) this can rise to over 90%. The bone marrow may contain 5-

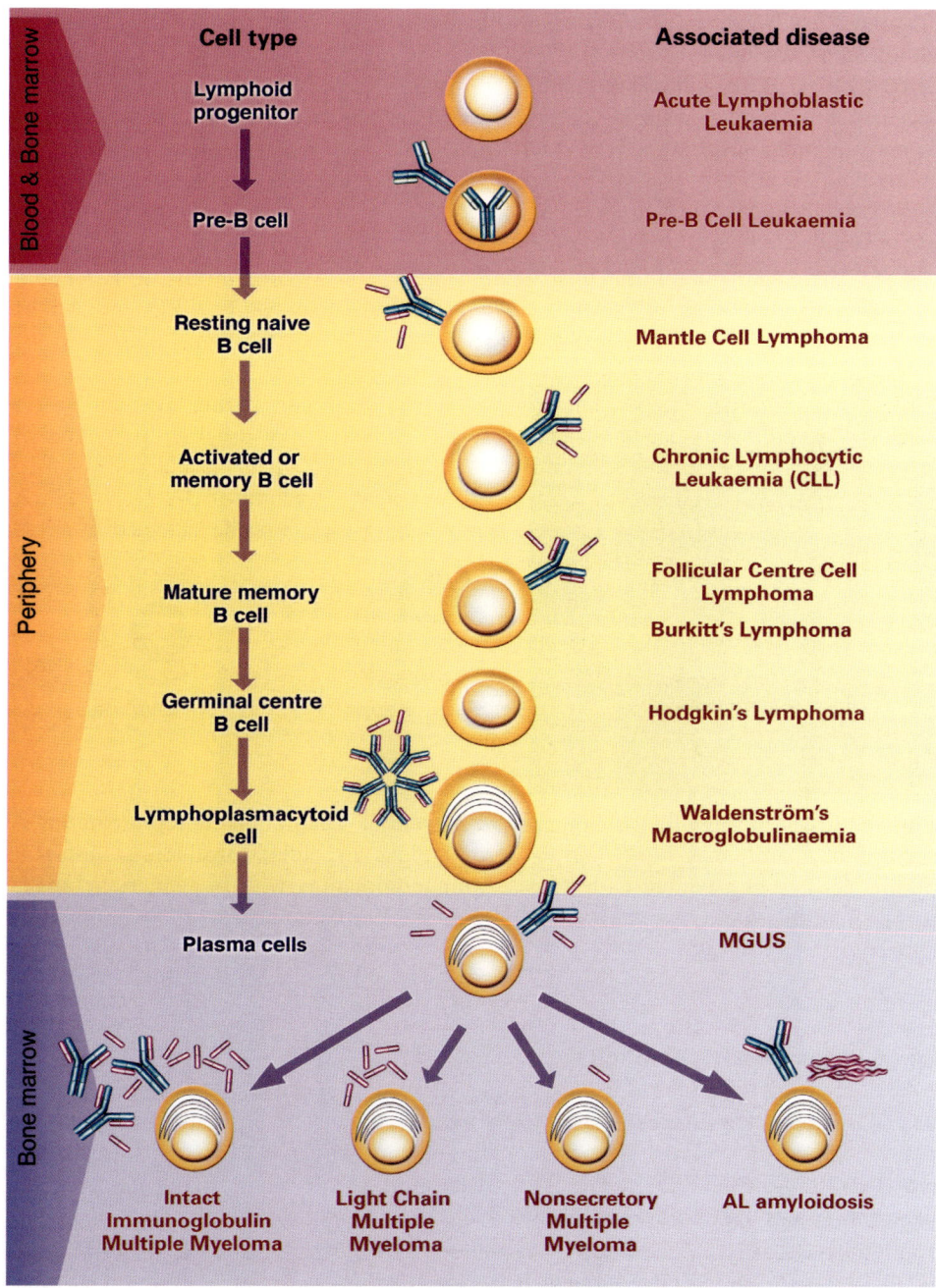

Figure 3.4. Development of the B-cell lineage and associated diseases.

10% plasma cells in chronic infections and autoimmmune diseases and this is associated with hypergammaglobulinaemia and corresponding increases in polyclonal sFLC concentrations. Bone marrow identification of monoclonal plasma cells by histology or flow cytometry is an essential part of MM diagnosis, and is frequently based on identifying intracellular κ and λ by direct immunofluorescence techniques *(Figure 3.5)*.

Plasma cells produce one of five heavy chain types, together with κ or λ molecules. There is approximately 40% excess FLC production over heavy chain synthesis, to allow proper conformation of the intact immunoglobulin molecules. As mentioned previously, there are twice as many κ-producing plasma cells as λ-producing plasma cells. κ FLCs are normally monomeric, while λ FLCs tend to be dimeric, joined by disulphide bonds, however higher polymeric forms of both FLCs may occur *(Figure 3.6)*.

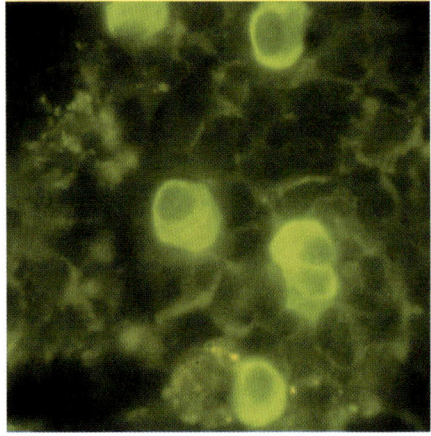

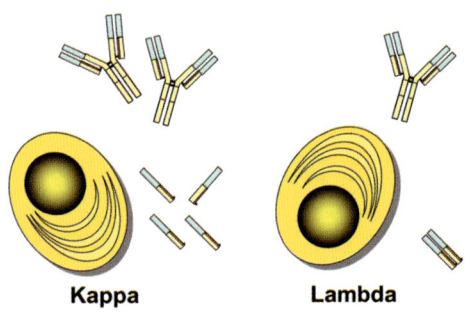

Kappa   **Lambda**

Figure 3.5. Immunohistochemical staining of κ producing bone marrow plasma cells from a patient with MM using fluorescein-conjugated, anti-κ antiserum. 5 plasma cells can be seen.

Figure 3.6. Diagrammatic representation of plasma cells producing intact immunoglobulins with monomeric κ and dimeric λ FLC molecules.

3.4. Clearance and metabolism

In normal individuals, sFLCs are rapidly cleared and metabolised by the kidneys depending upon their molecular size. Monomeric FLCs, characteristically κ, are cleared in 2-4 hours at 40% of the glomerular filtration rate. Dimeric FLCs, typically λ, are cleared in 3-6 hours at 20% of the glomerular filtration rate, while larger polymers are cleared more slowly. Removal may be prolonged to 2-3 days in MM patients in complete renal failure *(Chapter 13) (Solomon 1; Waldman; Miettinen)*. In contrast, IgG has a half-life of 21 days with minimal renal clearance *(Chapter 10)*.

Figure 3.7 shows a nephron, of which there are approximately half a million in each kidney. Each nephron contains a glomerulus with basement membrane fenestrations, which allow filtration of serum molecules into the proximal tubules. The pore sizes are variable with a restriction in filtration commencing at about 20-40kDa and being complete by 60kDa. Protein molecules that pass through the glomerular pores are then either absorbed unchanged (as with albumin) or degraded in the proximal tubular cells and absorbed or excreted as fragments *(Russo)*. This megalin/cubulin absorption pathway is important and is designed to prevent loss of large amounts of proteins and peptides into urine. It is very efficient and can process between 10-30g of small

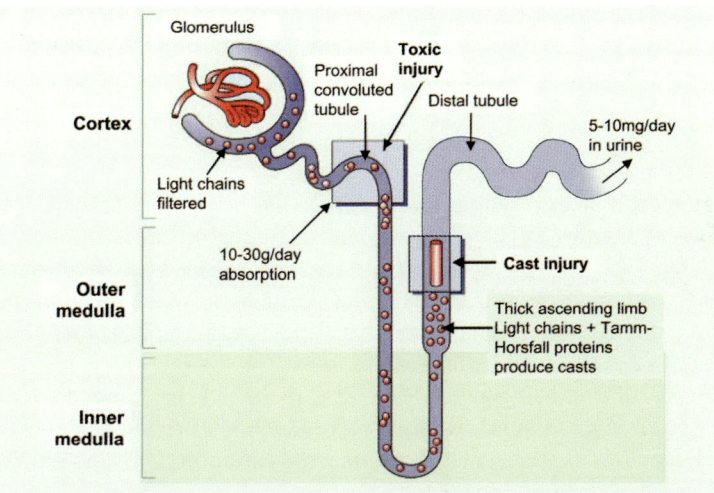

Figure 3.7. Nephron showing filtration, metabolism and excretion of FLCs. (Courtesy of R Johnson and J Feehally).

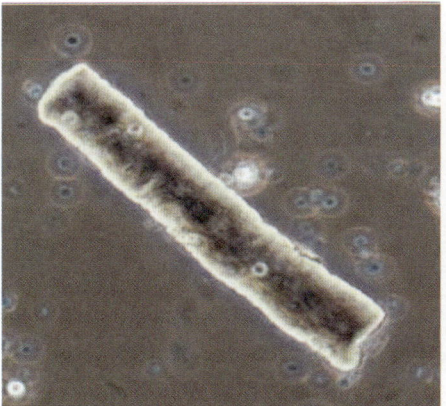

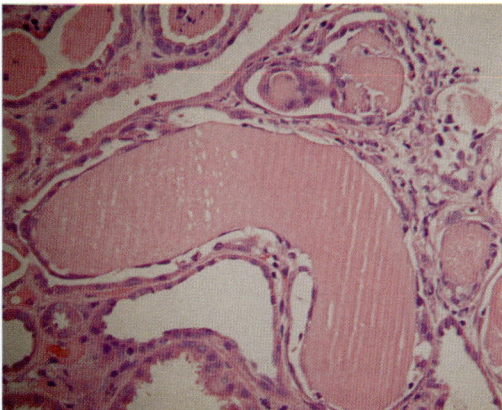

Figure 3.8. Waxy cast from the urine of a patient with multiple myeloma. (Courtesy of R Johnson and J Feehally).

Figure 3.9. Monoclonal FLC casts in the distal tubules of a patient with myeloma kidney. (Courtesy of C Hutchison).

molecular weight proteins per day, so under normal conditions, none passes beyond the proximal tubules *(Abraham GN; Wochner; Maack)*.

The distal tubule secretes large amounts of uromucoid (Tamm-Horsfall protein). This is the dominant protein in normal urine and is thought to be important in preventing ascending urinary infections. It is a relatively small glycoprotein (80kDa) that aggregates into polymers of 20-30 molecules. Interestingly, it contains a short peptide motif that specifically binds FLCs *(Ying)*. Together they form waxy casts that are characteristically found in acute renal failure associated with light chain multiple myeloma (LCMM) *(Figures 3.8 and 3.9) (Chapter 13) (Sanders 1, 2)*.

In normal individuals, 1-10mg of FLCs is excreted per day into the urine. Its exact origin is unclear but it probably enters the urine via the mucosal surfaces of the distal part of the nephrons and the urethra, alongside secretory IgA. This secretion is part of the mucosal defence system that prevents infectious agents entering the body. The 500mg of FLCs produced each day by the normal lymphoid system therefore flow through the glomeruli and is completely processed by the proximal tubules *(Russo)*.

If the proximal tubules of the nephrons are damaged or stressed (such as in hard exercise), filtered FLCs may not be completely metabolised and small amounts may enter the urine. Although important as markers of general renal function, particularly glomerular function, serum creatinine and cystatin C measurements cannot detect such minor degrees of nephron dysfunction.

As noted above, due to their smaller size κ FLC monomers are cleared 2-3 times faster than dimeric λ molecules *(Arfors)*. Although κ production rates are twice that of λ, their faster removal ensures that the actual serum concentrations are approximately 50%

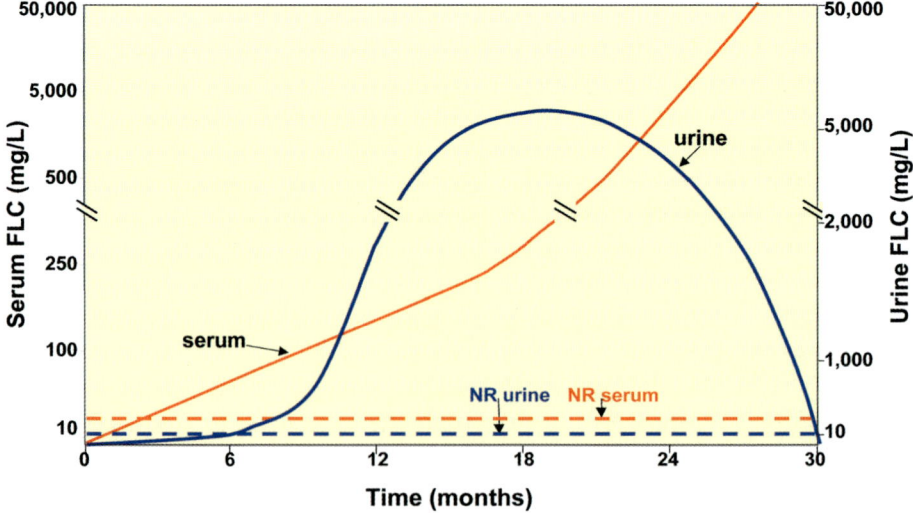

Figure 3.10. Changes in serum and urine free light chain concentrations during the evolution of a hypothetical patient with light chain multiple myeloma.

lower *(Chapter 5 and 20)*.

Because of the huge metabolic capacity of the proximal tubule, the amount of FLCs in urine, even when production is considerably increased, is more dependent upon renal function than synthesis by the tumour. As a consequence, serum and urine FLC concentrations may not be similar during the evolution of LCMM. This is shown in a hypothetical patient in Figure 3.10. The red line shows the steady increase in sFLCs as the tumour grows over the first 12 months. When synthesis of FLCs exceeds 10-30g/day (greater than 30 times normal) there is an overflow proteinuria and large amounts of FLCs enter the urine. It is normally at this point that patients with LCMM are identified.

When FLCs overwhelm the proximal tubules' absorption mechanisms, they enter the distal tubules and may cause inflammation or precipitate as casts. This can block the flow of urine causing affected nephrons to perish *(Figure 3.8 and 3.9) (Chapter 13 and 20)*. Rising concentrations of sFLCs are filtered by the remaining nephrons leading to a vicious cycle of accelerating renal damage with further increases in sFLC concentrations.

This process lengthens the serum half-life of FLCs so that concentrations rise rapidly as urine excretion decreases with the onset of terminal renal failure, falling to zero as the patient becomes aneuric *(Figure 3.10)*. Consequently, the amounts of FLCs in serum and urine diverge during disease progression. While increasing serum concentrations indicate worsening disease, falling urine concentrations may falsely suggest disease stabilisation or improvement. For example, Nowrousian et al. showed that urine FLC excretion decreased at high sFLC concentrations when there was significant renal impairment. In contrast, successful treatment of tumours may lead to a reduction of tubular casts, increased urine flow and more FLCs in urine.

Understanding the nephrotoxicity of FLCs and how it can influence FLC concentrations in serum and urine is important. The inevitable conclusion, from the physiological and pathological mechanisms described above, is that serum is preferable to urine for assessing FLC concentrations.

Test questions
1. *Is Bence Jones proteinuria "overflow", "glomerular" or "tubular" in origin?*
2. *What are the normal serum half-lives of IgG and FLCs?*
3. *Serum albumin concentrations are reduced in patients with nephrotic syndrome with gross proteinuria. Are serum FLC concentrations also reduced in these circumstances?*
4. *Which protein binds FLCs in the distal tubules?*
5. *Do urine FLC concentrations always increase alongside rising serum FLC concentrations?*

Answers
1. *Overflow proteinuria (Section 3.4, page 17).*
2. *IgG is 21 days and FLCs 2-6 hours (Section 3.4, page 14-15).*
3. *No, nephron damage from any cause increases serum FLC concentrations because glomerular filtration is always impaired (Section 3.4, page 16).*
4. *Uromucoid or Tamm-Horsfall protein (Section 3.4, page 16).*
5. *No. If there is significant renal impairment, urine FLC excretion falls (Section 3.4, page 16).*

4.1.	*Introduction*	*18*
4.2.	*Immunoassays for FLCs*	*20*
	A. Polyclonal antisera vs monoclonal antibodies	*20*
	B. Antisera specificity	*20*
	C. Accuracy and standardisation	*23*
	D. Assay sensitivity	*24*
	E. Assay ranges	*25*
	F. Antigen excess	*25*
	G. Precision	*27*
	H. Linearity	*27*
	I. FLC polymerisation	*29*
	J. Stability	*30*
	K. Serum and plasma comparisons for FLC assays	*30*
4.3.	*Comparison of FLC immunoassays on different instruments*	*31*
4.4.	*Comparison of different serum FLC immunoassays*	*32*
4.5.	*Quality control of FLC antisera and kits*	*32*
4.6.	*Scale-up of antiserum production*	*34*

Summary: FLC immunoassays have the following features:

1. Highly specific for serum and urine FLCs.
2. 1000 times more sensitive than serum electrophoretic tests.
3. Give better precision than electrophoretic tests.
4. Provide quantitative results.
5. Are performed on routine laboratory instruments.

4.1. Introduction

There are numerous assays for monoclonal free light chains (FLCs) in urine *(Table 4.1)*. Some, such as protein precipitation, are simple screening tests but are insensitive and non-specific, while others such as immunofixation electrophoresis (IFE) are more sensitive but laborious and may be difficult to interpret.

Some commonly used tests for FLCs are analysed in Table 4.1. Each assay has features that are satisfactory, to some degree or other, but all fail to detect FLCs in serum at concentrations within the normal range (3-25mg/L); or even at concentrations that are several times higher than the normal range. Dipstick assays, based upon dye uptake, are particularly unreliable for measuring cationic proteins such as FLCs and should not be used.

	Advantages	Disadvantages
Total urine protein	Simple, inexpensive, widely used	Sensitivity inadequate for FLC detection
Urine dipsticks	Simple, inexpensive, widely used	Sensitivity inadequate for FLC detection[1,2]
Serum protein electrophoresis (SPE)	Simple manual/semi-automated method Well established, inexpensive Monoclonal bands visualised Quantitative results with scanning	Insensitive (<500-2,000mg/L). Cannot detect FLCs at low concentrations Subjective interpretation of results
Urine protein electrophoresis (UPE)	Simple manual/semi-automated method Well established, inexpensive Monoclonal bands visualised Sensitive in concentrated urine (<10mg/L) Quantitative results with scanning	Subjective interpretation of results Urine may require concentration with possible protein loss[3,4] False bands from concentrating urine[5-8] Heavy proteinuria obscures results Cumbersome 24-hour urine collections[9]
IFE on serum and urine	Well established Good sensitivity for serum and very sensitive for concentrated urine (5-30mg/L)	Non-quantitative Serum sensitivity (~150mg/L) inadequate for normal serum FLC levels Manual/semi-automated technique Expensive use of antisera
Capillary zone electrophoresis (CZE)	Automated technology Quantitative	Less sensitive (~400mg/L) than IFE for serum FLCs (sFLCs) [10-13]
Total serum κ/λ assays	Automated immunoassay	Specificity inadequate for detecting many patients with light chain monoclonal gammopathies[14]

Table 4.1. Comparison of different assays for FLCs. (References: 1 (Fine), 2 (Penders), 3 (Ala-Houhala), 4 (Lindstedt), 5 (Bailey), 6 (Harrison), 7 (Hess), 8 (MacNamara), 9 (Brigden 1), 10-12 (Bossuyt 1-3), 13 (Katzmann 5), 14 (Boege).

An ideal test for FLCs would have the following characteristics:

- Sensitive diagnostic assay that identifies all patients producing monoclonal FLCs
- Quantitative measurement of FLCs and κ/λ ratios for monitoring patients
- High specificity with no interference from intact immunoglobulins
- Inexpensive to perform and easy to interpret
- Measures FLCs in serum (*see Chapter 3*)
- Can be performed on a variety of laboratory instruments *(Chapter 27)*

One modern solution to the problems of inadequate assay specificity and sensitivity is to use antibody-based methods. The following chapter describes the development of sFLC immunoassays and their validation on a variety of laboratory analysers.

4.2. Immunoassays for FLCs

A. Polyclonal antisera versus monoclonal antibodies

It is essential that FLC antibodies have high specificity and affinity. Early FLC immunoassays *(Chapter 5)* used polyclonal antisera but good specificity was difficult to obtain. Monoclonal antibodies (Mabs) seemed to be the obvious solution to the problem. However, considerable effort by my colleagues in the Immunology Department at The University of Birmingham, failed to produce antibodies that reliably recognised a full range of monoclonal FLCs. Other reported Mabs have also failed to have clinical impact, possibly for similar reasons. Furthermore, Mabs cannot be used in nephelometric or turbidimetric assays because they do not form immunoprecipitates, so enzyme immunoassays (or similar types) are required. These 3-stage assays need high serum dilutions and take considerable time to perform. Hence, they do not readily form part of the routine clinical testing alongside immunoglobulin assays.

To be practical, we therefore focussed on optimising polyclonal FLC antisera. The following description is an outline of the successful procedures involved in their development. Details of the many steps are propriety and are not available. However, in summary, sheep were immunised with κ or λ molecules that had been purified from urines containing Bence Jones proteins. The resultant antisera were adsorbed against purified IgG, A and monoclonal proteins and then affinity purified against mixtures of the respective FLCs that had been immobilised onto Sepharose. Antisera requiring further adsorption, as judged by the tests described below, were recycled through the adsorption and testing procedures until satisfactory.

B. Antisera specificity

Specificity is the most important aspect of the immunoassays and was evaluated using several techniques.

1. Immunoelectrophoresis

The antibodies were purified until they showed no cross-reactions by immunoelectrophoresis with the alternate FLC and intact immunoglobulin molecules *(Figure 4.1)*.

2. Western blot analysis

This sensitive technique was used to assess the reactivity of the antisera against immunoglobulin fragments and FLC polymers. The results showed that both κ and λ FLC antisera reacted strongly with two closely migrating bands at 25-30kDa and weakly with several larger and smaller molecular weight fragments. Similar staining patterns were observed using Mabs. The FLC antisera were readily able to detect monomers and dimers of both κ and λ molecules (*Figure 4.2*).

3. Haemagglutination assays

These assays are far more sensitive than immunoelectrophoresis and provide better assessment of specificity. Sheep red blood cells (RBCs) were sensitised with individual

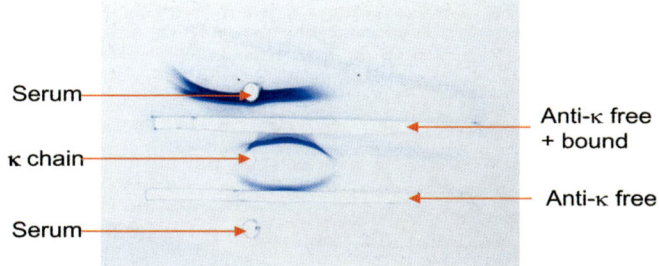

Figure 4.1. Immunoelectrophoresis showing the specificity of κ FLC antisera. The anti-κFLC antibody shows no cross reaction to proteins in normal human serum, including intact immunoglobulins (bottom well). Good anti-κ FLC activity is demonstrated by the presence of the arc against the purified κ chains. (Serum = normal human serum, κ chain = purified κ light chains).

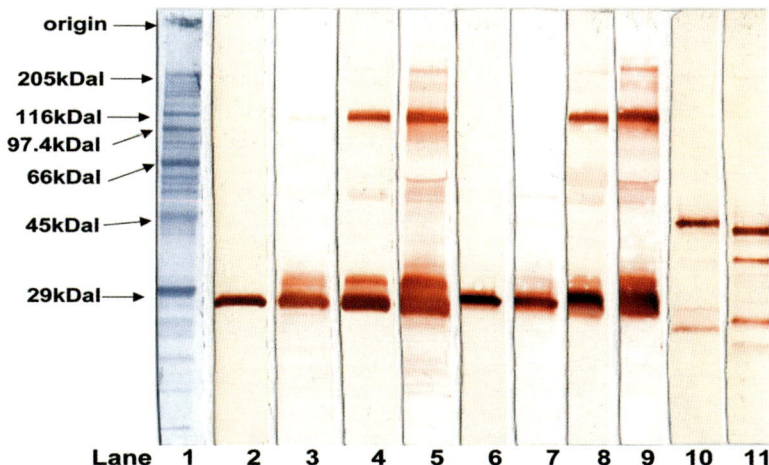

Figure 4.2. Western blots showing the specificity of the polyclonal FLC antisera compared with Mabs and the reaction of polyclonal antisera against FLC monomers and dimers separated by non-reducing sodium dodecyl sulfate - polyacrylamide gel electrophoresis (SDS-PAGE). (Lane 1, molecular weight markers. Lanes 2 & 3, urine containing κ FLCs and 4 & 5, normal serum all probed with mono- and polyclonal anti-κ. Lanes 6 & 7, urine containing λ FLCs and 8 & 9, normal serum probed with mono- and polyclonal anti-λ. Lanes 10 & 11, polyclonal FLC antisera reacting with monomers and dimers of κ and λ).

FLCs, and purified IgG, IgA and IgM and tested against the FLC antisera. The results showed that κ and λ FLC antibodies reacted with the appropriately labelled cells at >1:16,000 dilution and at <1:2 against cells coated with the alternate FLCs or intact immunoglobulins (*Figure 4.3*).

4. Nephelometry

Latex-conjugated FLC antisera were tested for specificity by nephelometry. Potentially interfering substances were added to serum containing known concentrations of FLCs and the changes in values indicated the effect on the assays *(Figure 4.4)*.

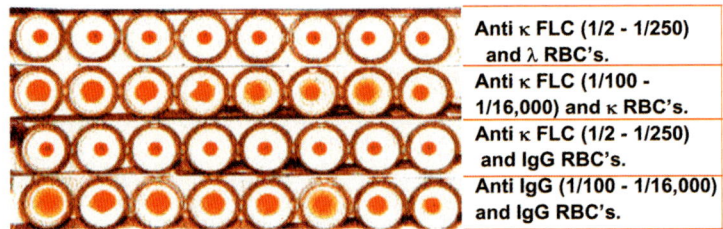

Anti κ FLC (1/2 - 1/250) and λ RBC's.
Anti κ FLC (1/100 - 1/16,000) and κ RBC's.
Anti κ FLC (1/2 - 1/250) and IgG RBC's.
Anti IgG (1/100 - 1/16,000) and IgG RBC's.

Figure 4.3. Haemagglutination assays showing the specificity of κ FLC antisera against RBC coated with purified FLCs and IgG.

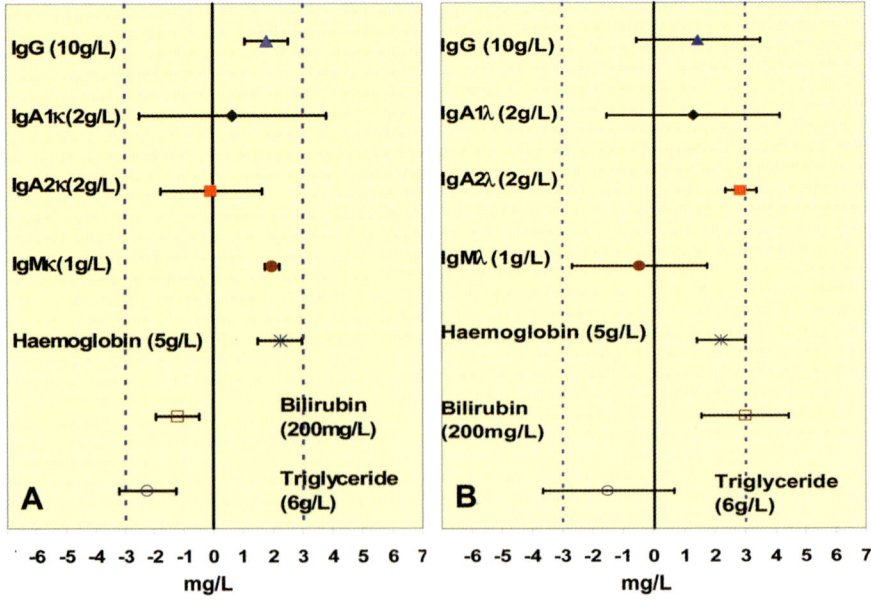

Figure 4.4. Specificity of (A) κ and (B) λ FLC antisera assessed by interference with the results of typical nephelometric assays upon the addition of various substances. Mean and 95% confidence limits for each added substance are shown.

Overall, the specificity assessments showed that FLC antisera had minimal reactivity with light chains on intact immunoglobulins and other potentially interfering substances (0.2-0.01%). These values are within the purity specification for FLC contamination in the tested interfering materials.

There have been no published independent specificity analyses of the nephelometric Freelite latex reagents. Nakano *(3)* et al. reported an evaluation but, in error, only tested FLC antisera that were manufactured for IFE, where specificity requirements are less demanding *(Robson 1)*.

C. Accuracy and standardisation

Accuracy is defined as the closeness of achieved results compared with their absolute values. Unfortunately, international standards do not exist for FLC measurements, so there are no reference points from which to assess the accuracy of results. Furthermore, each monoclonal light chain is unique, with its own special set of surface epitopes, so accurate measurements are difficult to obtain. Nevertheless, the light chain constant region domains have little structural variability, so they are good antibody targets.

In order to ensure accurate FLC immunoassays, a suitable basis for standardisation and calibration was required. It was considered that polyclonal FLCs should be used in order to minimise any potential problems that might arise from the use of unique monoclonal proteins *(Carr-Smith 1)*.

This was achieved in the following manner *(Figure 4.5)*:

1. Production and accurate quantification of pure polyclonal, *"primary"* FLC standards.
2. Production of secondary and *"working"* reference materials. Comparison with the primary standards.
3. Production of calibration materials for use in the FLC assays that were referenced against the *"working"* standards.
4. Analysis of a variety of normal and abnormal samples using a reference nephelometric method.
5. Comparison of results from other instruments with the reference method.

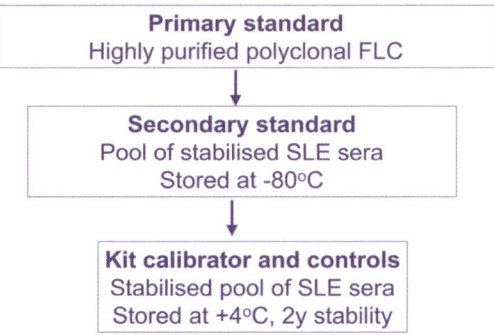

Figure 4.5. Flow chart for the production of standards for serum FLC immunoassays. (SLE: systemic lupus erythematosus, y: years)

The various materials were manufactured and purified in accordance with established procedures *(Bradwell 2)*. Each primary FLC preparation was found to be greater than 99% pure by silver-stained SDS-PAGE while the alternate FLC was not detected by haemagglutination inhibition and dot blot assays. The amino acid content of each primary standard was then determined in order to produce an accurate estimation of the protein content.

Secondary reference materials were prepared from pools of different monoclonal κ and λ proteins. These were not considered ideal for use as working calibrators, so additional reference materials were prepared from SLE sera that contained elevated polyclonal FLCs *(Figure 4.5)*.

The pure FLC preparations were used to assign κ and λ concentrations to the secondary reference preparations. Subsequently, κ and λ values were assigned to the serum pool by nephelometry. Each stage of the value transfer was completed at three dilutions and repeated three times. All protein preparations were stabilised and stored at -80°C until required. The final FLC values for the *"working"* reference preparations were 46.0mg/L for κ and 71.4mg/L for λ. These calibration values were used for all subsequent laboratory and clinical studies.

An alternative reference material based upon polyclonal FLCs extracted from urine has been proposed *(Nakano 1)*. However, sFLCs may differ chemically from urine FLCs and concentration measurements in the latter study were based upon inappropriate dye uptake comparisons with albumin rather than amino acid analysis. As yet, these two preparations have not been compared. Clearly, there is need for international agreement on standardisation with wide availability of a reference material for κ and λ FLCs.

D. Assay sensitivity

Assay sensitivity (the lowest limit of antigen detection) is somewhat dependent upon the types of sample measured. FLC assays can detect less than 0.1mg/L of κ and λ FLCs in undiluted, clear urines and CSF. Serum sensitivity is typically 5-fold worse because of interference from lipids and other light-scattering particles *(Table 4.2 and Chapter 6)*. Concentrations at 1-2mg/L are below the normal serum range so the FLC immunoassays allow extreme κ/λ concentration ratios to be established with considerable accuracy. This is important for monitoring patients under treatment because FLC concentrations may be below the normal range. Additional sensitivity for urine and CSF measurements *(Chapter 22)* can be obtained by adjusting concentrations of the reactants.

	Kappa	Lambda	Diagnostic requirement
SPE	500-2,000mg/L	500-2,000mg/L	monoclonal band
IFE	150-500mg/L	150-500mg/L	monoclonal band
FLCs	1.5mg/L	3.0mg/L	abnormal κ/λ ratio

Table 4.2. Representative routine sensitivity levels of frequently used FLC assays.

I. FLC polymerisation

FLC molecules are usually monomers or dimers but higher polymeric forms frequently occur *(Berggard; Diemert; Abraham 2; Sölling 2, 6; Heino; Émond)*. They act as multi-antigenic targets in immunoprecipitation assays which accelerates the formation of aggregates and leads to over-estimation of antigen concentrations. This occurs in patients with nonsecretory multiple myeloma (NSMM) who have undetectable concentrations of sFLCs by IFE but high concentrations by nephelometry *(Chapter 9)*. Sölling *(2 and 6)* indicated that with his assays, monomers and dimers were detected equally using antibodies against whole light chains, but antibodies directed only against FLC epitopes may preferentially detect dimers *(Heino)*.

In order to determine the effect of polymerisation on FLC quantification, FLC monomers, dimers and polymers were purified from myeloma sera and concentrations compared with total protein measurements. It was apparent that purified dimers were over-estimated by 1.5-fold and higher polymers by 1.5 to 3.5-fold *(Mead 4)*. However, in a study by Émond et al. a greater than 7-fold over-estimation was observed in two samples (when compared with CZE) in association with polymers of up to 200kDa.

An additional factor is that SPE tests can underestimate FLC concentrations. Variable polymerisation may cause *"smearing"* of monoclonal bands on the gels so that only a proportion of the monoclonal protein is measured *(Figure 9.2)*. FLCs also take up less protein stain than albumin so their concentration is underestimated *(Penders)*. It is likely that a combination of these factors causes the ten-fold or more over-estimations seen in the sera of some patients.

Whatever the explanation for the unexpected high results, it will be difficult to develop FLC assays that measure all molecular forms equally. Perfect quantification will be elusive.

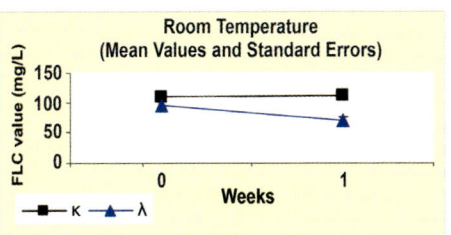

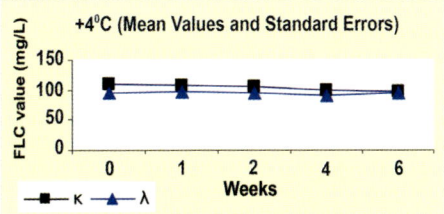

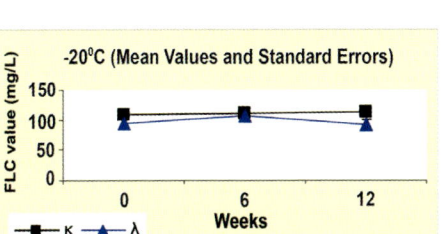

Figure 4.13. Stability of FLCs over different times and at different temperatures. (κ [square]; λ [triangle]. Standard deviations of results are mostly within the symbols).

J. Stability

FLC molecules are very stable in serum and urine *(Figure 4.13)*. Tencer *(1 and 2)* et al. showed that there was little variation in the concentrations of FLCs in samples stored at -20°C over a two year period. We have found similar concentrations of FLCs in both fresh and old normal sera stored for 20 years at -20°C. Clearly, these results indicate that FLCs do not dissociate from intact immunoglobulins, or fragment, over prolonged periods.

Stability of the FLC antisera is also important. *"Open-vial stability"* refers to the shelf-life of the antisera after their first use. This may be short because pipetting procedures can introduce FLC molecules into the vials that react with the remaining antisera and reduce its activity. Care should be taken not to contaminate antisera vials with sera from previously pipetted samples.

K. Serum and plasma comparisons for FLC assays

Some laboratories prefer to analyse blood proteins in plasma rather than sera. A study was performed to directly compare the concentrations of FLCs in plasma and serum. 50 paired serum and plasma samples from blood donors and 20 paired samples from patients with multiple myeloma (MM) were studied on a Beckman Immage, a Siemens Dade Behring BNII

	Beckman Coulter IMMAGE®		Siemens BN™II		Roche Hitachi 911	
	Kappa FLC	Lambda FLC	Kappa FLC	Lambda FLC	Kappa FLC	Lambda FLC
Slope	1.01	1.02	0.97	0.94	1.02	1.01
Intercept	-0.85	-1.29	0.63	4.37	-0.35	-1.71
Correlation	1.00	1.00	0.99	1.00	1.00	1.00

Table 4.3. Comparison of serum and ACD plasma FLC concentrations from 50 blood donors and 20 MM patients.

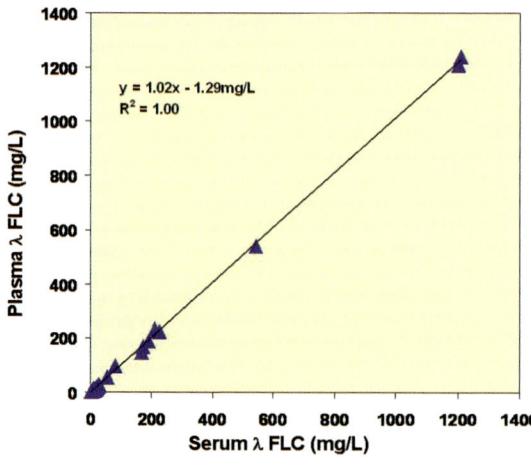

$y = 1.02x - 1.29 \text{mg/L}$
$R^2 = 1.00$

Figure 4.14. Comparison of serum and acid citrate dextrose plasma samples from 50 blood donors and 20 MM patients using a Beckman Immage.

and a Hitachi 911. Plasma samples were collected in acid citrate dextrose (ACD) or heparin and the sera were prepared by adding fibrin. Table 4.3, Figure 4.14 and Figure 4.15 show comparison data obtained on different instruments *(Smith LJ)*.

4.3. Comparison of FLC immunoassays on different instruments

sFLC kits are made for many laboratory instruments. A list of those currently available is given in Chapter 27. Comparison of FLC results using a Hitachi 911 and a Siemens Dade Behring BNII are shown in Figure 4.15. and four different instruments are shown in Table 4.4. Comparisons of normal ranges are discussed in Chapter 5. There are no significant differences between the results on different instruments *(Matters)*.

	Siemens BN™II	Beckman Coulter IMMAGE®	Roche Cobas Integra®	Binding Site SPAPLUS
Sensitivity (mg/L)	κ 0.30 (1/5) : λ 0.25 (1/5)	κ 3.0 (1/5) : λ 2.4 (1/5)	κ 0.6 (neat) : λ 1.3 (neat)	κ 0.4 (neat) : λ 0.5 (neat)
Precision (Intra-)	3.1-8.4%	2.0-8.1%	0.7-5.8%	1.6-3.4%
(CV) (Inter-)	4.7-8.4%	5.8-11.7%	0.7-2.7%	0.0-4.2%
Antigen excess	Good	Good	Good	Good
Analytical time	18 minutes	10 minutes	10 minutes	15 minutes
Time to run 20 normal/20 MM samples	52/127 minutes	40/84 minutes	33/75 minutes	33/68 minutes
Higher dilutions	Automatic	Off-line if high	Automatic & off-line if very high	Automatic & off-line if very high
Utility	Closed systems	Open system	Closed system, FLC channels on standard menu.	Closed system

Table 4.4. Summary of the characteristics of FLC assays on different instruments. (C.V. = coefficient of variation in precision studies. Numbers in brackets refer to sample dilutions). *See Chapter 27 for more details.*

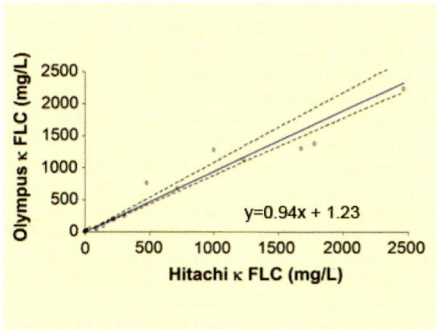
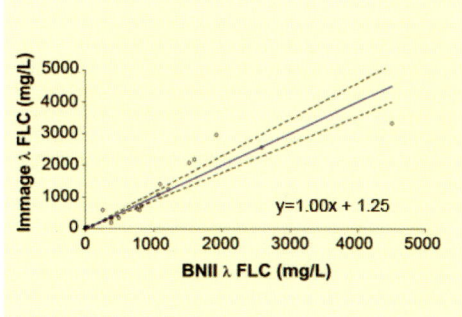

Figure 4.15. Comparison of serum samples on different platforms for κ (left) and λ (right).

4.4. Comparison of different serum FLC immunoassays

The Binding Site serum FLC assay (*Freelite™*) is the only one validated for serum measurements. New Scientific Company (NSC), from Italy, sells a urine assay for FLCs that they claim can be used for serum measurements. However, since the antisera are not latex-conjugated, there is insufficient sensitivity to measure normal sFLC concentrations. Furthermore, the antisera cross-react with intact immunoglobulins so that normal samples produce higher results than the Freelite assays. Consequently, sera containing elevated monoclonal FLCs may be classified as normal. In a study of 24 patients with monoclonal gammopathies, 15 were abnormal by *Freelite* but only 5 using NSC reagents. Of the 7 patients with AL amyloidosis in the cohort, all were positive by the *Freelite* assay but only 2 using NSC reagents. Furthermore, fluctuations of FLC concentrations were observed with *Freelite* reagents during monitoring that were undetectable by NSC reagents *(Ricotta; Morandeira)*. Thus, by scientific and clinical criteria, such unvalidated reagents should not be used for sFLC measurements.

4.5. Quality control of FLC antisera and kits

Maintaining batch-to-batch consistency is essential as the assays may be used for monitoring individual patients over many years. Effective quality control is ensured using a variety of techniques, two of which are briefly described here. External quality assurance schemes are described in Chapter 28.

Specificity is controlled by comparing a set of test results from each new batch of antiserum with results from previous batches. Typically, the panel of samples includes normal sera, sera with elevated polyclonal FLCs and myeloma sera. The results are compared using regression analysis and are considered acceptable when they fall within

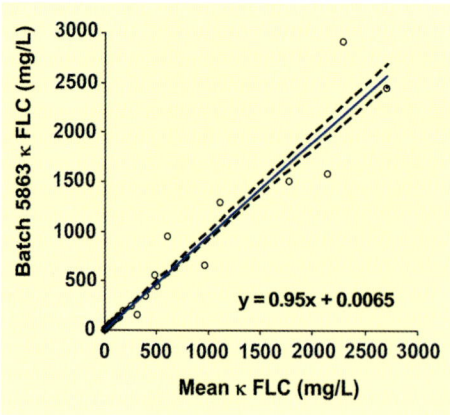
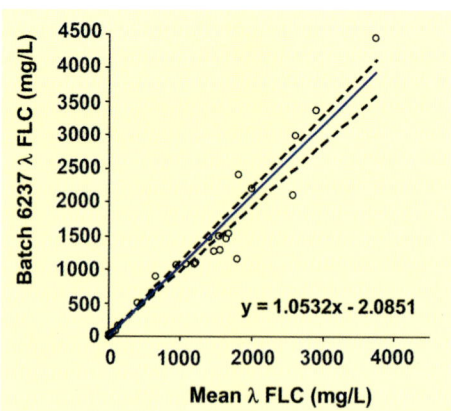

Figure 4.16. Comparison of results obtained from two batches of kits with the mean results from previous analyses. (The mean results were obtained from an average of 12 assays over 6 batches of reagent. The samples included 30 normal sera and 30 myeloma sera of various immunoglobulin types).

a defined set of criteria *(Figure 4.16)*.

Once suitable antisera have been selected and attached to latex particles, the other kit components are selected and assembled. Similar materials are used for calibrators and controls and comprise sera containing high concentrations of polyclonal FLCs. The first step is to assign a value to the kit calibrators using the working reference material. This is achieved using 100 separate assays and 10 separate calibration curves. The second step is to assign values to the control reagents using similar procedures.

Analytical comparisons are also made using normal sera. A typical evaluation on 30 normal samples produced the following results: mean κ = 10.8mg/L (range 4.5 - 17.1mg/L), mean λ = 18.0mg/L (range 8.2 - 31.7mg/L), mean κ/λ ratio = 0.58. Inter-instrument agreement is also important with results from 4 instruments shown in Table 4.4.

The long-term effect of technical improvements, together with more experience of kit production, customer feedback *(Figure 4.17)* and larger batch production, has improved quality. This is apparent from the considerable reduction in customer enquiry rates, per unit sold, since the kits were first introduced 9 years ago *(Figure 4.18) (Carr-Smith 6)*. Upward trends typically occur as larger reagent batches are manufactured. These are advantageous because there is more material that can be used for assignment of reference values, stability checks, calibrator checks, specificity and sensitivity analysis, etc. Downward trends are associated with the introduction of new kits such as for the SPA$_{PLUS}$™ instrument, in Autumn 2007. Newly introduced assays always have teething problems so it is of note that issues with the early kits have been largely resolved.

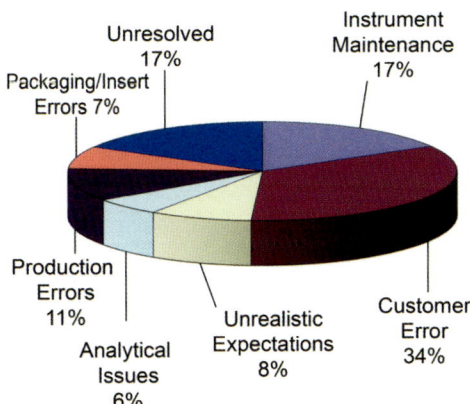

Figure 4.17. Analysis of outcome of customer enquiries.

Figure 4.18. Numbers of units (100 tests) sold per customer

4.6. Scale-up of antiserum production

Scale-up involves producing larger batches, both by pooling antisera from more sheep and by using larger and more automated laboratory equipment. The polyclonal reagents have to be adsorbed to complete specificity and then affinity purified by passage through large chromatography columns. This is controlled using an *AKTA biopilot system* that can automatically load, wash, elute and collect antibody peaks (*Figure 4.19*). Batch sizes have increased a thousand fold, from a few hundred millilitres to hundreds of litres. In the life-cycle of diagnostic tests, however, the FLC immunoassays are still in their youth. Improvements will be seen in sensitivity, specificity, reactivity, precision, utility, costs, etc.

Figure 4.19. AKTA biopilot system for controlling column

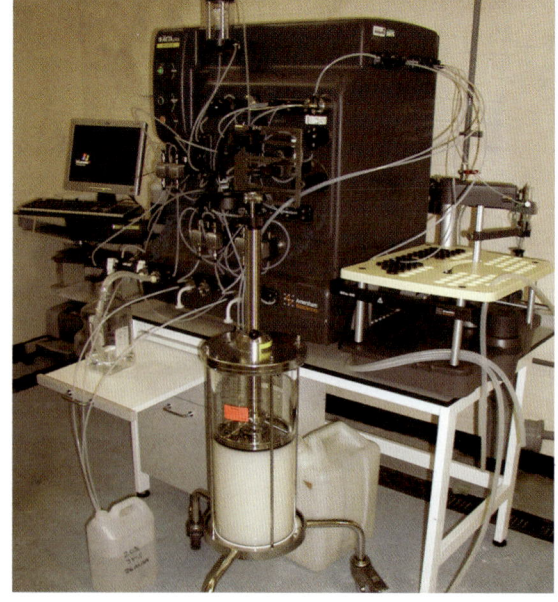

Test questions

1. Why do dye uptake tests for proteinuria fail to detect FLCs accurately?
2. How much more sensitive are immunoassays for FLCs than SPE?
3. Are sFLCs unstable?
4. What is a typical assay precision for FLC tests?

Answers

1. The dyes do not bind readily to cationic proteins such as FLCs (Section 4.1, page 18).
2. Approximately 1,000-fold (D. Assay Sensitivity and Table 4.2, page 24).
3. No. Analysis of frozen samples, even 20 years old, show normal light chain levels (J. Stability, page 30).
4. 5-10%. (G. Precision and Table 4.4, page 31).

5.1. *Serum free light chain (sFLC) normal ranges*   *36*
5.2. *Normal ranges in children*   *39*
5.3. *Variations in normal/reference ranges - hospital ranges*   *39*
5.4. *Borderline Results*   *41*
5.5. *Utility of sFLC assays for disease diagnosis*   *41*
5.6. *Urine free light chain (uFLC) normal ranges*   *42*

Summary: In sera from normal individuals:

1. sFLC concentrations and κ/λ ratios are maintained within narrow limits.
2. sFLC concentrations increase slightly with age due to reduced glomerular filtration.
3. Serum κ/λ ratios are raised slightly in unrecognised renal impairment.
4. sFLC concentrations are less variable than in urine.

5.1. Serum free light chain (sFLC) normal ranges

Normal range data for sFLCs have been published many times: those considered to be the most reliable are shown in Table 5.1. Serum results using polyclonal antibodies have varied considerably, indicating a degree of cross-reactivity with bound light chains in some of the assays. Perhaps unsurprisingly, assays using monoclonal antibodies have also produced varied results, the discrepancy between reports being more than tenfold *(Nelson; Abe; Nakano 2, 3)*. It is unclear whether the variations were due to specificity,

Publication	Date	κ sFLC	λ sFLC	κ/λ ratio
Abraham GN	1974	24.1 (19-36)[a]	17.4 (10-38)[a]	1.5 (range 0.9-2.4)
Sölling *(1)*	1975	13.2 (SD±3.8)[a]	10.6 (SD±3.1)[a]	1.2
Brouwer	1985	16.2 (CV 9%)[c]	41.4 (CV 10%)[c]	0.38
Axiak	1987	1.2 (0.8-7.5)[d]	-	-
Wakasugi *(1)*	1991	7.1 (SD±4.3)[c]	5.0 (SD±2.7)[c]	2.1 (SD±0.4)
Nelson	1992	10.0 (1.6-15.2)[d]	3.0 (0.4-4.2)[c]	3.3 (range 1.2-9.1)
Wakasugi *(2)*	1995	20.6 (SD±6)[ce]	16.2 (SD±8.6)[ce]	1.4 (SD±0.4)
Abe	1998	16.6 (SD±6.1)[d]	33.8 (SD±14.8)[d]	0.5 (range 0.25-1.0)
Bradwell *(2)*	2001	8.4 (4.2-13)[cef]	14.5 (9.2-22.7)[cef]	0.6 (range 0.36-1.0)
Katzmann *(6)*	2002	7.3 (3.3-19.4)[cef]	12.7 (5.7-26.3)[cef]	0.6 (range 0.26-1.65)
Nakano *(3)*	2004	3.1 (SD±1.2)[d]	3.3 (SD±1.4)[d]	0.9 (SD±1.3)
Nakano *(2)*	2006	43.5 (SD±12.0)[d]	55.2 (SD±17.9)[d]	0.9 (SD±0.23)

Table 5.1. Selected publications reporting normal sFLC concentrations (mg/L) and κ/λ ratios *(Bradwell 2; Nakano 2)*. ([a]polyclonal antibody against total light chains, [c]polyclonal antibody against FLCs, [d]monoclonal antibody against FLCs, [e]latex reagent, [f]95% range).

calibration or matrix differences, or a combination of these factors.

The most detailed study of FLC concentrations in normal individuals was published by Katzmann (6) et al. and showed similar results to those previously reported by Bradwell (2) et al. The same assay procedures were used in both studies but a wider age range of individuals was studied by Katzmann et al. Serum samples were obtained from 127 healthy blood donors (21-62 years) and 155 older, normal individuals (51-90 years). A summary of the median values and reference ranges in the Katzmann study is shown in Tables 5.1 and 5.2.

The results of κ and λ sFLC measurements for all individuals in the Katzmann study are shown in Figure 5.1. There were no significant differences between results obtained using either fresh or frozen sera but there was a trend towards higher concentrations in elderly people *(Figure 5.1 and Table 5.3)*. This occurred for both FLCs and also for the renal function marker, cystatin C, which was measured in the same samples. Calculations of sFLC/cystatin C ratios and κ/λ ratios on each sample normalised the elevated values *(Figure 5.1, c,e,f)*. Therefore, the higher sFLC values seen in older people can be explained by small reductions in glomerular filtration rate *(Deinum)*.

An alternative method of presenting normal range data is shown in Figure 5.2. This shows the serum κ and λ results for each person plotted on a logarithmic scale. Such data presentation inherently includes the κ/λ ratios and is useful for visualising results from individual patients. It also allows easy comparison between different disease groups and is used extensively in later chapters for clinical comparisons.

One discrepancy between the results of Bradwell et al. and Katzmann et al. and those of many earlier researchers is that serum κ concentrations were found to be lower than serum λ, whereas traditionally κ levels were higher than λ *(Table 5.1)*. Since there are nearly twice as many κ- as λ-producing lymphoid cells, this observation had seemed reasonable. Certainly, the total serum κ and λ levels reflect the normal ratio of κ to λ FLC synthesis *(Table 5.2)*.

A plausible explanation for the inverted serum κ/λ ratio relates to the kinetics of FLC clearance. Since κ molecules are normally monomeric (25kDa), their renal clearance is faster than dimeric λ molecules (50kDa); consequently they accumulate less in serum. Calculation of the κ/λ clearance rate from serum and urine FLC concentrations ([κ urine]/[λ urine] ÷ [κ serum]/[λ serum]) produced a result of 3.0 *(Bradwell 2)*. In a study on the movement of dextran polymers across capillary membranes, it was shown that molecules of 20kDa were cleared 3.2 times faster than 37kDa molecules *(Arfors)*. Admittedly, polysaccharides are different in charge, shape and flexibility from globular

	Free light chains	Total light chains
Kappa (95% range)	7.3mg/L (3.3-19.4)	2,520mg/L
Lambda (95% range)	12.7mg/L (5.7-26.3)	1,430mg/L
κ/λ ratio (100% range)	0.6 (0.26-1.65)	1.78 (mean)
κ/λ ratio (95% range)	0.6 (0.31-1.2)	N/A

Table 5.2. Median values and ranges for free and total light chain concentrations and κ/λ ratios in the sera of 282 normal individuals.

proteins of similar molecular weight, but differential glomerular filtration probably accounts for the inverse κ/λ ratios. This may not have been observed in some of the earlier FLC studies due to poor antisera specificity. Recent studies have shown more consistency in serum κ/λ ratios although Nakano *(2, 3)* et al. remain uncertain about actual FLC concentrations *(Table 5.1)*.

The κ/λ ratio is the most important parameter when distinguishing monoclonal from polyclonal increases in sFLCs. Polyclonal increases result from increased synthesis or decreased renal clearance of normal FLCs *(Chapter 20-21)*. These two processes increase both κ and λ concentrations equally, thereby maintaining a fairly constant κ/λ ratio. In patients with renal failure, the half-life of both FLCs is prolonged from a few hours to several days, so concentrations increase 20-fold or more. However, κ/λ ratios remain within fairly narrow limits. In contrast, in monoclonal gammopathies, only one of the FLC concentrations increases. Thus, κ/λ ratios distinguish monoclonal from polyclonal diseases.

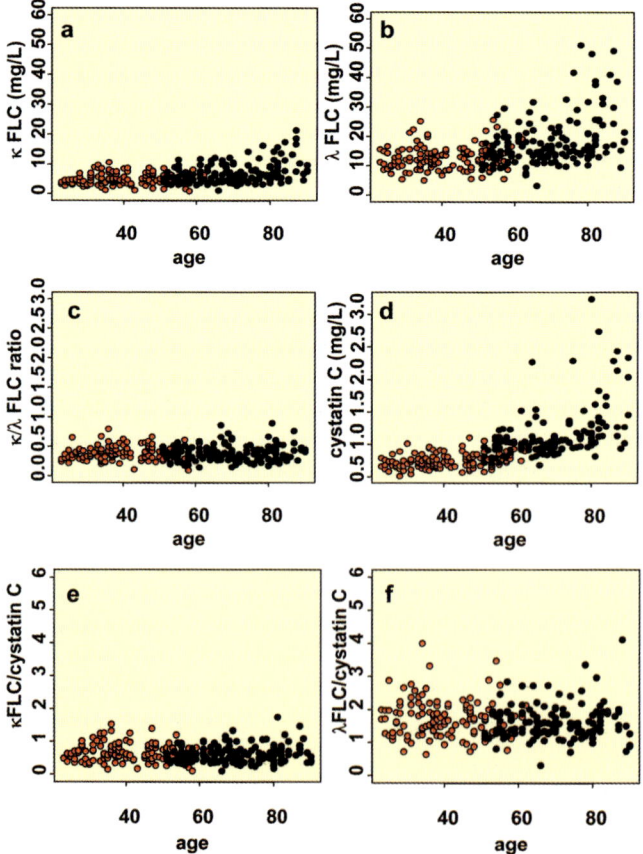

Figure 5.1. (a) κ, (b) λ sFLC concentrations and (c) κ/λ sFLC ratios versus age (years) in 282 normal serum samples together with (d) cystatin C results in the same patients. sFLC/cystatin C ratios (e) and (f) show no change with age, confirming renal deterioration as the cause of increased sFLC levels in elderly individuals. (red = fresh sera, black = frozen sera. *(Courtesy of JA Katzmann)*.

Age, years	κ FLC, mg/L	λ FLC, mg/L	FLC, κ/λ
20-29	6.3	12.4	0.49
30-39	7.2	13.6	0.55
40-49	7.5	12.8	0.58
50-59	6.4	11.3	0.59
60-69	6.9	11.8	0.70
70-79	8.0	11.9	0.65
80-90	9.1	15.1	0.64

Table 5.3. Median values for sFLCs and κ/λ ratios in different age groups.

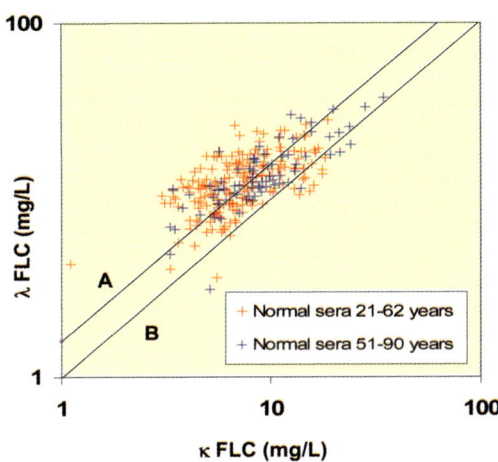

Figure 5.2. κ and λ sFLC concentrations in 282 normal sera. A: axis at the normal κ/λ ratio of 0.6. B: axis at a κ/λ ratio of 1.00.

5.2. Normal ranges in children

Only one study of sFLCs in children has been reported. This was conducted by Sölling (3), in 1977 (Figure 5.3), who noted that sFLC concentrations fall after birth and then increase with B cell development at six months. It is likely, therefore, that diseases such as agammaglobulaemia could be detected in the newborn from low sFLCs.

5.3. Variations in normal/reference ranges - hospital ranges

While decreasing renal function has little influence on sFLC κ/λ ratios, there is, nevertheless, a small but observable effect. As renal clearance falls the proportion of FLCs removed by other organs increases (Chapter 3.4). Since this is by pinocytosis of serum, there is no distinction between κ and λ removal rates. Relative to λ, κ FLC molecules are therefore removed more slowly than by the kidneys and concentrations rise. Hence, κ/λ ratios increase with deteriorating renal function and are highest in complete renal failure (Chapter 20). The data in Table 5.3 show the higher κ/λ ratios in normal elderly individuals, who typically have reduced glomerular filtration. This has practical importance for identifying multiple myeloma (MM) patients in screening studies. Thus, by using the Katzmann FLC ratio range of 0.26–1.65, assay specificity for patients with MM is 93%. When the renal reference range of 0.37–3.1 is used, the assay specificity increases to 99% *(Hutchison 5, 7; Vermeersch; Marshall; Abadie 2)*.

When screening symptomatic patients in a hospital setting, many ill patients have inflammatory conditions (causing increased polyclonal immunoglobulins and FLCs) with associated renal impairment and they show small increases in κ/λ ratios. An example of such borderline results, seen in the context of a screening study, is shown in Figure 5.4 *(Hill)*.

This is an important issue as some samples will be from patients who have no evidence of monoclonal gammopathy but have previously unrecognised renal impairment that should be investigated. Also, borderline elevated κ/λ ratios should be assessed in the context of an accurate glomerular filtration marker such as cystatin C when monitoring patients with changing renal function *(Chapter 20)*.

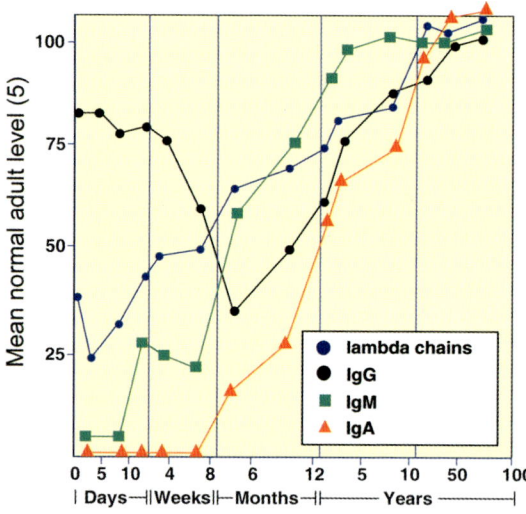

Figure 5.3. Changes in sFLC during childhood. Modified from Sölling *(3)* to allow comparison with adult concentrations. (Reproduced with permission from Scand J Clin Invest).

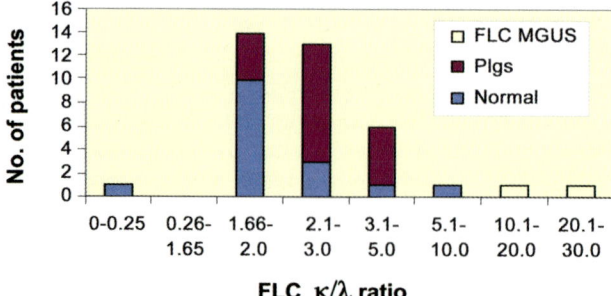

Figure 5.4. Frequency of borderline κ/λ sFLC ratios (37 in total) identified in a routine screening study for monoclonal gammopathies in 925 hospitalised patients. (PIgs: polyclonal immunoglobulins, FLC MGUS: FLC gammopathy of undetermined significance). (Courtesy of P Hill).

5.4. Borderline results

All tests may produce borderline results that should be considered in their clinical context. The diagnostic range for the serum κ/λ FLC ratio of Katzmann (6) et al. included all samples in a population of 282 individuals, so the extreme values are approximately 5 standard deviations from the median. Since this range is far greater than that used for most analytes (typically 2-3 standard deviations), borderline results for sFLC κ/λ ratios are more likely to be clinically significant. Variations in instrument performance, differences between instruments and variations in kit manufacture etc *(Chapter 4)* are correspondingly of less significance. In addition, FLC results are usually visualised on κ/λ ratio log plots. This is essential because of the huge range of clinical results; however, such a diagrammatic representation exaggerates differences between low concentration results compared with those at high concentrations. Consequently, undue emphasis may be placed on borderline analytical differences compared with huge differences in clinical results that are highly elevated.

From the clinical perspective, borderline elevated results are due to a variety of causes. Increases in FLC concentrations and κ/λ ratios from renal impairment are well documented *(Chapter 20)* as are those increases in sFLC concentrations associated with polyclonal B-cell activation in infections, autoimmune diseases and diabetes mellitus *(Chapter 21)*. Added to these are a variety of monoclonal diseases encompassing all the plasma cell dyscrasias, many lymphomas and leukaemias *(Chapter 18)*, AL amyloidosis *(Chapter 15)* and FLC monoclonal gammopathy of undetermined significance (MGUS) *(Chapter 19)*. Borderline abnormal results arising from these diseases may decrease the specificity of the test for light chain multiple myeloma (LCMM), for example, but need to be interpreted in clinical context and with the results of other laboratory tests. In screening studies, borderline elevated results may indicate a significant disease process. Secondary tests such as serum creatinine for assessing renal impairment are then necessary, while follow-up FLC tests will help determine whether a disease has progressed.

5.5. Utility of sFLC assays for disease diagnosis

Using the reference intervals and diagnostic ranges shown above, Katzmann *(6)* et al., assessed the utility of sFLC measurements for identifying monoclonal FLC gammopathies *(Table 5.4)*. Sensitivity, specificity, positive predictive value (PPV), negative predictive value (NPV) and accuracy were estimated for the sFLC κ/λ ratios on the basis of both the central 95% interval and a diagnostic range that included 100% of normal results. Accuracy was calculated as the proportion of individuals classified correctly. PPV and NPV were calculated assuming a 15% prevalence of monoclonal proteins in the samples submitted for monoclonal protein studies. The patients in these studies are described in Chapter 6 for immunofixation electrophoresis (IFE), and in the respective chapters for MM, AL amyloidosis, light chain deposition disease (LCDD) and polyclonal hypergammaglobulinaemia. In the small group of samples selected, sFLC measurements had a higher sensitivity than IFE for detecting low concentrations of monoclonal sFLCs *(Chapter 6)*.

The normal range data reported by Katzmann *(6)* et al. is the most detailed yet published and has been generally adopted *(See Chapter 26 for implementation)*.

	Reference Interval (0.3-1.2)		Diagnostic range (0.26-1.65)	
	Estimate	95% CL	Estimate	95% CL
Sensitivity %	98	91-100	97	89 - 100
Specificity %	95	92 -98	100	98 - 100
PPV %	78	65 - 89	100	91 - 100
NPV %	100	98 - 100	99	97 - 100
Accuracy %	96	93 - 98	99	98 - 100

Table 5.4. Comparison of reference intervals and diagnostic ranges for sFLCs and κ/λ ratios. (Sera from the 282 reference individuals and 25 polyclonal hypergammaglobulinaemia patients as well as 66 sera from AL amyloidosis, LCDD and MM patients were used to calculate utility. (CL: confidence limits).

5.6. Urine free light chain (uFLC) normal ranges

In general, FLC values in urine have been similar in all studies even when they have included random, 24-hour or early morning samples. Furthermore, κ concentrations were higher than λ in all publications apart from one recent study *(Nakano 2) (Table 5.5)*. This latter study also reported exceptionally high sFLC concentrations *(Table 5.1)* suggesting there were problems with antibody specificity. It is logical that urine contains more κ than λ molecules because of its faster renal clearance.

The overall similarity of uFLC results suggests that reported concentrations are reasonably accurate. This is in part because normal urine contains trivial amounts of intact immunoglobulin molecules that can interfere with accurate FLC quantification. The higher uFLC levels shown in the study by Bradwell (2) et al. for both κ and λ were explained by the use of early morning urine samples, which are typically 2-3-fold more concentrated than 24-hour urine samples. The mean ($\pm$SD) free κ concentration was 5.4 $\pm$ 4.95mg/L (n = 66; range, 0.36-20.3mg/L), and the mean ($\pm$SD) free λ concentration was 3.17 $\pm$ 3.3mg/L (n = 66; range, 0.81-17.3mg/L) *(Figures 5.5 and 5.6)*. The mean κ/λ ratio was 1:0.54 (95% confidence interval, 1:2.17-1:0.25). The mean normal uFLC excretion was 3.7 mg/g of creatinine for κ and 2.0mg/g of creatinine for λ. There was a positive but non-significant correlation of urine creatinine concentrations with κ (r = 0.22) and λ (r = 0.17) measurements.

As expected, the range of uFLC concentrations was much wider than for serum, and κ/λ ratios were more variable. Presumably, this reflects minor differences in renal handling, urine dilution and variations in mucosal secretion of FLCs. The wide range of normal uFLC concentrations is another argument in favour of serum measurements.

Publication	Date	κ uFLC	λ uFLC	κ/λ ratio
Peterson	1971	3.3 (1.1-6.7)[ab]	1.1[ab]	2.9 (1.7-3.5)
Hemmingsen (1)	1975	2.3[ab]	1.4[ab]	1.6
Sölling (1)	1975	3.2 (SD±1.2)[ab]	1.1 (SD±0.44)[ab]	2.9
Hemmingsen (2)	1977	1.2 (0.0-6.8)[bc]	0.8 (0.0-2.4)[bc]	1.5
Robinson	1982	3.1 (1.9-7.1)[bc]	1.45 (0.7-3.4)[bc]	2.4 (1.2-3.7)
Brouwer	1985	1.8 (0.2-7.5)[c]	0.8 (0.15-2.1)[c]	2.4 (0.75-4.5)
Ohtani	1997	1.6 (SD±2.5)[ceh]	0.8 (SD±1.8)[ceh]	2.5 (SD±2.1)
Abe	1998	2.96(SD±1.84)[d]	1.07 (SD±0.69)[d]	3.0 (1.4-4.4)
Bradwell (2)	2001	5.5 (0.39-15.1)[cefg]	3.17 (0.81-10)[cefg]	1.9 (0.46-4)[f]
Nakano (3)	2003	1.3 (SD±1.8)[dh]	0.5 (SD±0.7)[dh]	2.8 (SD±2.0)
Nakano (2)	2006	2.1 (SD±2.8)[dh]	2.8 (SD±3.5)[dh]	0.8 (SD±0.3)

Table 5.5. Selected publications reporting normal urine FLC concentrations in mg/L *(Bradwell 2; Nakano 2).* ([a]polyclonal antibody against total light chains, [b]uFLCs in mg/24 hours, [c]polyclonal antibody against FLCs, [d]monoclonal antibody against FLCs, [e]latex reagent, [f] 95% range, [g]early morning urine samples, [h]random urine).

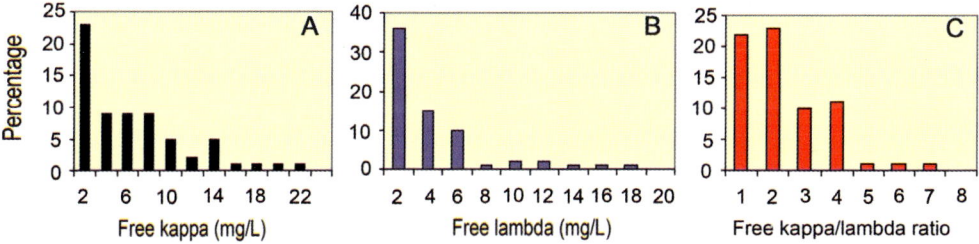

Figure 5.5. Frequency distributions of uFLC κ (A), λ (B) and κ/λ ratios (C) in 66 normal urines.

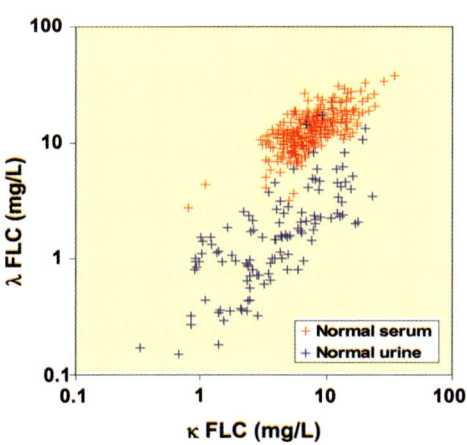

Figure 5.6. Comparison of FLC measurements in serum *(from Figure 5.2)* and early morning urine samples from healthy individuals.

Test questions

1. Why is the normal κ/λ ratio inverted in serum compared with urine?
2. How do κ/λ ratios correct for age-related increases in sFLC levels caused by decreases in glomerular filtration rates?
3. Is clonality preferably defined numerically, as κ/λ ratios, or by visualising a monoclonal band on electrophoretic gels?

Answers

1. Because the smaller monomeric κ molecules are cleared faster by the kidneys and enter the urine more readily than dimeric λ molecules (Section 5.1, page 37).
2. The non-tumour sFLC acts as a marker of glomerular filtration rate so the ratio eliminates the effect of deteriorating renal function (Section 5.1, pages 37-38).
3. Numerical κ/λ ratios are better than semi-quantitative, visualised bands particularly if they are faint (Chapter 6).

6.1. *Introduction*   45
6.2. *Serum and urine protein electrophoresis (SPE and UPE)*   46
6.3. *Serum immunofixation electrophoresis (sIFE)*   47
6.4. *Capillary zone electrophoresis (CZE)*   49
6.5. *Total κ and λ immunoassays*   50
6.6. *Urine tests for free light chains (FLCs)*   50
6.7. *Discordant serum and urine FLC results*   52

Summary: FLC immunoassays:

1. Are inherently much more sensitive than electrophoretic tests *(Figure 6.1)*.
2. Identify additional patients in all diseases associated with monoclonal gammopathies.
3. Provide quantitative κ/λ ratios compared with qualitative IFE.

Figure 6.1.
Sensitivity of assays for sFLC quantitation with error bars indicating the different claimed limits.

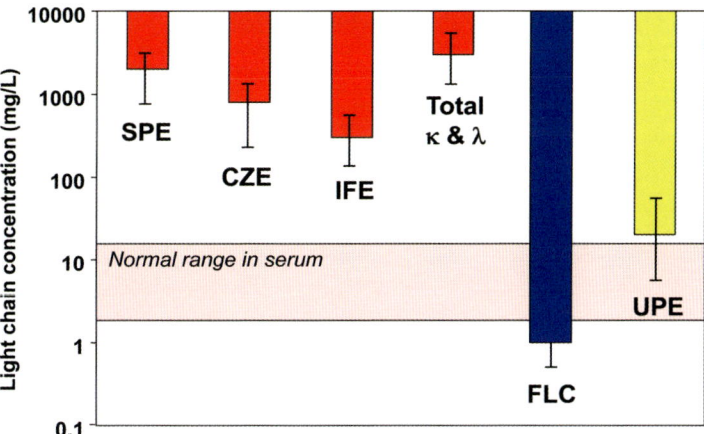

6.1. Introduction

Many factors need to be considered when deciding upon the most appropriate methods for measuring free light chains (FLCs). Requirements for the assays include sensitivity, accuracy, speed, cost, reliability, hands-on time, etc. A number of publications have compared different FLC assays, and many of their important features are discussed in other chapters. This chapter compares the sensitivity of routine laboratory assays for FLC detection in a clinical setting *(Figure 6.1)* - arguably the most important issue *(Kyle 2; Guinan)*.

6.2. Serum and urine protein electrophoresis (SPE and UPE)

SPE is the standard screening method for multiple myeloma (MM) and usually involves scanning agarose electrophoretic gels after the serum proteins have been separated, fixed and stained. A normal SPE is shown in Figure 6.2 and a monoclonal protein is shown in Figure 6.3.

The sensitivity of SPE for FLC detection is between 500mg/L and 2,000mg/L depending on whether or not the monoclonal protein migrates alongside beta proteins *(Bradwell 2)*. SPE is negative for FLCs in all patients with nonsecretory multiple myeloma (NSMM), the majority of patients with AL amyloidosis, and many patients with light chain multiple myeloma (LCMM) or other plasma cell dyscrasias *(Chapter 8, 9 and 15)*.

Examples of sera containing medium to high concentrations of FLCs are shown in Figure 6.4. Most show the typical electrophoretic abnormalities of monoclonal bands and hypogammaglobulinaemia. Sample No 2, however, appears normal yet the λ FLC concentration is elevated 50-fold. This is typical of many serum samples that are sent to the laboratory with a diagnosis of "possible multiple myeloma - please investigate". Current practice would require testing urine samples, but these are only available for 15-40% of the patients in most laboratories. Serum FLC (sFLC) assays would avoid the need for testing urine in all 9 patients presented *(Chapter 24)*. Screening sera by SPE and sFLCs is a simple and sensitive strategy for identifying new patients with monoclonal

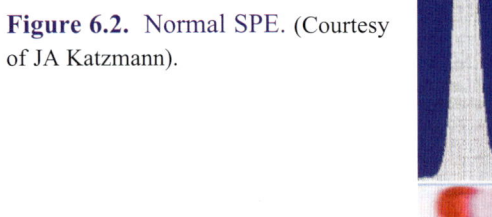

Figure 6.2. Normal SPE. (Courtesy of JA Katzmann).

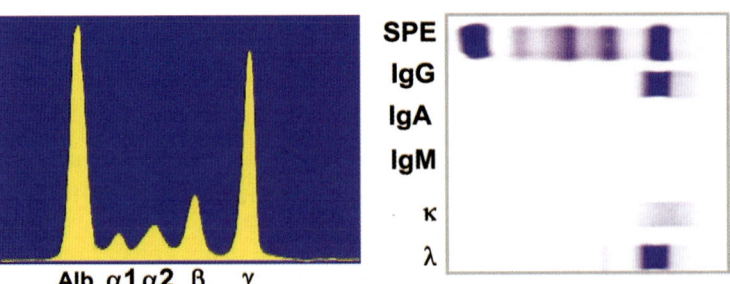

Figure 6.3. SPE and IFE of a serum IgGλ M-protein. Also identified is a small amount of polyclonal IgG with its associated κ and λ staining, and virtually no IgA or IgM. (Courtesy of JA Katzmann).

gammopathies and avoids the need for urine tests *(Chapter 23)*.

Murray et al. analysed the relationship between monoclonal protein quantification by SPE and immunonephelometry. Measurement of 2,095 IgG monoclonal proteins demonstrated nonlinearity of SPE at high monoclonal IgG concentrations. They concluded that changes in plasma cell populations may not be accurately reflected by changes in monoclonal protein values determined by SPE scanning densitometry and that clinicians should be aware of these limitations when quantifying IgG by SPE.

Urine protein electrophoresis (UPE) is much more sensitive than SPE since urine can be concentrated many times *(Figure 6.5)*. Thus, FLCs in urine can be detected at <10-20mg/L, although most laboratories claim a detection limit in the region of 40-50mg/L. In practice, high concentrations of background proteins and *"ladder banding"* caused by polyclonal FLCs prevent attainment of the ideal sensitivity. Furthermore, some patients with LCMM produce only small amounts of FLCs, so little, if any, passes the absorptive surface of the renal proximal tubules. These patients may, therefore, have no detectable levels of urine FLCs (uFLCs).

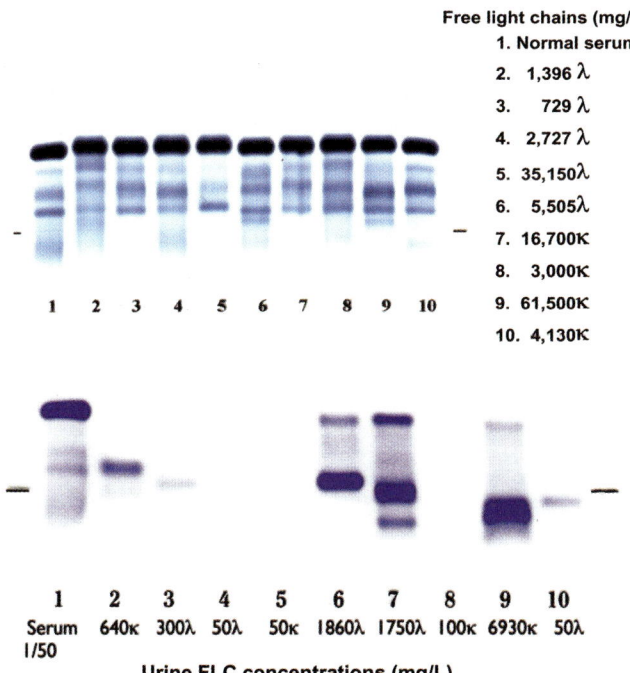

Free light chains (mg/L)
1. Normal serum
2. 1,396 λ
3. 729 λ
4. 2,727 λ
5. 35,150λ
6. 5,505λ
7. 16,700κ
8. 3,000κ
9. 61,500κ
10. 4,130κ

Figure 6.4. SPE in 9 patients with LCMM and one normal sample compared with the concentrations of sFLCs (mg/L). Some samples appear relatively normal by SPE but sFLC concentrations are grossly abnormal in all samples.

Figure 6.5. UPE in 9 patients with LCMM and one normal serum sample compared with measurements of uFLCs by immunoassay.

1　2　3　4　5　6　7　8　9　10
Serum　640κ　300λ　50λ　50κ　1860λ　1750λ　100κ　6930κ　50λ
1/50

Urine FLC concentrations (mg/L)

6.3. Serum immunofixation electrophoresis (sIFE)

sIFE is approximately 10-fold more sensitive for FLC detection (150-500mg/L) than SPE but still considerably less sensitive than FLC immunoassays. One particular disadvantage of IFE is that it cannot be used to quantify monoclonal immunoglobulins because of the presence of the precipitating antibody. IFE is also rather laborious to perform and visual interpretation may be difficult. A typical result on a sample containing a substantial amount of IgGλ monoclonal protein is shown in Figure 6.3.

In a study from The Mayo Clinic *(Katmann 6)*, it was shown that all of 46 serum samples with low concentrations of monoclonal FLCs were correctly identified by FLC immunoassay *(Figure 6.6)*. IFE was less sensitive and detected FLCs in some of the sera only after multiple assays and at different sample dilutions. The study included serum samples from patients in whom IFE was only positive for uFLCs. In addition, one sample shown in Figure 6.6 was negative by sIFE and urine IFE (uIFE) and sFLC immunoassays.

The high clinical sensitivity of FLC assays is dependent upon assessing the individual FLC concentrations and the κ/λ ratios. Tumour suppression of the normal plasma cells in the bone marrow reduces the concentrations of the alternate FLC and, thereby, enhances the sensitivity of the κ/λ ratio. The alternate FLC concentrations, and hence the κ/λ ratios, are important aspects of the diagnostic accuracy of sFLC immunoassays.

Figure 6.6 highlights the poor correlation that exists between sFLC concentrations and the detection of monoclonal proteins by IFE. This may be due to polymerisation of the FLCs, which prevents the formation of visible, narrow monoclonal bands during electrophoresis. This has been observed in patients with NSMM *(Chapter 9)* and probably occurs in most MM sera. However, rare serum samples from patients with AL amyloidosis and light chain deposition disease (LCDD) may be negative by sFLC assays but positive by IFE. This indicates that sFLC assays and IFE should be used as complementary diagnostic tests in patients with AL amyloidosis *(Chapter 15)*.

Some reports have suggested that sFLC analysis is not as sensitive as IFE for

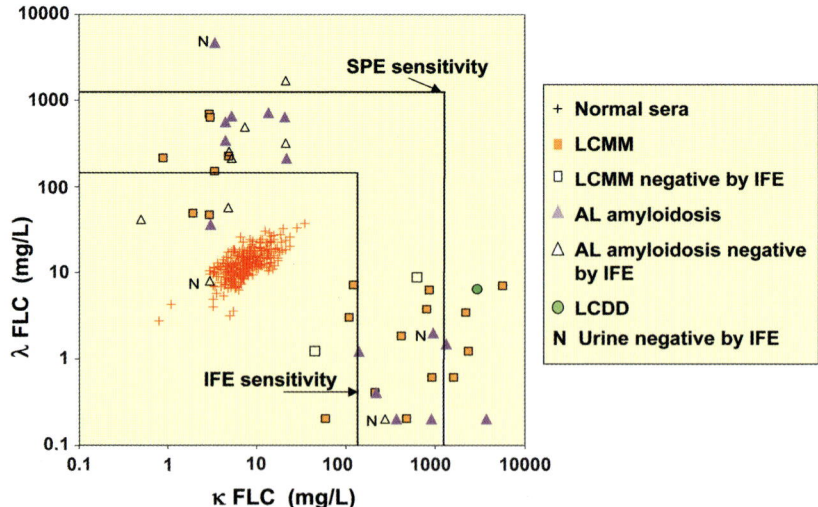

Figure 6.6. FLC concentrations in sera containing monoclonal FLCs that were *"difficult to detect"* by IFE.

detecting some monoclonal gammopathies (Jaskowski; Mehta). This is true, since there are many patients who produce monoclonal intact immunoglobulins alone, which cannot be detected by sFLC analysis. Such reports indicate some confusion by the authors about the specificity of the assays rather than their sensitivity.

6.4. Capillary zone electrophoresis (CZE)

CZE is used in many clinical laboratories for serum protein separation and is able to detect most monoclonal proteins. However, compared with IFE, CZE fails to detect monoclonal proteins in 5% of positive samples (Katzmann 5; Bossuyt 2; Bakshi). These so-called "false negative" results encompass low-concentration and "hidden" monoclonal proteins (e.g. in the transferrin peak).

Marien et al. compared the sensitivity of sFLC assays and CZE for the detection of low concentration monoclonal immunoglobulins. Frozen sera from 55 patients, previously shown by IFE to contain monoclonal proteins, but which were negative by CZE, were assessed by immunoassays for FLCs. They included both intact immunoglobulin and FLC monoclonal proteins.

The results showed that all 21 samples from patients with LCMM had abnormal FLC results *(Figure 6.7)*. In addition, 13 out of 33 samples from patients with intact monoclonal immunoglobulins had abnormal FLC ratios (not shown). These patients had monoclonal gammopathy of undetermined significance (MGUS) and B-cell derived malignant diseases. FLC immunoassays were therefore more reliable than CZE for detecting LCMM, and identified many abnormal samples in patients from other disease groups. Similar findings were reported from Bakshi et al., and by others in monoclonal protein screening studies *(Chapter 23)*.

Figure 6.7 indicates that several samples with FLC concentrations greater than 1,000 mg/L were not detected by CZE. Under ideal conditions the sensitivity of this technique may be better but it is still substantially less sensitive than IFE. However, CZE does at least provide a quantitative measure of the monoclonal proteins, whereas IFE does not.

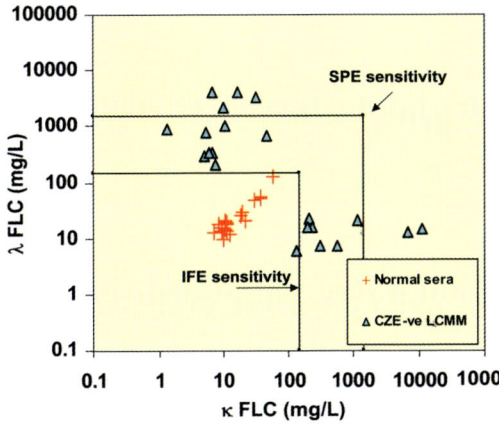

Figure 6.7. Serum κ and λ FLC concentrations in 20 control samples and 21 samples that were normal by CZE but had monoclonal proteins by IFE. (Courtesy of X Bossuyt).

6.5. Total κ and λ immunoassays

Unfortunately, immunoassays for total κ and λ are sometimes used to try and identify patients with LCMM. This is in spite of warnings by The College of American Pathologists and others that the method is not sensitive enough for routine clinical use *(Kyle 2)*. Indeed, samples containing many grams of FLCs may be completely missed using this technique.

The sensitivity of sFLC assays and total κ and λ assays were compared in a study by Marien et al. Sixteen serum samples from patients with LCMM were investigated (samples were from the CZE study described earlier). Total κ and λ concentrations were measured using Beckman-Coulter reagents on the IMMAGE® nephelometer and sFLC concentrations were measured by Freelite™ assays from The Binding Site. All samples were abnormal by FLC assays *(Figure 6.8)*. This compared with only 5 of the 16 samples by total κ and λ assays, and one λ patient was misclassified as κ. Total κ and λ assays do not have a role in the clinical laboratory.

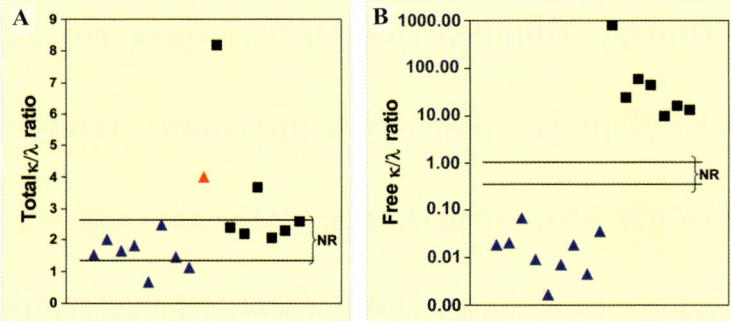

Figure 6.8. Comparison of total serum light chains (A) and sFLC κ/λ ratios (B) for identifying patients with κ and λ LCMM. κ patients: black squares; λ patients: blue triangles. The red triangle is a misclassified sample. Normal range limits are shown.

6.6. Urine tests for free light chains (FLCs)

There are many methods for detecting uFLCs, and some of these are outlined in Chapter 4. Serum immunoassays are preferable because urine is a poor fluid for assessing FLC concentrations *(Chapter 3)*. As shown by Nowrousian et al. significant amounts of sFLCs are necessary to cause FLC proteinuria in patients with MM. In that study, median monoclonal serum κ and λ concentrations were 113 mg/L and 278mg/L respectively, before Bence Jones proteinuria occured. Further clinical comparisons of serum and urine sensitivity are described in Chapter 24.

Herzog and Hoffman compared the sensitivity of 3 different urine protein tests *(Table 6.1)* with serum and urine FLC immunoassays in 33 patients with MM who had Bence Jones proteinuria as assessed by IFE. Five of the patients had LCMM while 28 patients had intact monoclonal immunoglobulins in the serum with additional uFLC excretion. Fifteen of the latter patients also excreted intact monoclonal immunoglobulins into the urine. Results showed that sFLC κ/λ ratios provided the most sensitive analysis and better than uFLC κ/λ ratios *(Table 6.1)*. Total protein tests for uFLCs were poor. The reference method for this study was uIFE and positive results were the basis for sample selection.

Analytical technique	Normal ranges	Sensitivity
sFLC κ/λ ratio	κ/λ >0.26 to < 1.65	33 out of 33
uFLC κ/λ ratio	κ/λ >2.04 to <10.37	29 out of 33
Urine total κ/λ ratio + levels	κ/λ >1 to <5.2 & Tκ+λ<10mg/L	29 out of 32
Urine albumin/total protein	albumin<0.3 & TP <300mg/L	14 out of 31
Urine total protein	total protein <100mg/L	26 out of 32

Table 6.1. Sensitivity of different analytical techniques for FLC detection in 33 patients with MM *(Herzog)*. (Tκ+λ = total κ + λ concentrations; TP = total protein).

Van Hoeven et al. compared the sensitivity of three tests for FLCs (sFLC, uFLC und UPE) with reference to uIFE in 98 urine samples. sFLC assays were more sensitive than UPE for the detection of monoclonal FLCs. The authors also showed that the sensitivity of uFLC was greater than UPE (75% versus 44%) in samples with an abnormal uIFE. The difference between uFLC and UPE was even more pronounced for positive uIFE samples (89% versus 52%).

Fulton et al. compared sFLC with uIFE when screening for monoclonal gammopathies. In 314 samples from 142 patients the sensitivity of the sFLC ratio and uIFE was 91% and 81%, respectively, with a specificity of 100% and 99%. Other studies have shown that in a clinical setting sFLC tests are considerably more sensitive than uIFE for identifying patients with residual disease *(Chapters 8, 9, 10, 11, 12 and 24)*.

A comparison of uIFE and uFLC immunoassays for Bence Jones protein detection was made by Viedma et al. While FLC immunoassays correctly identified many of the positive and negative samples, high background polyclonal FLCs (arising from renal damage) obscured correct interpretation of κ/λ ratios. Hence, uIFE was more accurate for monoclonal protein identification. Similar results have been observed by Katzmann et al. (personal communication).

Snyder et al. assessed whether measuring uFLCs and/or urine total light chains was useful in addition to UPE for monitoring patients with a urine monoclonal protein. Using 336 uIFE-positive urine samples (that were also SPE positive), they showed that the diagnostic sensitivities of uFLC and urine total light chain measurement were 80% and 70%, respectively. In samples that were SPE negative, diagnostic sensitivities decreased substantially. They concluded that uFLC and urine total light chain assays are not useful in addition to SPE in diagnostic testing. Urine total light chain values were in closer agreement with serum monoclonal protein values measured by SPE and provided a quality check on measurements of urinary monoclonal protein values in disease monitoring.

Le Bricon et al. analysed the sensitivity of different urine assays for Bence Jones proteins in 20 patients with MM. The urine samples contained monoclonal intact immunoglobulins and/or monoclonal FLCs. Comparisons were made between assays for total protein (Pyrogallol Red), sodium dodecyl sulfate-agarose gel electrophoresis (SDS-AGE: sensitivity 50mg/L in unconcentrated urine) and FLC immunoassays (sensitivity <1mg/L). The results confirmed the superior sensitivity of the FLC immunoassays. Six of the 20 patients were abnormal by the total protein test, 11 by SDS-AGE and 16 by FLC immunoassays. Details of the 6 samples that were only abnormal by FLC immunoassays are shown in Table 6.2. The immunoassays were particularly helpful for identifying monoclonal FLCs when these were present in low concentrations. There was a reasonable correlation between the quantitative results for the different methods. Herzum et al. also compared FLC immunoassays with other urine tests for FLCs. The FLC assays showed good precision (<10%), linearity and correlation between different instruments (Hitachi and BNII). However, the assays overestimated the amounts of Bence Jones protein present in urine samples. 37 samples were compared using the following methods: benzethonium chloride, Biuret, modified Biuret and FLC immunoassays. Measurements using immunoassays produced the highest results in 26 of the samples. The Biuret methods, which measure peptide bonds and are considered to be good general assays for proteins, had the best correlation with FLC immunoassays. This provides supporting evidence for the accuracy of FLC tests.

Lueck et al. assessed the utility of antibodies specific for FLCs in uIFE using the Sebia electrophoresis system. They found that antibodies against total light chains detected FLC bands more effectively than FLC-specific antibodies and concluded that such antibodies were of little clinical use. However, no comparison was made in this study of the latex-enhanced FLC nephelometric immunoassays on the urine samples.

Serum M-protein	Urine protein (mg/L)	Urine FLCs		Urine SDS-AGE
		κ & λ (mg/L)	κ/λ ratio	
IgGκ	50	92:10	9.2	Negative
IgAκ	40	28:6	4.7	Negative
IgAλ	50	7:94	0.07	Negative
IgAλ	960	66:749	0.09	+/- polyclonal
IgDλ	100	3:21	0.14	Negative
FLC	1,990	60:21	0.29	+/- polyclonal

Table 6.2. Comparison of urine tests in 6 patients who were negative for monoclonal urine proteins by SDS-AGE but were abnormal by FLC *(Le Bricon)*. There was good overall correlation, but sample 6 probably contained albumin.

6.7. Discordant serum and urine FLC results

Quantitation of monoclonal FLCs by electrophoretic methods often produces lower results than FLC immunoassays. H. Zitterberg (Goteborg University, Sweden: Personal communication) reviewed the laboratory records from 22 newly diagnosed LCMM patients who presented over a 2-year period. By quantitative electrophoresis, none had more than 3g/L of sFLCs and 19 were below 1g/L. This was in contrast to 50% that were over 3g/L in the study by Bradwell *(1)* et al. using sFLC immunoassays. Part of the explanation is that uFLCs and sFLCs are frequently polymerised, which leads to high results by immunoassay *(Chapter 4)*. Also, protein dye tests underestimate the concentrations of FLCs to a variable extent.

Discordant sFLC and uFLC results are also caused by inaccuracies in 24-hour urine testing. Siegel *(2)* et al. evaluated 623 24-hour urine samples that had unexpectedly increased creatinine clearance (CrCl) for 24-hour protein, uIFE and UPE monoclonal proteins. Unexpectedly, abnormally increased CrCl was found in 19% of the samples, which was accompanied by increased urinary monoclonal protein and total protein, but with no sFLC increase in the corresponding serum samples. Thus, results requiring 24-hour urine collections were highly susceptible to error, whereas sFLC analysis was unaffected by these errors.

Although not quantitative, IFE is considered to be the "gold standard" for identifying monoclonal uFLCs. Nevertheless, the detection of minor monoclonal bands by this method can be misleading. There are many reasons:

1. Precipitating antibodies against one or other FLC may not be completely specific and may cross-react with other urine proteins. This can produce a false positive band with the appearance of a monoclonal FLC.
2. A narrow protein precipitate band can occur at the sample application site on the gel and appear like a monoclonal band.
3. Restricted *"ladder banding"* in concentrated samples can give the appearance of monoclonality.
4. Heavy proteinuria containing polyclonal FLCs may produce confusing background staining, which obscures correct interpretation of monoclonal FLC bands.
5. An inadequate antibody to one of the FLCs (usually λ) gives the impression that only the alternate FLC is being excreted. Although there may be a broad band, it can suggest monoclonality. This is a fairly frequent occurrence.
6. After transplantation, oligoclonal bands are produced for many months. These may produce confusing results in urine samples and appear to be monoclonal: sFLC ratios, on the other hand, may be normal.
7. IFE is not quantitative so it is difficult to compare one result with another. The majority of the staining intensity (80% or more) is due to the secondary antibody so it is not possible to assess the amounts of monoclonal protein present.

Even when a monoclonal uFLC band is visually convincing, it should be evaluated alongside other clinical and laboratory data because it may be inconsequential. As discussed elsewhere, clinically significant monoclonal uFLCs are exceptional when sFLCs are normal. In a study of 110 patients with AL amyloidosis *(Katzmann 3)*, all but one had serum monoclonal proteins and the remaining patient was negative by all tests including urine analysis *(Table 6.3)*. Isolated, minimal uFLC excretion is usually clinically insignificant *(Chapter 19.4, Figure 19.5)* although rarely patients may have AL amyloidosis *(Palladini 5; Chapter 15)*.

uIFE may correctly identify monoclonal FLCs when serum analysis is normal, under the following unusual circumstances:

1. When there is damage to isolated nephrons allowing leakage of monoclonal FLCs into urine. Serum levels, in contrast, may not be raised sufficiently to produce abnormal κ/λ ratios as defined by the normal range *(Chapter 5)*.
2. Intact immunoglobulin molecules that enter the proximal tubules may be partially metabolised by tubular cells. FLCs and other immunoglobulin fragments can subsequently be released and enter the urine *(Chapter 10)*.
3. When monoclonal FLCs are truncated and rapidly cleared. The short serum half-life prevents accumulation in serum but significant levels could be present in urine. This must be a very rare occurrence since it has not been observed in any study to date.
4. Abnormal amino-acid sequences could change the shape of the *"hidden"* epitopes on the FLC constant regions. This might prevent detection by antisera specific for FLCs. However, IFE antisera against the whole FLC structures would detect some remaining, undistorted epitopes and produce positive results *(Coriu)*. This is speculative and has not been observed to date.
5. When the serum and urine samples are collected at different times, results may be discordant. Normally, serum samples are collected at the clinic but the urine is either collected earlier or is subsequently sent in by post. If the samples are separated by a significant time, even a few days, results may be quite different because of the short serum half-life of FLCs *(Chapter 10)*.

In spite of the possibility that uIFE is more sensitive for monoclonal FLCs under rare circumstances serum assays provide a reliable basis for assessing true FLC production by tumours. Detailed physiological, technical and clinical reasons for using sFLC assays rather than urine assays are discussed in Chapter 24.

Table 6.3. Comparison of serum and urine tests for identifying 110 patients with AL amyloidosis *(Katzmann 3)*. Urine tests provided no additional

Test	Sensitivity
FLC κ/λ ratio	91%
sIFE	69%
uIFE	83%
FLC κ/λ ratio and uIFE	91%
FLC κ/λ ratio and sIFE	99%
sIFE and uIFE	95%
All three tests	99%

Conclusions

Since normal sFLC concentrations are considerably lower than the detection limit of all serum electrophoretic methods, some samples will always be misidentified as negative using these techniques. The numbers of patients and types of diseases missed by serum electrophoretic tests are shown diagrammatically in Figure 6.9. Although reports vary in their claims for the levels of sensitivity achieved with electrophoretic tests, the benefit of the increased sensitivity of sFLC measurements is clearly evident.

Hence, sFLC measurement has now been included in the international guidelines of the International Myeloma Working Group, details of which are presented in Chapter 25 *(Dispenzieri 7)*.

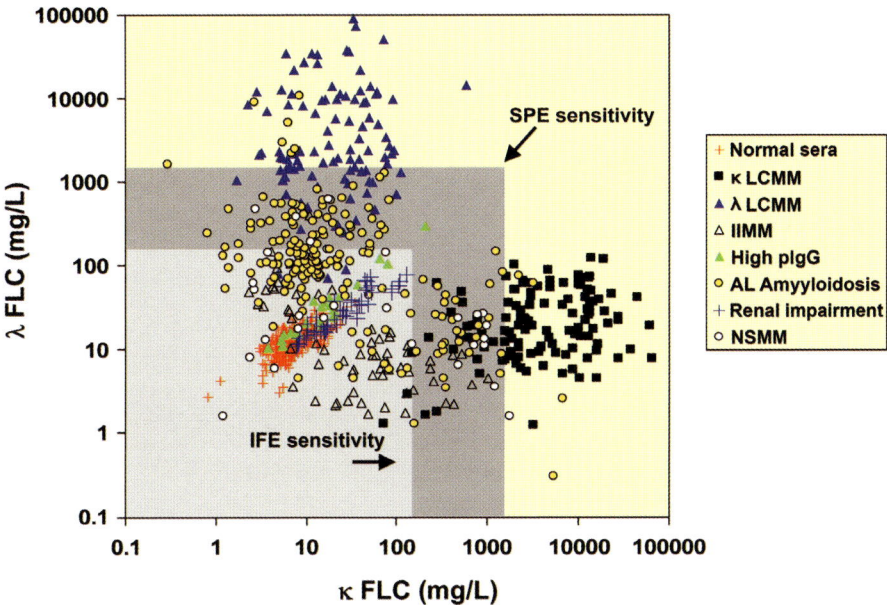

Figure 6.9. κ/λ logarithmic plot of sFLCs showing samples that would be mis-identified as negative using SPE and sIFE. (High pIgG: polyclonal hypergammaglobulinaemia, IIMM: intact immunoglobulin multiple myeloma. *For the data on the patient groups see the relevant chapters).*

Test questions
1. What is the origin of ladder banding in UPE?
2. Which is more sensitive, CZE or SPE?
3. Do total serum light chain tests have a useful role in the laboratory?
4. Under what circumstances is uIFE unreliable?
5. Which is more specific for monoclonal proteins - uFLC immunoassays or uIFE?

Answers
1. Oligoclonal FLCs focus separately during electrophoresis (Section 6.2, page 47).
2. CZE (Section 6.1 and figure 6.1, page 45).
3. No. FLC immunoassays are better in all respects (Section 6.5, page 50).
4. There are several reasons listed in Section 6.7, page 52.
5. uIFE (Section 6.6, page 51).

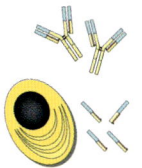

SECTION 2 - Free light chains and monoclonal gammopathies

Section 2A. (Chapters 7-14). *Multiple myeloma (MM)* **59**
Section 2B. (Chapters 15-17). *Diseases with monoclonal light chain deposition* **130**
Section 2C. (Chapters 18-19). *Other diseases with monoclonal free light chains* **154**

The main clinical applications of serum FLC (sFLC) measurements are for patients with monoclonal gammopathies *(Chapter 29)*. Figure 7.0 shows the associated diseases seen at the Mayo Clinic in 2005 *(Kyle 11)*. These data are from a specialist referral centre; general hospitals see different patterns of disease referral with a higher percentage of MM and monoclonal gammopathy of undetermined significance (MGUS) and fewer AL amyloidosis patients.

Table 7.0 shows the approximate annual incidence of the more common monoclonal gammopathies reported in the USA. Over the past 9 years, there have been publications covering many aspects of FLC usage in light chain multiple myeloma (LCMM), nonsecretory multiple myeloma (NSMM) and AL amyloidosis while its benefits in MGUS are now well established. Other clinical applications have recently been described including in solitary plasmacytomas, asymptomatic (smouldering) multiple myeloma (ASMM), Waldenström's macroglobulinaemia (WM) and B-CLL all of which are described in their respective chapters.

Dispenzieri *(3)* et al. recently suggested that these observations on sFLCs represent an important step in understanding monoclonal gammopathies and add to previous definitions. 1st, there was the separation of polyclonal and monoclonal hypergamma-globulinemia. 2nd, there was the realization that premalignant MGUS existed. 3rd, the concept of ASMM became an accepted principle. 4th, there was an international consensus on the definitions for the spectrum of MGUS to ASMM to active MM. Each

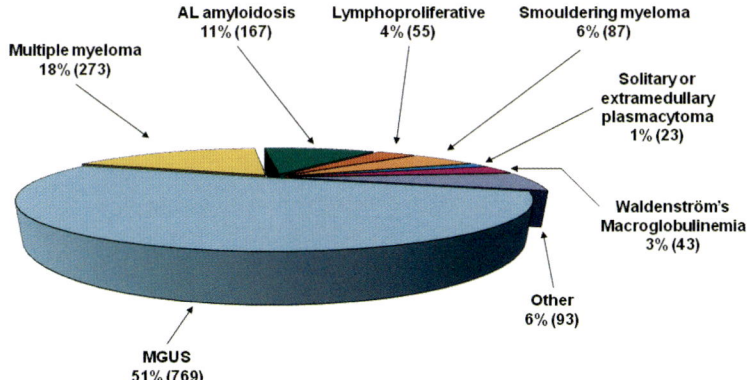

Figure 7.0. Monoclonal gammopathies diagnosed at the Mayo Clinic during 2005 *(Kyle 11)*.

of these benchmark decisions were based upon observational studies that incorporated the following four variables to reach conclusions and recommendations:

1. M-protein; **2.** Bone marrow plasmacytosis; **3.** Symptom status; **4.** Time.

They proposed that sFLC κ/λ ratios should be introduced as another variable that may better define these clinical entities. It is a simple test that provides information about underlying plasma cell biology and baseline sFLC κ/λ ratios provide valuable prognostic information. Furthermore, all recent studies indicate that in the spectrum of disease development from MGUS, through ASMM to symptomatic MM, increasing monoclonal FLC production becomes progressively more probable. This accumulating evidence has led to new international guidelines (Chapter 25) where sFLC analysis is recommended in patient diagnosis and prognosis, as well as the management of a number of the individual conditions *(Dispenzieri 7)*.

The following chapters review these studies and suggest investigations that may be of clinical value in the future. In addition, there are many other studies showing the importance of sFLCs when screening for monoclonal diseases, replacement of urine testing, their role in patient monitoring, etc. *(Section 4)*. There is even the probability of reversing myeloma kidney damage with the combination of novel *"high cut-off"* haemodialysis and chemotherapy *(Chapter 13)*.

	Annual Incidence	*Monoclonal proteins	sFLC abnormal	Median survival	**Utility of sFLC
IIMM	15,500	100%	96%	3-4 years	Important
LCMM	3,200	100%	100%	3-4 years	Important
NSMM	650	0%	>70%	3-4 years	Important
Plasmacytoma	1,000	50%	~80%	>10 years	Important
AL Amyloidosis	2,500	75-90%	>95%	1-2 years	Important
LCDD	100	90%	90%	3-4 years	Important
Waldenström's	1,000	100%	97%	5 years	Interesting
MGUS	1,000,000	100%	30-60%	>12 years	Important
ASMM	250	100%	~90%	5-10 years	Important
#B-CLL	10,000	~50%	~35%	~5 years	Interesting

Table 7.0. Incidence of diseases producing monoclonal proteins in the USA *(Malpas)*. *% showing monoclonal proteins by traditional electrophoretic methods. ** Based on current publications. #Other B-cell lymphoid malignancies also produce monoclonal proteins. IIMM: intact immunoglobulin multiple myeloma, LCDD: light chain deposition disease, B-CLL: B-cell chronic lymphocytic leukaemia.

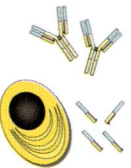

Chapter 7. *Multiple myeloma (MM) - Introduction*   *59*
Chapter 8. *Light chain multiple myeloma (LCMM)*   *61*
Chapter 9. *Nonsecretory multiple myeloma (NSMM)*   *69*
Chapter 10. *Intact immunoglobulin multiple myeloma (IIMM) - theoretical considerations*
     of sFLC and Ig measurements   *77*
Chapter 11. *Intact immunoglobulin multiple myeloma - sFLCs at diagnosis*   *86*
Chapter 12. *Monitoring patients with IIMM using sFLCs*   *95*
Chapter 13. *The kidney and monoclonal free light chains*   *111*
Chapter 14. *Asymptomatic (smouldering) multiple myeloma (ASMM)*   *126*

Chapter

7

SECTION 2A - Multiple myeloma

Multiple myeloma - Introduction

Multiple myeloma (MM) is the 2nd most common form of haematological malignancy after non-Hodgkin lymphoma. In Caucasian populations the incidence is approximately 35 per million, per year, it increases with age and there is a slight male preponderance. In the UK there are 3,000-3,500 new cases per year, almost 20,000 in the USA and 86,000 world-wide, with a median survival of 3 years *(Katzel)*. At any one time, there are 10,000 to 15,000 patients with the disease in the UK and nearly 250,000 world-wide. For those who enter clinical trials, the median 5-year survival is approximately 50%. The incidence of MM in different populations is shown in Table 7.1. Furthermore, there is a slowly rising incidence and survival is increasing *(Figure 7.1) (Katzel; Kumar 3)*.

Diagnosis is based on the presence of excess monoclonal plasma cells in the bone marrow, monoclonal proteins in serum or urine and related organ or tissue impairment such as hypercalcaemia, renal insufficiency, anaemia or bone lesions *(Chapter 25)*. Normal plasma cell content of the bone marrow is about 1% while in MM the content is typically greater than 30% but may be over 90%. Concentrations between 5-10% are equivocal since plasma cell distribution in MM may be patchy or there may be other causes for the plasmacytosis such as chronic infections. Identification of the monoclonal cells is preferably achieved using immunohistochemical staining for κ and λ light chains *(Figure 3.5)*.

Location	Men	Women	Location	Men	Women
Argentina	30	21	Japan	18	12
Australia	30	26	Thailand	5	4
Canada	40	27	UK	35	25
Germany	30	25	USA (black)	72	68
India	13	8	USA (white)	40	26

Table 7.1. Annual age-standardised incidence of MM by location per million *(Malpas)*.

The osteolytic lesions of MM, classically seen in skull X-rays *(Figure 7.2)*, are not a constant feature. Patients may have osteosclerotic bone lesions or no detectable abnormalities. Also, osteolytic lesions may occasionally be seen in other diseases.

The immunoglobulin classes of monoclonal proteins produced by the plasma cell clones reflect their normal frequency in the body. This is shown in Figure 7.3 and is based on more than 2,500 patients entered into the UK MRC MM trials from 1980 to 1998.

Figure 7.1 Improving survival in MM resulting from novel treatments. (This research was originally published in Blood *(S Kumar 3)* © the American Society of Hematology).

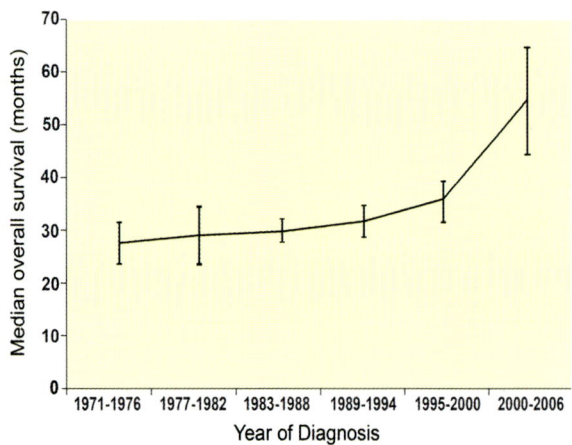

Figure 7.2. Osteolytic lesions in the skulls of 2 patients with MM.

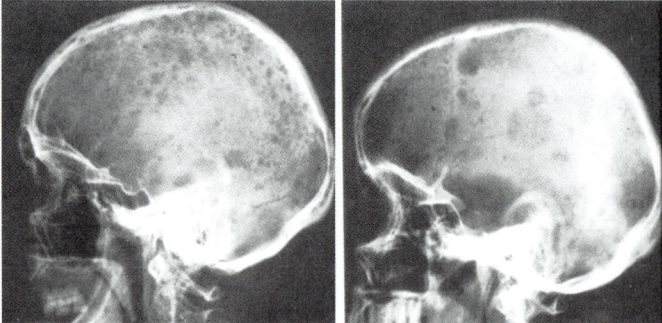

Figure 7.3. Classification of MM based upon M-protein production. (UK, MRC MM trials).

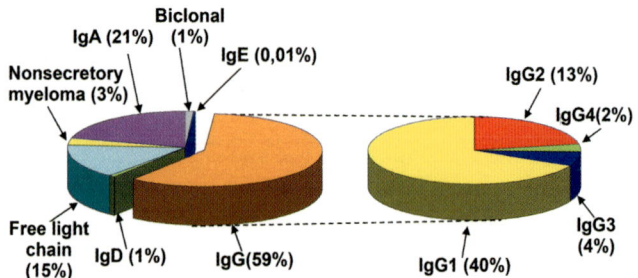

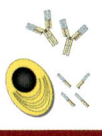

8.1.   *Diagnosis of LCMM using serum free light chains (sFLCs)*   **61**
8.2.   *Monitoring LCMM using sFLCs*   **64**
         Clinical case history No 1   **66**
         Clinical case history No 2   **68**

Summary: In patients with LCMM, sFLC concentrations are:

1. Always elevated when urine free light chains (uFLCs) are elevated.
2. Diagnostically more sensitive than serum immunofixation electrophoresis (sIFE).
3. A better indicator of minimal residual disease than urine measurements.
4. Able to indicate disease changes more accurately than urine measurements.
5. More easily measured than in urine.
6. Able to provide numerical results compared with immunofixation electrophoresis (IFE).
7. Able to provide additional information about renal and bone marrow function from the alternate FLC concentrations and κ/λ ratios.
8. Useful in the differential diagnosis of patients with bone pain and fractures, unexplained renal disease and other clinical features of multiple myeloma (MM) when urine samples are not available.

8.1. Diagnosis of LCMM using sFLCs

The typical clinical features of LCMM, such as bone pain, fractures, renal failure and anaemia alert the physician to the diagnosis. Bence Jones protein in the urine, in the absence of intact monoclonal immunoglobulins in the serum, alongside a positive bone marrow biopsy, confirms the diagnosis. On occasions, symptoms and signs can be obscure so that urine tests for LCMM are not considered for some time. This leads to delays in diagnosis as indicated by the occasional published case report *(Brigden 2; van Zaanen)* illustrating the difficulties facing the clinicians *(see Clinical Case Histories 1 and 2)*.

Commonly, the initial screening test for LCMM is serum protein electrophoresis (SPE). This demonstrates a monoclonal FLC band in approximately 50% of patients and others may show hypogammaglobulinaemia. sIFE demonstrates monoclonal bands in most patients but ultimately a urine test is required to identify and quantitate the monoclonal FLCs. Figure 8.1 shows some typical electrophoretic test results.

Since immunoassays for sFLCs are more sensitive than electrophoretic tests *(Chapter 6),* could urinalysis be stopped when the diagnosis of LCMM is being considered? The answer to this question was published in *The Lancet* in 2003 *(Bradwell 1)* as a clear, "*Yes*". The study was based on archived sera from 224 patients with LCMM entered into the UK MRC Myeloma trials between 1983 and 1999. The clinical diagnosis of

MM had been established using bone marrow plasma cell content, the presence of monoclonal FLCs in the serum or urine (without intact monoclonal immunoglobulins) and the presence of lytic bone lesions.

The results showed that at the time of diagnosis, all patients had abnormal concentrations of the appropriate sFLC *(Figure 8.2)* and abnormal κ/λ ratios. The concentrations were clearly different from FLCs measured in the serum of 282 blood donors *(Chapter 5)*. The non-tumour FLCs, produced by normal plasma cells, were also

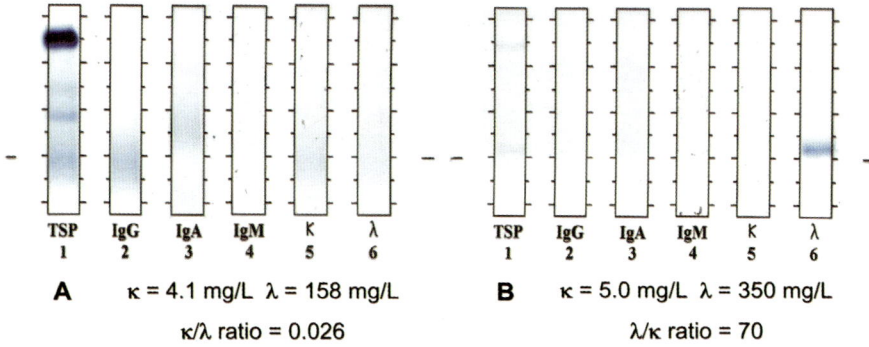

Figure 8.1. Serum and urine IFE from a patient with λ LCMM compared with FLC immunoassay results. SPE and IFE of (A) serum with no abnormality visible and (B) urine showing a λ FLC band.

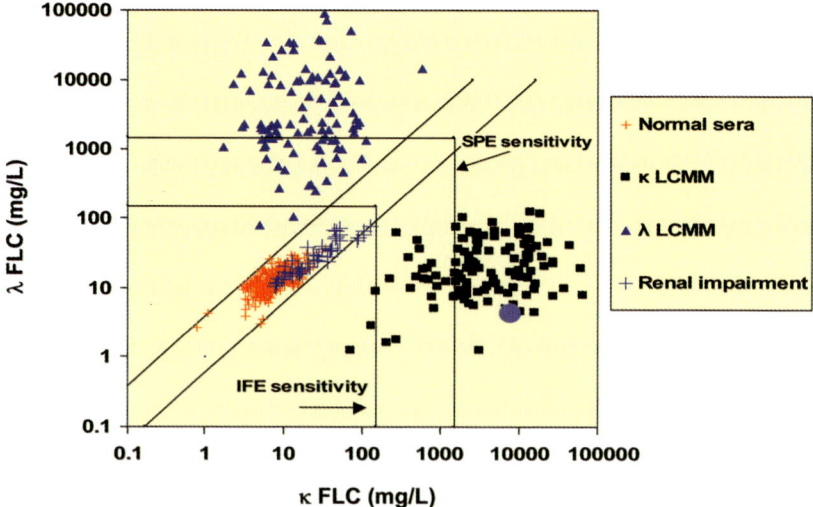

Figure 8.2. sFLC concentrations in normal individuals, patients with LCMM and patients with renal impairment but no plasma cell dyscrasia *(Chapter 20)*. The diagonal lines separate monoclonal from polyclonal diseases. ● Clinical case No 1.

abnormal in a high proportion of the patients. Some were elevated as a result of renal impairment while others were low because of bone marrow suppression. Comparison of results was made with 31 patients who had renal impairment from causes other than monoclonal gammopathies. Characteristically, they had elevations of both κ and λ FLC types with normal κ/λ ratios (*Chapter 20*).

Similar results have been found in all studies and have included over 600 patients (*Table 8.1*). No patient with untreated LCMM has yet been identified who was positive by urine tests but negative by sFLC tests. These include patients with renal failure when urine samples may be difficult to obtain because of poor urine flow (*Hutchison 5 and Chapter 13*).

In *The Lancet* study mentioned above, all the patients had elevated concentrations of both serum and urine monoclonal FLCs, so it might be expected that the two variables

Study (Year)	Patient Numbers	κ	λ	κ/λ Ratio Abnormal
Bradwell (1) 2003	224	123	101	100%
Abraham (3) 2002	28	9	19	100%
Nowrousian 2005	17	na	na	100%
Kang 2005	23	14	9	100%
Wolff 2006	5	na	na	100%
Mosbauer 2007	9	5	4	100%
van Rhee 2007	49	na	na	100%
Hutchison (5) 2008	13	5	8	100%
Piehler 2008	7	4	3	100%
Harding 2009	7	4	3	100%
Drayson (3) 2009	223	na	na	100%

Table 8.1. sFLC measurements in studies of LCMM. (na: not available).

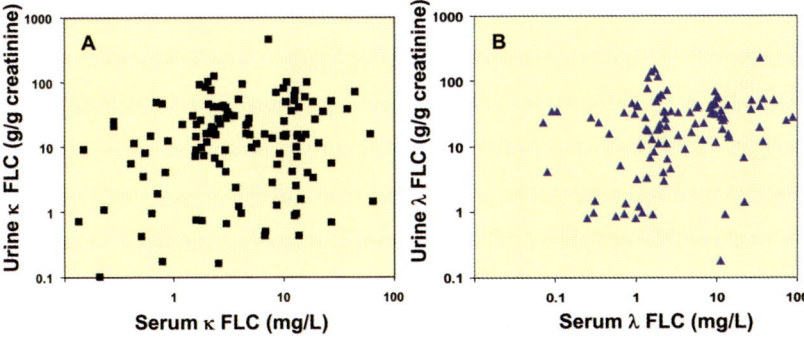

Figure 8.3. Relationship between serum and urine FLCs in 224 patients with LCMM at the time of diagnosis. (A) κ: r=0.29; p=0.0012. (B) λ: r=0.13; p=0.183. uFLCs were measured by immunoassay and corrected for urine dilution using creatinine concentrations.

would be highly correlated. In fact, there was only a modest association *(Figure 8.3)*. This may seem rather surprising, at first glance, but it can be explained by the effect of the renal tubular metabolism of FLCs. The amounts of FLCs observed in urine are highly dependent upon renal function. When renal function is normal, the proximal tubules remove 20-30g of protein per day *(Chapter 3)*. Hence, serum and urine FLC concentrations are frequently discordant, with serum being more representative of tumour mass than urine concentrations. The amounts of sFLCs required to produce abnormal urine tests is discussed in Chapter 24.

Similar results were found in a study of 66 patients with MM by Nowrousian et al. Nearly all patients who had uFLCs by IFE had abnormal sFLC κ/λ ratios. Furthermore, more than 60% of patients under treatment, with negative urine tests for Bence Jones protein, had abnormal sFLCs, indicating the additional sensitivity of the serum test. The study concluded that assessment of sFLCs was much more sensitive than urine IFE (uIFE) for identifying Bence Jones protein in patients with MM. This sensitivity is reflected in the international guidelines *(see Chapter 25 and Dispenzieri 7)*.

8.2. Monitoring LCMM using sFLCs

Assays that are useful in disease diagnosis are typically useful for disease monitoring. This is particularly true for FLC assays. Not only are FLC immunoassays inherently quantitative but also their precision is considerably better than measurements of monoclonal bands on electrophoretic gels.

Figure 8.3 shows that there is little correlation between serum and urine FLC concentrations at the time of disease diagnosis. In contrast, changes in serum and urine FLC concentrations observed during the course of the disease show a good correlation. This is illustrated in Figure 8.4 in 2 patients from the Mayo Clinic *(Abraham 3)*. In both patients the concentrations of FLCs in serum and urine fell following chemotherapy (although in the first patient this was not in parallel, possibly due to inadequate 24-hour urine collections). In an expanded study of 71 patients *(Abraham 3)*, a good correlation

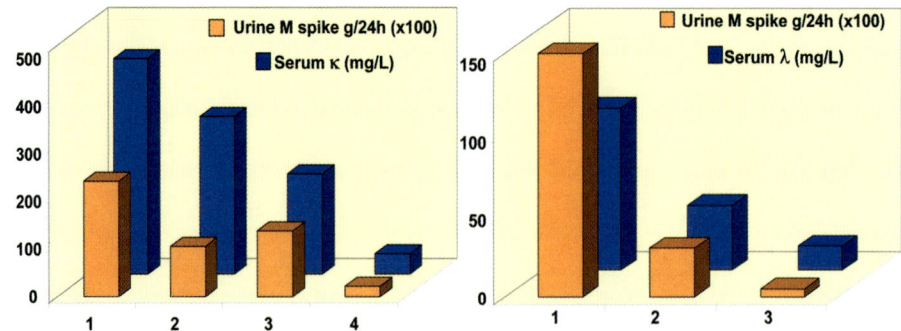

Figure 8.4. Comparison of 24-hour urinary monoclonal-(M-) protein concentrations and sFLCs in 2 patients with LCMM, measured at different times after chemotherapy. (Courtesy of RA Kyle and JA Katzmann).

was found between changes in serum and urine FLC concentrations *(Figure 8.5)*. The authors concluded that sFLC measurements provided a satisfactory alternative to 24-hour urine collections for monitoring patients with LCMM.

In *The Lancet* study *(Bradwell 1)*, changes in sFLC concentrations were assessed as indicators of responses to treatment. The results showed that 99% of patients (81/82) had reductions in sFLCs compared with 95% (78/82) for the corresponding uFLCs. This indicated a marginally better sensitivity for serum tests when initial responses to chemotherapy were evaluated. However, there were considerable differences when assessing rates of remission. 32% (26/82) of the patients were considered to be in complete remission as assessed by normal uFLC concentrations but this compared with only 11% (9/82) from sFLC levels. In the same clinical trials, 10% of patients with intact immunoglobulin multiple myeloma (IIMM) (117/1189) had complete serological remission. Since the serum responses to chemotherapy were similar in IIMM and LCMM, the results indicated that uFLC measurements were relatively insensitive for assessing residual disease. This has been substantiated in many other studies *(Chapter 12)*.

The discrepancy between serum and urine results is mainly due to renal metabolism of FLCs *(Chapter 3)*. Other factors are the greater sensitivity of the immunoassays and errors in collecting and measuring urine samples *(Chapter 24)*. While changes in serum and urine FLC concentrations generally occur in parallel, sFLCs remain abnormal in many patients when urine is normal. Clinical responses and assessment of residual disease are better judged from sFLC measurements.

Comparisons of serum and urine assays for monitoring LCMM are illustrated for two patients in Figures 8.6 and 8.7. In both patients the uFLC measurements became normal while serum tests remained abnormal.

The results indicate that sFLC measurements have an important role in identifying and managing these patients. FLC tests are included in international diagnosis and response guidelines for MM *(Chapter 25)*. A detailed comparison of the use of serum versus urine FLC tests is given in Chapter 24.

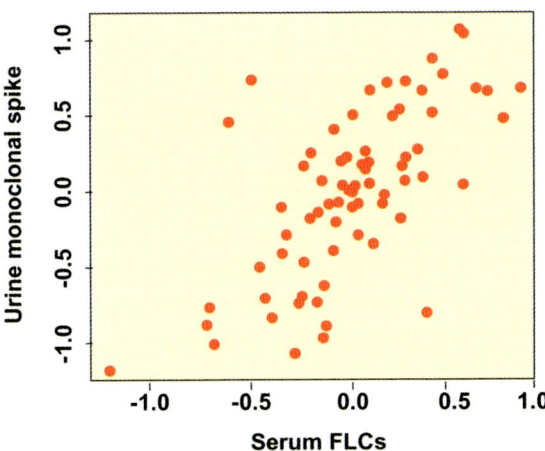

Figure 8.5. Correlation between changes in serum and urine FLCs during the evolution of their disease in 71 patients with LCMM. Initial measurement values were changed to zero and transformed logarithmically for comparison purposes (P=0.0001). (Courtesy of RA Kyle and JA Katzmann).

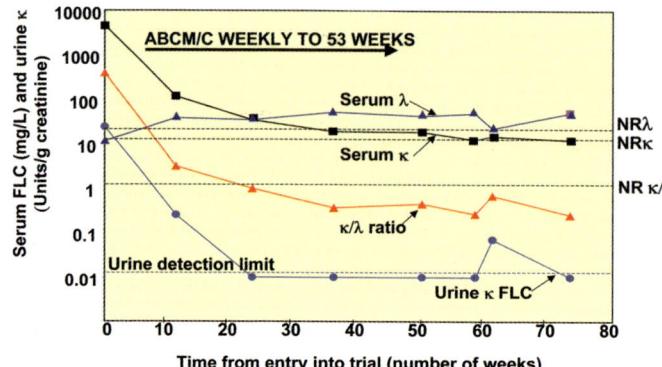

Figure 8.6. Serum and urine FLC levels in a patient with κ LCMM during treatment. (NR: normal range, ABCM: adriamycin (doxorubicin), busulphan, cyclophosphamide, melphalan).

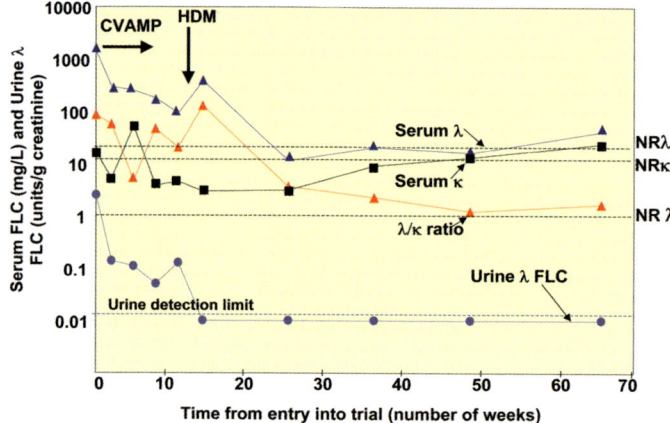

Figure 8.7 Serum and urine FLC levels in a patient with λ LCMM during treatment. (CVAMP: cyclophosphamide and vincristine, adriamycin, melphalan, methyl-prednisolone, HDM: high dose melphalan).

Clinical case history No 1. Unusual clinical features in a patient with LCMM.

A 63-year-old woman attended hospital with severe pain in her right shoulder. An X-ray showed minor erosive changes in the shoulder joint while blood tests were normal apart from a marginally elevated serum calcium at 2.69 mmol/L (NR: 2.08-2.67). An orthopaedic surgeon recommended a hemiarthroplasty. At operation, the head of the humerus was eroded and the glenoid was almost entirely replaced by 'extremely soft bone' suggestive of malignancy or a metabolic cause. The operation was aborted and the tissue was sent for histological examination.

Re-examination of the patient failed to identify any additional clinical features. Investigations showed a normal chest X-ray while the serum calcium had increased further to 3.41mmol/L and she was hypophosphataemic at 0.55mmol/L (NR: 0.67-1.54). Other results included a raised alkaline phosphatase at 234 IU/L (NR: 30-115) and an increased parathyroid hormone related peptide at 9.5 pmol/L (NR: 0.7-1.8) while parathyroid hormone levels were low at 9ng/L (NR: 10-60). A metastatic tumour deposit was considered the most likely cause of her shoulder disease. The search for a primary tumour, however, was unsuccessful: CT scans of the abdomen and thorax were normal, as were the serum cancer markers, CA-199, CA-125 and CEA.

The possibility of MM was considered. Bone histology from the surgically resected specimen showed osteoarthritis and osteopenia and there was no excess of plasma cells. A skeletal survey showed only osteoporotic bone with no discrete lytic lesions and no features of MM. Serum and urine electrophoretic tests showed no monoclonal gammopathy but assessment of immunoglobulins by nephelometry revealed hypogammaglobulinaemia: IgG 3.20g/L (NR: 5.3- 16.5), IgA 0.15g/L (NR: 0.8- 4.0) and IgM 0.26g/L (NR: 0.5- 2.0).

In view of the diagnostic difficulties, sFLC measurements were requested, with the following results:- κ 7,840mg/L (NR: 3.3 - 19.4); λ 4.4mg/L (NR: 5.7 - 26.3); κ/λ ratio 1,782 (NR: 0.26-1.65) (*Figure 8.8*). uFLC concentrations by immunoassay were as follows: κ 371mg/L; λ 4.7mg/L; κ/λ ratio 79.

Subsequent bone-marrow aspiration of the iliac crest indicated a high concentration of abnormal plasma cells. This established the diagnosis as κ-secreting LCMM with production of excess parathyroid hormone related peptide leading to hypercalcaemia.

Several months after her initial clinical presentation, the patient was finally treated with chemotherapy. This resulted in a satisfactory reduction of the serum κ FLC concentrations and normalisation of serum calcium. Since then, her shoulder has remained unstable although pain free.

Comment. The diagnosis of LCMM can be difficult. In this patient the clinical features were atypical and did not provide a clear lead to the diagnosis, so tests for MM were not considered for some time. When electrophoretic analysis of serum and urine were eventually performed the results showed no monoclonal proteins. Reassessment of the original urine sample, at a later date, showed a low concentration of monoclonal κ by IFE that had been overlooked on the earlier analysis. The final diagnosis was LCMM with limited plasma cell infiltration of the bone marrow.

The response of the tumour to chemotherapy was good. The serum κ FLC concentrations fell for 45 weeks with a half-life of approximately four weeks (*Figure 8.8*). The patient is likely to have a good clinical course in the medium term.

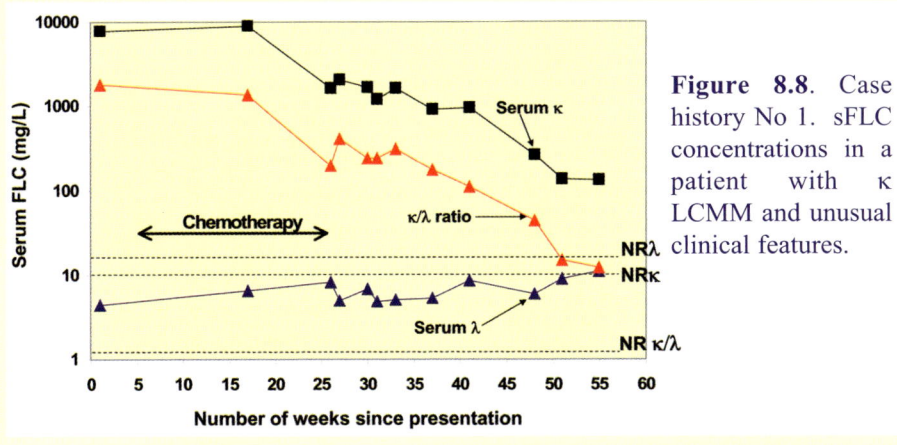

Figure 8.8. Case history No 1. sFLC concentrations in a patient with κ LCMM and unusual clinical features.

Clinical case history No 2. Free light chain breakthrough during relapse of MM.

A 56-year-old man was followed for refractory MM grade IIIB (Durie and Salmon classification). The IgGλ M-protein was 69 g/L, at diagnosis, on the scanned SPE gel. Following a bone marrow allograft, SPE and IFE showed a reduction and stabilisation of the IgGλ band at 3.0 g/L.

Three months after the allograft the patient was clinically deteriorating with no change in the IgGλ M-protein band *(Figure 8.9)*. SPE showed a peak in the gamma-region while IFE showed a monoclonal IgGλ but no monoclonal FLC band.

sFLC analysis showed: κ 1.3mg/L, λ 242mg/L and κ/λ ratio of 0.005, clear evidence of the tumour that was not apparent from the monoclonal IgGλ level. The diagnosis was relapse of the MM with FLC breakthrough. The patient died one month later.

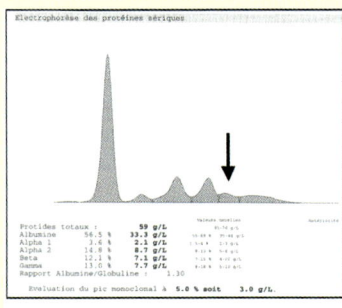 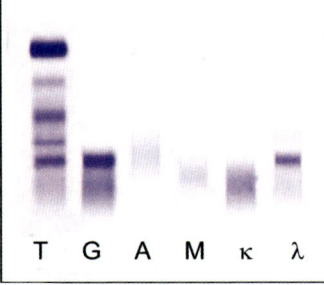

Figure 8.9. Case history No 2. SPE scan and sIFE of the patient during clinical relapse. (Courtesy of L Guis)

Test questions
1 How many patients with LCMM will be missed if urine tests for Bence Jones protein are abandoned for sFLC tests?
2. Is the correlation between serum and urine concentrations of monoclonal FLCs good, medium or poor?
3. What is the median concentration of serum monoclonal λ FLCs required to overflow into the urine and produce positive urine FLC tests?

Answers
1. *None (Section 8.1, page 61).*
2. *Poor (Figure 8.3).*
3. *Approximately 300mg/L (Chapter 24, Figure 24.2).*

Nonsecretory multiple myeloma (NSMM)

9.1.   *Introduction*   **69**
9.2.   *Diagnosis of nonsecretory multiple myeloma (NSMM)*   **70**
9.3.   *Monitoring NSMM*   **72**
          Clinical case history No 3   *73*
          Clinical case history No 4   *75*

Summary: In patients with NSMM, serum free light chain (sFLC) measurements:

1. Are important for diagnosis.
2. Identify relapses and responses to treatment earlier than other tests.
3. Allow monitoring without the need for repeated bone marrow biopsies or radiological scans.
4. Allow patients to be included in clinical trials from which they were previously excluded.

9.1. Introduction

Nonsecretory multiple myeloma (NSMM) accounts for 1-5% of all multiple myeloma (MM) patients. The disease is characterised by the absence of monoclonal proteins in serum and urine using electrophoretic tests *(Drayson 2; Bladé 3; Dreicer)*.   Nevertheless, monoclonal proteins can usually be demonstrated in the bone marrow plasma cells by immunohistochemical staining. Using high sensitivity tests such as isoelectric focusing, monoclonal proteins have been detected in the sera of some patients *(Sheeham)*. Other patients have tumour cells that produce but do not secrete monoclonal immunoglobulins into the blood. Finally, 10-15% of NSMM patients are true *"non-producers" (Raubenheimer).* In these patients the tumour plasma cells contain no detectable immunoglobulins. As tests for monoclonal proteins have become more sensitive, fewer patients are now classified as NSMM. Yet, even in expert hands, 2-3% of patients with MM have undetectable serum or urine M-proteins by immunofixation electrophoresis (IFE) (*Bladé 3; Reilly; Katzmann 3; van Rhee).*

From a logical standpoint, such patients cannot be producing significant amounts of intact monoclonal immunoglobulins. IgG molecules accumulate in serum with a half-life of 3-4 weeks so production from even small clones of plasma cells is visible as monoclonal bands on serum protein electrophoresis (SPE) gels. In contrast, free light chains (FLCs) have a serum half-life of only 2-6 hours, 100-200 fold less. Clonal production of monoclonal FLCs, therefore, needs to be correspondingly that much greater to produce similar serum levels to those found in IgG producing MM. Hence, more sensitive techniques are required to detect FLC producing clones when they are small or when production is inefficient.

Investigations on urine samples may also be unhelpful because patients with NSMM usually have normal renal function. The modest increases in monoclonal FLC production, typically seen, may not be sufficient to damage or overwhelm the reabsorption capacity of the kidneys and enter the urine *(Chapter 3).*

9.2. Diagnosis of nonsecretory multiple myeloma (NSMM)

The above arguments suggest that sensitive assays for sFLCs might detect monoclonal proteins in a proportion of patients with NSMM. The results from a large study are shown in Table 9.1 *(Drayson 2)*.

Archived sera were obtained from patients studied in the UK, MRC MM trials between 1983 and 1999. Out of 2,323 patients, 64 (2.8%) were diagnosed with NSMM and of these, 28 were selected for study because they had complete clinical records and the appropriate stored serum samples. In all patients, serum concentrations of κ and λ were compared with results from SPE and IFE tests. The results showed that 19 of the 28 sera had elevated κ or λ sFLC concentrations and abnormal κ/λ ratios. A further 4

Classification based on sFLCs		κ FLC (mg/L)	λ FLC (mg/L)	κ/λ ratio	Bone marrow plasma cells %	Other Results
Normal sera ranges		3.6-16	8.1-33	0.36-1.0		
12: elevated free κ and increased κ/λ ratio	1	1754	1.6	1096	85	IFE κ +/-
	2	1201	3.6	333	82	BJP κ +/-
	3	935	11	85	70	ND
	4	487	6.6	74	20	ND
	5	931	13.2	71	>90	IFE κ +/-
	6	730	11.1	65	35	ND
	7	978	19.4	50	65	ND
	8	920	26.3	35	14*	ND
	9	789	25.6	31	>50	BJP κ +/-
	10	480	23.8	20	30	ND
	11	151	11.5	13	66	Hist κ+ve
	12	79.8	30.8	2.6	50	ND
7: elevated free λ with reduced κ/λ ratio	13	11.2	196	0.057	20	IFE λ+
	14	2.7	50.9	0.053	74	ND
	15	2.6	61	0.043	6*	ND
	16	17.8	624	0.029	8	IFEλ, BJP+/-
	17	3.8	144	0.026	70	ND
	18	7.7	389	0.019	60	ND
	19	2.8	481	0.005	29	IFE λ+
4: suppression of either κ, λ or both free light chains	20	4.5	6	0.75	21	ND
	21	1.2	1.6	0.75	55	ND
	22	2.4	8.1	0.296	34	ND
	23	3.6	13.1	0.274	70	ND
5: κ or λ normal or elevated and κ/λ ratios normal or borderline	24	16.2	23.4	0.692	67	ND
	25	20.7	33	0.627	73	ND
	26	77	142	0.543	18	IFE λ+
	27	8.3	17.4	0.477	9*	ND
	28	8.6	25.2	0.341	80	ND

Table 9.1. sFLC concentrations in 28 patients with NSMM *(Drayson 2)*. Hist: immunohistochemical confirmation of MM; IFE+/-: weak diffuse bands; IFE+: weak narrow band; BJP+/-: low concentrations of urine FLCs; ND not detected; *trephine biopsy +ve for MM.

samples showed abnormally low levels of one or both FLCs. sFLC concentrations in the remaining 5 samples were substantially normal (*Figure 9.1*).

Careful repeat testing of the sera by IFE, using optimal sensitivity (*Table 9.1*), showed monoclonal sFLCs in 6 of the 28 sera but the monoclonal bands were mostly weak and diffuse. Rather surprisingly, in 9 of the 28 patients no monoclonal bands were seen using IFE even though the immunoassays indicated sFLC concentrations of >200mg/L, and some were considerably higher. In many of these samples, the elevated FLC concentrations should have been easily detectable by IFE.

IFE gels applied with sera from 5 of the samples containing high concentrations of κ FLCs are shown in Figure 9.2. These are compared with 3 κ samples from patients with typical light chain multiple myeloma (LCMM). The sFLCs in the NSMM samples failed to focus into the same narrow monoclonal bands seen in the LCMM sera.

Two sera from (NSMM) patients with substantial concentrations of sFLCs (980mg/L and 1,700mg/L), were subjected to size-separation gel chromatography and found to contain highly polymerised FLCs (40-200kDa) (*Figure 9.3*). This suggested that variable polymerisation caused the monoclonal bands to smear on the SPE gels (*see polymerisation in Chapter 4*) and this could account for their absence or diffuse appearance. Such large polymers would have minimal renal clearance compared with monomeric FLCs. Good renal function would be maintained (typical of these patients) and little FLC would enter the urine. These observations concur with other reports that describe polymerised or structurally abnormal FLCs in some patients with MM (*Abdalla IA; Coriu; Émond*).

Of additional interest, it was found that diffuse bands were more common in κ producing patients (*Table 9.1*). Hence, λ patients with low FLC production are more likely to produce discrete monoclonal bands and be classified as *"secretory"* LCMM. At one time, this dearth of λ patients led to the suggestion that such patients may not exist. Moreover, the

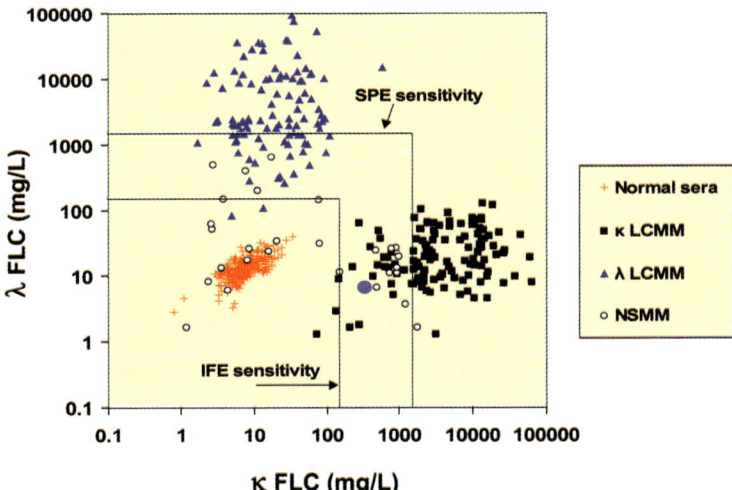

Figure 9.1. sFLC concentrations in patients with NSMM compared with normal individuals and patients with LCMM. ● Clinical case No 3.

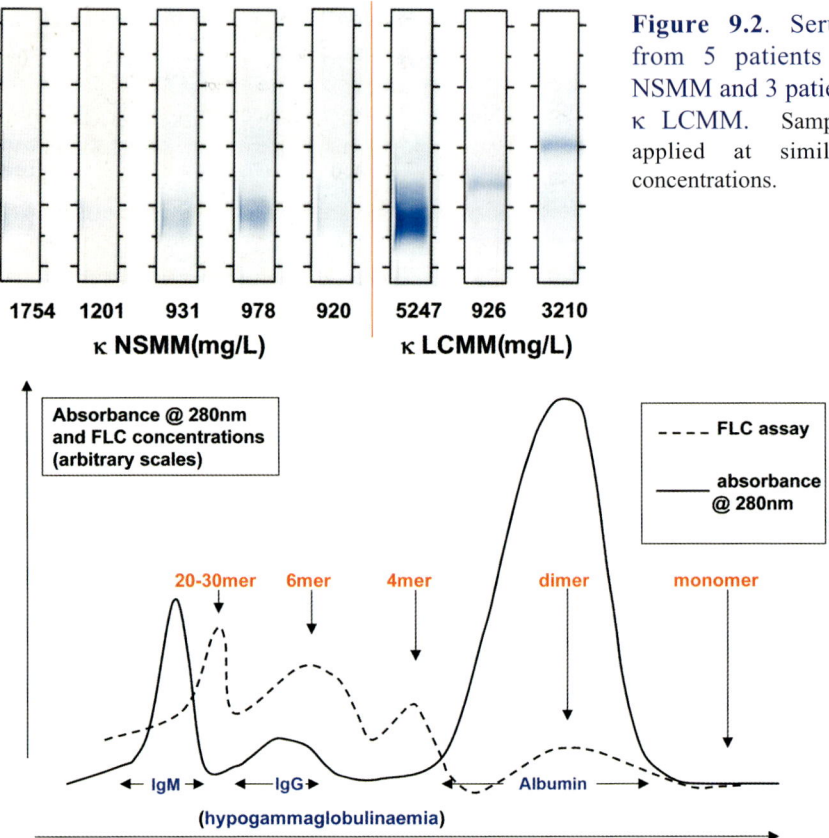

Figure 9.2. Serum IFE from 5 patients with κ NSMM and 3 patients with κ LCMM. Samples were applied at similar FLC concentrations.

Figure 9.3. Size-separation gel chromatography showing the FLC size variation in a serum sample from a patient with NSMM. The sample contained 1,754 mg/L of κ FLCs by immunoassay but was negative by SPE and IFE *(Figure 9.2)*.

observed higher frequency of κ polymerisation probably explains the 4:1 ratio of κ to λ NSMM patients reported in the literature *(Bladé 3)*.

There have been no other large studies on sFLC measurements in NSMM but there have been reports confirming the above observations in smaller groups of patients. Katzmann *(3)* et al. reported sFLCs in 5 patients with NSMM at diagnosis and all were abnormal. Six others had received high dose therapy and were in clinical remission, a finding supported by the FLC results. Similarly, van Rhee et al. reported 4 of 5 NSMM patients with abnormal FLC concentrations. The advent of the sFLC immunoassays has made true NSMM even rarer than previously observed, perhaps only 1 in 200 myeloma patients *(Shaw; Rutherford)*. sFLC testing should improve clinical outcome *(Kumar 4)*.

9.3. Monitoring NSMM

sFLC concentrations are also important for monitoring disease progress and are recognised as such in international guidelines *(Chapter 25)*. In an initial study, samples from 6 patients showed elevated sFLC levels at clinical presentation, reduced levels during plateau phase, and increased levels at relapse *(Figure 9.4)*. One of the patients showed a discordance between clinical features and sFLC concentrations (No.2). While the patient was in remission from a clinical viewpoint, rising concentrations of FLCs indicated imminent disease relapse.

Many patients with NSMM have been studied prospectively since the assays first became available. Two examples are described below.

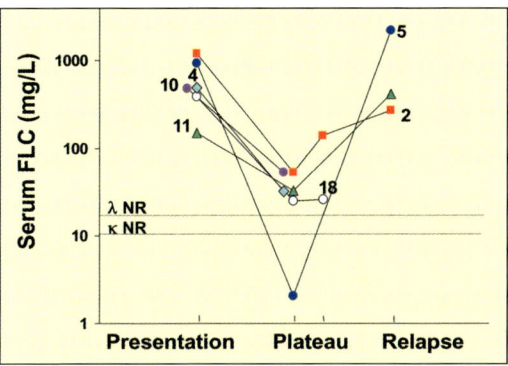

Figure 9.4. Changes in sFLCs and clinical status in 6 patients with NSMM. (Numbers refer to patients in Table 9.1).

Clinical case history No 3. NSMM with "difficult to assess" symptoms during clinical relapse.

A 38-year-old woman, presented with a fractured rib following mild trauma. Over the following months, the pain subsided but non-specific symptoms including breathlessness, vague chest pains and tiredness persisted. During this time, full blood counts, erythrocyte sedimentation rate (ESR) and biochemistry were all normal as were chest X-rays and lung function tests. In the absence of a diagnosis, the general practitioner considered a psychiatric assessment.

Seven months after the initial presentation, she remained symptomatic and was re-investigated, whereupon bone scans and X-rays showed extensive osseous lesions. Immunoglobulin measurements showed immune paresis, but no serum monoclonal protein was detected. She was noted to have hypercalcaemia (2.85 mmol/L - NR: 2.08-2.67) but had normal renal function. In view of the absence of monoclonal immunoglobulins, MM was still considered unlikely. However, a skull X-ray and CT scan showed osteolytic lesions *(Figure 9.5)* so a skull biopsy was performed which was reported as 'plasmacytoma/NSMM'. She was given chemotherapy comprising ABCM for the following 8 months that resulted in clinical remission.

Seven months later and over 2 years after the initial presentation, she re-attended hospital because of chest pains and breathlessness. Again, clinical examination was normal, as were routine biochemistry and haematology tests. Immunology tests showed reduced immunoglobulins but no detectable monoclonal protein by serum electrophoresis. A bone marrow biopsy showed 5% plasma cells that were morphologically normal. Chest X-ray, a ventilation perfusion scan and lung function tests revealed no evidence of pulmonary disease. Blood tests were requested for FLCs, the results of which were: κ 330mg/L, λ 6.5mg/L and κ/λ ratio 51, suggesting recurrence of NSMM *(Figure 9.1 and Figure 9.6 at week 67)*. Doubt was expressed regarding the validity of the results so FLC measurements were repeated 2 and 3 weeks later and showed κ increases to 470mg/L and then 525mg/L with a rising κ/λ ratio, confirming recurrence of the disease.

FLC concentrations were assessed retrospectively from archived samples and then the patient was monitored prospectively. Figure 9.6 shows that the κ FLC concentrations had increased rapidly during the tumour recurrence, with an apparent doubling time of 30 days as indicated by the κ/λ ratio. Serum κ concentrations

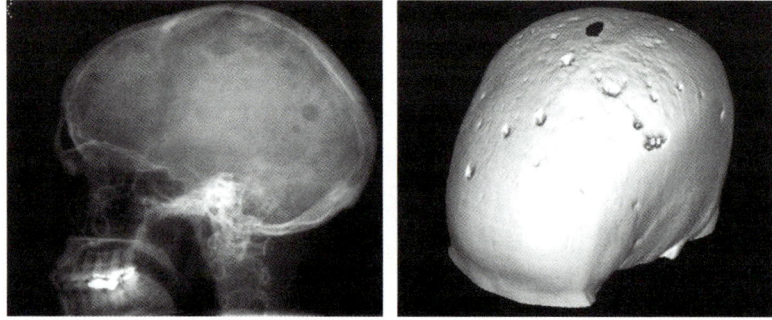

Figure 9.5. X-ray and computerised tomography (CT) scans of the skull in NSMM.

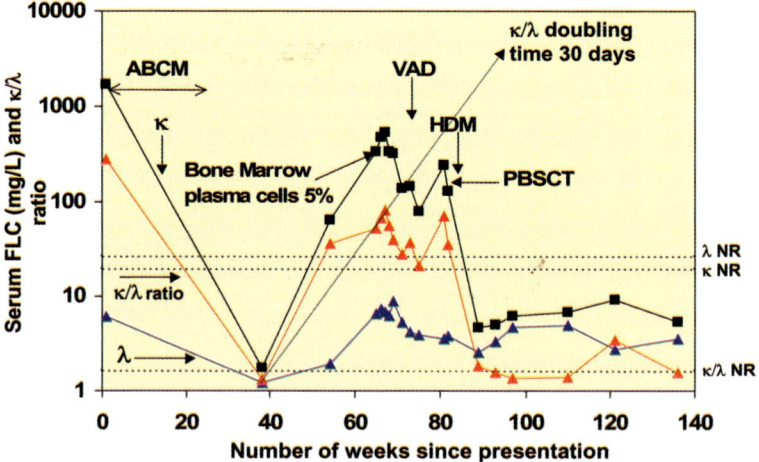

Figure 9.6. sFLC concentrations in Patient No 3 during the course of the disease. The changing κ/λ ratio is related to the tumour growth. (ABCM: adriamycin (doxorubicin), busulphan, cyclophosphamide, melphalan; VAD: vincristine, adriamycin (doxorubicin), dexamethasone; HDM: high dose melphalan; PBSCT: peripheral blood stem cell transplant).

subsequently reduced during VAD chemotherapy prior to high dose melphalan and PBSC rescue. During the period of relapse, the alternate FLC increased in concentration suggesting deteriorating renal clearance of FLCs from impaired glomerular filtration. Serum κ and λ concentrations and the κ/λ ratio returned towards normal, post-transplant, as the patient went into clinical remission. For 2 years following the transplant the patient remained completely well.

Discussion: NSMM is rare so it is not normally considered when a patient first presents with symptoms but rather when other diseases have been excluded. Even then, normal serum and urine electrophoretic tests for monoclonal proteins tend to deceive the diagnostician. When MM is finally thought of, the patient is subjected to a painful bone marrow biopsy, which is not undertaken lightly. Clearly, since sFLC measurements are important they should be requested when the diagnosis of MM is first considered.

While monitoring these patients, repeated serum tests should generally replace other forms of investigation. At present, skeletal surveys using X-rays, magnetic resonance imaging (MRI) or positron emission tomography (PET) may be used, together with repeated bone marrow biopsies. None of these tests is likely to provide results that are as representative of the changing total tumour burden as sFLC tests. sFLC concentrations assess FLC production from all of the bone marrow and extramedullary sites. They are likely to be a better reflection of overall tumour activity than bone marrow aspirations or skeletal surveys.

Of additional interest in these patients is the κ/λ ratio. This is a more accurate measure of changing monoclonal FLC production than individual FLC concentrations since the alternate FLC compensates for alterations in glomerular filtration rate. This was apparent during the relapse phase in this patient. After week 40, the alternate FLC gradually increased which suggested impaired renal function from renal deposition of tumour-produced FLCs.

It is also of note that normalisation of the κ/λ ratio had not occurred even 2 years after treatment. This suggests either that the clone of tumour cells persists or that bone marrow function has not completely returned to normal. It is likely that the patient has residual disease *(Chapter 12)*.

Clinical case history No 4. A patient with NSMM/plasmacytoma excluded from clinical trials.

A 37-year-old man with pelvic pain was found to have a solitary plasmacytoma located in the right iliac crest. Bone marrow biopsy of the opposite iliac crest was normal and no monoclonal protein was identified in serum or urine. Treatment comprised surgical resection followed by irradiation (5,000Gy). Subsequently, he remained asymptomatic, but 5 years later a routine skeletal survey showed a thoracic spine lesion at T2 that was irradiated. Over the following 7 years further painful lesions developed. These were identified using different scanning techniques (particularly PET) and were treated with irradiation or melphalan and prednisolone.

Throughout this period, and in spite of repeated testing, no monoclonal protein was identified by SPE and UPE. Finally, 12 years after the initial presentation FLC immunoassays became available and showed; κ 7.5mg/L, λ 632mg/L and a κ/λ ratio 0.01. These results identified a λ producing tumour with no associated suppression of the κ FLC *(Figure 9.7)*. One month later, λ concentrations had increased to 700mg/L, prompting treatment with thalidomide (50mg/day) and dexamethasone (40mg weekly). Over the subsequent 7 months, serum λ gradually fell to 33mg/L and the κ/λ ratio began to normalise. Based on the FLC results, dexamethasone was reduced to 12 mg per week and he remained well and in complete remission.

Figure 9.7 shows the changes in sFLC concentrations over a 12 month period. The effectiveness of the drugs and the doses required can all be monitored during this period of therapy. This has produced clear benefits for the patient and avoided costly scans and painful bone marrow biopsies. Furthermore, the patient can be entered into clinical trials of new treatments when absence of a disease marker had previously led to his exclusion. The patient has been monitored successfully using sFLC assays for many years since the original tests were performed.

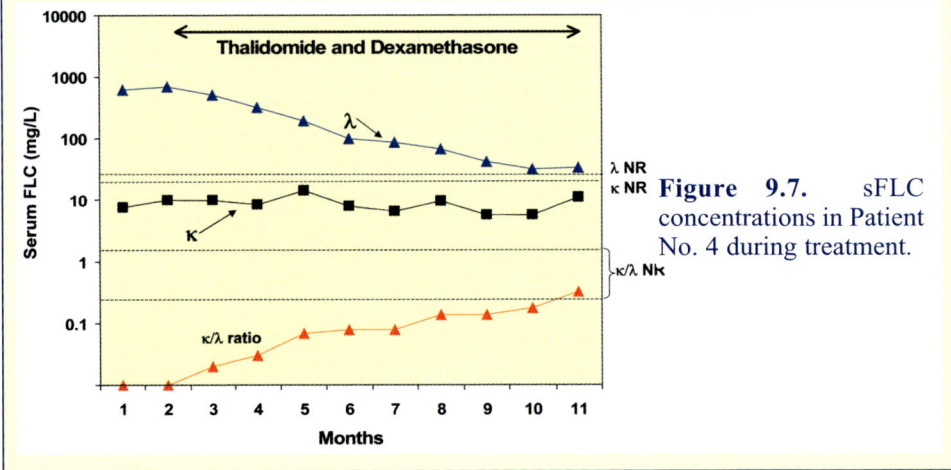

Figure 9.7. sFLC concentrations in Patient No. 4 during treatment.

Test Questions
1. Which is more sensitive, serum or urine IFE, for detecting NSMM?
2. Why do most so called "NSMM" patients have excess monoclonal FLCs rather than excess monoclonal intact immunoglobulins?
3. Are sFLC tests more accurate than bone marrow assessments of tumour responses?

Answers
1. *By definition, neither test is positive in NSMM (Section 9.1, page 69).*
2. *IgG accumulates in serum 100-200 times more than FLCs, for the same tumour production when it is readily detected by insensitive techniques such as SPE (Section 9.1, page 69).*
3. *sFLC tests are sometimes the most accurate method of assessing tumour responses to therapy. sFLC concentrations are a consequence of production from all the tumour deposits, not just the part of the tumour that is sampled in a biopsy (Clinical case history No. 3, page 73)*

Chapter

10

Intact immunoglobulin multiple myeloma (IIMM) - theoretical considerations of sFLC and Ig measurements

10.1.	*Introduction*	*77*
10.2.	*Half-life of sFLCs*	*78*
10.3.	*Half-life of IgG, IgA and IgM and recycling by FcRn receptors*	*79*
10.4.	*Tumour killing rates and half-lives of immunoglobulins*	*80*
10.5.	*Blood volume changes in monoclonal gammopathies*	*82*
10.6.	*Changes in sFLCs and immunoglobulins during treatment*	*83*

Summary:

1. sFLC concentrations are low because of fast renal clearance.
2. sFLC κ/λ ratios inherently compensate for changing clearance rates.
3. Serum IgG half-life is both prolonged and variable because of FcRn recycling.
4. Changes in haematocrit and blood volume alter immunoglobulin concentrations.
5. The 2-6 hour half-life of sFLCs allows tumour cell killing rates, early responses, residual disease and early tumour relapse to be quickly observed during treatment.

10.1. Introduction

Serum concentrations of free light chains (FLCs) and intact monoclonal immunoglobulins (Igs) reflect the balance between their production and clearance rates. Production rates vary, not only between different patients, but also in individual patients as their tumours progress or respond to treatment. Salmon and Smith in 1970 showed that for IgG myeloma, average synthesis rates in different patients ranged from 12,500 to 85,000 molecules of IgG per minute per myeloma cell. However, this was constant over time which means that changes in total synthesis rates reflect changes in tumour mass for an individual patient. This is presumably also true for monoclonal FLC synthesis.

Clearance rates are more complicated. Under normal circumstances, sFLCs are cleared rapidly via renal glomeruli, but in renal failure clearance is by general pinocytosis resulting in half-lives of between 2 and 3 days, depending upon the degree of renal impairment. For IgA and IgM molecules, half-lives appear to be constant at around 5 to 6 days, with clearance by pinocytosis, while for IgG, clearance is prolonged to 21 days by saturable recycling receptors (Brambell receptors; more recently called neonatal receptors [FcRn]). These various influences on serum concentrations of FLCs and immunoglobulins mean that both isolated and serial measurements may not reliably relate to tumour size or changing tumour size.

These variable clearance rates apply equally to monoclonal immunoglobulins and their normal counterparts; hence the value of sFLC κ/λ ratios, that inherently compensate for varying clearance rates as renal function changes. The same rationale applies to IgG monoclonal immunoglobulins, with ratios of monoclonal Igs to normal immunoglobulins (eg. IgGκ/IgGλ) numerically compensating for varying removal rates. This is discussed in detail in Chapter 32.

In addition, other factors need to be considered. Large molecules such as IgM, and to a lesser extent IgA and IgG, are affected by changes in blood volume. In contrast, the smaller FLC molecules, which are 70-80% extravascular, are affected less. Furthermore, sFLC κ/λ ratios automatically compensate for volume changes.

The purpose of this chapter is to analyse the theoretical basis for the utility of sFLC measurements in multiple myeloma (MM) patients who produce both intact monoclonal immunoglobulins and FLCs.

10.2. Half-life of sFLCs

As indicated in Chapter 3, the predominant clearance mechanism for sFLCs is filtration through the renal glomerular fenestrations. At 25kDa, monomeric FLC molecules (usually κ) have a half-life of 2 hours while dimeric molecules (usually λ) have a half-life of between 4 and 6 hours *(Figure 10.1)*. Serum FLCs that are more highly polymerised have longer half-lives. The 200- to 300-fold shorter serum half-lives of FLCs compared with IgG (21 days) allows a much more sensitive evaluation of changing monoclonal protein production and hence tumour size, during treatment *(Bidart)*.

As noted earlier, the half-lives of sFLCs become prolonged to 2-3 days when there is renal failure. This is particularly relevant at high sFLC concentrations as they directly cause renal damage, further increasing sFLC concentrations. The relationship between sFLCs, urine FLCs and tumour progression is shown in Figure 3.10. Accelerating rises in sFLC concentrations suggest accelerating prolongation of the half-life as a result of renal failure rather than rapid tumour growth. Serum FLC κ/λ ratios correct for varying clearance rates because both types of FLC are removed by the same mechanisms. Such measurements are helpful when renal function is abnormal or changing.

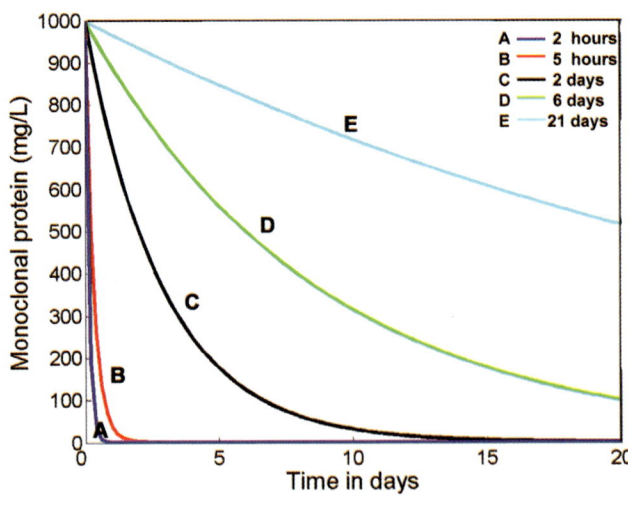

Figure 10.1.
Calculated serum half-life curves for different immunoglobulin molecules.
(A) monomeric κ; (B) dimeric λ; (C) monomeric κ with renal failure; (D) IgA and (E) IgG. (Courtesy of N Evans and M Chappell).

10.3. Half-life of IgG, IgA and IgM and recycling by FcRn receptors

Under normal circumstances, serum proteins that are too large for renal filtration (>65kDa) have a half-life of between 2 and 3 days and are removed by pinocytosis. This process occurs in all nucleated cells as they obtain their essential nutrients from plasma. However, for IgG (and albumin) the half-life is prolonged to 20-25 days by FcRn receptors *(Kim; Akilesh; Anderson CL)*. These proteins have a structure similar to Class I MHC molecules with a heavy chain of 3 domains and a single domain light chain comprising β_2-microglobulin *(Figure 10.2)*. They are the same receptors that transport IgG from the pregnant mother to the developing foetus in the last trimester of pregnancy.

The heterodimeric FcRn molecules protect both IgG and albumin from acid digestion and recycle them back to the cell surfaces to be released into the slightly alkaline environment of the blood. This process occurs many times under normal circumstances, resulting in the half-lives of both IgG and albumin extending from 3 to 21 days *(Figure 10.3)*. Interestingly, IgG and albumin molecules do not compete for the same sites on the receptor although the exact mechanism and sites of binding are unknown.

When there are no functioning FcRn receptors, as in patients with familial hypercatabolic hypoproteinaemia (a disease associated with a genetic deficiency of β_2-microglobulin), the half-lives of IgG and albumin are only 3 days. Such patients have hypogammaglobulinaemia, not from failure of production, but simply from excessive

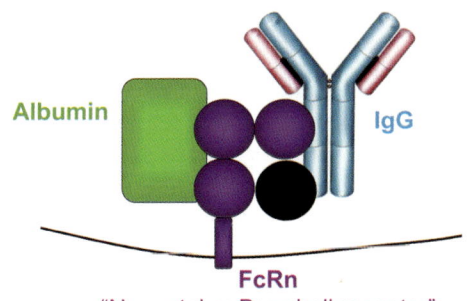

Figure 10.2. Diagram of FcRn structure with binding of IgG and albumin molecules. Albumin and IgG are bound non-competitively (Courtesy of J Hobbs).

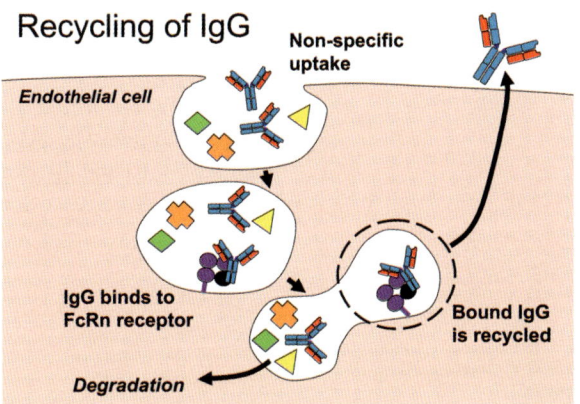

Figure 10.3. Recycling of IgG molecules by FcRn receptors. (Courtesy of J Hobbs).

catabolism. FcRn receptors are functional in most nucleated cells including renal podocytes, and presumably affect urine IgG and uFLC concentrations *(Chapter 24) (Tryggvason; Haymann; Kobayashi; Chaudhury; Akilesh 2; Roopenian).*

10.4. Tumour killing rates and half-lives of immunoglobulins

Due to the long serum half-lives of intact immunoglobulins, in particular IgG, their respective serum concentrations are slow to reflect changes in tumour production. This is evident from calculations of serum half-life curves for the different Igs and sFLCs in relation to tumour cell killing rates. Figure 10.4 shows calculated half-life curves for serum monomeric κ and serum IgG when the tumour cells are killed at different rates. The half-life of sFLCs adds only a few hours to the changing tumour production rates such that tumour responses and complete remission can be rapidly identified. In contrast, the 21-day serum half-life of IgG is so long that it largely obscures differences in rates of tumour destruction. Furthermore, complete remission can only be identified many months later using IgG levels rather than FLCs. For the first time in MM, a short half-life marker is available that allows reliable assessment of tumour cell killing rates and earlier identification of tumour remission *(Pratt 2).*

Production of both FLCs and intact immunoglobulins usually recommence when MM relapses *(Chapter 12).* When tumour recurrence is associated with synthesis of both types of molecule, they typically become detectable again at approximately the same time. (In practice, re-synthesis is not always synchronous so one type of immunoglobulin molecule may reappear before the other - see FLC breakthrough – *Chapter 12.7).* If tumour re-growth occurs before all the monoclonal IgG has disappeared, then FLC measurements will be more sensitive for detecting tumour recurrence *(Figure 10.5).* This is because the rising concentrations of monoclonal IgG from tumour relapse are superimposed upon falling concentrations from the initial tumour response, thereby obscuring an early IgG increase. In contrast, sFLCs normalise early, so increasing production by the relapsing tumour is not masked by residual falling

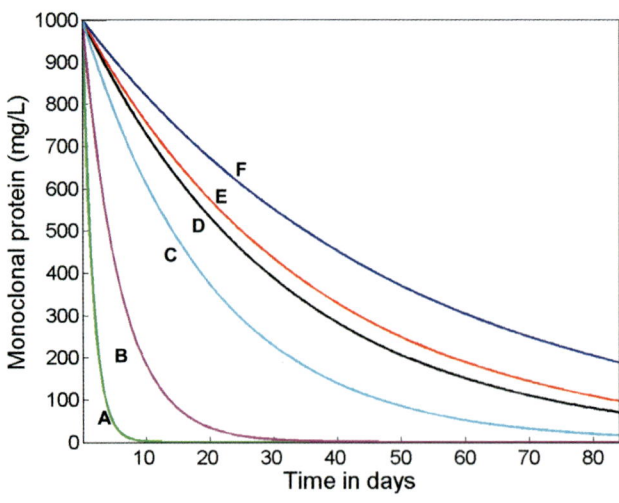

Figure 10.4. Calculated serum half-life curves for monomeric κ (A,B,C) and IgGκ (D,E,F) when the tumour cells are killed at: 50% per day (A, D); 50% in 4 days (B, E) and 50% in 2 weeks (C, F). (Courtesy of N Evans and M Chappell).

the tumour sFLC fell and then rapidly recovered within 10-20 days of a treatment cycle. These patients were monitored over multiple treatment cycles and showed repeated falls and rises in sFLC co-incident with treatment *(Figure 10.10)*. When evident, the relapse of sFLC was rapid, with a doubling time of less than 10 days. This may correspond to the biological half-life of proteosome inhibition and recovery, rather than to tumour killing and regrowth. By comparison, intact immunoglobulin monoclonal proteins did not show the same peaks and troughs. Generally the sFLC levels indicated disease response earlier than the immunoglobulin measurements. They concluded that monitoring patients with sFLCs provided an opportunity to follow the kinetics of tumour kill, which is obscured by the slow clearance of intact monoclonal immunoglobulins. They added that sFLC measurements indicated early tumour responses that could facilitate relevant changes of treatment strategy. This may have a major bearing on the costs of treatment and utilisation of resources. Similar observations have been made by others using bortezomib *(Kyrtsonis 1; Robson 3)*.

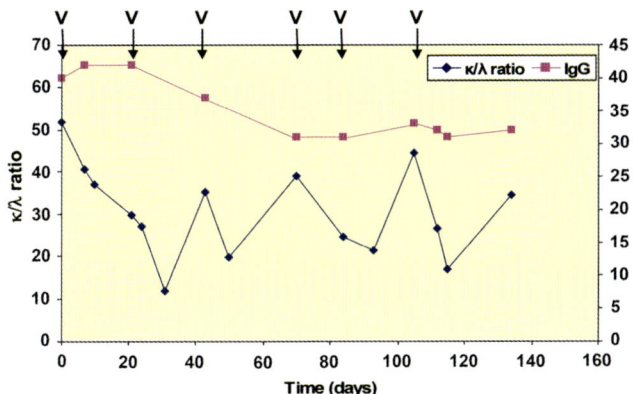

Figure 10.10. Changes in sFLC concentrations during 6 cycles of Bortezomib (V) showing rapid responses to treatment and subsequent relapses. (Courtesy of GP Mead).

Test questions
1. *What is the ideal serum half-life of a tumour marker?*
2. *What is the mechanism of slow clearance of IgG?*
3. *Why do IgG and FLC half-lives vary with concentration?*
4. *Do blood volume changes affect sFLC concentration ratios?*

Answers
1. *The shorter the half-life the better, provided sufficient amounts remain for accurate detection (Section 10.2, page 78).*
2. *Recycling receptors bind IgG and protect it from acid degradation in pinocytotic vesicles (Section 10.3, page 79).*
3. *IgG receptors saturate at high IgG concentrations so the half-life shortens (Section 10.3, page 79). FLCs damage the kidney so filtration slows (Section 10.2, page 78).*
4. *No (Section 10.8, page 82)*

Chapter

11

Intact immunoglobulin multiple myeloma (IIMM) - sFLCs at diagnosis

11.1. *Introduction*   **86**
11.2. *Myeloma cell diversity: production of monoclonal immunoglobulins (Igs) and FLCs* **86**
11.3. *sFLC concentrations at disease presentation*   **88**
11.4. *Disease stage and sFLCs*   **91**

Summary: In patients with intact immunoglobulin multiple myeloma (IIMM), monoclonal serum free light chains (sFLCs):

1. Are produced by clones of cells with considerable diversity.
2. Are abnormal in 95% of patients at disease presentation.
3. Show no correlation with intact monoclonal immunoglobulin concentrations.
4. Are markers of disease stage.
5. Are markers of tumour outcome.
6. Are independent of albumin and β_2-microglobulin for disease staging.

11.1. Introduction

Approximately 80% of patients with multiple myeloma (MM) produce intact immunoglobulin monoclonal proteins *(Figure 7.3)*, of whom ~50% have excess monoclonal FLCs in urine by immunofixation electrophoresis (IFE) *(OR McIntyre)*. Serum protein electrophoresis (SPE) and sIFE tests for FLCs are less frequently positive because sFLC concentrations are below the detection limits of these assays.

The first attempt to measure sFLCs IIMM was by Sölling *(5)* in 1982. Using column chromatography, FLCs were first separated from bound light chains and then measured using antibodies against whole light chains. Using this technique monoclonal sFLCs were present in 86% of IIMM patients. Recently, using sensitive sFLC immunoassays, many authors have reported even higher prevalences of monoclonal FLCs. These studies, together with their clinical relevance will be described in this chapter.

11.2. Myeloma cell diversity: production of monoclonal Igs and FLCs

Myeloma plasma cells have huge genetic and mophological diversity, so mixed clones occur at clinical presentation. Subsequent evolutionary selection by chemotherapy leads to escape mutants with different protein expression profiles. Predictably, myeloma cell populations with mixed monoclonal protein expression arise from this diversity, although perhaps surprisingly, this was only recently confirmed by histology. Using a double immunofluorescence staining method, Ayliffe et al. estimated the incidences of different types of plasma cells in bone marrow biopsies *(Figure 11.1)*. In 82% of patients, single populations of cells were present that contained either intact

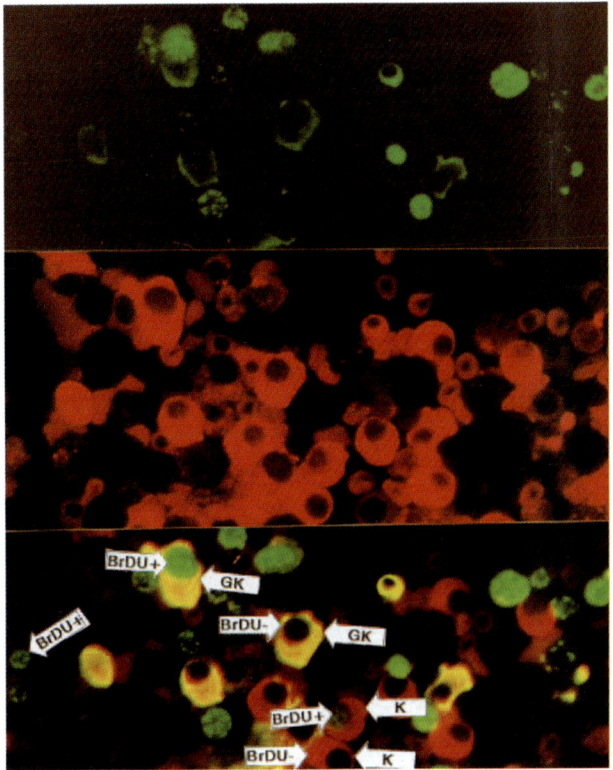

Figure 11.1. The same microscope field of an IgGκ bone marrow sample showing dual populations stained with anti-IgG fluorescein isothiocyanate (FITC, green) and anti-κ tetramethylrhodamine isothiocyanate (TRITC, red). In addition, anti-bromo deoxyuridine (-BrdU) FITC (green) staining of nuclei has been performed to show cells in S-phase of the cell cycle. Upper panel shows anti-IgG FITC cytoplasmic and anti-BrdU FITC nuclear staining, middle shows the same field with anti-κ TRITC staining and lower shows a double exposure of the 2 upper plates superimposed to demon-strate double stained IgGκ cells (yellow), κ only cells (red) and S-phase cells with green nuclei. Arrows indicate BrdU+ and BrdU- intact immunoglobulin + cells, BrdU+ and BrdU- κ only cells together with non-plasma cells in S-phase. (Reproduced with permission from Haematologica *(Ayliffe)* and M Ayliffe).

Figure 11.2. Classification of 95 cases according to the presence of FLCs in the urine (Bence Jones protein (BJP) positive (+) or negative) and cell populations demonstrated in the marrow (monoclonal intact immunoglobulins (M-Igs) and/or FLC only) together with their incidence, immuno-chemical findings and survival since diagnosis. (Reproduced with permission from Haematologica *(Ayliffe)* and M Ayliffe).

Cellular Type	Single population M-Ig + BJP Neg	Single population M-Ig + BJP +	Dual population M-Ig + & FLConly BJP+		Single population FLConly BJP+
Marrow plasma cells					
& extra-cellular products	M-Igs with or without free light chains	M-Igs with free light chains	M-Igs with free light chains	Free light chains only	Free light Chains only
Number of cases in group (%)	30 (32%)	40 (42%)	17 (18%)		8 (8%)
M-Ig in Serum	+	+	+ in 7/17 (42%) Neg in 10/17(56%)		Negative
FLCs in urine	Negative	+	+		+
Median survival since diagnosis	86 months	38 months	22 months		22 months

monoclonal immunoglobulins (with or without monoclonal FLCs [74%]) or monoclonal FLCs alone (8%). However, 18% of the samples contained a mixture of both cell populations *(Figure 11.2)*. Progression from cells making intact monoclonal immunoglobulins to cells restricted to FLC production alone was also shown to occur in some patients during the course of their disease. Furthermore, the presence of *"FLC-only"* cells was associated with shortened survival. This indicates that monoclonal FLC production at a cellular level is an adverse prognostic marker both at clinical presentation and during relapse of patients with IIMM. The relationship between monoclonal FLC production and poor survival has been observed in a series of studies that are described below.

11.3. sFLC concentrations at disease presentation

Using archived samples from the UK MRC MM trials, Mead *(1)* et al. assessed sFLC concentrations at the time of presentation in a series of patients: 314 with IgG MM, 142 with IgA MM, 36 with IgD MM and 5 with IgE MM. Overall, 88% had elevated sFLCs with the following breakdown: IgG 84%, IgA 92%, IgD 94%, and all 5 of the IgE MM patients. Some of the remaining patients had normal or reduced concentrations of FLCs but their κ/λ ratios were abnormal, indicating monoclonality in association with bone marrow suppression. In total, 96% of all MM patients (including light chain [LCMM] and nonsecretory [NSMM] disease) had abnormal FLC concentrations or abnormal κ/λ ratios *(Figures 11.3 to 11.6)*. This percentage is higher than previously reported *(OR McIntyre; Sölling 5)*, and reflects the increased sensitivity of the FLC immunoassays and in particular, the use of the suppressed alternate FLC to identify abnormal κ/λ ratios. It is also of note that there was complete concordance between the monoclonal FLC type identified by κ/λ ratios and that obtained by IFE *(Figure 11.3)*: this provided an important specificity validation of the FLC immunoassays.

Other large studies of MM at disease presentation have shown similar results. Orlowski et al. studied sFLCs in 487 patients and noted that 94% had abnormal sFLC

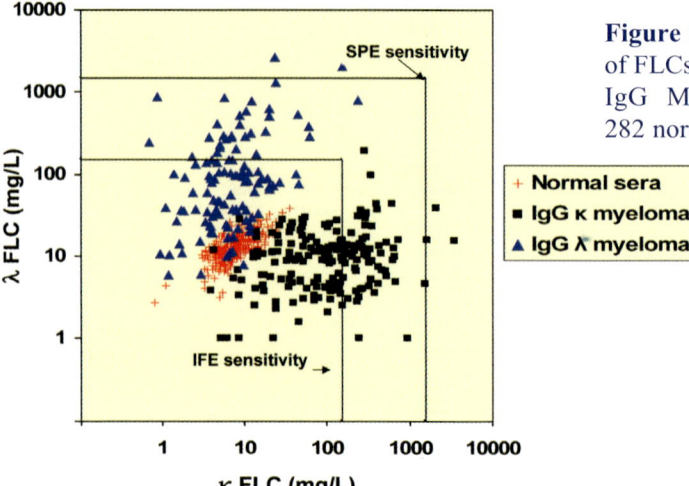

Figure 11.3. Concentrations of FLCs in 314 patients with IgG MM compared with 282 normal sera.

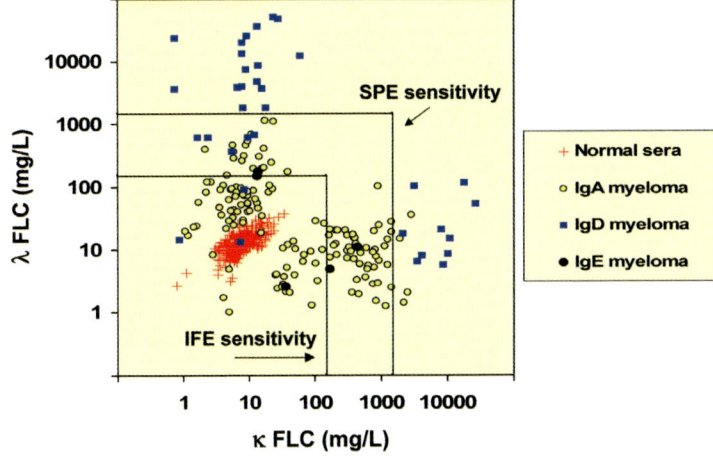

Figure 11.4. Concentrations of sFLCs in 142 IgA, 36 IgD and 5 IgE MM patients compared with 282 normal sera.

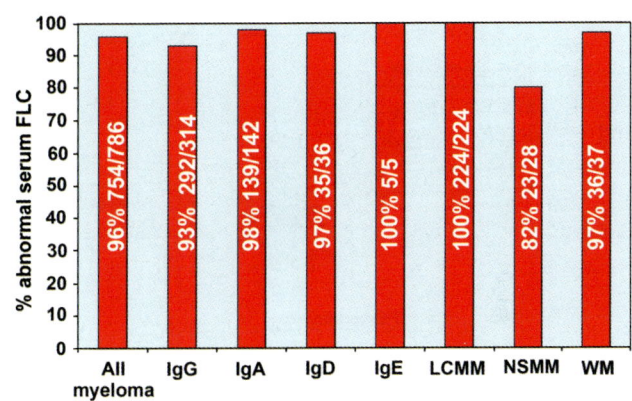

Figure 11.5.

Frequency of abnormal sFLC concentrations in patients with different types of MM and Waldenström's macroglobulinaemia (WM).

κ/λ ratios. Snozek *(2)* et al. reported that monoclonal sFLCs were present in 95% of 576 MM patients at disease presentation, a figure identical to that observed by Owen *(1)* et al. in 207 patients entered into the UK MRC trials. Dispenzieri *(6)* et al. observed abnormal sFLC κ/λ ratios in 96% of 399 patients and, most recently, Katzmann *(8)* et al. observed abnormal ratios in 96.8% of 467 patients with MM.

As sFLC concentrations are normal in some patients with IIMM, it is clear that serum electrophoretic tests are essential for MM diagnosis. In contrast, sFLC assays are more sensitive for the identification of FLCs in LCMM and NSMM. Therefore, when MM is suspected, the optimum laboratory practice should be to test sera by both SPE/IFE and sFLC assays.

In the study by Mead *(1)* et al. sFLC concentrations were higher in IgA than IgG patients but highest in IgD patients (similar to LCMM patients *(see Figure 11.4)*. High levels of urine FLC (uFLC) excretion and excess of λ compared with κ are typical of IgD MM *(Bladé 3; Kanoh)*. Five patients with IgE MM are included in Figure 11.4. Although the number of patients is small, the FLC results are presumably representative of the disease.

A subgroup of 69 patients with IIMM (55 IgG and 14 IgA), each with no detectable uFLC excretion (less than 40mg/L), is shown in Figure 11.6. Serum analysis showed that 95% of the patients had abnormal FLC κ/λ ratios. It is of note that none of the patients had increased concentrations of the alternate sFLC, indicating there was insignificant renal impairment. At clinical presentation there was no correlation between serum creatinine and sFLC levels (as for LCMM - *Figure 8.3)*.

It was also apparent that sFLC measurements showed no significant correlation with serum levels of intact monoclonal immunoglobulins (by Pearson correlation coefficient r). Results for IgGκ and IgGλ are shown in Figures 11.7 and 11.8.

IgGκ vs κFLC: r = -0.0145 (n=150). IgGλ vs λFLC: r = 0.0037 (n=116).
IgAκ vs κFLC: r = 0.212 (n=68). IgAλ vs λFLC: r = -0.0330 (n=64).

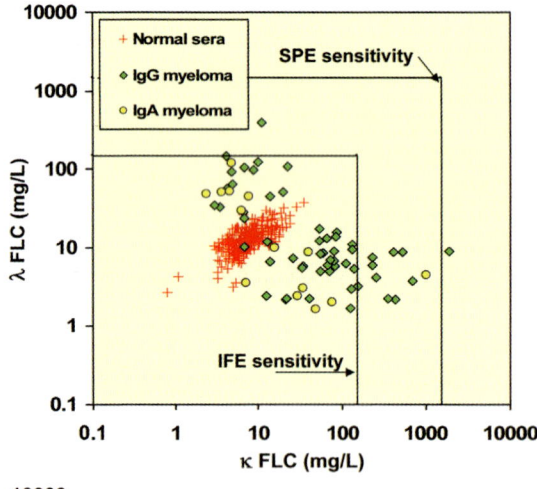

Figure 11.6. sFLCs in 55 IgG and 14 IgA MM patients who had no urine FLC excretion compared with 282 normal sera.

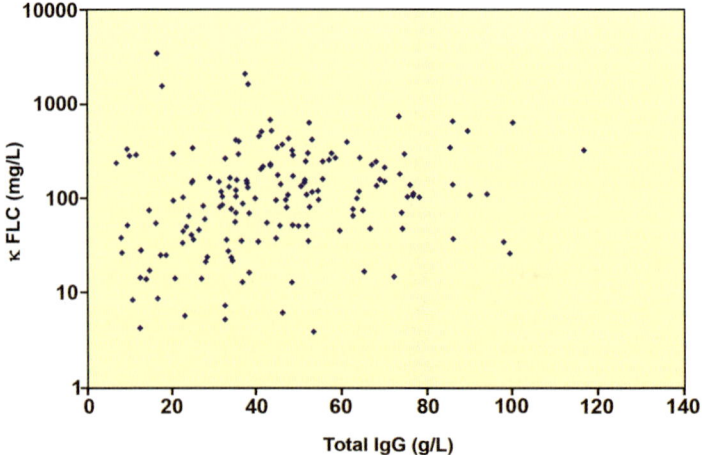

Figure 11.7. Serum monoclonal IgGκ measured by scanning densitometry, and sFLCκ concentrations in 150 IgGκ MM patients. (Pearson rank correlation r = -0.015).

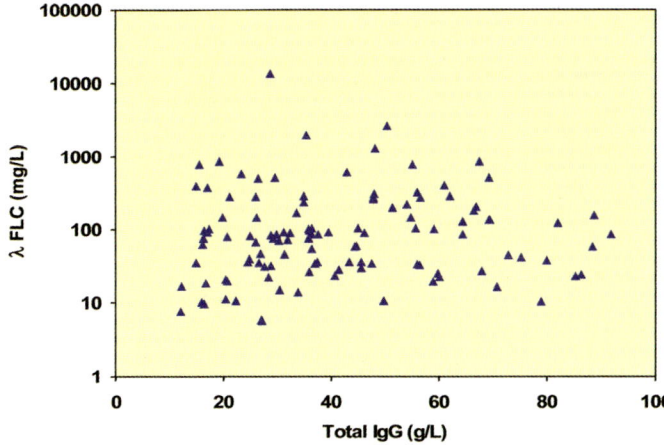

Figure 11.8. Serum monoclonal IgGλ measured by scanning densitometry, and sFLCλ concentrations in 116 IgGλ MM patients. (Pearson rank correlation r = -0.0037).

This lack of correlation is an important issue because it indicates that sFLC concentrations are independent markers of the disease process in IIMM. sFLCs provide additional disease information, both at initial presentation and when monitoring patients.

11.4. Disease stage and sFLCs

It has been known for some time that in MM, high monoclonal uFLC excretion at clinical presentation is predictive of poor survival, and hence, disease stage. For example, the outcome for 351 patients with LCMM was compared with 1,512 IgG and 717 IgA patients entered into the UK MRC myeloma multicentre trials *(Drayson 1)*. LCMM patients had the shortest median survival of 1.9 years (P<0.001) compared with 2.3 years for IgA patients and 2.5 years for IgG patients. It follows that sFLC measurements should be of more value than urine FLCs, due to their greater clinical sensitivity and analytical accuracy.

Several recent studies have assessed the relationship between sFLCs at disease presentation and subsequent outcome. Kyrtsonis *(3)* et al. investigated the prognostic value of baseline sFLC κ/λ ratios in 94 MM patients. The median baseline ratio for κ MM (κ/λ ratio) was 3.57 and for λ MM (λ/κ ratio) was 45.1. sFLC ratios above the observed median values correlated with elevated serum creatinine and lactate dehydrogenase, extensive marrow infiltration and κ or λ light chain type of the intact monoclonal immunoglobulins. Importantly, the 5-year disease specific survival was 82% in patients with sFLC κ/λ ratios lower than the median compared with 30% for κ/λ ratios equal to or greater than the median (P<0.001) *(Figure 11.9A)*.

In current myeloma practice, patients are categorised using the International Staging System (ISS) based upon serum albumin and β$_2$-microglobulin measurements alone. Kyrtsonis *(3)* et al. assessed these parameters alongside sFLC analysis and showed that κ/λ ratios were an additional independent prognostic factor. Combination of the ISS with sFLC ratios in newly diagnosed MM patients for time to progression and survival *(Table 11.1)* showed significantly worse survival with abnormal sFLC ratios (P<0.0001) *(Figure 11.9B)*.

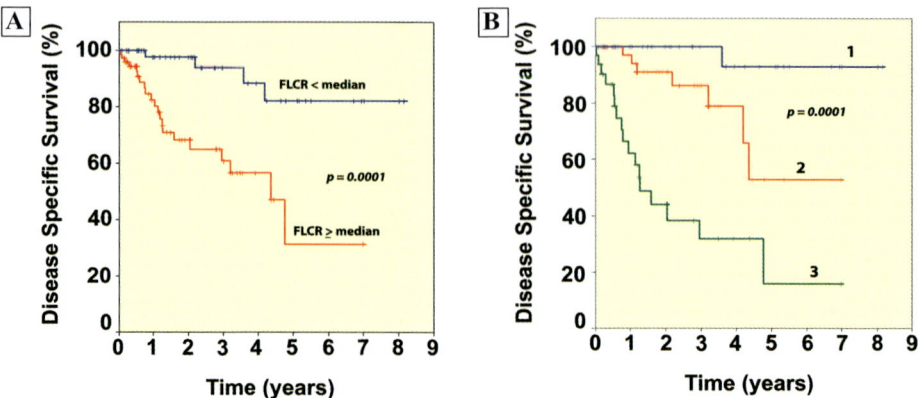

Figure 11.9. **A**. Disease-specific survival of MM patients according to baseline sFLC κ/λ ratios. **B.** Disease-specific survival of MM patients according to abnormalities of baseline sFLC κ/λ ratios, serum albumin or β$_2$-microglobulin as used in the ISS. (Courtesy of M-C Kyrtsonis).

In a similar study of 576 patients at the Mayo Clinic, abnormal sFLC ratios at presentation were again important independent markers of outcome. Snozek *(2)* et al. showed that sFLC ratios assessed alongside the ISS in patients with 0, 1, 2 or 3 risk factors (sFLC κ/λ ratios <0.03 or >32, β$_2$-microglobulin >3.5g/L and albumin >35g/L) had median overall survival times of 51, 39, 30 and 22 months, respectively (P<0.001). Because these data provided additional outcome information, it was suggested that sFLC ratios should be incorporated into the ISS to provide a new risk stratification model.

van Rhee et al. studied the relationship between sFLCs and outcome in 301 patients undergoing intensive treatment *(Figure 11.10)*. They observed that baseline top-tercile sFLC levels >750 mg/L were associated with inferior overall (p=0.005) and event-free survival (p=0.007). Baseline concentrations of other serum or urine immunoglobulins (excluding β$_2$-microglobulin and albumin) did not identify prognostic subgroups. In an associated investigation, Cavallo et al. indicated that sFLC concentrations and abnormal κ/λ ratios were both highly correlated with cytogenetic abnormalities and MM staging.

While results from these different studies show that both sFLC concentrations and sFLC ratios are prognostic, it is unclear which is preferable. Should it be sFLC κ/λ ratios,

Patient Subgroup	Pts (%)	3-yr DSS (%)	5-yr DSS (%)
Low sFLCR and ISS <3	61 (29)	95	90
Either High sFLCR or ISS=3	96 (46)	82	56
High sFLCR and ISS=3	50 (24)	37	24

Table 11.1. Disease specific survival (DSS) in 207 newly diagnosed patients with MM according to the combined sFLC κ/λ ratios and and the ISS comprising serum albumin and β$_2$-microglobulin. (Courtesy of M-C Kyrtsonis).

absolute values or subtracted FLC concentrations (involved minus uninvolved)? This question was recently addressed by Dispenzieri (6) et al., when analysing the outcome of 399 MM patients at clinical presentation. Patients were divided into terciles, and regardless of the baseline variable used (i.e. FLC ratios, involved FLCs or subtracted FLCs), patients with the lowest tercile of FLC had the best outcomes when compared with the two higher terciles. The outcomes for the two higher terciles were nearly identical *(Figure 11.11)*. The respective median overall survivals were 49.4, 41.7 and 42.1 months, and the median progression-free survivals were 34.9, 28.7 and 29.5 months.

Since high sFLCs frequently cause renal damage because of their inherent toxicity, reduced survival might be related to renal-associated mortality. In a study of sFLC concentrations in patients presenting with renal failure, patients with the highest concentrations were more likely to have cast nephropathy and hence a higher mortality rate *(Chapter 13)*. However, such patients are not normally included in routine clinical trials so it is unclear what role renal impairment has in determining outcome in the above studies.

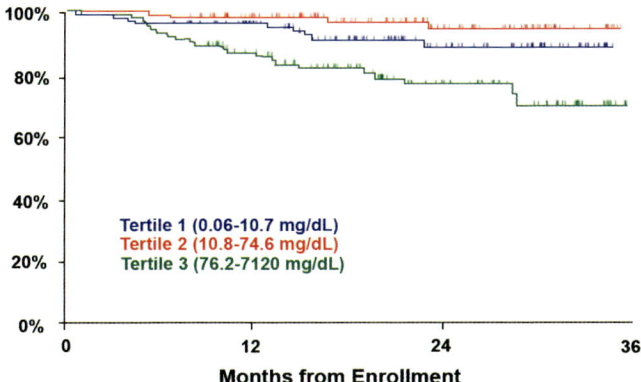

Figure 11.10. Kaplan-Meier overall survival plot according to terciles of baseline concentrations of sFLCs. Outcome was inferior among patients with top-tertile sFLC baseline levels. (This research was originally published in Blood *(van Rhee)* © the American Society of Hematology).

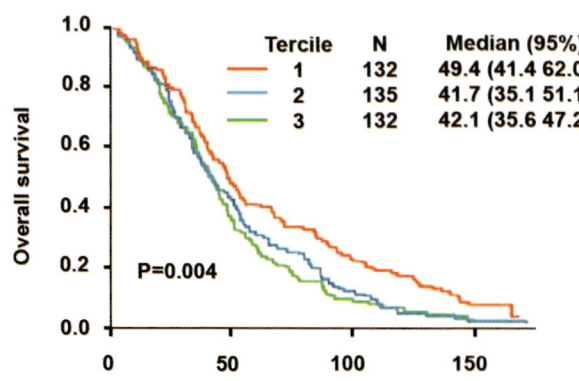

Tercile | N | Median (95%)
1 | 132 | 49.4 (41.4 62.0)
2 | 135 | 41.7 (35.1 51.1)
3 | 132 | 42.1 (35.6 47.2)

P=0.004

Figure 11.11.
Kaplan-Meier overall survival plot according to terciles of baseline concentrations of sFLCs. Outcome was superior among patients with bottom-tercile sFLC baseline levels. (This research was originally published in Blood *(Dispenzieri 6)* © the American Society of Hematology).

Test Questions

1. Do MM plasma cells produce more than one type of M-protein?
2. How often are sFLCs abnormal in IgA MM?
3. Are sFLCs independent of the ISS as risk factors in MM?
4. Why are high sFLCs an adverse risk factor?

Answers
1. Yes, there is considerable diversity in bone marrow myeloma cells (Section 11.2, page 86).
2. In 98% of patients at clinical presentation (Section 11.3, page 88).
3. Yes, they are independent of serum albumin and β₂-microglobulin. (Section 11.4, page 91).
4. Because their expression is increased as tumours become genetically more diverse, more aggressive (Section 11.4, page 91) and patients develop renal impairment (Chapter 13).

12.1.	Introduction	95
12.2.	Overview of sFLCs for monitoring patients with IIMM	95
12.3.	Bone marrow responses and short sFLC half-life	97
12.4.	Speed of response using sFLCs	98
12.5.	Complete response and residual disease	100
12.6.	sFLCs during disease relapse	102
12.7.	Free light chain escape (breakthough)	104
12.8.	sFLCs following high dose melphalan (HDM) and bone marrow transplant	105
12.9.	Long term variations in sFLC levels after high dose therapy	107
12.10.	Drug selection using rapid responses in sFLCs	108
12.11.	Utility of sFLCs in IIMM - comments on utility	109

Summary: sFLCs are useful for monitoring IIMM because they:

1. Correlate better with bone marrow biopsy data than intact immunoglobulins.
2. Have a short serum half-life that allows detection of early responses or lack of responses to treatment.
3. May be abnormal in remission when other tests are normal and thereby indicate remaining tumour with shorter time to progression and reduced survival.
4. May show disease relapse earlier than other markers.

12.1. Introduction

Serum free light chain (sFLC) measurements are being widely used for monitoring patients with IIMM. Monoclonal sFLCs are produced in ~95% of patients, are independent of intact monoclonal immunoglobulin production and have a short half-life *(Chapters 10 and 11)*. Therefore, they are of use in all stages of the disease; during early treatment, assessing response rates, for indicating remission and residual disease, and during relapse. This chapter analyses numerous publications that have addressed these various issues. Many other studies support these results but not all can be discussed in detail for lack of space. Therefore, publications containing the largest series of patients or the most important results have been selected. A full list of publications is shown in Chapter 31.

12.2. Overview of sFLCs for monitoring patients with IIMM

The various attributes of sFLC for monitoring patients are best demonstrated using some patient examples. Detailed analysis of the various aspects of the sFLC responses are provided in the subsequent sections.

Figure 12.1 shows a patient with multiple myeloma (MM) producing monoclonal IgGκ and monoclonal κ serum FLCs (but no urine FLCs). During the initial therapy, κ concentrations rapidly fell with a half-life of approximately 20 days, while IgGκ, quantified from serum protein electrophoresis (SPE) gels, only gradually returned to normal. The apparent half-life of IgGκ was between 100 and 200 days. (This depended, in part, upon which densitometric scanning device was used: white light scanners may under-read at high protein concentrations). The observed response comprised the tumour kill half-life plus the half-life of sFLCs or serum IgG. A similar pattern of sFLC and IgG responses were seen during a subsequent relapse and treatment period.

A further example is given in Figure 12.2 in which the IgGκ half-life was 50 days. In contrast, sFLCs were well within the normal range at the first measurement point

Figure 12.1.
Monitoring of a MM patient using IgGκ and κ sFLCs. SPE gels are shown for each sample.

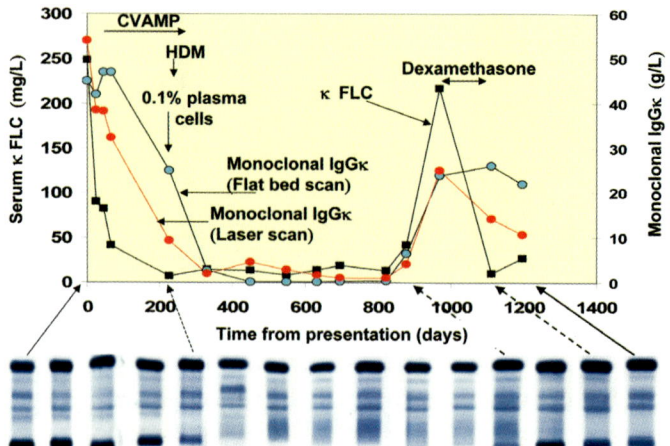

Figure 12.2.
Monitoring of MM using IgGκ and κ sFLCs. SPE gels are shown for each sample. IFN = interferon. κNR: upper limit of κ FLC normal range.

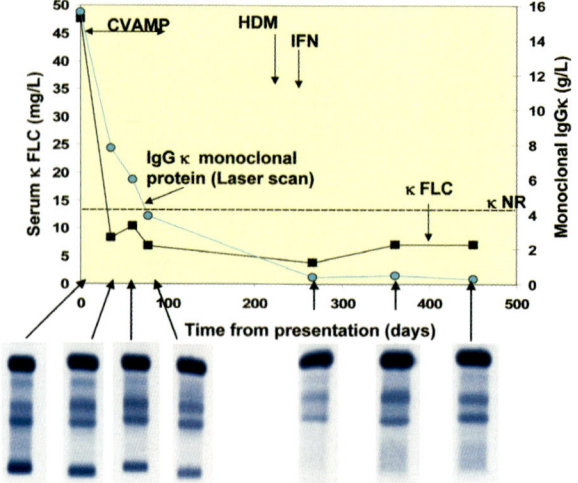

following treatment, and presumably earlier; yet chemotherapy was continued for many weeks longer. In highly vulnerable patients it may be of no harm to stop chemotherapy if sFLC levels have normalised in order to minimise drug side effects.

12.3. Bone marrow responses and short sFLC half-life

The value of sFLCs rather than IgG levels for accurately assessing rates of response in IIMM is apparent from bone marrow sampling. During treatment, monoclonal plasma cell counts correlate better with changes in sFLCs (and serum β_2 microglobulin) than intact monoclonal IgG *(Figure 12.3)*. In a study of 51 patients by Mead *(3)* et al., *(Table 12.1)* and others *(Tang)*, sFLC concentrations were better than both urine free light chains (uFLCs) and SPE/serum immunofixation electrophoresis (sIFE) for assessing disease status. Many patients with normal bone marrow biopsies had elevated monoclonal immunoglobulins by SPE, reflecting slow clearance rates of IgG. 4 patients had abnormal bone marrow biopsies but normal sFLC levels (non-FLC producers). Interestingly, 5 patients had abnormal sFLCs but normal marrows suggesting that the biopsies had been taken from the wrong part of the bone marrow. In a disease with patchy distribution, a serum test that measures protein production from all tumour cells is likely to be highly sensitive for residual disease in some patients.

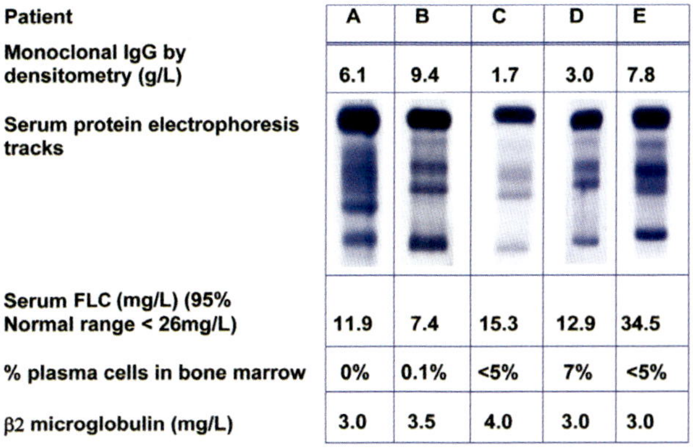

Patient	A	B	C	D	E
Monoclonal IgG by densitometry (g/L)	6.1	9.4	1.7	3.0	7.8
Serum protein electrophoresis tracks					
Serum FLC (mg/L) (95% Normal range < 26mg/L)	11.9	7.4	15.3	12.9	34.5
% plasma cells in bone marrow	0%	0.1%	<5%	7%	<5%
β2 microglobulin (mg/L)	3.0	3.5	4.0	3.0	3.0

Figure 12.3.
Accuracy of different blood tests for assessing bone marrow plasma cell volume in MM. Because of slow catabolism, IgG concentrations lag behind reductions in bone marrow plasma cell content during treatment.

Table 12.1.
Comparison of bone marrow plasma cell counts in MM using different tests for monoclonal proteins.

		Bone marrow	
		Normal	**Abnormal**
Serum free light chains	Normal	19	4
	Abnormal	5	47
Urine free light chains	Normal	21	21
	Abnormal	3	30
Serum IFE	Normal	10	5
	Abnormal	14	46

12.4. Speed of response using sFLCs

A rapid decrease in concentration of sFLCs compared with IgG has been shown in many studies. It is most apparent with drugs that produce a fast tumour response. Thus, after HDM, reduction in sFLC concentration is seen within days *(also see sections 12.8 and 12.9 below)*. Dispenzieri *(5)* et al. showed that in an intensive testing study in which FLCs were tested on most days post-transplant, a 90% reduction by day 7 predicted complete response.

Bortezomib (Velcade) is a rapidly acting chemotherapeutic agent in which particularly fast tumour responses occur. Das et al. showed that 6 of 8 patients had responses of which 3 showed repeated falls and rises of sFLCs co-incident with treatment cycles *(Chapter10, Figure 10.10)*. The relapse of sFLC was very rapid with doubling times of less than 10 days. Such rapid changes presumably correspond with the biological half-life of proteosome inhibition and recovery rather than tumour killing and regrowth. By comparison, the intact monoclonal immunoglobulins did not show the same peaks and troughs. Similar observations have been made by others using Bortezomib *(Kyrtsonis 1; Robson 3)*.

These patterns of response can only be observed with the use of tumour markers that are rapidly cleared from the serum, together with short sampling intervals. Generally, the sFLC levels indicated disease response earlier than the immunoglobulin assays. The authors concluded that monitoring patients with sFLC provided an opportunity to follow the kinetics of tumour killing, which is otherwise obscured by the slow clearance of intact monoclonal immunoglobulins. They added that sFLC measurements rapidly indicated tumour responses that could allow relevant changes of treatment strategy. This may have major bearing on cost of treatment and utilisation of resources.

In a study by Patten et al., the short serum half-life of FLCs was used to assess early treatment responses to Actimid (a thalidomide analogue) in relapsed, refractory MM. 12 patients were treated while being observed for changes in monoclonal immunoglobulins *(Table 12.2)*. A beneficial change in the sFLC κ/λ ratio of greater than 50% by day 7 or 28 predicted clinically responding or stable disease. Changes of less than 25% by day 28 predicted clinical progression or stable disease. There was a relatively poor correlation between parameters but in responding patients, reductions in sFLC concentrations predicted outcome earlier than intact monoclonal immunoglobulin measurements. The study concluded that sFLC κ/λ ratios allow early risk stratification and should allow tailoring of therapy in this difficult group of patients. In addition, faster dose escalation studies for drugs with serious side-effect profiles might be possible when using sFLC measurements.

The fast response of sFLCs to chemotherapy was studied by Hassoun *(2,3)* et al. They found that normalisation of the FLC κ/λ ratio after the first or second cycle was highly predictive of outcome. *"Since the aim of therapy is to achieve a complete response, assessment of sFLCs after 1 or 2 cycles may be an important milestone in the decision-making for these patients".* Under other circumstances, sFLC measurements may indicate that a longer duration or a change of chemotherapy is appropriate.

Cavallo et al. studied 140 patients, comparing early sFLC responses with subsequent outcome. They found that normalization of sFLCs increased the probability of subsequent complete remission when assessed both at 30 days (p=0.002) and prior to peripheral blood stem cell transplant (PBSCT) (p=0.001) *(Figure 12.4)*. They also observed that sFLCs were highly correlated with cytogenetic abnormalities and MM staging - the 2 most important prognostic factors for event-free survival and overall survival. They concluded that *"sFLCs can be expected to have a major independent predictive power for outcome with longer follow-up"*. Comparison was also made between sFLC responses and functional positron emission tomography. There was a high correlation between the two measurements for predicting outcome (p<0.002) but the PET scans normalised faster *(Walker)*.

Patient	Actimid	Immunoglobulin (g/L)		Free light chain κ/λ (or λ/κ) ratios		
	mg	pre	day 28	pre	day 7	day 28
&1	1	87	90	115	75	150
*2	1	30	22	38	5	5
&3	1	0	3	7	10	-
*4	1	22	22	50	35	28
*5	5	25	12	53	1	0.9
*6	10	69	52	157	50	-
&7	5	42	39	0.6	1	0.8
&8	5	37	31	8	6	7
&9	2	30	24	17	7	8
*10	2	3	3	1283	1547	1360
*11	2	69	60	37	27	7
*12	2	28	23	6	4	3

Table 12.2. sFLCs and monoclonal intact immunoglobulin concentrations in 12 patients treated with the thalidomide analogue, Actimid. *Early response seen in sFLCs. &FLC responses at 7 or 28 days predicted relapsing or stable disease. (Courtesy of S Schey)

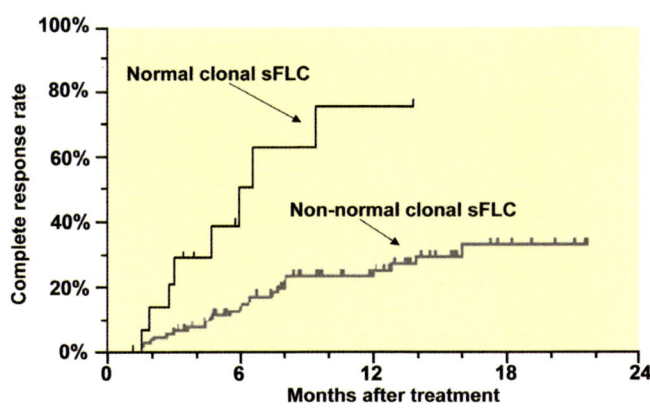

Figure 12.4. Normalisation of sFLCs prior to PBSCT in MM is a good indication of subsequent complete response after treatment. (Courtesy of G Tricot).

Mösbauer et al. evaluated the sensitivity of sFLCs in 26 MM patients for early detection of remission. 3 patients converted to IFE negativity for monoclonal immunoglobulins during follow-up. However, the corresponding sFLC concentrations became normal at a median of 128 days earlier (range 110-144 days). They concluded that sFLCs detected remission earlier than IFE *(Figure 12.5)*.

Orlowski et al. studied the largest cohort of patients for sFLC response rates. In 487 patients (with at least one previous relapse), they assessed the long term impact of sFLC ratio normalisation after each 21-day cycle of bortezomib and doxorubicin. At baseline, 6% had normal ratios but this increased to 12% after the first cycle, 17% after the second cycle and eventually 23%. Importantly, time to progression after cycle 1 was 345 days if the ratio normalised, versus 225 days if the ratio remained abnormal (p<0.0005). After cycle 2, time to progression fell to 325 days in the normalised group versus 224 days for the abnormal group (p<0.001). Similarly, early sFLC ratio normalisation corresponded with earlier complete responses (p<0.001) and partial responses (p<0.0001) after each cycle compared with residual abnormal sFLCs. This large study confirms the beneficial impact of sFLCs as a response marker that tracks the tumour killing effectiveness of chemotherapy.

Dispenzieri *(5)* et al. and Nakorn et al. showed similar results. Serum FLC responses after 2 months of therapy were superior to measurements of intact monoclonal immunoglobulins for predicting overall responses.

12.5. Complete response and residual disease

Absence of intact monoclonal immunoglobulins by IFE after chemotherapy is well described as prognostic for good survival. However, their slow clearance and the relatively poor sensitivity of IFE suggests that sFLC analysis should be a better prognostic marker, or at least additive to IFE. This is supported by the above observations that early normalisation of sFLCs is indicative of good responses. Hence, some patients in complete remission by IFE may be reclassified as having residual disease by the more sensitive sFLC tests. This has led to the classification of stringent complete response (sCR) in the International Uniform Response Criteria for MM and the IMWG recommendation of sFLC measurements in all patients who have achieved a CR to determine whether they have attained a sCR *(Chapter 25)*. Several studies, described

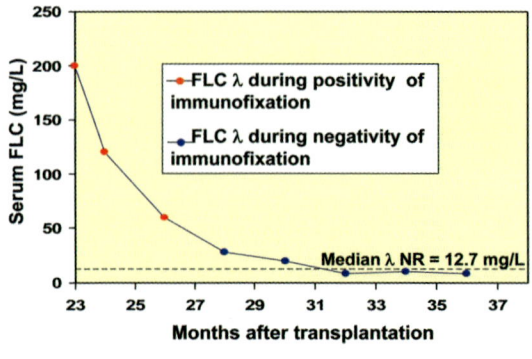

Figure 12.5. Serum FLCs and IFE during the evolution of patients with MM into remission. (Courtesy of N Kroger).

below, have confirmed the high sensitivity and value of sFLCs for assessing complete responses.

Sirohi et al. studied sera for sFLCs from 107 MM patients who were in complete remission as assessed by electrophoretic tests. Many of these patients had abnormal concentrations of sFLCs (*Figure 12.6*). Patients with ratios outside the 95% of the normal range had a 1.3-fold increased risk of progression (p<0.05) and patients with results outside the 100% range had a 2.7-fold increased risk (p<0.01) (Fisher's exact test). Patients with both FLCs elevated (but normal κ/λ ratios) were considered to have impaired renal function *(Chapter 20)* or disordered immune reconstitution and showed no increased risk of relapse. Thus, even though intact monoclonal immunoglobulins were undetectable by IFE in all patients, sFLCs were frequently abnormal and predicted worse outcome.

Reid *(2)* et al. measured sFLCs in 36 patients with light chain multiple myeloma (LCMM) who were clinically in complete remission or had stable disease *(Figure 12.7)*. 17 of the patients had no detectable FLCs in either serum or urine by IFE (Sebia system).

Figure 12.6. sFLC concentrations in normal individuals and 107 patients with MM in complete remission, both clinically and by serum and urine IFE.

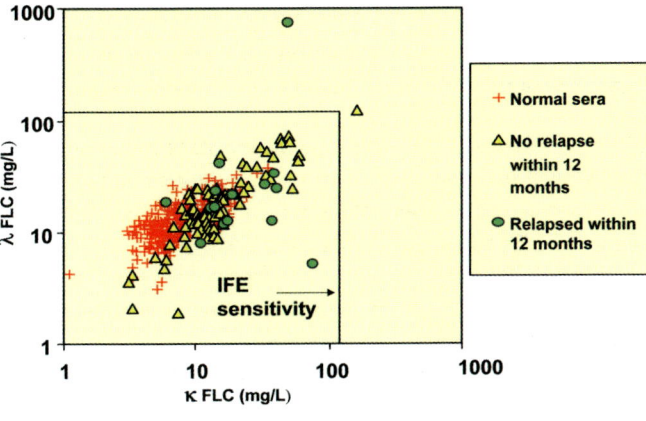

Figure 12.7. sFLC concentrations in 36 patients with LCMM in complete remission or with stable disease. CR: complete response.

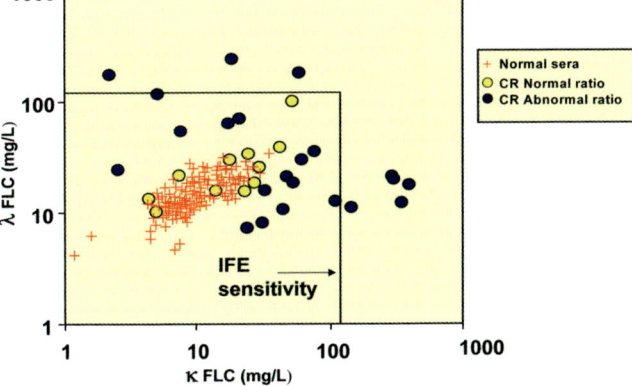

However, 11 of these patients had abnormal κ/λ ratios with increased FLC concentrations. In contrast, all 19 patients with abnormal IFE results had abnormal sFLC concentrations. In a further data set of 61 patients with IIMM in complete remission by conventional criteria *(Reid 2)*, 17 had abnormal sFLC κ/λ ratios *(Figure 12.8)*. Survival data was available for 44 of the 97 patients in these 2 groups and showed that abnormal FLC ratios were highly significant for reduced overall survival (P<0.007).

More results from the UK have recently been reported and further support the utility of sFLC studies for defining response rates. A study of 207 patients in the MRC Myeloma IXth trial, by Owen *(1)* et al. showed that normal sFLC results at the end of induction therapy (prior to HDM) predicted attainment of an IFE negative complete response. Thus, 5% of patients had normal sFLCs at presentation and this increased to 46% at the end of induction therapy and 79%, 100 days after HDM. The sFLCs after induction therapy predicted IFE normality at day 100, i.e, with normal sFLCs, 70% became IFE negative whereas with abnormal sFLCs, only 30% became IFE negative. Lebovic et al. showed similar findings in a smaller study. Both studies considered the sFLC measurements to be a useful tool for defining response rates.

These results establish the use of sFLCs for assessing residual tumour but the benefit of extra treatment for patients needs to be determined in a trial setting. A few patients with residual disease, as identified by abnormalities of sFLCs alone, have been given additional high dose therapy and PBSCT. An example of a patient treated only on the basis of residual sFLCs is shown in Figure 12.9. The patient was normal by SPE and UPE but had high λ sFLC concentrations. A successful response of sFLCs to thalidomide was observed over a six-month period.

12.6. sFLCs during disease relapse

Relapse with monoclonal FLCs only may occur for the following reasons:-

1. Since sFLC assays are intrinsically more sensitive than IFE, they will be abnormal earlier if the tumour relapses simultaneously with monoclonal FLCs and intact immunoglobulins. Since FLCs are produced by 95% of patients with IIMM *(Figure*

Figure 12.8.
sFLC concentrations in 61 patients with IIMM whilst in complete remission, both clinically and by serum and urine IFE. CR: complete response.

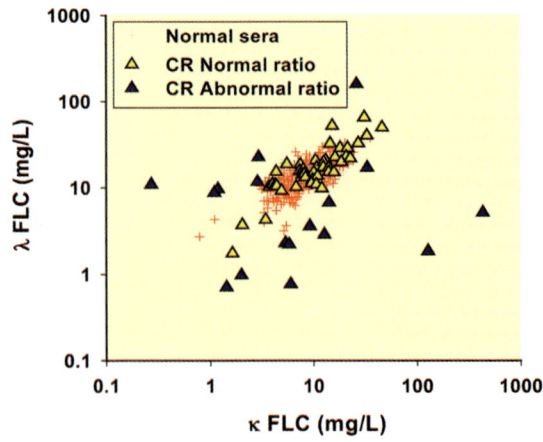

11.5), this is likely to be a common occurrence with frequent serum sampling.

Mösbauer et al. evaluated the sensitivity of sFLCs in 9 MM patients in complete remission by IFE and sFLCs for early detection of relapse. 8 of the 9 patients showed relapse by sFLC ratios earlier than IFE positivity by a median of 98 days (range 35-238 days) *(Figure 12. 10).* Earlier studies *(Tate 2,4),* with lower FLC sensitivity for relapse (10-15%) may be explained by less frequent serum sampling (typically 3 monthly) *(Mead 2).*

2. When monoclonal immunoglobulin production by the tumour changes to FLCs only or a relative increase in FLCs. FLC escape/breakthrough (perhaps more appropriately called sero-conversion) may occur in 10-15% of patients with modern intensive treatment *(see section 12.7 below).*

3. When relapse occurs within a few months of successful chemotherapy and intact monoclonal immunoglobulins are still abnormal by IFE. In this situation, the falling concentrations of IgG or IgA hide the early increases from the tumour relapse. In

Figure 12.9. A patient treated with thalidomide to normalise λ sFLCs after the IgG monoclonal protein had disappeared. (Courtesy of MR Nowrousian).

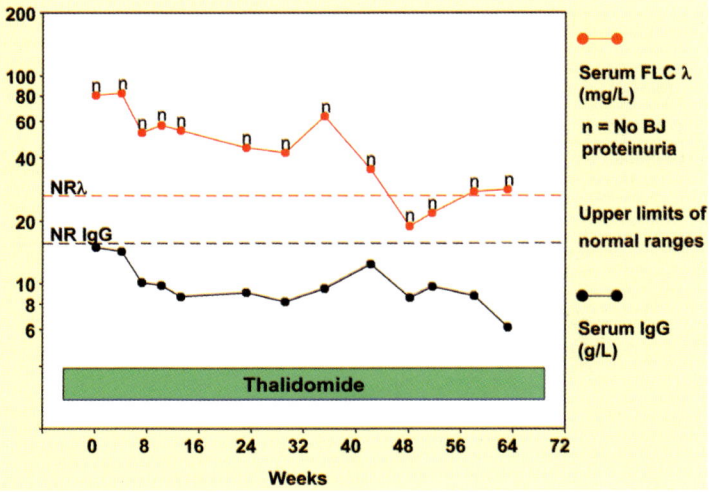

Figure 12.10. Serum FLCs and IFE during relapse of patients with MM. (Reproduced with permission from Haematologica *(Mösbauer).*

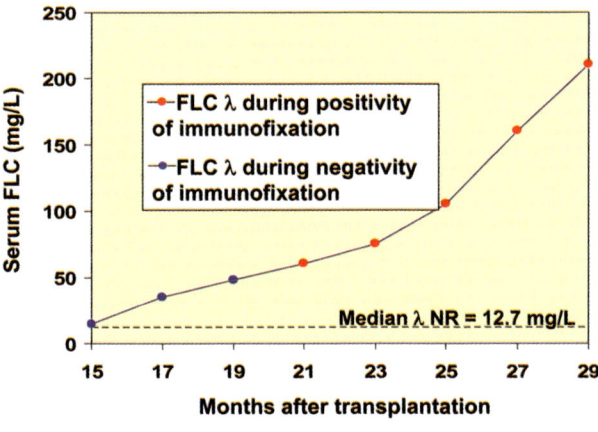

contrast, sFLCs have already normalised because of their short half-life, so the relapse occurs from lower base-line levels *(Figure 12.11 and Chapter 10)*.

12.7. Free light chain escape (breakthrough)

The first observation of FLC breakthrough was in 1971 by JR Hobbs when monitoring patients in the first UK myeloma trial. He noted that patients producing intact monoclonal immunoglobulins relapsed with only Bence Jones proteinuria in 5% and with a relative increase in 35%. While this has since been generally accepted as a not infrequent phenomenon, it has not been reliably documented. A typical example is shown in Figure 12.12.

Kühnemund *(1, 2)* et al. recently reported 2 patients with IgA MM who relapsed with monoclonal FLCs only *(Figure 12.13)*. Dawson et al. reported 3 patients with FLC escape, 2 of which were IgA. They noted fulminant relapses with extra-medullary features and suggested that this may be a more frequent occurrence with modern and more aggressive chemotherapy.

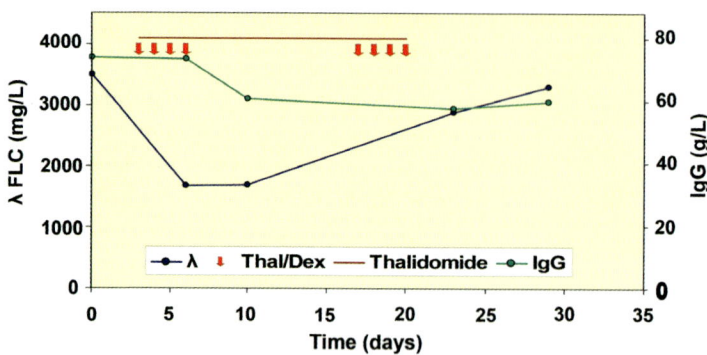

Figure 12.11. Early tumour recurrence identified from rising λ sFLC levels while IgGλ continued to fall. Tumour relapse occurred before IgGλ had stabilised.

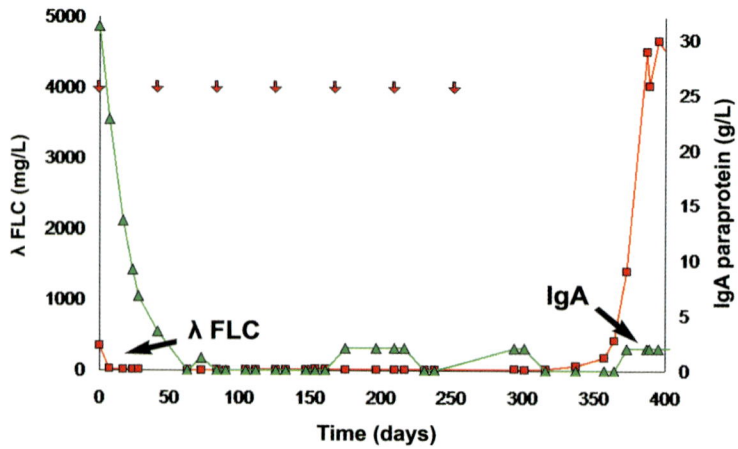

Figure 12.12. Serum λ FLCs and IgAλ in a patient showing FLC escape. Free light chain escape was apparent as a conversion from IgA production to FLC only. (Courtesy of E Liakopoulou)

In order to better establish the incidence of this phenomenon we reviewed 30 IgG and 36 IgA patients randomly selected from the UK, MRC VIIth myeloma trial *(Hobbs JAR 1, 2)*. 4 of the patients (6%) had FLC escape only (1 IgG and 3 IgA). A further 7 (3 IgG and 4 IgA) had a relative increase in FLCs during relapse compared with intact monoclonal immunoglobulin production (partial FLC escape). In total, 17% of patients (11/68) showed FLC escape and more in the IgA group (7/11). Although the patient numbers were small, there was no bias to the intensive treatment arm of the trial. It was noted that urine tests did not reliably detect FLCs in 7 of the patients. This suggests that the incidence of FLC escape is higher than evidence derived from urine studies.

The importance of this phenomenon is two-fold. First, the serological diagnosis of disease relapse cannot be relied upon by measuring intact monoclonal immunoglobulins alone. Second, high concentrations of monoclonal sFLCs cause renal failure. Continuous observations of sFLC during patient monitoring will provide early indications of renal impairment and any risk of renal failure. Chemotherapy can then be used to specifically reduce the load of toxic FLCs on the kidneys and lessen the incidence of renal failure in relapsing patients *(Chapter 13)*.

12.8. sFLCs following high dose melphalan and bone marrow transplant

Pratt *(2)* et al. analysed the detailed changes in sFLC concentrations after PBSCT in 19 patients. Before transplant (but after induction chemotherapy), 11 of the patients had elevated levels of the tumour-produced FLC with abnormal κ/λ ratios. After transplantation, in all patients with associated intact monoclonal immunoglobulins, the tumour-produced FLC concentrations fell within 48 hours (median half-life 4.3 days) and faster than the monoclonal paraprotein (median half-life 14 days). The rate of fall and range of reduction of FLCs varied between individual patients indicating different tumour killing rates/chemosensitivity *(Figures 12.14 to 12.16)*. Figure 12.16 shows a patient treated with chemotherapy in whom there was an incomplete tumour response.

Figure 12.13. Serum λ FLCs and IgAλ in a patient during relapse. Free light chain 'escape' was apparent as the IgA clone disappeared (Courtesy of M Engelhardt).

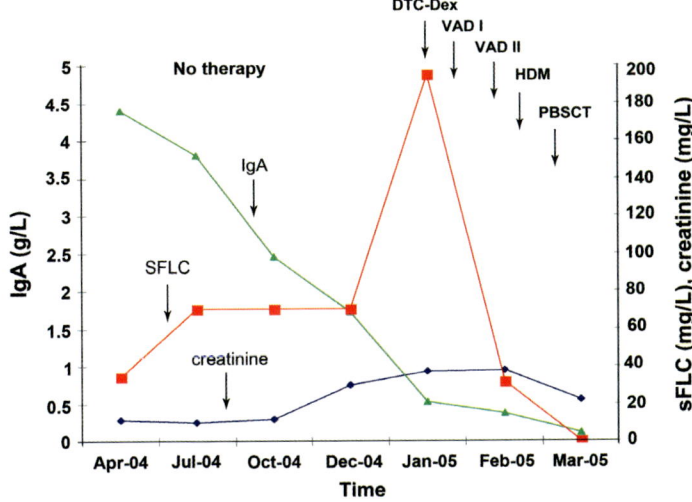

This was evident earlier from the failure of λ concentrations to normalise quickly.

Figures 12.14 and 12.15 illustrate bone marrow engraftment with functioning plasma cells after HDM and PBSCT. In Figure 12.14, concentrations of both FLCs fell until day 10 and then increased rapidly. Three days later, total IgG began to rise. This indicated bone marrow engraftment and occurred a few days after the return of platelet and neutrophil production. This appears to be the normal pattern of FLC response as engraftment takes place.

Minimum levels of FLCs during treatment were 3-5mg/L. This indicated that some FLC production remained. Presumably, there were long-lived plasma cells that were resistant to the drugs. Since κ/λ ratios were normal, the producing cells were probably polyclonal rather than monoclonal tumour cells. For the patient in Figure 12.14, the early stages of engraftment showed greater numbers of κ plasma cells being generated. This produced a *"spike"* in the κ/λ ratio around day 12. Total IgG concentrations fell

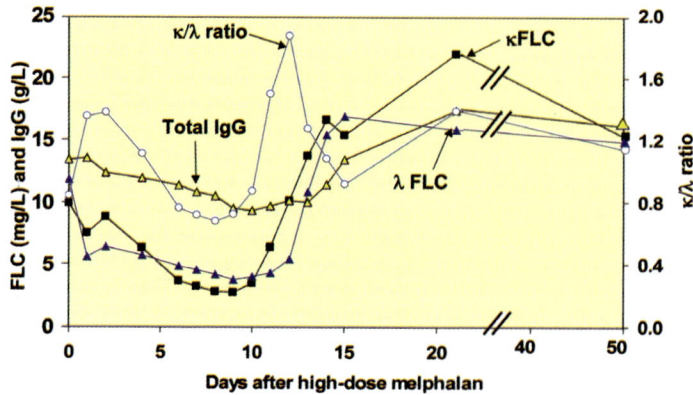

Figure 12.14. Changing immunoglobulin concentrations in a MM patient with IgGκ and FLC κ after high-dose melphalan treatment (day 0) and PBSCT. (Courtesy of G Pratt).

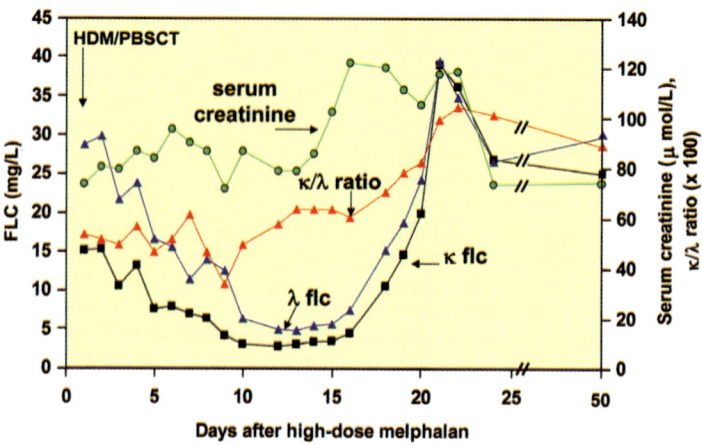

Figure 12.15. Changes in FLCs and creatinine while monitoring a patient with MM after high-dose melphalan and PBSCT. (Courtesy of G Pratt).

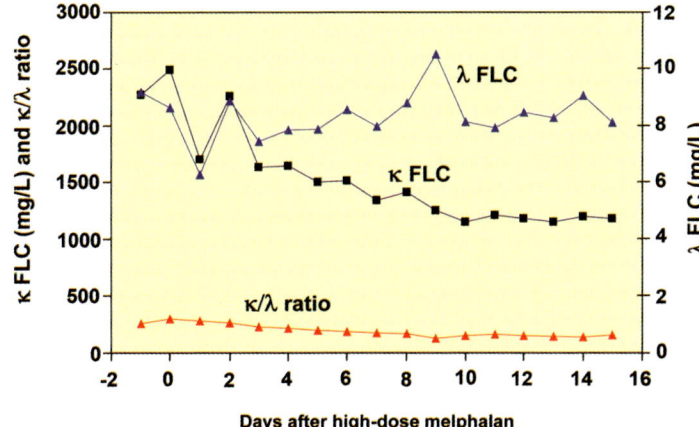

Figure 12.16.
Concentrations of sFLC after high dose melphalan identifying a poor treatment response. (Courtesy of G Pratt).

more slowly after treatment and increased later than FLCs during engraftment. By day 50, FLCs and immunoglobulins were within their normal concentration ranges.

Figure 12.15 illustrates similar features to Figure 12.14 but there is the added complication of renal impairment from gastrointestinal bleeding at day 14. This markedly affected the concentrations of sFLCs. Initially, after HDM, serum λ half-life was 3 days, compared with 20 days for IgG. FLC concentrations increased from day 15 and exceeded normal levels by day 21. They returned to normal, alongside serum creatinine levels on day 24 when normal renal function was restored.

FLC recovery occurred either after or around the time of neutrophil engraftment in all the patients *(Figures 12.14 and 12.15)*. Since early lymphocyte recovery is associated with a more favourable outcome, the time to FLC normalisation may be a useful prognostic marker.

Overall, FLC measurements provided a sensitive monitor of changes in the numbers of tumour cells after PBSCT and pre-dated the intact immunoglobulin response. Further follow-up is required to ascertain whether differences in the kinetics of the FLC responses have any prognostic value. Similar findings were reported by others *(Dispenzieri 5)*.

12.9. Long term variations in sFLC levels after high dose therapy

When bone marrow myeloma cells have been eradicated with intensive therapy, normal haematopoesis may not return for many months. This may show as minor fluctuations of sFLC concentrations around the normal range. Caution should, therefore, be taken when interpreting the data. Figures 12.17 and 12.18 show patients in whom there was a gradual change in κ/λ ratios over many months. It is unknown whether this is indicative of adverse outcome. However, when assessing residual disease or complete remission the relevance of minor fluctuations in sFLC levels needs to be considered carefully *(see MM response criteria guidelines in Chapter 25)*.

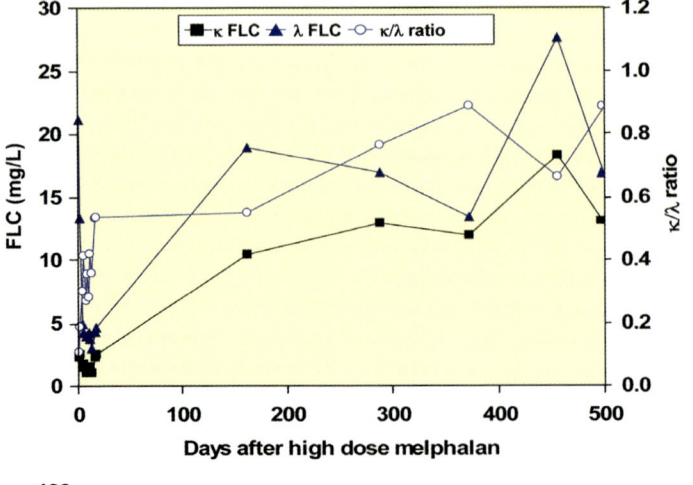

Figure 12.17.
Long term evaluation of sFLC concentrations in a patient following high dose melphalan and PBSCT.

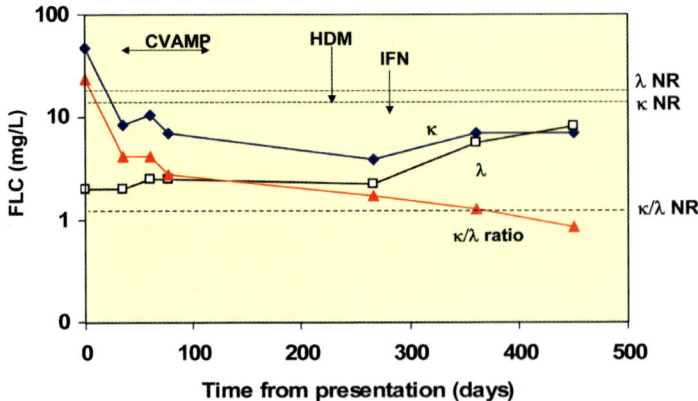

Figure 12.18.
Changes in sFLC levels and κ/λ ratios during long-term tumour monitoring. IFN: interferon.

12.10. Drug selection using rapid responses in sFLCs

The short half-life of sFLCs may be useful for identifying effective drugs, and equally, ineffective drugs, particularly during the later stages of the disease. With each relapse, and as the disease becomes more refractory to treatment, patients are given more drugs, with many toxic side effects. The selection of a drug or drug combinations and the necessary doses can be based upon short-term responses in sFLC concentrations. When using IgG levels for such purposes it may be several weeks before it is appreciated which drug is effective *(Bradwell 4)*.

An example of rapid response to drugs, identified using FLCs, is shown in Figure 12.19. The patient was diagnosed with LCMM and given dexamethasone 24 hours before vincristine and adriamycin and monitored frequently for sFLCs. 10-12 hours after the first dose of dexamethasone (apoptosis time), sFLC concentration fell by 40%. Further falls followed subsequent doses. In contrast, there was no clear evidence of vincristine or adriamycin responses. These observations suggest that serial monitoring of FLCs, whilst introducing drugs one-by-one, may allow active components to be identified.

Figure 12.19. Changes in κ sFLC concentrations in a patient with LCMM at initial treatment. The effect of dexamethasone is clearly detectable within 24 hours. Dex = dexamethasone. (Courtesy of G Galvin)

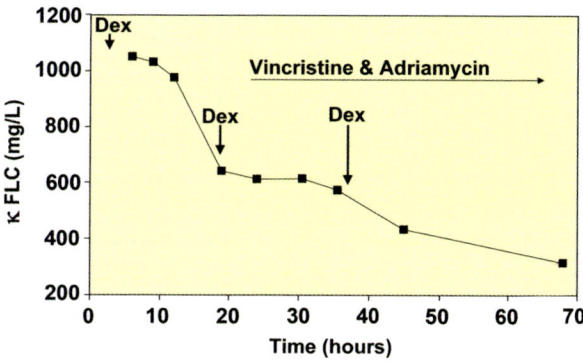

12.11. Utility of sFLCs in IIMM - comments on utility

Utility	Comments
Independent marker of disease	Differential production of FLCs and Igs
Prognostic at presentation	Identifies risk in addition to other markers
Better correlation with bone marrow than IIMM	Due to short serum half-life
Useful fast response marker	Due to short serum half-life
Earlier indicator of remission	Due to short serum half-life
Stringent remission criteria	FLC producing clones eliminated
Early relapse	When Igs still falling but patient relapses
FLC escape	Increasing rate with intensive treatments
Renal damage	Treat FLC levels to prevent renal damage
Urine test replacement	In international guidelines
For quantifying residual disease	More sensitive than IFE
Risk assessment of asymptomatic smouldering MM progression	Relationship to IgH gene translocation

Test Questions

1. Why can sFLC measurements be used as an early marker of relapse in IIMM?
2. Why might the short half-life of sFLCs be clinically useful?
3. Can serum FLC measurements help when assessing residual disease in MM?
4. Can serum FLC tests be normal when IFE is abnormal in patients going into complete remission?

Answers

1. *There are 3 ways. They may be detected before Ig's rise because sFLC assays are more sensitive, there may be FLC conversion, or FLCs may relapse from normality while Igs have not normalised (section 12.6, page 102).*
2. *It allows an early assessment of treatment response or failure (section 12.4, page 98).*
3. *Some patients are normal by serum and urine electrophoretic tests but have abnormal sFLC concentrations. Statistically, these patients are more likely to have residual disease and shorter survival (section 12.5, page 100).*
4. *Yes. Because of the short serum half-life of serum FLCs they normalise earlier than sIFE for intact immunoglobulin monoclonal proteins (section 12.4, page 98).*

Chapter

13

The kidney and monoclonal free light chains

AR Bradwell, P Cockwell and C Hutchison

13.1. *Introduction*   *111*
13.2. *Normal FLC clearance and metabolism*   *112*
13.3. *Nephrotoxicity of monoclonal FLCs*   *113*
13.4. *Diagnosis of myeloma kidney using sFLC analysis*   *116*
13.5. *Removal of FLCs by plasma exchange*   *117*
13.6. *Model of FLC removal by plasma exchange and haemodialysis*   *117*
13.7. *Removal of FLCs by haemodialysis*   *119*
13.8. *Recovery from renal failure following FLC removal by haemodialysis*   *121*

Summary: Monoclonal sFLCs:

1. Cause renal impairment in approximately 30% of patients with multiple myeloma (MM) and dialysis dependent renal failure in 10%.
2. Should be measured in all MM patients to identify those at risk of renal damage.
3. Are not adequately removed by plasma exchange.
4. Can be removed by haemodialysis using *"high cut-off"* dialysers, leading to renal recovery.

13.1. Introduction

Renal failure is a major cause of morbidity and mortality in patients with MM. At initial presentation, up to 50% of patients have renal impairment (serum creatinine >1.5mg/dL or >130μmol/L); 12 to 20% have acute renal failure (ARF) and 10% become dialysis dependent. They represent 2% of the dialysis population and there are approximately 5,000 new patients, worldwide, each year. Furthermore, there is a 50% mortality within 6 months of diagnosis *(Kyle 7; Hutchison 2)*

While reversible factors such as dehydration, hypercalcaemia and medication are frequently involved, monoclonal FLCs are the most potent cause of irreversible renal failure. Large amounts of sFLCs readily pass through the glomerular fenestrations and overwhelm the absorptive capacity of the proximal tubules. On entering the distal tubules, they co-precipitate with Tamm-Horsfall protein to form waxy casts that both block the flow of urine and cause interstitial inflammation *(Herrera 1)*. Furthermore, high concentrations of FLCs are directly toxic to tubular cells *(Sanders 1,2; Basnayake)*.

Studies have analyzed renal recovery rates after FLC removal by plasma exchange. This is a logical approach, but results have been disappointing. Although an early report was optimistic *(Zucchelli)*, the largest and most recent controlled trial (97 patients) showed no clinical benefit *(Clark WF)*. A subsequent editorial in the Journal of the American Society of Nephrology (JASN) listed the shortcomings of this study, including the failure to monitor either serum or urine FLC concentrations *(Ritz)*. It was noted,

"This resembles anti-hypertensive treatment without measuring blood pressure." Clearly, the efficiency of plasma exchange for FLC removal could not be judged.

Because FLCs are relatively small protein molecules (κ~25 kDa: dimeric λ~50 kDa) they are present in similar concentrations in serum, extravascular compartments and tissue oedema fluid *(Takagi)*. Thus, the intravascular compartment may contain only 15 to 20% of the total amount. A series of 3.5 litre plasma exchanges that removed 65% of intravascular FLCs on each occasion would have little overall impact, particularly if production were not reduced at the same time by chemotherapy. An alternative approach is to remove FLCs by haemodialysis. Although this is not possible with routine dialyzers because of their small pore sizes (12-15kDa), a new generation of *"high cut-off"* dialyzers, allows FLC removal *(Hutchison 2)*. By using extended dialysis, large amounts of FLCs can be removed without the attendant clotting and deproteination problems that may limit the extended use of plasma exchange.

This chapter discusses the normal renal handling of FLCs, their role in renal failure in MM, clinical case studies and current management strategies such as plasma exchange. A mathematical model of FLC removal is presented and plasma exchange is compared with the utility of haemodialysis. Finally, clinical evidence for the beneficial use of *"FLC removal haemodialysis"* is presented.

13.2. Normal FLC clearance and metabolism

In normal individuals, sFLCs are rapidly cleared by the kidneys depending upon their molecular size *(Figure 13.1 and Chapter 3)*. Monomeric FLCs, characteristically κ, are cleared in 2-4 hours at 40% of the glomerular filtration rate (GFR). Dimeric FLCs, typically λ, are cleared in 3-6 hours at 20% of the GFR, while larger polymers are cleared more slowly. Removal is prolonged to 2-3 days in MM patients who are in complete renal failure when FLCs are removed by the liver and other tissues. In contrast, IgG has a normal serum half-life of 21 days that is not affected by renal impairment.

Figure 13.1. Renal injury caused by FLCs. (This figure was published in Acute renal failure: Myeloma kidney. Winearls CG. In: Johnson RJ and Feehally J, eds. Comprehensive clinical nephrology, Mosby: Page 238, Figure 17.5 © Elsevier (2003)).

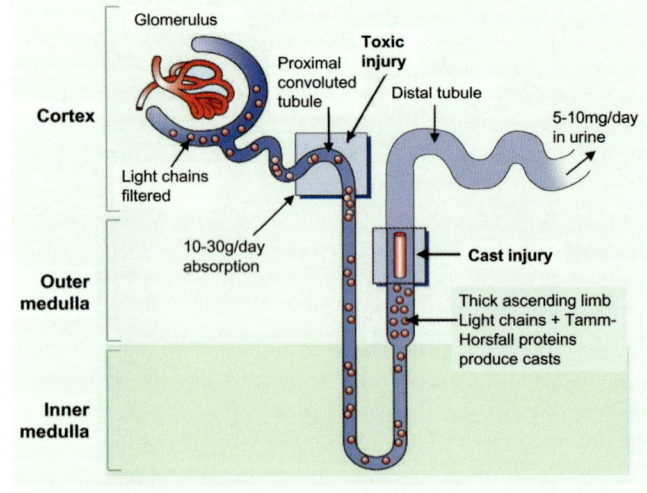

After filtration by the glomeruli, FLCs enter the proximal tubules and bind to brush-border membranes via low-affinity, high-capacity receptors called cubulins and megalins *(Johnson RJ)*. Binding provokes internalisation of the FLCs, subsequent proteolysis into smaller peptides and finally their excretion into the urine flow. The concentration of FLCs leaving the proximal tubules, therefore, depends upon the amount in the glomerular filtrate, competition for binding uptake from other proteins and the absorptive capacity of the tubular cells. A reduction in GFR, due to loss of nephrons, increases sFLC concentrations so that more is filtered by the remaining functioning nephrons. Subsequently, and with increasing renal failure, hyperfiltering glomeruli leak albumin and other proteins which compete with FLCs for absorption thereby causing more to enter the distal tubules.

FLCs entering the distal tubule can bind to Tamm-Horsfall protein (uromucoid). This is the predominant protein in normal urine and is thought to be important in preventing ascending urinary infections. It is a glycoprotein (85kDa) that aggregates into high molecular weight polymers of 20-30 units. Interestingly, it contains a short peptide motif that has a high affinity for FLCs *(Ying)*.

13.3. Nephrotoxicity of monoclonal FLCs

The main renal pathology in the context of MM and acute renal failure is myeloma kidney (cast nephropathy). This is caused by precipitation of FLCs with uromucoid as waxy casts and is characteristically found in ARF associated with MM *(Figures 13.2 and 13.3) (Johnson RJ)*. The casts obstruct tubular fluid flow, leading to disruption of the basement membrane and interstitial damage. Rising concentrations of sFLCs are filtered

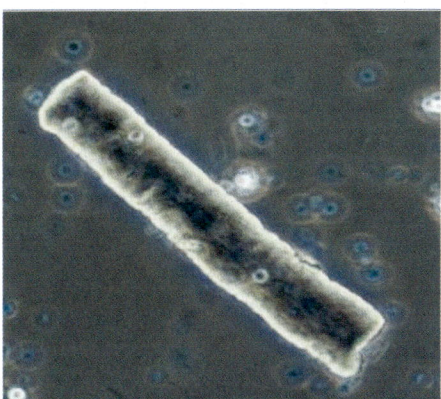

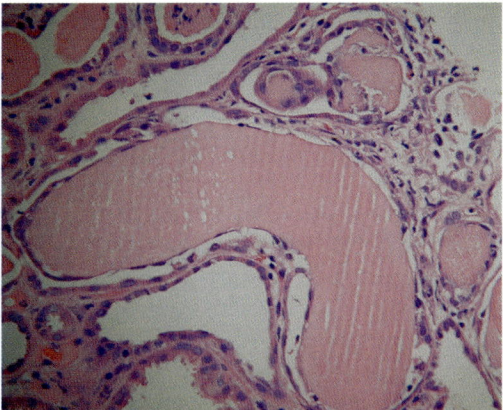

Figure 13.2. Waxy cast from the urine of a patient with multiple myeloma. (This figure was published in Investigation of renal disease: Urinalysis. Fogazzi GB. In: Johnson RJ, Feehally J, eds. Comprehensive clinical nephrology, Mosby: Page 41, Figure 4.3B). © Elseivier (2003)).

Figure 13.3. Classic casts in the distal tubules of a patient with light chain multiple myeloma. (Courtesy of C Hutchison).

by the remaining functioning nephrons which become blocked, leading to a vicious cycle of further increases in sFLC concentrations and progressive renal damage. This may explain why some MM patients, without apparent pre-existing renal impairment, suddenly develop catastrophic and irreversible renal failure. The process is aggravated by other factors such as dehydration, diuretics, hypercalcaemia, infections and nephrotoxic drugs.

Monoclonal FLCs cause renal impairment by several mechanisms, a variety of which may contribute to both acute and chronic renal failure.

Mechanisms of renal FLC toxicity
1. Activation of inflammatory mediators in the proximal tubule epithelium.
2. Proximal tubule necrosis.
3. Fanconi syndrome (renal tubule acidosis) with FLC crystal deposition.
4. Cast nephropathy.
5. AL amyloidosis *(Chapter 15)*.
6. Light chain deposition disease *(Chapter 17)*.

In MM, monoclonal sFLC concentrations are abnormal in >95% of patients *(Chapter 11)* and can have a wide range of concentrations. Importantly, their toxicity also varies considerably. This was elegantly shown by Sanders and Brooker *(1,2)* using isolated rat nephrons and is, in part, related to variable affinity for Tamm-Horsfall protein. However, in spite of much effort to show otherwise, particular molecular charge and κ or λ type are not now considered relevant to FLC toxicity. Furthermore, highly polymerised FLCs (a frequent finding in MM - *Chapter 4*) are probably not nephrotoxic because they cannot pass through the glomerular fenestrations. This may partly account for the lack of renal damage in some patients who have very high sFLC concentrations.

The amount of sFLCs necessary to cause renal impairment was studied by Nowrousian et al. They showed that the median serum concentrations associated with overflow proteinuria (and hence potential for tubular damage) was 113mg/L for κ and 278mg/L for λ. These are approximately 5-10 fold above the normal serum concentrations and presumably relate to the maximum tubular reabsorption capacity of the proximal tubules. Since the normal daily production is ~500mg, increases to ~5g/day are likely to be nephrotoxic in most patients.

As renal impairment develops, progressive increases occur in both the monoclonal sFLCs and the polyclonal non-tumour sFLCs *(Chapter 20)*. Concentrations of monoclonal sFLCs below 300mg/L are rarely associated with renal impairment as judged by the associated normal levels of the non-tumour FLCs *(Figure 8.2)*. These concentrations are somewhat higher than those observed by Nowrousian et al. in patients with renal impairment, but are in the same general range.

There have been several urine studies that have related urine FLC excretion rates to renal impairment. Typically, the associated renal impairment rises with increasing urine FLCs. One study showed that 7%, 17% and 39% of patients had renal impairment with excretions rates of <0.005g/day, 0.005-2.0g/day, and >2g/day, respectively *(Bladé 1)*. However, FLC excretion is an indicator of renal damage in addition to its cause.

The causal relationship between sFLC concentrations and renal impairment is illustrated in 3 patients below *(Figures 13.4 to 13.6) (Abdalla SH)*. In each case, urine FLC concentrations were low and not indicative of the underlying renal deterioration.

The patient shown in Figure 13.4 had developed MM expressing IgAλ and λ FLCs, 3 years earlier. She had a good initial response to vincristine, adriamycin (doxorubicin), dexamethasone (VAD) but was frail. Over a 7-month period, her serum λ FLC levels increased from 1,500 to 2,800 mg/L, then 4,500 mg/L and finally 15,000 mg/L in month 9, while serum creatinine had increased slightly to 120 µmol/L. Her medication was not changed from thalidomide, cyclophosphamide and pulsed dexamethasone on account of her frailty. Unfortunately, the sharp rise in sFLCs led to ARF which was apparent one month later (as shown by the high creatinine concentration). She failed to respond to therapy and died shortly after.

Figure 13.5 illustrates a patient who developed ARF that responded to chemotherapy and haemodialysis. As in Figure 13.4, the sFLC concentrations were a sensitive indicator of impending renal failure and concentrations appeared to reach a threshold before ARF developed. Although renal function was deteriorating at the 4th month, the patient felt well and declined treatment, only to present in renal failure one month later. The potential seriousness of the high FLC concentrations had not been fully appreciated. She was given VAD and put on haemodialysis for 6 weeks. This produced a good response of the MM but only partial renal recovery. It is of note that sFLC concentrations were, consistently, more sensitive than serum creatinine as a marker of falling GFR.

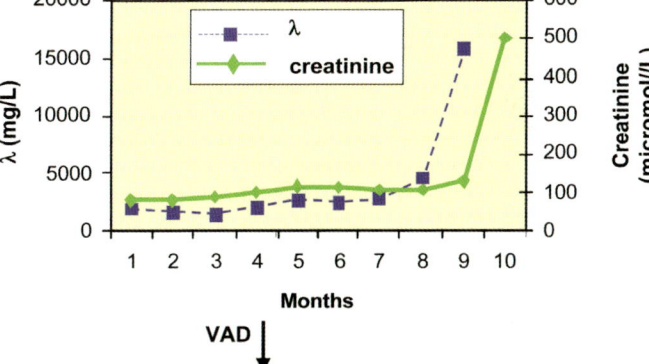

Figure 13.4. Serum λ FLC concentrations in a patient with IgAλ MM during development of ARF. (Courtesy of S Abdalla).

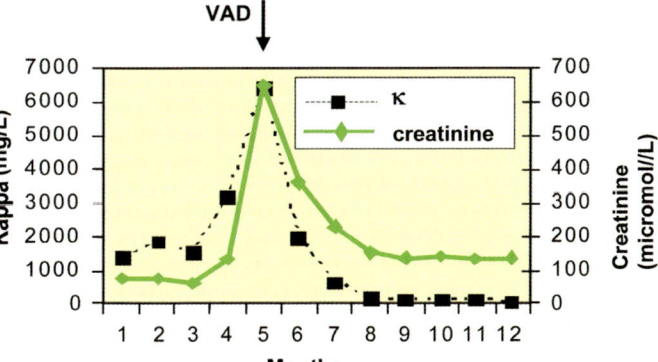

Figure 13.5. sFLC concentrations in a patient with LCMM during the development of ARF and subsequent response to VAD (arrow). (Courtesy of S Abdalla).

Figure 13.6. Serum κ FLCs causing renal impairment in a patient who appeared to have stable disease from IgG measurements. Satisfactory improvement in renal function followed chemotherapy. VED: vincristine, epirubicin and dexamethasone. (Courtesy of MR Nowrousian).

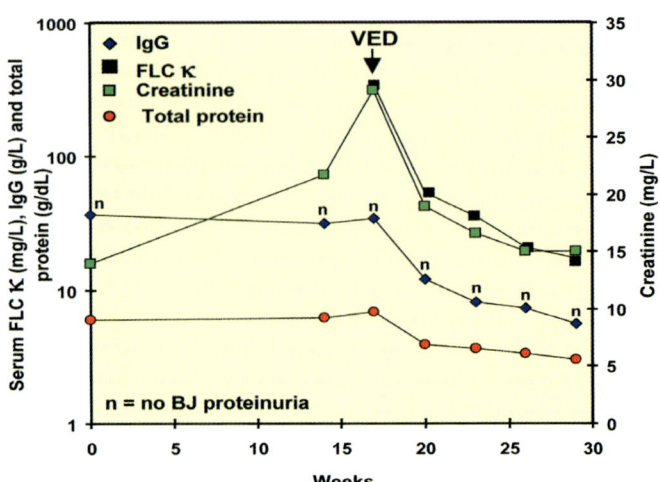

Figure 13.6 shows sFLCs, immunoglobulins and creatinine results in another patient over a 30-week period. Total protein and IgG tests indicated stable disease during the first 18 weeks of monitoring. However, the serum creatinine increased from 1.2 to 3.0mg/dL (130 to 240 μmol/L). sFLC measurements showed unexpectedly high κ concentrations at 500 mg/L with a κ/λ ratio of 4.04, in spite of no FLCs in the urine. It was considered that the high FLC concentrations might be causing renal damage so chemotherapy was commenced. Over the ensuing 2 months the serum κ and serum creatinine almost returned to normal concentrations.

13.4. Diagnosis of myeloma kidney using sFLC analysis

The diagnosis of renal failure from MM has been based upon testing for FLCs in the urine together with serum protein electrophoresis (SPE) and serum immunofixation electrophoresis (sIFE). Positive results are followed by a renal biopsy. Since sFLC tests are more sensitive than urine tests, it is logical to commence with sFLC analysis alone. SPE and sIFE are irrelevant since they test for intact monoclonal immunoglobulins, which do not cause myeloma kidney.

To confirm the diagnostic accuracy of sFLC analysis we reviewed an unselected cohort of 142 patients who had presented with dialysis-dependent ARF of unknown cause. Of the 41 patients with monoclonal gammopathies, 40 had MM with abnormal sFLC ratios and 1 had monoclonal gammopathy of undetermined significance (MGUS) *(Figure 13.7)*. Receiver operating characteristic (ROC) curve analysis showed that a slightly modified sFLC ratio range increased the specificity from 93% to 98% with no loss of sensitivity *(Chapter 20)*. By comparison, urine testing for monoclonal FLCs was negative in one of the 24 MM patients that were assessed *(Hutchison 5)* and 17 had inadequate urine samples.

These results indicate that sFLC κ/λ ratios are a sensitive and specific, yet simple method for identifying monoclonal FLC production in patients with MM and ARF. Rapid diagnosis in these patients will allow early initiation of renal biopsies for

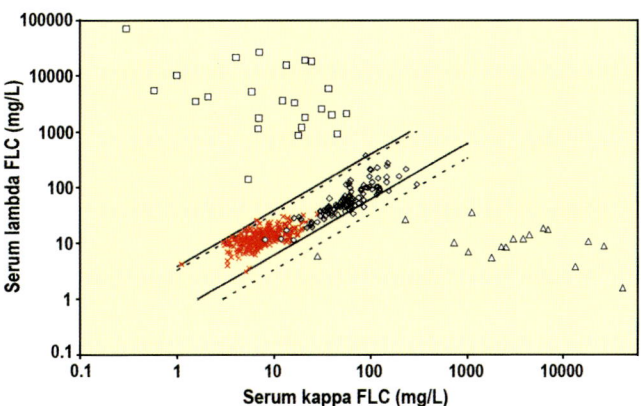

Figure 13.7. sFLCs in 142 patients presenting with dialysis-dependent, ARF. 41 patients with κ (triangles) and λ (squares) MM are distinct from blood donors (red crosses): P<0.001) and from patients with ARF from other causes (diamonds: P<0.001).

confirmation of myeloma kidney followed by *"high cut-off haemodialysis"* and the appropriate chemotherapy *(Figure 13.8)*.

13.5. Removal of FLCs by plasma exchange

Clearly, the pre-renal load of sFLCs is an important factor in renal toxicity. It seems logical that renal recovery might occur if sFLC concentrations were lowered rapidly. This hypothesis is supported by the work of Leung *(2)* and colleagues who studied patients with biopsy proven myeloma kidney and serial FLC measurements. A reduction of greater than 50% in sFLC concentrations was associated with renal recovery. Historically, it has been suggested that plasma exchange might improve the patient outcomes by FLC removal.

Several early studies indicated renal function improvement using plasma exchange. For example, in 1988, Zucchelli et al. compared MM patients on peritoneal dialysis (control group) with plasma exchange (some patients also on haemodialysis). Only 2 of 14 in the control group had improved renal function compared with 13 of 15 in the plasma exchange arm, and survival was better (P<0.01). However, peritoneal dialysis is less effective than haemodialysis for removing FLCs *(Chapter 20)* so the comparison was not ideal.

This early success was not repeated in subsequent controlled trials. WJ Johnson et al. in 1990, compared 10 patients on forced diuresis with 11 who had forced diuresis and plasma exchange and found no difference in outcome. Most recently, a large study was reported by WF Clark et al. (2005). Patients were treated with chemotherapy and were randomly allocated to receive additional plasma exchange. Again, there was no statistically significant benefit for plasma exchange.

13.6. Model of FLC removal by plasma exchange and haemodialysis

In order to understand the effectiveness of plasma exchange we developed a compartmental mathematical model that was applicable to patients being treated for MM and renal failure *(Hutchison 2)*. The following parameters were considered:- sFLC concentrations at clinical presentation; monomeric κ and dimeric λ clearance rates with and without renal failure; partition of sFLC between vascular and extravascular compartments (including oedema fluid); flow of FLCs between compartments; half-life of sFLC in renal failure, sFLC production rates, and tumour killing rates with chemotherapy. Data from patients with MM were fitted to the model to analyse rates of sFLC removal. Simulations were conducted to compare 6 and 10 plasma exchange treatments (of 3.5L each over 12 days) with 5 different haemodialysis protocols. Chemotherapeutic tumour killing rates that were considered included 100% on the first day; 10%, 5%, and 2% per day and no killing.

In renal failure the half-life of sFLCs is approximately 3 days. Assuming a starting concentration of 10,000mg/L, it would take approximately 20 days for the FLCs to be metabolised assuming complete tumour killing with the first chemotherapy dose *(Figure 13.8: 1)*. More realistically, 10% of the tumour might be destroyed per day by aggressive chemotherapy resulting in a slower FLC reduction *(Figure 13.8: 2)*. Addition of plasma exchange procedures increased removal rates by approximately 25% but

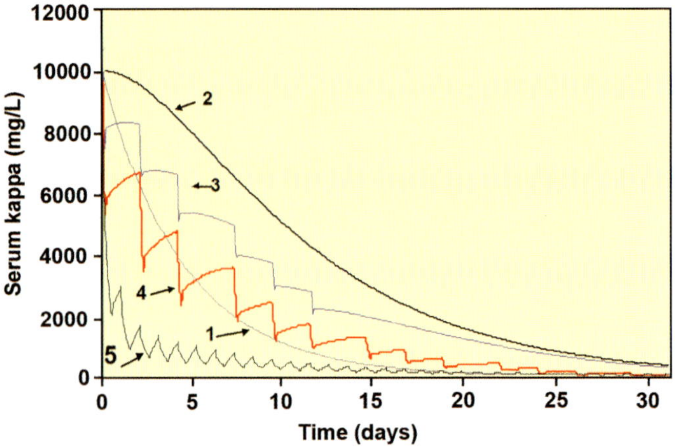

Figure 13.8.
Calculated reduction of κ sFLCs in a patient with LCMM by 100% (immediate) tumour killing (1) and 10% per day (2). The effect of 6 x 3.5 litre plasma exchanges over 12 days is indicated (3) based on 10% kill per day (2) plus the addition of haemodialyis for 4 hours, x3 per week (4) and 8 hours/day (5).

Figure 13.9. Failure of 16 plasmapheresis sessions (arrows) to lower sFLCs (circles) but decline in response to Bortezomib (B), Cyclophosphamide (C): Dexamethasone (Dex) and Thalidomide (Thal). (Reproduced with permission from Transfusion *(Cserti)* and Cserti).

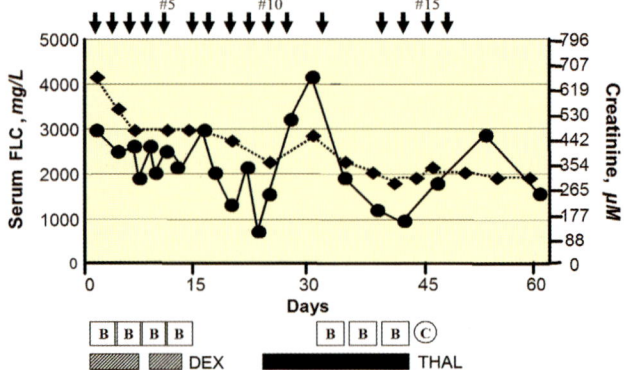

concentrations were not reduced below toxic levels (500mg/L) at 4 weeks. The rapid reductions in sFLC concentrations during the procedure and their subsequent re-entry from the extra-vascular to the intra-vascular compartment can be seen.

Results from the model calculations suggest that plasma exchange is not effective unless used intensively. One report on 3 patients undergoing plasma exchange suggested no benefit *(Figure 13.9) (Cserti),* while another report, on one patient, considered it was helpful *(Pillon).* Lack of success can be explained by on-going high production rates and re-entry of FLCs from extravascular compartments (including oedema fluid).

It should be noted that peritoneal dialysis does not remove FLCs efficiently *(Chapter 20).* Presumably, this is because the volume of exchanged fluid is much lower than in haemodialysis.

13.7. Removal of FLCs by haemodialysis

As an alternative to plasma exchange, sFLCs can be removed more effectively by haemodialysis, provided the pore sizes of the membranes are large enough. Conventional dialysers have a molecular weight cut-off around 15-20kDa so the

Figure 13.10. Photomicrograph of pores in a Gambro high *"cut-off"* (HCO™) membrane and a high-flux membrane (Polyflux™). (Courtesy of H Goehl, Gambro Dialasoren GmbH).

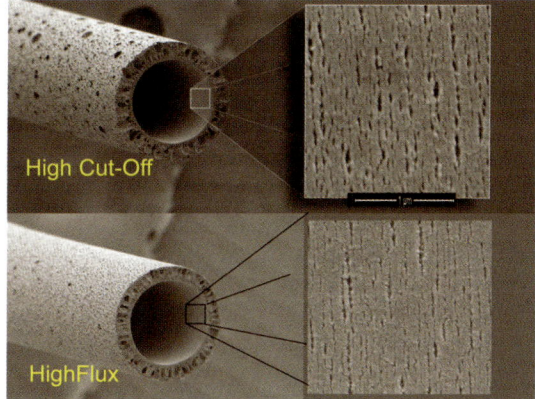

Figure 13.11. Calculated clearance of serum kappa FLCs with 10% tumour killing per day (2) and the addition of haemodialyis for 8 hours/day (5), 12 hours per day (6), 8 hours on alternate days but no tumour killing (7) and no tumour killing with no haemodialysis (8). (See Table 13.1 for calculation details and Figure 13.8).

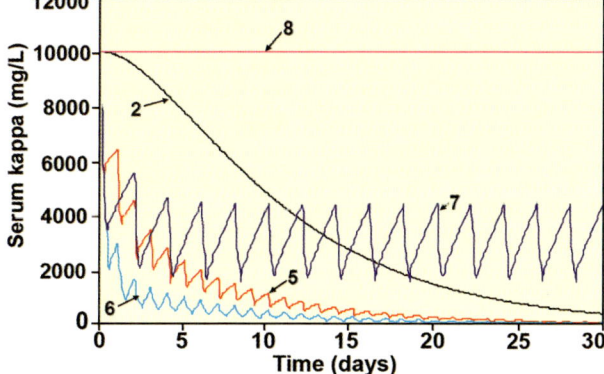

filtration efficiency for FLCs is very low. However, some of the new *"protein-leaking"* dialysers have much larger pores, particularly the Gambro *"high cut-off dialyser"* *(Figure 13.10) (Hutchison 2)*. Furthermore, haemo-diafiltration is more effective at removing small protein molecules than haemodialysis. Indeed, there are sporadic reports of patients with AL amyloidosis and end-stage renal failure who appear to have improved survival when haemo-diafiltration is instigated. But, there is currently inadequate clinical data to provide a clear conclusion in amyloidosis patients.

Model calculations *(Figures 13.8 and 13.11 and Table 13.1)* indicate that the

Method of FLC removal	Percent therapeutic FLC removal over 3 weeks and reduction time from 10g/L to 0.5g/L (days) with chemotherapy tumour killing rates/day of 100% to 0%				
	100%	10%	5%	2%	0%
None	NA(14)[1]	NA(30)[2]	NA(52)	NA(121)	NA(10g/L)[a 8]
PE x 6 in 10 days	29(10)	24(29)[3]	17(52)	9(121)	3(10g/L)[a]
PE x 10 in 10 days	40(8)	34(29)	25(52)	13(121)	4(10g/L)[a]
HD 4 hrs x 3/week	60(7)	54(19)[4]	53(31)	51(73)	50(3.6g/L)[a]
HD 4hrs daily	76(4)	73(13)[5]	72(23)	71(55)	70(1.9g/L)[a]
HD 8hrs daily	87(3)	85(7)	84(14)	83(29)	82(1.0g/L)[a]
HD 8hrs alt.days	79(4)	73(13)	72(19)	70(47)	69(1.5g/L)[a 7]
HD 12hrs daily	91(2)	89(5)[6]	89(8)	88(16)	88(0.7g/L)[a]
HD 18hrs daily	93(2)	93(3)	93(4)	92(8)	91(0.6g/L)[a]

Table 13.1. Model calculations of the efficiency of therapeutic removal of sFLCs. Numbers are the additional percentage of FLC removed by intervention compared with normal metabolism. Numbers in parentheses are the time in days for sFLC concentrations to reduce to 5% of the starting concentrations (from 10g/L to 0.5 g/L). For superscript numbers 1 to 8, the simulations are shown in Figures 13.8 and 13.11. [a]sFLC concentrations at day 150 for simulations in which reductions to 0.5 g/L were not achieved. PE: plasma exchange. HD: hemodialysis. NA: not applicable.

Figure 13.12. Concentrations of λ FLCs in serum and dialysate fluid of a patient with ARF due to LCMM undergoing dialysis with the Gambro *"high cut-off"* dialyser. The dialyser was renewed (arrows) as FLC leakage slowed.

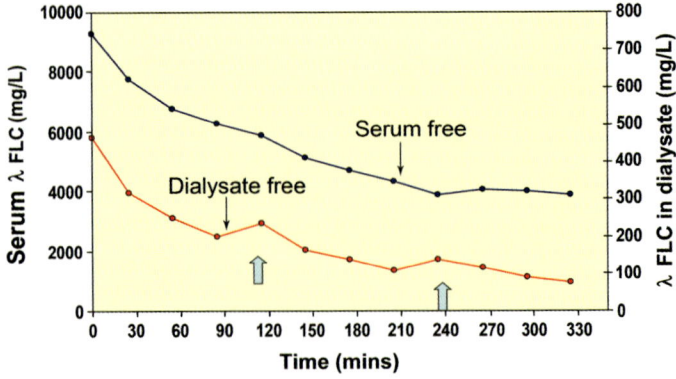

prolonged use of *"high cut-off"* dialysers could reduce κ sFLC concentrations to less than 0.5g/L in 2-3 days with ~95% of the sFLC removed. Dimeric λ molecules are considerably larger so they are removed more slowly. Based upon these modelling results, we assessed the utility of various dialysers for sFLC removal in patients with MM. The Gambro high cut-off membrane with pores that are large enough to filter FLCs was the most efficient by a considerable margin. Figure 13.12 shows the rapid removal of λ FLCs in a patient with MM using the Gambro HCO™ dialysis membrane over a 6 hour period. Also shown are the large amounts of FLCs in the dialysate fluid. A total of nearly 40 grams of FLCs were removed in a single dialysis session.

13.8. Recovery from renal failure following FLC removal by haemodialysis

Intensive application of FLC removal haemodialysis has been applied to 17 patients with cast nephropathy by Hutchison *(6)* et al. in Birmingham, UK. Figures 13.13 to 13.15 show examples of the clinical response in 3 patients who presented with ARF. The first of the 3 figures shows good FLC reductions with renal recovery *(Figure 13.13)*. The second failed to respond to the initial chemotherapy but responded to bortezomib

Figure 13.13. sFLCs in a patient presenting with MM and ARF who responded to haemodialysis with renal recovery. Pre- and post-dialysis concentrations are shown. Numbers indicate grams of FLCs removed (and hours of dialysis) per session. Dialysis was over 22 days leading to renal recovery that has been maintained for over 2 years.

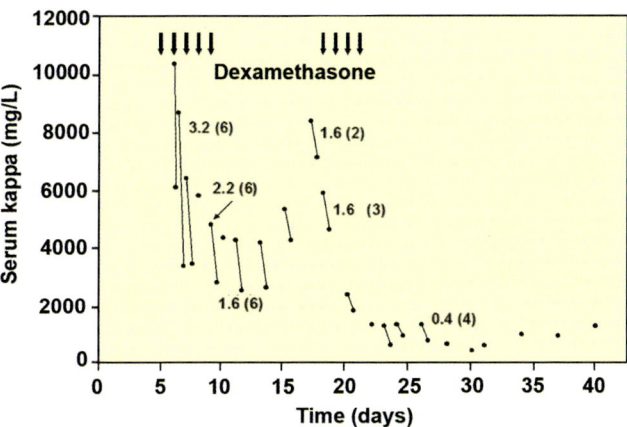

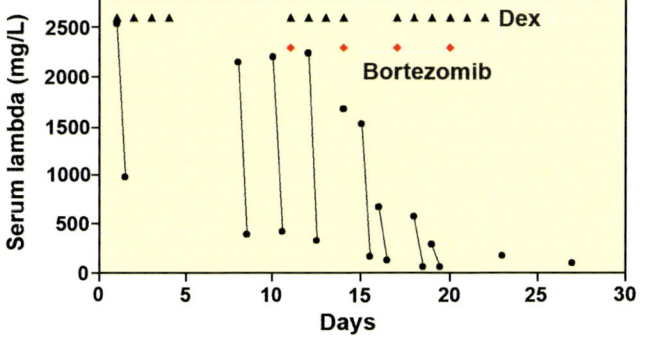

Figure 13.14. sFLC in relapsing MM and ARF who recovered with chemo-therapy and haemodialysis. Pre- and post-dialysis sFLC measurements are shown. The patient received dialysis over 20 days then had renal recovery with GFR rising to 50. At 1 year the patient remained in complete MM remission.

(Figure 13.14). Haemodialyis was unable to produce sustained FLC responses without accompanying satisfactory chemotherapy. The third patient failed to achieve renal recovery *(Figure 13.15).* He developed sepsis and chemotherapy was stopped. In spite of multiple, prolonged periods of haemodialysis and removal of 1.7Kg of FLCs, there was no recovery of renal function. sFLC concentrations oscillated during treatment in much the same manner as shown from model calculations shown in Figure 13.11 (line 7).

Figure 13.15. sFLCs in a patient presenting with relapsing MM and ARF who remained on dialysis because of inadequate chemotherapy, in spite of removing 1.7Kg of FLCs (shown per 10 day period).

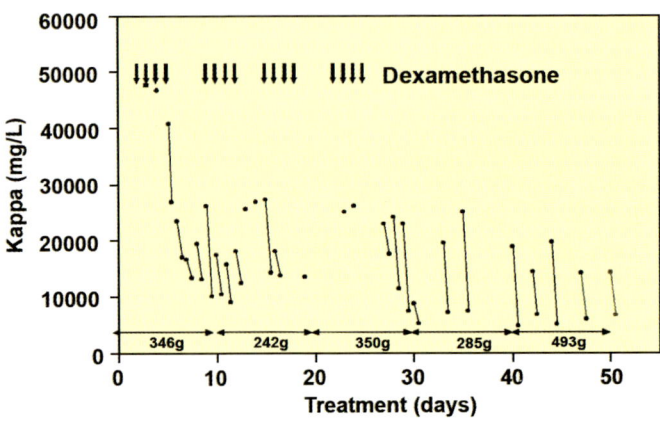

Figure 13.16. Kaplan-Meier plot of renal recovery in 17 patients with MM and ARF treated by chemotherapy and sFLC removal haemodialysis. The control group comprised 17 patients with MM treated over the previous 6 years with similar chemotherapy but using conventional haemodialysis 3 times per week.

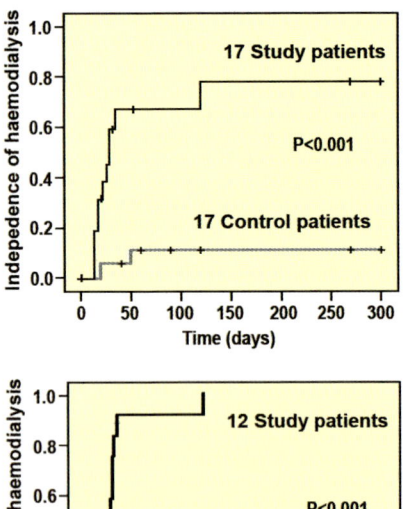

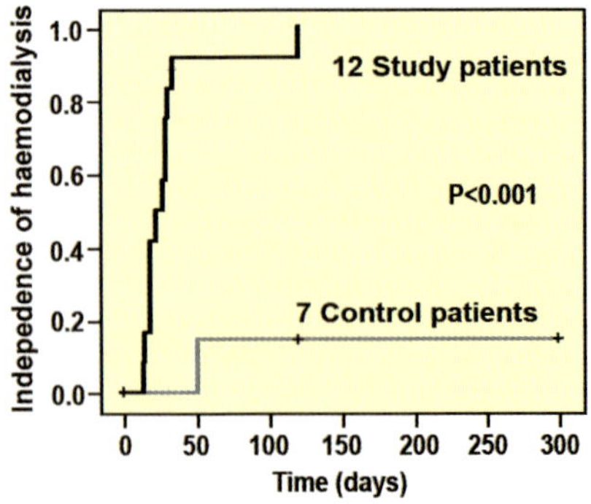

Figure 13.17. Kaplan-Meier plot of the 12 patients with ARF who had recovery of MM following chemotherapy (from Figure 13.6 above). The control group comprised 7 patients with MM treated over the previous 6 years who went into remission but who were treated with conventional haemodialyis 3 times per week.

It should be emphasised that all patients needed both effective chemotherapy to switch off FLC production together with rapid reduction of the pre-renal load, for renal recovery to occur *(Figure 13.18)*. Of the 19 patients studied, 6 had early infective complications resulting in their chemotherapy being withheld. The patients did not achieve an early FLC reduction and only one subsequently became independent of dialysis. The remaining 13 patients had chemotherapy and dialysis and all had an early reduction in serum FLCs and subsequent renal recovery.

The renal recovery rate of 17 of these patients is shown in Figure 13.16, alongside historical recovery rates in a case matched population at the same hospital. Clearly, the patients receiving intensive FLC removal had better renal recovery rates. Of these two populations, when only those with documented tumour responses to chemotherapy were assessed, all 12 patients treated with FLC removal haemodialysis had renal recovery *(Figure 13.17)*. This was in contrast to the control group, where only 1 patient had renal recovery out of 7 who responded to chemotherapy. This suggests an additional benefit of dialysis over chemotherapy alone.

Figure 13.19 analyses patient survival in the 19 patients treated with FLC removal haemodialysis. The 5 patients in the treated group who died were those that failed to regain renal function. All 12 patients who became independent of renal replacement therapy remained alive at 6 months. GFR continued to improve for several months after

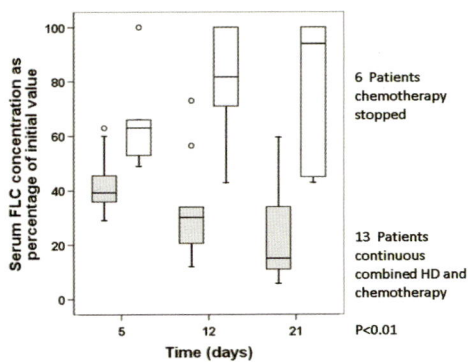

Figure 13.18. sFLC levels over time in 19 patients receiving FLC removal by dialysis. Six patients (clear boxes) had chemotherapy withheld due to infections and 13 patients (grey boxes) had chemotherapy and dialysis and subsequent renal recovery.

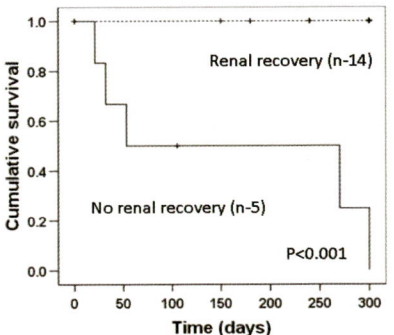

Figure 13.19. Kaplan-Meier plot of survival time in 17 patients with MM and ARF treated by FLC removal haemodialysis. Those with renal recovery had longer survival than those who remained on haemodialysis.

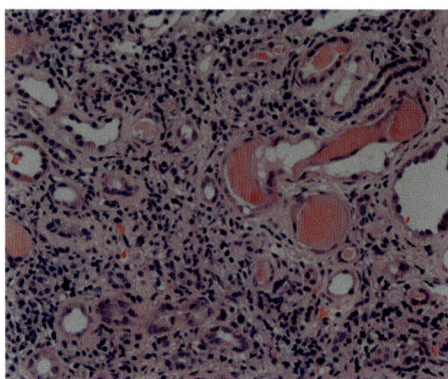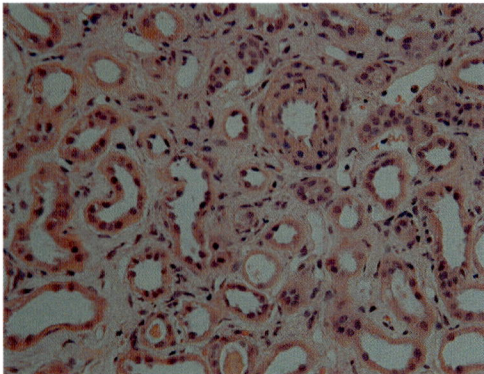

Figure 13.20. Renal biopsies of a patient **A**, at presentation and **B**, 6 weeks after *"high cut-off"* haemodialysis showing resolution of FLC casts in the distal tubules. (Courtesy of K Basnayake).

dialysis independence was achieved. Furthermore, this was associated with complete disappearance of the casts in the one patient in whom serial renal biopsies were available *(Figure 13.20)*.

Two German groups have obtained similar results using the same dialysis protocols *(Bachmann; Kleeberg)*. Overall, these exceptional results suggest that renal recovery is possible in most patients with MM presenting with ARF from cast nephropathy. Alternative strategies for FLC removal by haemodialysis not using a high cut-off membrane have been proposed, including utilising the combination of haemodiafiltration and adsorption *(Testa)* The capacity of these strategies for FLC removal needs to be directly compared with high cut-off haemodialysis.

Other groups have suggested that high rates of renal recovery are possible with modern chemotherapy alone, particularly using combinations that include bortezomib. Ludwig et al. showed recovery in 40% of patients, but none had renal biopsies and few were dialysis dependent. Even better recovery rates, at 73% of patients, were reported by Kastritis et al. but again patient selection was unsatisfactory *(Bergner)*. The control group in the plasma exchange study of WF Clark et al. also had 40% renal recovery and were in a similar clinical state. This suggests that few patients with cast nephropathy that are dialysis dependent will recover with chemotherapy alone. Supporting evidence comes from a retrospective study that showed only 25% recovery in dialysis dependent patients with cast nephroapthy given bortezomib and conventional haemodialysis *(Chanan-Khan)*. However, for patients with non-dialysis dependent renal impairment several studies now support the use of a chemotherapy regime containing a novel agent (bortezomib or thalidomide) in preference to historic regimes to improve renal outcomes *(Li; Roussou; Dimopoulos 2)*.

The clinical challenge will be to identify and quickly treat patients after diagnosis. Traditionally, ARF due to MM has been identified by serum and urine studies for monoclonal FLCs. Our data suggests that all patients with MM can be identified using sFLC tests alone (Fig 13.7) and concentrations are usually greater than 500mg/L

(Hutchison 5). Renal biopsy is also essential to specifically identify cast nephropathy. Other causes of ARF in MM do not have huge concentrations of sFLCs and do not require intensive, *"high cut-off,"* haemodialysis for renal recovery.

Return of higher GFR rates is likely to be better with rapid reduction of sFLCs to normal concentrations. We have experience of 2 patients whose FLC removal haemodialysis was not commenced until one month after initial presentation with ARF. Recovery to dialysis independence was slow with barely satisfactory GFR at 6 months, in spite of complete MM remission.

A German and UK, multi-centre, prospective, randomised controlled trial comparing the Gambro HCO™ dialyser with conventional haemodialysis is currently underway. It is planned to recruit 90 MM patients presenting with ARF and cast nephropathy. They will receive bortezomib, dexamethasone and doxorubicin with randomisation to intensive FLC removal haemodialysis or conventional haemodialysis.

Test questions
1. *What percentage of patients with MM have renal impairment at presentation?*
2. *What protein binds FLCs in the distal tubules?*
3. *What is the benefit of plasma exchange in acute myeloma kidney?*
4. *Are haemodialysis membranes porous to FLCs?*
5. *Is renal recovery possible from light chain cast nephropathy?*

Answers
1. *40-50% (Section 13.1, page 111).*
2. *Tamm-Horsfall or uromucoid protein (Section 13.2, page 112).*
3. *Clinical trials suggest there is minimal benefit (Section 13.5, page 117).*
4. *Normal membranes are impermeable at 10-15kDa but the Gambro high cut off dialyser has pore sizes of ~50kDa and rapidly clears FLCs (Section 13.7, page 119).*
5. *Normal recovery rates are 10-20% but can be up to 80% if effective chemotherapy for MM is combined with FLC removal haemodialysis (Section 13.8, page 121).*

Summary: In ASMM, sFLC κ/λ ratios:
1. Are abnormal in approximately 90% of patients.
2. When abnormal are associated with an increased risk of progression.
3. Can be used to produce a risk model in combination with other factors.

Patients with asymptomatic multiple myeloma (ASMM) satisfy two of the criteria for MM:- monoclonal protein concentrations greater than 30g/L and/or 10% or more monoclonal plasma cells in the bone marrow, but no features of end-organ damage such as bone fractures (CRAB criteria) *(Kyle 10)*. Such patients were classified in the Durie and Salmon staging system as Grade 1 *(Section 25.1)*. The time to disease progression is typically 2-4 years so there is no need for immediate treatment but patients should be monitored on a regular basis *(Kyle 12; Rosinol)*. Since the presence of urine free light chains (uFLCs) is an adverse prognostic indicator *(Weber DM)*, serum free light chain (sFLC) levels may also relate to outcome. Extrapolation from the data of Rajkumar *(5)* et al., showing the prognostic importance of monoclonal FLCs in monoclonal gammopathy of undetermined significance (MGUS), would further suggest their utility in ASMM.

Augustson *(2)* et al. studied 43 patients with ASMM who had been registered on the UK, MRC multiple myeloma trials between 1980 and 2000. They found that abnormal FLC κ/λ ratios were present in 36 (84%) of the patients while 7 (16%) were normal. In

Figure 14.1. Kaplan-Meier plot of survival in 43 patients with ASMM with (blue) and without (green) abnormal κ/λ ratios.

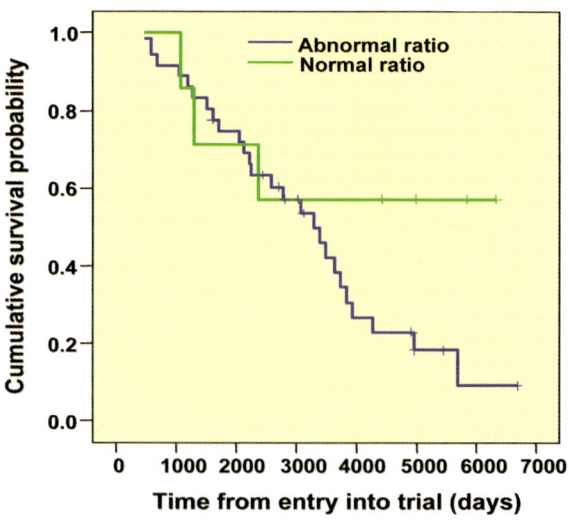

the 26 patients with abnormal ratios who progressed, the median time period was 713 days. This compared with 1,323 days in the 6 patients who progressed who had normal κ/λ ratios. There was no significant difference in survival between the two groups (p<0.13) but patient numbers were inadequate for reliable statistical power *(Figure 14.1)*.

The largest study of sFLCs in patients with ASMM has recently been reported from the Mayo Clinic. Dispenzieri *(3)* et al. showed clearly that abnormal sFLCs indicated an increased risk of progression to MM. Baseline serum samples obtained within 30 days of diagnosis were available from 273 patients. At a median follow-up of 12.4 years, transformation to active disease had occurred in 59%. Abnormal sFLC ratios were present in 90% at baseline and were associated with adverse outcome. The degree of ratio abnormality was independent of other ASMM risk factors including the number of bone marrow plasma cells (BMPC) and quantity of serum intact monoclonal immunoglobulins (M-protein).

The study concluded that abnormal sFLC ratios were an important additional determinant of clinical outcome. An increasingly abnormal sFLC ratio was associated with a higher risk for progression to active MM. Patients with a normal (0.26 to 1.65) or near normal ratio (0.25 to 4) had a rate of progression of 5% per year, while patients with markedly abnormal ratios either <0.0312 (1/32) or > 32 had a rate of progression

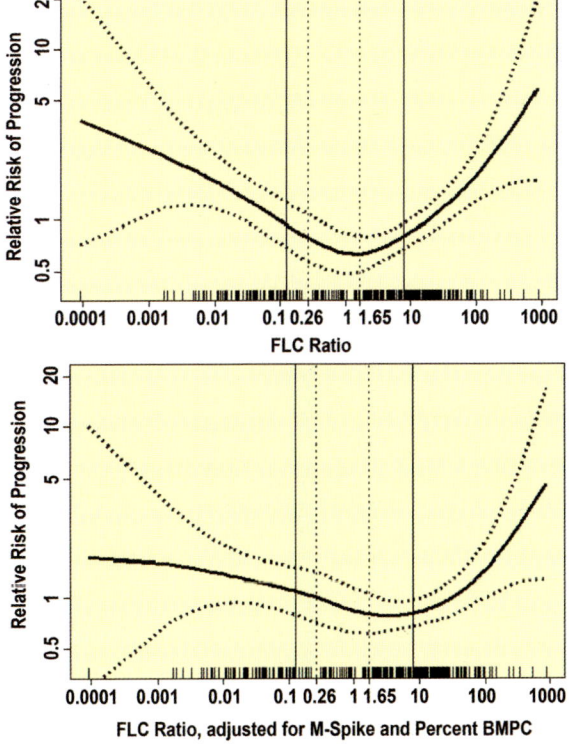

Figure 14.2. Effect of increasing abnormal sFLC ratios on the relative risk of progression of ASMM to MM or related disorders. BMPC: bone marrow plasma cells. (This research was originally published in Blood *(Dispenzieri 3)* © the American Society of Hematology).

of 8.1% per year *(Figure 14.2)*. This increase persisted after adjusting for the competing causes of death. The best cut-off point for progression was sFLC ratios of <0.125 or greater than 8, giving a hazard ratio for progression to active MM of 2.3 times that of patients with sFLC ratios between 0.125 to 8 *(Figure 14.3)*.

Incorporation of sFLC κ/λ ratios with the two factors of BMPC and M-protein concentration produced a highly significant risk model. 5 year progression rates in high- (BMPC >10% and serum M-protein >30g/L), intermediate - (BMPC >10% and serum M-protein <30g/L), and low-risk groups (BMPC <10% and serum M-protein >30g/L) were 76%, 51%, and 25%, respectively. The cumulative probability of progression at 10 years was 50% in patients with 1 risk factor; 65% for 2 risk factors; and 84% for 3 risk factors *(Figure 14.4)*. Corrected for death as a competing risk, the 10 year rates were 35%, 54%, and 75%, respectively (P< 0.001). Use of urinary M-protein of 50 mg/24h could not substitute for the sFLC ratio in this model indicating the value of serum rather than urine for FLC analysis.

The authors noted that unlike MGUS, in which the rate of progression remains constant over time *(Chapter 19)*, the overall risk of progression in ASMM was greatly

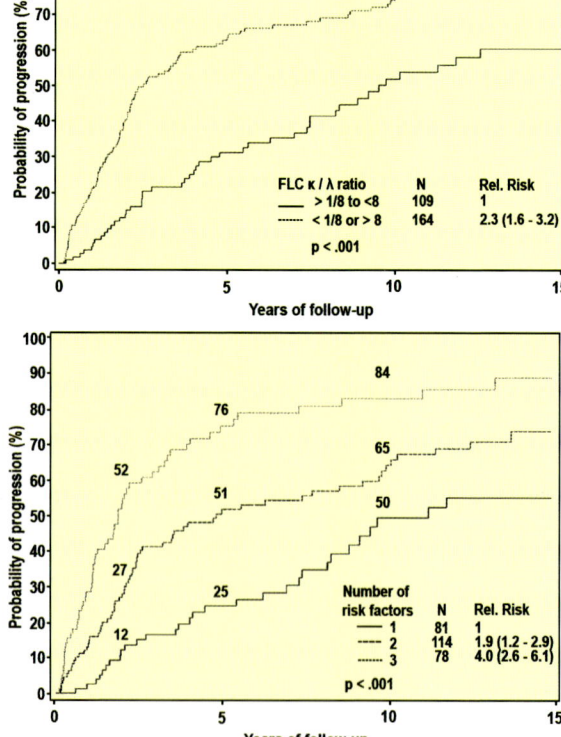

Figure 14.3. Risk of ASMM progressing to MM using 2 different levels of sFLC κ/λ ratios. (This research was originally published in Blood *(Dispenzieri 3)* © the American Society of Hematology).

Figure 14.4. Risk of ASMM progressing to MM with 1, 2 or 3 high risk factors comprising; abnormal sFLC κ/λ ratios, BMPC >10% and serum M-protein >30g/L* (This research was originally published in Blood *(Dispenzieri 3)* © the American Society of Hematology).

influenced by the length of time from diagnosis, with the highest rates in the first few years *(Figure 14.3 and 14.4)*. This was most notable in the high-risk group, in whom the probability of progression was about 26% per year for the first 2 years but slowed to 8% per year for the next 3 years. In contrast, the low-risk group progression rates were 6% per year for the the first 2 years and about 4%, per year, subsequently. Maybe, some patients classified as ASMM are biologically identical to MGUS, and with increasing follow-up the cohort becomes enriched with such patients, resulting in progressively decreasing rates of progression. Why abnormal FLC κ/λ ratios should predict a worse outcome in ASMM is unclear, but the authors speculated that these patients might have immunoglobulin heavy chain translocations or other genetic disruptions associated with disease progression *(Kumar 1)*. The importance of finding adverse risk factors for ASMM is discussed in a recent review *(Bladé 7)*. It is also highlighted in guidelines *(Chapter 25)* where it is recommended that sFLC should be assessed at baseline in ASMM *(Dispenzieri 7)*, patients should be risk stratified to guide on follow up *(Kyle 16)* and participation in clinical trials should be considered for patients at high risk.

Test Questions
1. How frequently are sFLCs abnormal in ASMM?
2. What is the rate of progression of ASMM with highly abnormal sFLC ratios?

Answers
1. In approximately 90% of patients (pages 126-127).
2. 26% in the first 2 years if BMPC and M-protein are also highly abnormal; 2.3 times the progression rate of patients with more normal sFLC ratios (page 128).

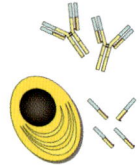

SECTION 2B - Diseases with monoclonal light chain deposition

Chapter 15 *AL Amyloidosis*   *130*
Chapter 16 *Localised amyloid disease*   *148*
Chapter 17 *Light chain deposition disease (LCDD)*   *149*

Chapter	SECTION 2B - Diseases with monoclonal light chain deposition
15	**AL Amyloidosis**

15.1.	*Introduction*	*130*
15.2.	*sFLCs at disease presentation*	*132*
15.3.	*Initial sFLC measurements correlate with survival*	*137*
	Clinical case history No 5	*137*
15.4.	*Monitoring patients with AL amyloidosis*	*138*
	Clinical case history No 6	*142*
15.5.	*Stem cell transplantation and sFLCs*	*143*
15.6.	*Solid organ transplantation and sFLCs*	*144*
15.7.	*Brain naturetic peptide (NT-proBNP) and sFLCs in AL amyloidosis*	*144*
15.8.	*sFLCs in renal failure complicating AL amyloidosis*	*145*
	Clinical case history No 7	*145*

Summary: In patients with AL amyloidosis, sFLC concentrations:

1. Are elevated in over 95% of patients at disease presentation.
2. Provide a quantitative assessment of circulating fibril precursors.
3. Correlate with changes in amyloid load during treatment and are predictive of clinical outcome.
4. Are helpful in measuring early responses to treatment and identifying disease relapses before other markers.
5. Are useful for monitoring patients undergoing stem cell and solid organ transplantation.
6. May be directly cardiotoxic.
7. Increase during renal failure, independently of FLC synthesis.

15.1. Introduction

AL amyloidosis (primary systemic amyloidosis) is a protein conformation disorder characterized by the accumulation of monoclonal free light chains (FLCs), or their fragments, as amyloid deposits (*Figure 15.1A*). Typically, these patients present with heart or renal failure but the skin, peripheral nerves and other organs may be involved (*Figures 15.1B - 15.2*) (*Gertz 2; Merlini; Kyle 6*).

Median survival used to be little more than 18 months but with improved chemotherapy and better monitoring techniques (particularly using sFLCs), median survival is now 6-8 years in patients with good responses to treatment. A slowly growing clone of plasma cells secretes the monoclonal FLCs that are typically λ type (κ to λ frequency: 1:2). It is of interest that clonal plasma cells from patients with renal deposits more commonly have the *6α Vλ* light chain, variable region genes, while those with cardiac and multisystem disease have the *1c, 2α2 and 3r Vλ* genes, although associations with other genes may be found *(Commenzo)*.

AL amyloidosis is one-fifth as common as multiple myeloma (MM) with an annual incidence of 9 per million, i.e., there are approximately 600 new patients per year in the UK and 2,500 in the USA. The median age of presentation is 70 years of age and it is rare before 40. Men represent between 60% and 65% of patients and less than 10% have associated MM.

The presence of a monoclonal protein in the serum and urine of patients is an important diagnostic feature and is a common finding. However, the underlying monoclonal gammopathies can be subtle and are undetectable in 5-20% of patients depending upon the sensitivity of the electrophoretic method. Figure 15.3 shows the serum abnormalities in a patient with AL amyloidosis. Serum protein electrophoresis (SPE) indicates a typical nephrotic pattern (low albumin, elevated α2 and low γ

Figure 15.1.
AL amyloidosis showing (A) formation of amyloid fibrils from FLC domains and (B) classic facial features with periorbital purpura. (Courtesy of PN Hawkins).

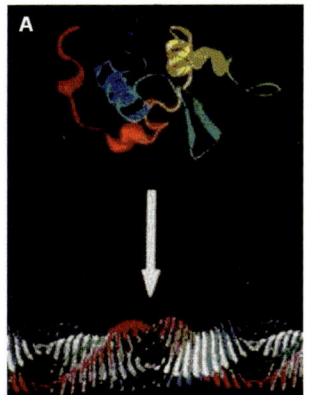 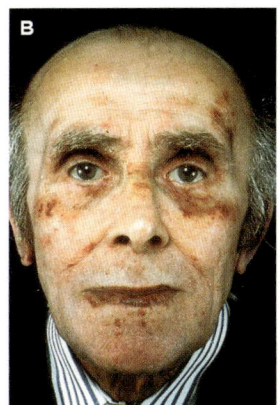

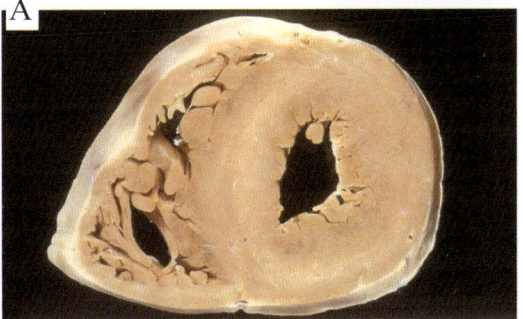 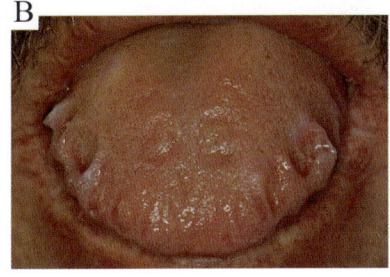

Figure 15.2. A. AL amyloidosis in the heart showing thickening of the left ventricular walls leading to heart failure. **B**. AL amyloidosis showing macroglossia that occurs in 20% of patients. (Courtesy of PN Hawkins).

fraction), but there is no observable monoclonal spike. Serum immunofixation electrophoresis (sIFE) shows some polyclonal immunoglobulin in the γ fraction and a small monoclonal λ protein that migrates in the β/γ region. This band is too small to be quantified by scanning densitometry of the SPE gel since it is undetectable against the background proteins. Figure 15.4 shows the urine protein electrophoresis (UPE) from the same patient. It contains a considerable amount of protein, particularly albumin and there is a small monoclonal spike. IFE indicates a monoclonal λ protein against a background of polyclonal κ and λ FLCs. The monoclonal band is difficult to quantify by UPE and is of modest use for the purpose of disease monitoring.

15.2. sFLCs at disease presentation

Abnormal sFLCs are not a diagnostic test for AL amyloidosis, that is the preserve of tissue biopsies and DNA tests. However, FLCs are frequently abnormal and act as a useful initial screening tool alongside sIFE and if positive, are essential for monitoring patients. If negative, the diagnosis of AL amyloidosis should be questioned.

The incidence of abnormal sFLC measurements has been assessed by many centres. The first study was from the National Amyloidosis Centre in London, UK. In a retrospective analysis of stored serum, 98% of 262 patients had abnormal sFLC

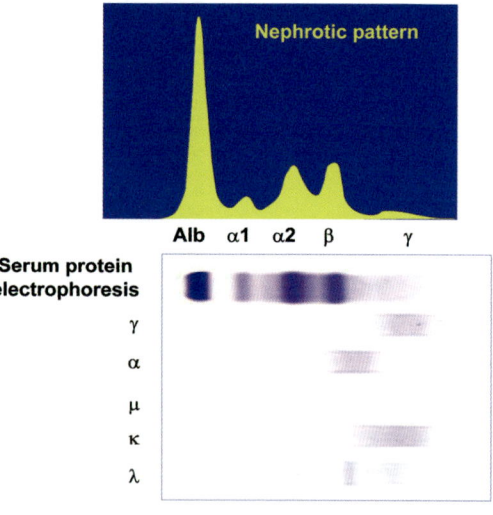

Figure 15.3.   Serum from a patient with AL amyloidosis showing a nephrotic pattern on SPE (top) and a small (non-quantifiable) monoclonal λ protein in the β/γ region by IFE (below). (Courtesy of RA Kyle and JA Katzmann).

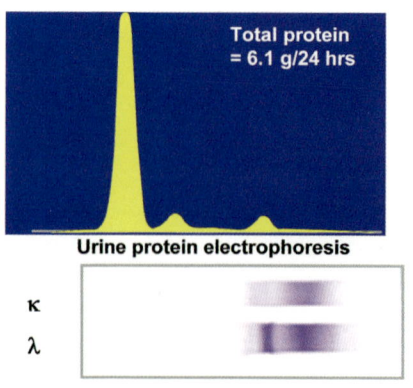

Figure 15.4. UPE and urine IFE from the same patient as in Figure 15.3, showing a monoclonal λ protein band. (Courtesy of RA Kyle and JA Katzmann).

concentrations at the time of clinical presentation *(Lachmann)*. In contrast, only 3% of patients had sufficient serum concentrations of monoclonal FLCs to be quantitated by SPE. Many patients had elevated FLCs in the urine but as discussed in Chapter 3, serum measurements are preferable. Comparisons of the results with electrophoretic tests are shown in Figures 15.5 and 15.6.

Comparison of the sFLC results from individual patients is shown in Figure 15.7 alongside normal serum samples. Concentrations of FLCs in the AL amyloidosis patients are similar to those observed in nonsecretory multiple myeloma (NSMM) but lower than those found in light chain multiple myeloma (LCMM). Classification into κ or λ types by FLC immunoassays agreed fully with IFE and bone marrow phenotyping (in the 207/262 samples that were available). In most patients, the concentrations of monoclonal FLCs were within the range of 30-500 mg/L. There was no correlation of sFLC concentrations with intact monoclonal immunoglobulins, when present.

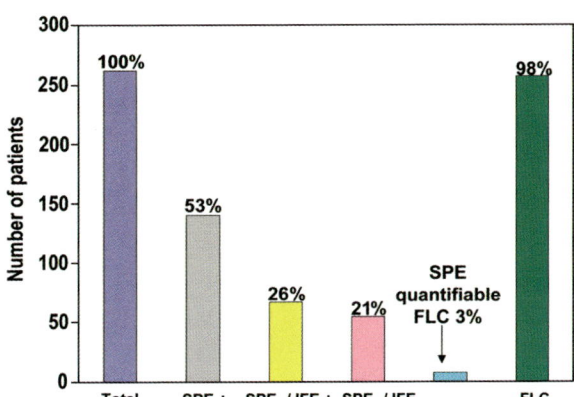

Figure 15.5. Comparison of electrophoretic tests and sFLC immunoassays in 262 patients with AL amyloidosis studied at the UK National Amyloidosis Centre. Only 3% were quantifiable by SPE.

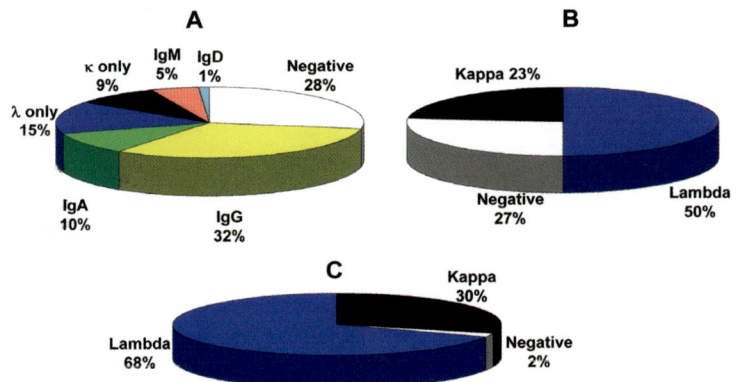

Figure 15.6. Diagnostic accuracy of different assays in AL amyloidosis. Electrophoretic test results in serum (A) and urine (B) were based on 430 patients studied at The Mayo Clinic. Serum FLCs (C) are from 262 patients at The UK National Amyloidosis Centre.

The diagnostic accuracy of traditional serum and urine tests has been compared with sFLC assays in a study from The Mayo Clinic *(Katzmann 6)*. Samples from 95 patients with AL were selected, based upon whether serum or urine tested positive or negative for monoclonal proteins by IFE and bone marrow immunohistochemistry. For samples that were serum and urine positive by IFE, the sensitivity of sFLC immunoassays was marginally lower *(Figures 15.8 and 15.9)*. In those patients whose sera were IFE negative for κ or λ, the sFLC results showed a sensitivity of 95% and 100% respectively. In patients who were negative by serum and urine IFE (but confirmed by bone marrow tests), sFLCs had a sensitivity of 86%. In the bone marrow negative group of 4 patients,

Figure 15.7. sFLCs in 262 patients with AL amyloidosis at diagnosis, 282 normal sera, 224 patients with LCMM and 28 patients with NSMM.

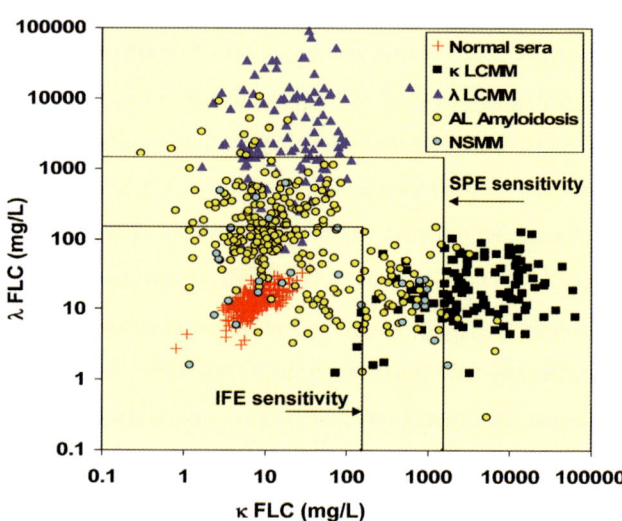

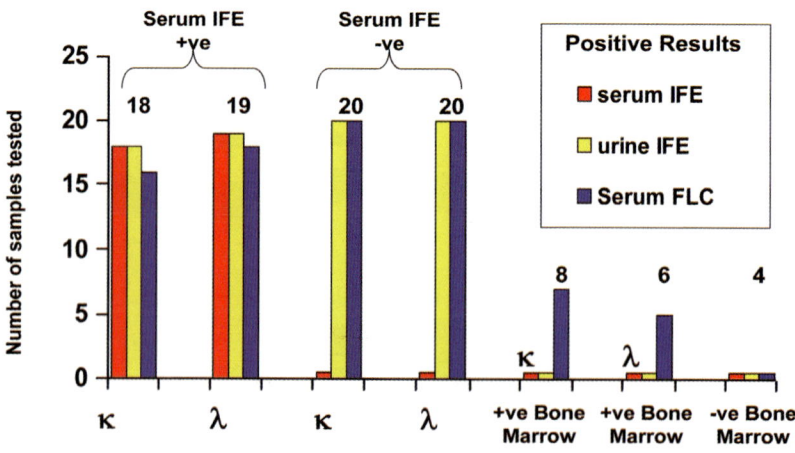

Figure 15.8. Diagnostic sensitivity of serum FLCs and IFE in 95 patients with AL amyloidosis. Numbers refer to the samples in each category. (Courtesy of RA Kyle and JA Katzmann).

sFLCs were not diagnostic.

The study above was based on samples from deliberately selected diagnostic groups. A further study was undertaken to evaluate FLC assays in 34 unselected patients with AL amyloidosis who had undergone peripheral blood stem cell transplant (PBSCT) *(Abraham 4)*. Of the 34 patients, at the time of transplantation, 26 were abnormal by serum IFE, 28 by urine IFE and 24 by both serum and urine IFE. However, only 19 patients had a monoclonal serum and 17 a monoclonal urine protein that could be quantified by electrophoretic tests. Overall, only about half of the patients who had undergone transplantation could be evaluated by serum or urine monoclonal proteins. In contrast, changes in sFLCs could be used for assessing all 34 patients although, in 4 patients, the concentrations were within the normal range.

Similar results were found by Akar *(1,2)* et al. in 169 patients with AL amyloidosis that were evaluated for the utility of sFLC immunoassays. Elevated concentrations of κ or λ FLCs were found in 96% and 94% respectively of their respective types, higher than any other individual test. However, monoclonality, as defined by κ/λ ratios was less sensitive - 89% for κ patients and 73% for λ patients. For the κ patients, sensitivity was higher than other tests but for λ patients serum and urine IFE were more sensitive (79% and 92% respectively). The conclusions were similar to those from Katzmann *(6)* et al. described above: 1) sFLC measurements are a useful screening test and supplement other tests, 2) The quantitative nature of the FLC immunoassays has value in monitoring patients, 3) FLC tests are complementary to IFE and other tests used in AL amyloidosis. Indeed, this is reflected in recent guidelines specifically addressing the use of sFLC assessment *(Dispenzieri 7)*, *(Chapter 25)*.

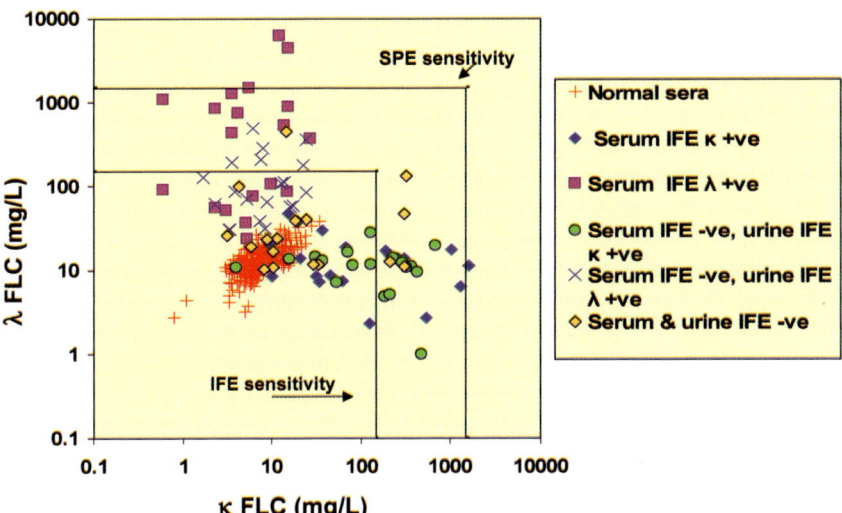

Figure 15.9. sFLCs in 95 patients with AL amyloidosis and 282 normal serum samples. The patients are divided into the diagnostic categories shown in Figure 15.8. (Courtesy of RA Kyle and JA Katzmann).

An audit of the utility of different diagnostic tests in AL amyloidosis, during the year 2003, was published by Katzmann (3) et al. *(Table 15.1)*. In 110 new patients, FLC κ/λ ratios were the most sensitive test and in combination with serum IFE identified 109 of the patients. Urine tests failed to identify the remaining patient that was negative by serum tests.

Many other groups have reported similar high rates of sFLC κ/λ ratio abnormalities. Palladini (3) et al. reported 86% abnormal in 116 patients, Morris et al. reported 97% in 31 patients, and Bochtler et al. reported 87% in 133 patients. A recent and detailed screening study *(Katzmann 8)* showed a detection rate of 88.3% of the sFLC assay for AL amyloidosis. Combination of sFLC, SPE and sIFE showed a combined sensitivity of 97.1%. Addition of urine studies to the serum screening panel elevates this detection to 98.1%. This illustrates that in some rare instances, AL amyloidosis might be picked exclusively by urine studies as previously reported by Palladini and why 24h-urine IFE is additionally recommended by the IMWG guidelines *(Dispenzieri 7)* when screening for AL amyloidosis *(Chapter 25)*. All authors agreed that the sFLC assays were most useful for subsequent patient monitoring.

In the studies described above, there was little correlation between the concentrations of sFLCs detected by immunoassays and SPE or IFE *(Figure 15.9)*. Some patients had surprisingly high concentrations of FLCs by immunoassay but IFE was normal. It is possible that the FLCs were polymerised in some of the samples and this may have inhibited the formation of narrow bands on the electrophoretic gels. This has been observed in sera from patients with NSMM *(see Figure 9.2)*.

Discordant results:

In 3-15% of patients, sFLCs are normal by immunoassay but detectable by serum IFE. Theoretical explanations include:

1. The FLCs may be missing epitopes so they are not detected by FLC antibodies but are detected by antibodies used whole light chains used in IFE.

2. The FLCs may be truncated so that they pass quickly into the urine. Accumulation cannot occur in the serum so concentrations are in the normal range, but FLCs may be detectable in the urine. Indeed, exceptionally rare patients produce monoclonal proteins that comprise only the variable domains of the light chains.

3. Structurally aberrant molecules occasionally produce antigen excess conditions even at low concentrations and produce normal FLC results. Dilution of the samples normally allows the molecules to be detected accurately and thereby produce more reliable

Table 15.1. Sensitivity of different diagnostic tests and their combinations in 110 patients with AL amyloidosis at the time of disease diagnosis.

Test	Sensitivity
FLC κ/λ ratio	91%
Serum IFE	69%
Urine IFE	83%
FLC κ/λ ratio and urine IFE	91%
FLC κ/λ ratio and serum IFE	99%
Serum IFE and urine IFE	95%
All three tests	99%

assessments of their concentrations.

While these three issues should be considered when curious results are seen, in practice, they have not been the cause of any observed discrepancies between properly measured results of sFLC and sIFE tests.

In rare patients both sFLCs and serum IFE are normal. Explanations include:
1. Some FLC molecules may have a high affinity for the amyloid deposits so any circulating FLCs would be rapidly removed. Possibly, these patients are particularly resistant to treatment.
2. In a similar manner, patients with extensive amyloid deposits might have a huge capacity for FLC removal. Any newly synthesised molecules would be cleared rapidly by a combination of binding to the amyloid mass and glomerular filtration, thereby preventing the accumulation of FLCs in serum.
3. The amyloid may be due to the deposition of a different protein.
4. The reference range for serum FLCs includes some borderline abnormal results because it has been made too wide *(Chapter 5)*.

Whatever the reason for normal sFLC results in some AL amyloidosis patients, in the vast majority these assays provide an important diagnostic and monitoring tool.

15.3. Initial sFLC measurements correlate with survival

In multiple myeloma, the baseline concentrations of sFLC correlate with outcome. The same is apparent for AL amyloidosis. In a study of 119 patients being treated with PBSCT, Dispenzieri *(4)* et al. showed that sFLC concentrations above the median value of 152mg/L were associated with shorter survival than values below the median *(Figure 15.10)*. A four parameter staging system has recently been reported *(Kumar 5)*.

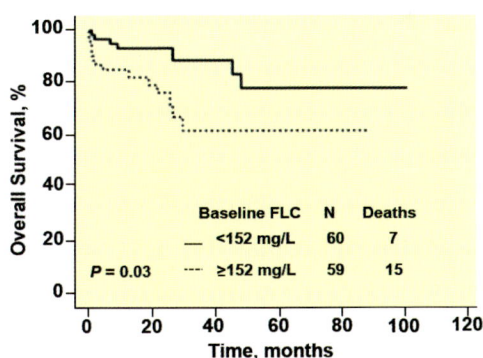

Figure 15.10. Overall survival according to baseline sFLCs in 119 patients with AL amyloidosis. Median values of sFLCs - 152mg/L. (This research was originally published in Blood *(Dispenzieri 4)* © the American Society of Hematology).

Clinical case history No 5. AL amyloidosis identified by FLC analysis when electrophoretic tests were doubtful *(Guis)*.

A 40-year-old woman, with spontaneous bruises, asthenia, abdominal pains and a possible cardiomyopathy, was investigated for suspicion of AL amyloidosis. Abdominal fat biopsy showed Congo Red positivity. SPE showed hypogammaglobulinaemia but no monoclonal proteins.

IFE showed a weak λ band without a corresponding intact immunoglobulin (*Figure 15.11*). A weak λ arc was also visible by serum immunoelectrophoresis. Quantitative immunoglobulin measurements were: IgG 4.9g/L; IgA 1.02g/L and IgM 0.32g/L indicating hypogammaglobulinaemia. sFLC analysis showed: κ 7.8mg/L; λ 210 mg/L and κ/λ ratio 0.04.

Nephelometric FLC quantification was, thus, clearly abnormal and provided a measurable parameter for subsequent disease monitoring. In contrast, FLCs were barely detectable by conventional electrophoretic assays.

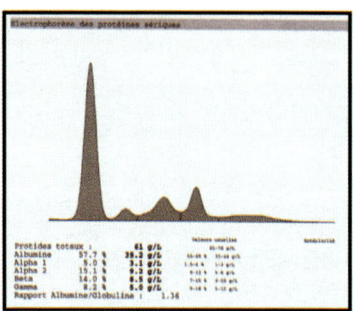 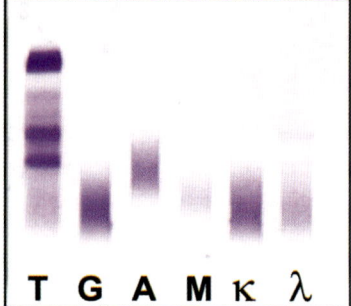

Figure 15.11. Clinical Case history No 5. SPE scan and IFE of the patient's serum. A weak λ band is visible. (Courtesy of L Guis).

15.4. Monitoring patients with AL amyloidosis

"The introduction of the serum immunoglobulin free light chain assay has revolutionized our ability to assess hematological responses in patients with low tumor burden......." Dispenzieri A, Gertz MA, Kyle RA. Blood 2004.

"The Freelite serum free light chain assay represents a landmark advance in the management of AL amyloidosis....." Wechalekar AD, Hawkins PN, Gillmore JD. Br J Haem 2008.

The aim of therapy in AL amyloidosis is to suppress the monoclonal plasma cell clone that produces the amyloidogenic FLC. When the supply of amyloid-forming protein is reduced, the balance between amyloid deposition and clearance may be favourably altered. Although complete suppression of the clonal plasma cells is desirable, reduction in the amyloidogenic sFLC concentrations by 50-70% is often sufficient to lead to stabilisation or regression of amyloid deposits.

Using SPE, the depositing monoclonal FLCs can rarely be quantified in serum. In the study by Lachmann et al. only 3% of patients had sufficiently high concentrations of monoclonal sFLCs to be quantitated by SPE *(Figure 15.5)* so patients could not be reliably monitored *(see Figure 15.18 as an example)*. This had previously led to the use of I[123] labelled serum amyloid P scans (SAP scans) as an alternative method of assessment.   Uptake of the radiolabelled protein into amyloid deposits allows identification of the affected organs and the amounts deposited. Furthermore, uptake varies with time, in parallel with changes in clinical status. This is seen in patients during treatment with chemotherapy and is compared with the concentrations of sFLCs in Figures 15.12 and 15.18.

Investigations by Lachmann et al. in 137 patients with AL amyloidosis confirmed the important relationship between amyloid deposits seen on the scans and serum FLC concentrations. Patients were divided into 3 groups dependent upon whether the SAP

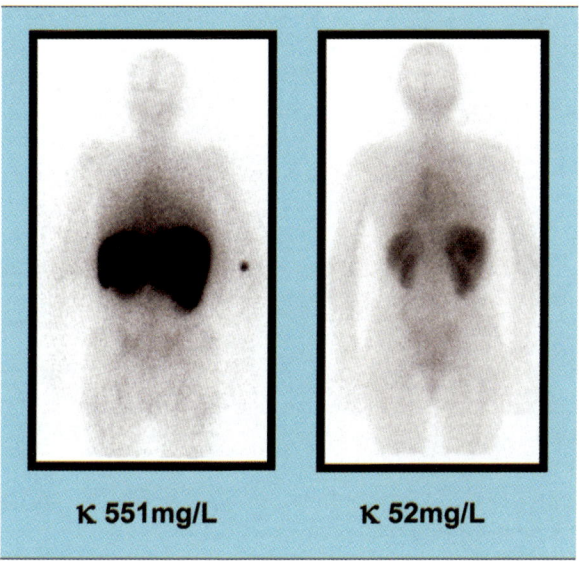

Figure 15.12. I[123] labelled serum amyloid P scans in a 52-year-old woman, viewed posteriorly. Reduction of AL deposits in the liver and spleen after one year of chemotherapy can be seen. Serum κ FLCs reduced from 551mg/L to 52mg/L over the same period. (Courtesy of PN Hawkins).

Figure 15.13. Comparison of disease status from serum amyloid P scans and serum FLCs in 127 patients with AL amyloidosis before and 12 months after commencing chemotherapy.   The mean percentage of remaining FLCs in each group are indicated (Kruskal-Wallis test: P<0.0001). (Courtesy of PN Hawkins).

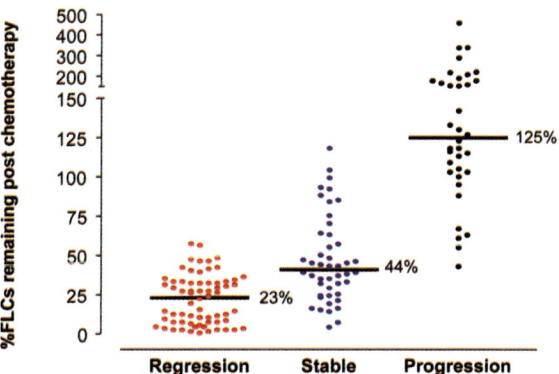

scans of the amyloid deposits showed regression, no change, or progression following chemotherapy. There was a good correlation with changes in sFLC concentrations during the same period (*Figure 15.13*). This indicated that sFLC measurements could provide a simple measure of changes in disease status in patients with AL amyloidosis.

The importance of sFLC measurements for assessing chemotherapy in AL amyloidosis has been repeatedly demonstrated. The first study was by Lachmann et al. in a retrospective analysis of 164 patients. Each patient received high-dose melphalan, intermediate or low-dose therapy and all were monitored with SAP scans and blood tests on a six-monthly basis. Stored blood samples were analysed for sFLCs in the 137 patients who survived at least 6 months. Results showed that a reduction in the amyloidogenic FLCs by 50% or more, following chemotherapy, was associated with a 10-fold survival advantage. There was an 88% probability of survival for 5 years if the sFLCs had fallen by more than 50%, compared with 39% if the sFLCs had reduced by less than 50% (P <0.001) (*Figure 15.14*). Median survival was 15 months in the 27 patients whose sFLCs showed no response (P < 0.0001). A greater than 50% fall in sFLC was more significantly related to good outcome than any other clinical or biochemical measure, while reductions by >90% were associated with the best survival.

As a result of these observations, the authors made several recommendations on the use of sFLC measurements, most of which were incorporated into UK and international guidelines *(Chapter 25):-*

1. In order to minimise the toxicity associated with chemotherapy, measurements of the sFLC should be made approximately 2 weeks after each course of chemotherapy. This should assist with decision-making about continuation of treatment.

2. It may be appropriate to discontinue chemotherapy at an early stage if the amyloidogenic FLC concentrations have:-

    a. Fallen to within the normal range.

    b. Fallen to a plateau level for at least a month.

    c. Fallen by 50-70% and toxicity or adverse effects are deemed to render further chemotherapy undesirable.

    d. Not fallen significantly after three courses of treatment, suggesting that alternative treatments should be considered.

Other large studies have confirmed the utility of sFLC for monitoring responses to chemotherapy. Wechalekar *(1)* et al. used an approach of risk adapted therapy using cyclophosphamide, thalidomide and dexamethasone in 65 patients with advanced disease. Responses of sFLCs were divided into complete, partial (>50% reduction) and no response. The differences in survival between the groups was clear (p<0.001) *(Figure 15.15)*.

Tan et al. reviewed 81 patients who were unsuitable for PBSCT and had received melphalan and dexamethasone alone. A 25% reduction in sFLCs (difference between involved and uninvolved) predicted prolonged overall survival that was not reached at 30 months compared with 11.3 months median survival for non-responders (P<0.0001) *(Figure 15.16)*. More recently, Goodman *(3)* et al. reviewed 204 patients treated with VAD like regimens and showed the prolonged survival benefits of sFLC responses

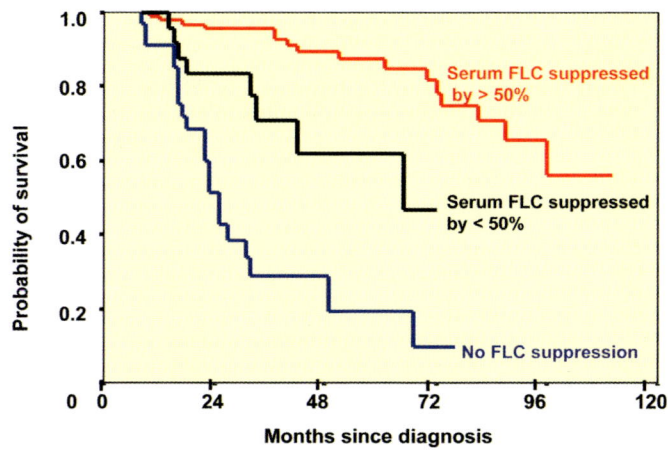

Figure 15.14. Kaplan-Meier probability of survival in 137 patients with AL amyloidosis showing that a reduction of sFLCs by greater than 50% following chemotherapy was associated with increased survival. (Courtesy of PN Hawkins).

Figure 15.15. Kaplan-Meier probability of survival in 65 patients with advanced AL amyloidosis treated with CTD based upon haematological responses. (This research was originally published in Blood *(Wechalekar 1)* © the American Society of Hematology).

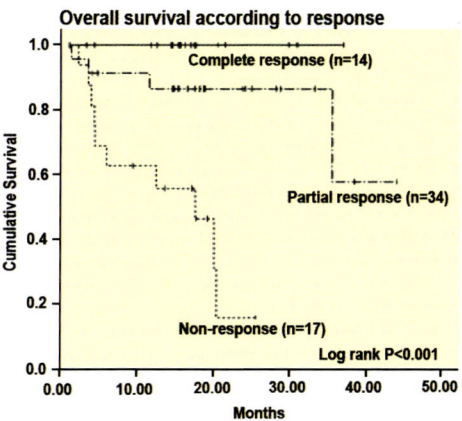

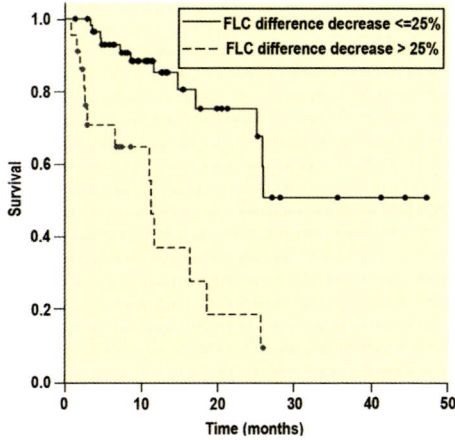

Figure 15.16. Kaplan-Meier probability of survival in 81 patients with severe AL amyloidosis treated with melphalan and dexamethasone. 25% reduction with sFLCs was associated with increased survival. (This research was originally published in Blood *(Tan)* © the American Society of Hematology).

(P<0.0001) *(Figure 15.17)*. While complete suppression of the clonal plasma cells is desirable, reduction in the amyloidogenic sFLC concentrations by 50-70% is often sufficient to lead to stabilisation or regression of amyloid deposits. Continuation of toxic chemotherapy when FLC concentrations have responded may be unnecessary or even harmful. Risk-adapted therapy, based upon early changes in FLCs, may become common practice.

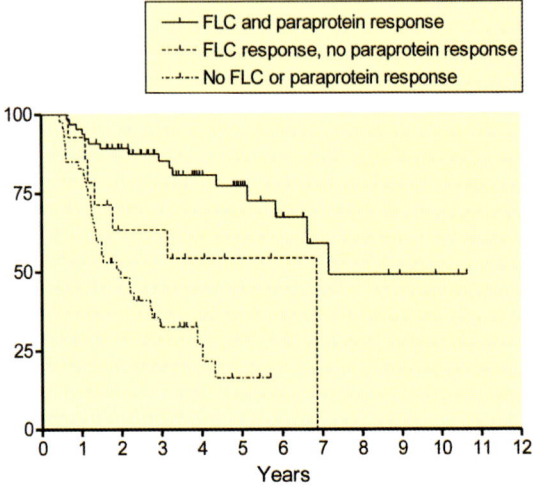

Figure 15.17. Kaplan-Meier probability of prolonged survival in 204 patients with AL amyloidosis treated with VAD regiments according to sFLC responses. (Courtesy of PN Hawkins).

Clinical case history No 6. Use of sFLCs to monitor a patient with AL amyloidosis.

A 49 year old man presented with congestive cardiac failure. After establishing a diagnosis of AL amyloidosis, he was given a heart transplant. He was subsequently treated with melphalan and prednisolone for a year, but then gradually developed increasing autonomic neuropathy with gastrointestinal symptoms, weight loss, hypotension and proteinuria. A cardiac biopsy showed evidence of amyloid in the graft. Two years after his initial presentation, he was given high dose melphalan and a PBSCT. This was successful as judged by diminishing proteinuria from 5.5g to 2.3g per day over the following months and more stable blood pressure. The patient regained some weight, returned to jogging and was relatively well for the following few years.

During his 6th year of illness, he gradually became short of breath, lost weight and renal function worsened. Deterioration continued with an episode of aspiration pneumonia followed by syncopal episodes. End-stage renal failure finally developed and he died seven and a half years after the initial presentation. Throughout his illness, he had a low level of monoclonal IgGκ protein in his serum, detectable only by IFE. Changes in its concentration had not been sufficient to act as a useful clinical marker *(Figure 15.18)*.

Retrospective analysis of serum samples showed that a monoclonal κ FLC had been present at different stages of his disease. It was present in greatly elevated concentrations at presentation but fell following the PBSCT and was undetectable for several years. It then recurred, as minor symptoms developed. Investigations at that time were normal and it was considered that the amyloidosis remained under control. In retrospect, rising FLC concentrations indicated otherwise.

Subsequently, symptoms progressed in parallel with rising κ sFLC levels but the monoclonal IgGκ, detectable by IFE, remained unchanged. Development of progressive renal and cardiac failure indicated the terminal phase of the illness and he became too ill to be treated with chemotherapy. Perhaps, if FLC results had been available before the final illness, earlier treatment with chemotherapy could have produced a favourable outcome.

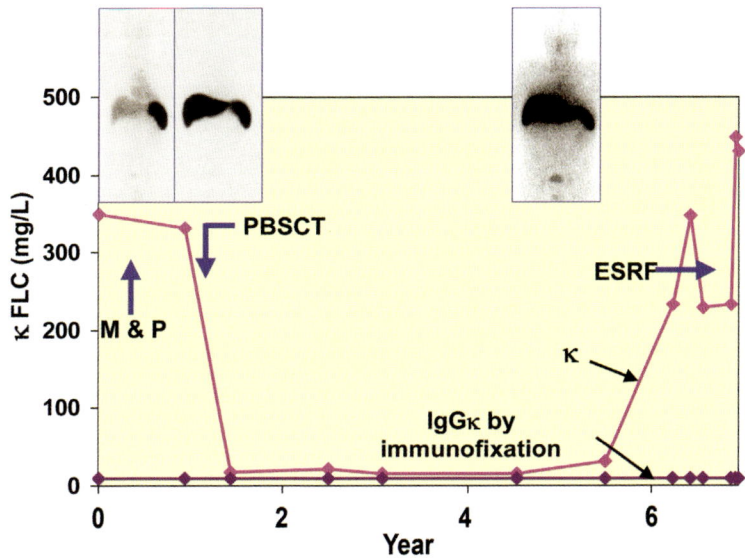

Figure 15.18. Changes in SAP scans and serum monoclonal proteins during the disease course of a patient with AL amyloidosis. M&P: melphalan and prednisolone; ESRF: end stage renal failure. (Courtesy of PN Hawkins).

15.5. Stem cell transplantation and sFLCs

The first study was from The Mayo Clinic *(Abraham 4)*. 34 patients with AL amyloidosis were assessed following PBSCT. 75% could be evaluated for clinical responses by serum or urine electrophoretic tests, whereas all could be assessed using sFLC concentrations. sFLCs were either the only marker for measurement, or decreased before the other tests in 21 of the 34 patients. Overall, changes in FLC concentrations showed a better correlation with changes in organ function than changes in electrophoretic tests.

A subsequent study by the same authors *(Dispenzieri 4)* showed that there was a higher risk of death with high baseline sFLC *(see Section 15.3 above)*. Furthermore, reduction post-transplant, was a more powerful predictor of survival than complete haematological responses *(Figure 15.19)*. Other studies have shown similar results when monitoring reductions in sFLC ratios or concentrations *(Cohen 1; Goodman 2)*. All agree that FLC quantification should be standard practice for monitoring AL amyloidosis patients undergoing PBSCT.

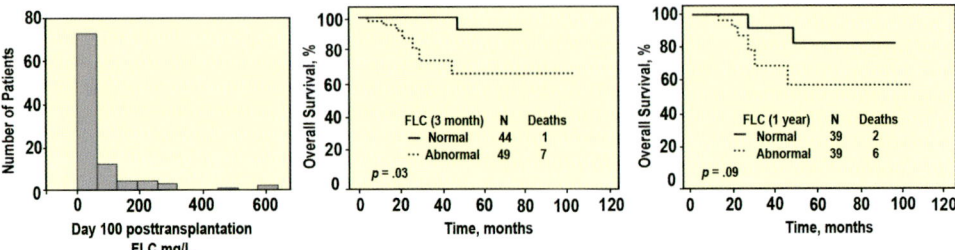

Figure 15.19. Absolute value of sFLCs after PBSCT.* (A. Distribution of levels 100 days after PBSCT, B. overall survival by sFLCs at day 100 and C. at 1 year. (This research was originally published in Blood *(Dispenzieri 4)* © the American Society of Hematology).

15.6. Solid organ transplantation and sFLCs

Renal transplanation. There are no substantial publications on the role of sFLCs in patients with AL amyloidosis and renal failure. Since the kidneys are the most common site of amyloid deposition it is likely that serum rather than urine measurements of FLCs would be helpful in monitoring and predicting organ damage. Furthermore, renal transplantation is helpful in patients with advanced renal disease. Wechalekar et al. *(4)* reported on 16 patients who had renal transplantation and indicated that the median survival was at least 7.4 years despite a very poor haematological response to chemotherapy in many. sFLC measurements for assessing disease relapse with damage to the transplant are likely to be helpful.

Cardiac transplantation. This is a controversial procedure in AL amyloidosis because of recurrence of amyloid in the graft and shortage of donors. However, it has been used when amyloid deposits in other organs are not too severe. Gillmore et al. reported on 5 patients with heart and stem cell transplantation and showed substantial survival benefits *(for an example see Figure 15.18)*. Mignot et al. reported successful suppression of monoclonal λ sFLCs by 80% using melphalan and dexamethasone followed by cardiac transplantation. In spite of initial severe cardiac failure, recovery of organ function and AL amyloidosis control remained good at 2 years with normal sFLC concentrations *(Figure 15.20)*. The early and rapid fall in monoclonal sFLCs was a useful guide to the success of the subsequent transplantation.

15.7. Brain naturetic peptide (NT-proBNP) and sFLCs in AL amyloidosis

An important link between cardiac dysfunction in AL amyloidosis and falling sFLC concentrations was observed by Palladini *(1,2)* and colleagues. 51 AL amyloidosis patients with symptomatic myocardial involvement were given chemotherapy and monitored for sFLCs and the amino-terminal fragment of naturetic peptide type B (NT-proBNP), a sensitive marker of myocardial dysfunction in AL amyloidosis. During

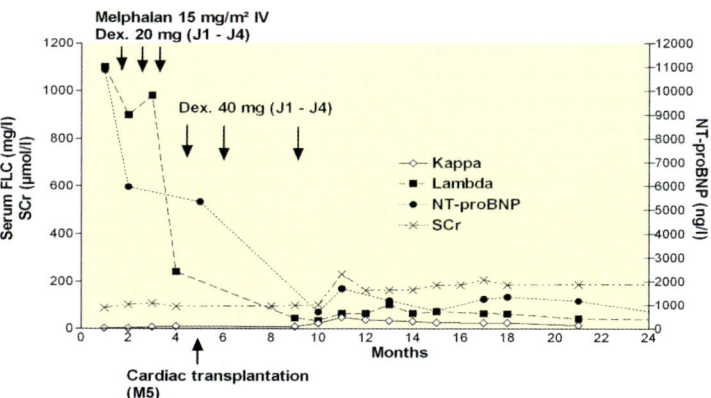

Figure 15.20. Sustained fall in sFLCs following chemotherapy and cardiac transplantation. (Reproduced with permission of Haematologica *(Mignot)* and A Mignot).

treatment, 22 patients had a reduction of sFLCs by more than 50%, including 9 patients who had disappearance of monoclonal immunoglobulins by IFE and there was a corresponding reduction of NT-proBNP levels (p < 0.001). Survival was better in responders than non responders (p<0.001). In some patients with sFLC reductions, the heart failure resolved without concommitant reduction of wall thickness at echocardiography.

These observations suggest that the amyloidogenic precursors equate to the circulating monoclonal FLCs. Of particular interest was the rapid improvement in heart failure, supporting a toxic effect. Further evidence has come from *in-vitro* studies of cardiac myocytes, where direct toxicity of amyloidogenic FLCs has been observed. It further emphasises the utility of the FLC assays and the importance of rapidly reducing their concentrations in patients with heart failure. In addition, although correlated, changes in sFLCs and NT-proBNP levels were independent markers of survival.

15.8. sFLCs in renal failure complicating AL amyloidosis

Renal failure frequently complicates AL amyloidosis and inevitably leads to increased serum concentrations of the amyloidogenic FLC even when the disease is not progressing. However, increases also occur in the concentrations of the non-depositing FLC *(Chapter 20.1)*. Under these circumstances, the κ/λ ratio is a good determinant of disease progression. Furthermore, changes in the alternate FLC can be used to assess changes in renal function. This is illustrated in a patient with AL amyloidosis caused by a κ-secreting plasma cell clone *(Figure 15.21)*.

It should be noted that the normal range for κ/λ ratio changes slightly in patients with renal failure. As renal function deteriorates, less FLCs are cleared through the glomeruli and more are removed by pinocytosis in other tissues. This reduces the clearance rate of κ more than λ, so the median value for the normal κ/λ ratio increases slightly *(Chapter 20)*.

Clinical case history No 7. Use of the κ/λ ratio to assess AL amyloidosis in a patient with renal impairment.

A patient with AL amyloidosis was treated with cyclophosphamide and vincristine, adriamycin (doxorubicin), methyl-prednisolone (C-VAMP), after which serum κ FLC concentrations increased and then remained stable for six months before increasing further (*Figure 15.21*). Serum creatinine concentrations changed in a similar manner suggesting that a reduction in renal clearance could account for the rise in κ concentrations. However, the κ/λ ratio steadily increased over the same period, indicating a clonal increase in κ concentrations. The κ/λ ratio then increased sharply which suggested that the deteriorating renal function was due to increased amyloid deposition. A renal transplant was then successfully performed.

These results illustrate a typical dilemma facing the clinician managing these patients. The deteriorating renal function could be due to one or more factors. However, the steadily rising κ/λ ratio indicated a recurrence of the FLC clone leading to renal deposition of amyloid.

Clearly, the κ/λ ratio should be carefully monitored, but it is not clear whether this is the only consideration. As renal function deteriorates, a greater proportion of the sFLC are due to failure of renal clearance rather than de-novo synthesis. The additional circulating amounts of depositing FLCs might accelerate amyloid formation. The depositing FLC concentrations can be estimated by subtraction of the non-depositing FLC concentrations (polyclonal). Indeed, it has been suggested that this should replace κ/λ ratios when monitoring patients with MM. Guidelines for the diagnosis and monitoring of AL amyloidosis are described in Chapter 25.

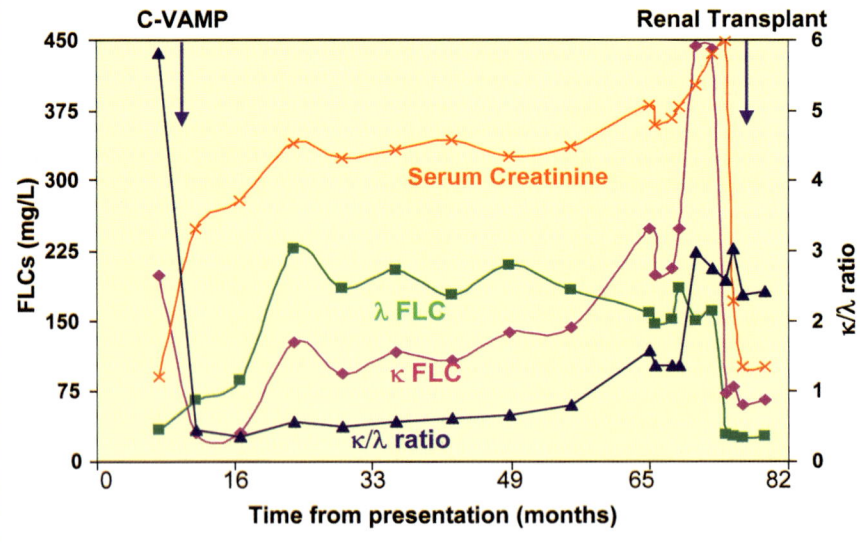

Figure 15.21. Changes in serum FLCs and renal function in a patient with monoclonal κ AL amyloidosis. (Courtesy of PN Hawkins).

Test questions

1. What is the frequency of abnormal sFLCs in AL amyloidosis?
2. Why are some patients with AL amyloidosis negative for sFLCs?
3. How frequently should patients undergoing treatment for AL amyloidosis be assessed for sFLCs?
4. What is the median survival of AL amyloidosis patients who have normalisation of sFLCs following chemotherapy?

Answers

1. 90-95% *(Section 15.2, page 132).*
2. *There is no clear explanation but possibly the sFLCs are being cleared very quickly into the amyloid deposits (Section 15.2, page 132).*
3. *Response or lack of response to treatment can be detected within 1-2 weeks using sFLC measurements (Section 15.4, page 138).*
4. *> 10 years (Section 15.4, page 138).*

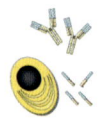

Although normally systemic, amyloid deposition may also be limited to single organs. The distribution depends upon the biochemical nature of the amyloid fibril protein and as in AL amyloidosis, light chain fragments may be involved. Serum free light chains (sFLCs) have been evaluated in patients with localised amyloid disease attending the UK National Amyloidosis Centre, as presented in Table 16.1. An enlarged series of 235 cases was reported by the same group in 2005 *(Goodman 1)*.

Overall, elevated levels of sFLCs are less commonly observed in localised amyloidosis than in systemic AL amyloidosis and, even when present, the concentrations are lower. sFLC concentrations may, therefore, assist in distinguishing the different types of amyloid disease and also systemic from localised light chain amyloid disease.

Site of amyloid deposits	Number of patients	Monoclonal proteins*	Abnormal κ/λ ratios
Bone	7	3 (43%)	6 (88%)
Bladder	25	1 (4%)	3 (12%)
Bowel	10	4 (40%)	3 (30%)
Bronchial	13	2 (15%)	1 (8.0%)
Nodular pulmonary	13	3 (23%)	3 (23%)
Laryngeal	22	0	1 (4.5%)
Nasopharynx	16	1 (6%)	2 (13%)
Skin	18	2 (11%)	2 (11%)
Occular	10	2 (20%)	2 (20%)
Lymph node	16	3 (19%)	5 (31%)
Miscellaneous	6	0	1 (17%)

Table 16.1. Frequency of monoclonal proteins in patients with localised amyloid disease. *Serum monoclonal proteins or light chain proteinuria identified by electrophoretic tests. (Courtesy of PN Hawkins).

17.1. Introduction   *149*
17.2. Diagnosis of LCDD using sFLC assays   *149*
       Clinical case history No 8   *150*
17.3. Monitoring LCDD using sFLC assays   *151*
       Clinical case history No 9   *152*

Summary: sFLC measurements are important in LCDD because:

1. They are abnormal in 90% of patients at the time of diagnosis.
2. They are useful for monitoring disease progress.
3. They may identify patients that were previously unrecognised.

17.1. Introduction

In light chain deposition disease (LCDD), monoclonal serum free light chains (sFLCs) are precipitated on the basement membranes of cells in the kidneys and other organs. As with AL amyloidosis, the disease is progressive and leads to organ failure of the kidneys, heart or liver and has a poor prognosis *(Solomon 2; Buxbaum 1, 2)*. This rare disease differs from AL amyloidosis by being more frequent in younger women (30-50 years) and renal failure is a common presenting feature. The deposits usually contain κ FLCs (*Vκ1 and Vκ4*) without amyloid P component. Some of the patients have serum and urine monoclonal proteins detectable by electrophoretic tests.

17.2. Diagnosis of LCDD using sFLC assays

Although biopsy is necessary for the diagnostic work up of a patient with LCDD, the clinical utility of adding sFLC analysis to diagnostic screening panels has clearly been demonstrated. This was shown most recently where 18 LCDD patients were included as part of a larger screening study aimed at evaluating several serum and urine based screening algorithms *(Katzmann 8)*. The analysis showed that sFLC testing alone, a panel of SPE and sFLC were both as sensitive (77.8%) for LCDD detection as a panel of SPE, serum IFE and urine IFE. The most recent IMWG guidelines *(Dispenzieri 7)* recommend sFLC in combination with SPE and serum IFE as a clinically effective screening panel for monoclonal gammopathies, with the exception of AL amyloidosis which additionally requires a 24h-urine IFE. Proposed screening algorithms are discussed further in Chapter 23 and guidelines are detailed in Chapter 25.

The study by Katzmann *(8)* et al., mentioned above, supports previous studies evaluating the clinical utility of the sFLC tests in LCDD. Briefly, sFLC concentrations were measured in 19 patients with LCDD by Katzmann *(6)* et al. and were abnormal in 17 (*Table 17.1 and Figure 17.1*). One sample was falsely negative by sFLC analysis but positive by serum immunofixation electrophoresis (IFE). In a subsequent publication by the same authors, 7 further patients were studied and all had raised sFLC concentrations *(Katzmann 3)*. Clinical case history No 8 illustrates the clinical sensitivity of the FLC tests compared with conventional serum and urine electrophoretic assays *(Figure 17.2)* *(Guis)*.

Figure 17.1. sFLCs and serum and urine electrophoretic tests in 19 patients with LCDD. BM = bone marrow. (Courtesy of RA Kyle and JA Katzmann).

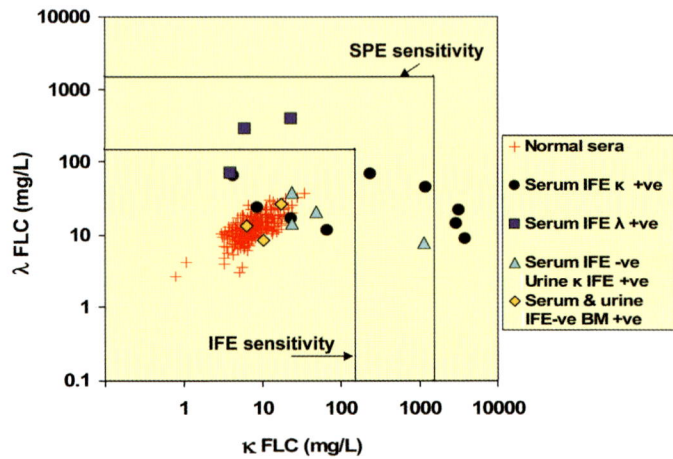

Table 17.1. Detection rates by sFLCs in 19 LCDD patients. BM: bone marrow.

Classification	Elevated FLC	Abnormal FLC κ/λ ratio
Serum IFE κ +ve	8/9	8/9
Serum IFE λ +ve	3/3	3/3
Serum IFE -ve; Urine: IFE κ +ve	4/4	4/4
Serum and uIFE -ve. BMPCs κ +ve	1/3	2/3
Total abnormal for sFLCs	**16**	**17**

Clinical case history No 8. Light chain deposition disease undetectable by conventional electrophoretic assays.

A 66-year-old man suffering from asthenia and anaemia was investigated for serum protein abnormalities. Serum protein electrophoresis (SPE) and serum and urine IFE tests showed no evidence of monoclonal immunoglobulins *(Figure 17.2)*. Serum immunoglobulins were normal/low: IgG 8.5g/L; IgA 0.4g/L and IgM 0.2g/L. However, sFLC concentrations were highly abnormal: κ 294mg/L; λ 71.6mg/L and κ/λ ratio 4.1. These results indicated a monoclonal gammopathy and renal impairment. FLC quantification allowed the depositing FLC to be easily identified and supported the clinical diagnosis of LCDD obtained by renal biopsy.

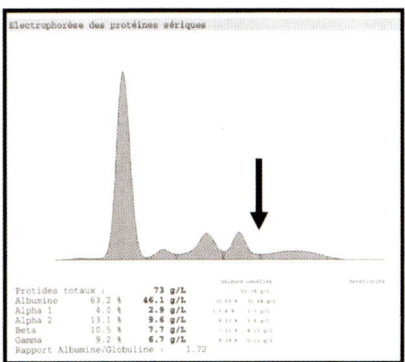 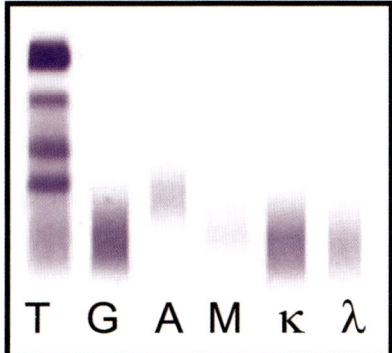

Figure 17.2. LCDD showing normal SPE (scanning densitometry) and IFE, but sFLCs were highly abnormal (κ 294mg/L: λ 71.6mg/L and κ/λ ratio: 4.1). T: Protein stain. (Courtesy of L Guis).

17.3. Monitoring LCDD using sFLC assays

It is logical to monitor these patients using sFLC assays and initial reports indicate that changes in concentrations are as helpful as in patients with AL amyloidosis. Case history 9 illustrates the utility of sFLC analyses in a patient who was difficult to monitor by other methods *(Brockhurst)*.

A series of 17 patients with biopsy-proven LCDD was studied by Wechalekar *(3)* et al. sFLCs were abnormal with a clonal bias in 15 (88%): 11 (64%) had κ excess, 4 (23%) had λ excess while 2 (11%) had polyclonal increased FLCs. The median κ levels were 317mg/L (range 8.5-2,260) while the median λ levels were 64mg/L (range 17-10,700).

A total of 10 patients received systemic chemotherapy for the underlying plasma cell dyscrasias as follows: vincristine, adriamycin (doxorubicin), dexamethasone (VAD) - 4, cyclophosphamide and vincristine, adriamycin, melphalan, methyl-prednisolone (C-VAMP) - 1, VAD followed by autologous stem cell transplant - 2, melphalan and prednisone - 1 and intermediate dose melphalan - 2. 8 (80%) patients had sFLC responses with a median decrease of 63% (range 31 - 95%) compared with pre-treatment values. One had no change in sFLC levels (which did not show clonal bias pre-treatment) but had a very good partial response of the intact monoclonal immunoglobulin. Only 2 patients had complete normalization of FLC levels. Renal function improved in 2, remained unchanged in 5 (including 3 patients with end-stage renal failure) and worsened in 1 patient. Both patients with abnormal liver function and cardiac involvement showed improvement. The median overall survival was 59 months.

The authors concluded that measurement of sFLCs detected 33% more patients with LCDD than SPE and IFE. The assay was also useful for monitoring response to treatment. Detection of abnormal sFLCs may shorten time to diagnosis in patients without monoclonal intact immunoglobulins. Measurements of FLCs was recommended as a useful addition to the screening tests for patients with suspected LCDD and also for monitoring responses to chemotherapy *(Kuypers)*.

Hassoun *(1)* et al. reported on 5 patients with LCDD, one with light and heavy chain DD and one with light chain crystal DD. All had abnormal sFLCs at diagnosis. Patients were given high dose melphalan and peripheral blood stem cell transplant (PBSCT) with good responses that could be monitored with sFLCs.

It is important to note that patients with chronic kidney disease due to deposition of monoclonal FLCs are difficult to identify and monitor *(Sanders 3; Possi)*. It is likely that many patients have detectable monoclonal FLCs in serum but are undiagnosed from urine studies. This issue is discussed in detail in Chapter 20.5.

Clinical case history No 9. Light chain deposition disease monitored with sFLC assays. (Courtesy of I Brockhurst, Leicester, UK).

A 49-year-old Caucasian male presented to the nephrologists with flu-like symptoms, hypertension and face, hand and leg swelling. Serum electrolytes were normal, creatinine clearance was 140mL/min and urinary protein quantification was 2.4g/24 hours. A renal biopsy demonstrated normal histology and immuno-fluorescence tests. He was managed with a 120mg daily dose of frusemide and antihypertensives. Follow-up was initially uneventful with renal function remaining stable.

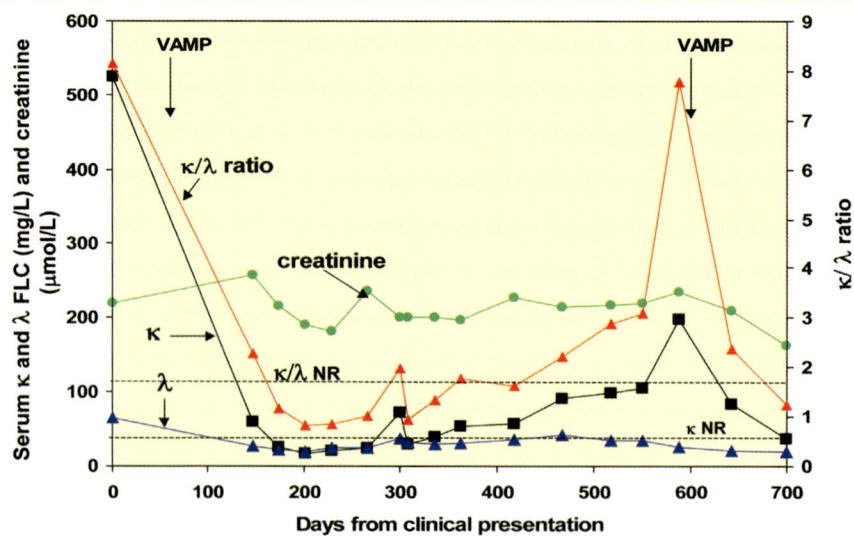

Figure 17.3. Monitoring of a patient with LCDD using sFLC assays (Courtesy of Ian Brockhurst, Leicester, UK).

10 years later he presented with nephrotic syndrome. Serum biochemistry showed: creatinine 165μmol/L (NR 60-120μmol/L), albumin 33g/L (NR >40g/L), cholesterol 8.7mmol/L (NR <5.5mmol/L) and urinalysis revealed 3+ proteinuria. A further renal biopsy showed nodular glomerulosclerosis with evidence of LCDD on electron microscopy. Congo red staining was negative. SPE, immunoglobulin levels and urinary Bence Jones protein assays were all normal.

He was referred to the haematology department to rule out an underlying B-cell clonal disorder. Bone marrow aspirate and trephine revealed normal cellular marrow with no morphological or immunophenotypic evidence of MM and, again, Congo Red staining was negative. An iodine[123] labelled, serum amyloid P scan showed no evidence of amyloid deposition. Serum was tested for sFLCs with the following results: κ 526.0mg/L (normal range 3.3 - 19.4mg/L), λ 64.6mg/L (normal range 12.7 - 26.3mg/L) and κ/λ ratio 8.14 (normal range 0.26 - 1.65) (*Figure 17.3*). Subsequently he developed atrial fibrillation. A 24-hour tape showed irregularities in the atrial chamber and intermittent disruption of AV node conduction. He had a dual chamber pacemaker fitted and cardiac biopsy performed, which showed no evidence of amyloid or light chain deposition.

Within 2 months his renal function had deteriorated further with a serum creatinine of 210μmol/L, creatinine clearance of 67mL/min and a 24 hour urine protein leakage of 13.8g. In order to delay the need for dialysis he was treated with 3 cycles of VAMP chemotherapy (vincristine 0.4mg/day for 4 days, doxorubicin 9mg/m2/day for 4 days and methylprednisolone 1g/m2 for 5 days per cycle). Subsequent to the chemotherapy, renal function improved and this was also observed in the sFLC levels and κ/λ ratio. 3 months after the chemotherapy, 24 hour urinary protein excretion was 0.1g/L.

For the following year renal function remained stable but then the κ/λ ratio and serum creatinine concentrations began to increase again. He was treated with a further 3 cycles of VAMP and similar improvements in renal function and sFLC levels were seen. The patient has remained reasonably well since.

Test question
1. What proportion of patients with LCDD have raised sFLCs?

Answer
1. ~90% (Section 17.2 page 149).

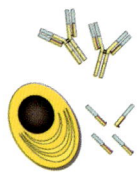

Chapter 18. *Other malignancies with monoclonal free light chains*   *154*
Chapter 19. *Monoclonal gammopathies of undetermined significance (MGUS)*   *166*

Chapter

18

SECTION 2C - Other diseases with monoclonal free light chains

Other malignancies with monoclonal FLCs

18.1. *Solitary plasmacytoma of bone*   *154*
18.2. *Extramedullary plasmacytoma*   *156*
18.3. *Multiple solitary plasmacytoma (+/- recurrent)*   *156*
18.4. *Plasma cell leukaemia*   *157*
18.5. *Waldenström's macroglobulinaemia*   *157*
18.6. *B-cell, non-Hodgkin lymphomas*   *159*
        B-cell, non-Hodgkin lymphoma complicated by AL amyloidosis   *162*
18.7. *B-cell, chronic lymphocytic leukaemia*   *162*
18.8. *POEMS syndrome*   *164*
18.9. *Cryoglobulinaemia*   *165*

18.1. Solitary plasmacytoma of bone

These bone tumours represent 3-5% of plasma cell neoplasms and are twice as common in women than men *(Figure 18.1)*. Approximately 50% progress to multiple myeloma (MM) over 3-4 years while 30-50% are alive at 10 years. The criteria for the disease are shown below *(Kyle 5)*.

Criteria for the diagnosis of solitary plasmacytoma of bone:
- Low concentration or no monoclonal (M)-protein in serum and/or urine
- Single area of bone destruction due to clonal plasma cells
- Bone marrow not consistent with MM
- Normal skeletal survey (and magnetic resonance imaging (MRI) of spine and pelvis if done)
- No related organ or tissue impairment (no end organ damage other than a solitary bone lesion)

Immunofixation electrophoresis (IFE) of serum and/or concentrated urine shows a small monoclonal protein in approximately 50% of patients. When present, this is useful for guiding therapy and persistence is associated with worse outcome.

The potential use of serum free light chains (sFLCs) has recently been investigated in two studies. In the first report, 13 patients with solitary plasmacytoma were assessed at diagnosis and during progression to MM *(Leleu 2)*. By conventional electrophoretic tests, 5 patients had IgG, 2 had IgA and 3 had FLC only monoclonal proteins, while 2 were nonsecretory. In total, 5 of the 13 had monoclonal κ FLCs detectable by electrophoretic methods. However, using the more sensitive sFLC immunoassays, 7 patients had κ and 2 had λ monoclonal proteins, including one of the nonsecretory plasmacytomas that was serum κ positive. There was complete concordance between the κ and λ types identified by the FLC assays and the bound light chain type on the intact monoclonal immunoglobulin identified by IFE. After radiotherapy, 7 patients showed reductions in sFLC concentrations. Three patients who progressed to MM showed no reductions in FLC levels.

In a larger study from the Mayo Clinic, 116 patients were retrospectively investigated *(Dingli 1)*. At the time of the analysis, 43 had progressed to MM with a median time of 1.8 years. sFLC ratios were abnormal in 54 (47%) patients at diagnosis and this was associated with a higher risk of progression (P=0.039), ie., 44% at 5 years versus 26% with normal sFLC ratios *(Figure 18.2)*, and the group had a shorter survival time *(Figure 18.3)*. A risk stratification model was then constructed of concentrations of more or less than 5g/L of monoclonal immunoglobulin together with normal or abnormal sFLC ratios. Low, intermediate and high risk groups corresponded to none, 1 or 2 positive risk factors and these gave a progression rate at 2 years of 13%, 26% and 62% respectively

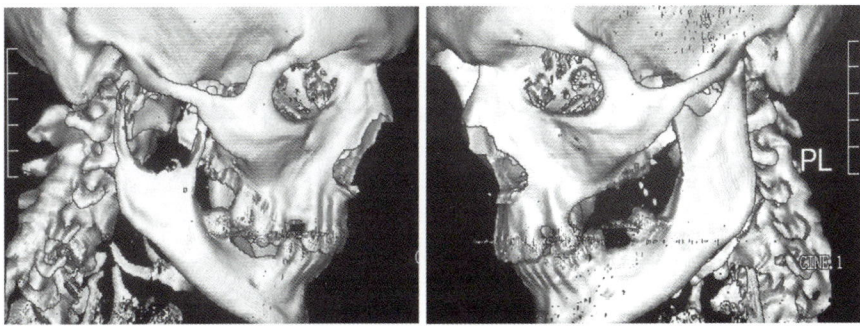

Figure 18.1. Solitary plasmacytoma of the right ramus of the mandible (Courtesy of Ade Olujohungbe).

Figure 18.2. Kaplan-Meier plots for time to progresssion to multiple myeloma in 116 patients with solitary bone plasmacytoma and normal (62) or abnormal (54) sFLC ratios. (This research was originally published in Blood *(Dingli 1)*, © the American Society of Hematology).

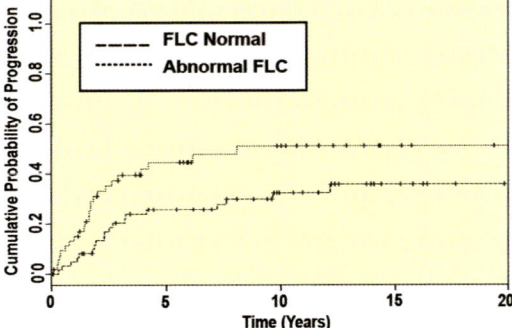

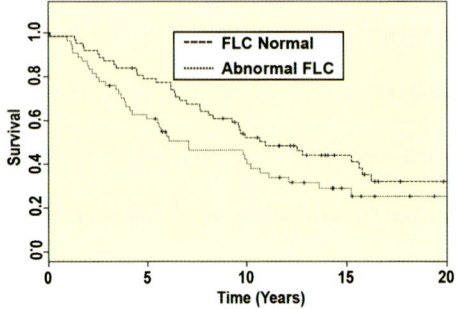

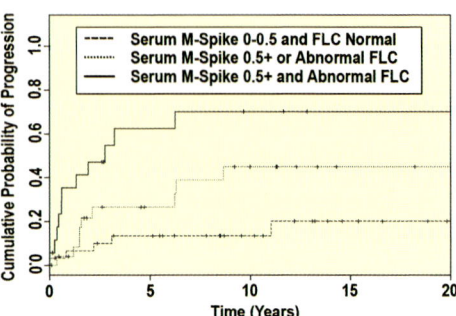

Figure 18.3. Kaplan-Meier plots for survival in 116 patients with solitary bone plasmacytoma and normal (62) or abnormal (54) sFLC ratios. (This research was originally published in Blood *(Dingli 1)*, © the American Society of Hematology).

Figure 18.4. Risk of progression in solitary plasmacytoma of bone using sFLCs and serum monoclonal immunoglobulins. Graphs A, B and C indicate high, intermediate and low risks of progression. (This research was originally published in Blood *(Dingli 1)*, © the American Society of Hematology).

(Figure 18.4). It is of interest that urine studies of FLC excretion also showed a correlation with outcome. The authors commented that sFLC analysis provided an important new prognostic indicator.

18.2. Extramedullary plasmacytoma

This is a plasma cell tumour that arises outside the bone marrow and can occur in any organ, although it is found particularly in the upper respiratory tract. Local tumour irradiation is the treatment of choice and only 15% progress to MM. When present, the monoclonal protein is typically IgA *(Kyle 5)*. As with solitary plasmacytoma of bone, sFLC measurements may be helpful in managing some of these patients.

Criteria for the diagnosis of extramedullary plasmacytoma

- Low concentration or no M-protein in serum and/or urine
- Extramedullary tumour of clonal plasma cells
- Normal bone marrow
- Normal skeletal survey
- No related organ or tissue impairment (no end organ damage including bone lesions)

18.3. Multiple solitary plasmacytoma (+/- recurrent)

Up to 5% of patients presenting with solitary plasmacytomas develop multiple lesions in the bone or elsewhere, without evidence of MM *(Kyle 5)*. As with solitary plasmacytoma, sFLC measurements may be helpful in managing some of these patients.

Criteria for the diagnosis of multiple solitary plasmacytomas (± recurrent)

- Low concentration or no M-protein in serum and/or urine
- More than one localised area of bone destruction or extramedullary tumour of clonal plasma cells which may be recurrent
- Normal bone marrow
- Normal skeletal survey and MRI of spine and pelvis if done
- No related organ or tissue impairment (no end organ damage other than the localised bone lesions)

18.4. Plasma cell leukaemia

High concentrations of plasma cells in the blood (>20% of nucleated cells and >2.0 x 10^9/L) define plasma cell leukaemia. It may occur without evidence of MM or may develop from leukaemic transformation of pre-existing myeloma *(Kyle 5)*. Monoclonal proteins are present in some patients and there is one report of a patient with monoclonal sFLCs *(Kruger)*. The patient was monitored effectively with sFLCs and plasma cell counts during treatment with Bortezomib and an allogeneic stem cell transplant.

18.5. Waldenström's macroglobulinaemia

Waldenström's macroglobulinaemia (WM) is a low-grade, lymphoproliferative disorder that is associated with the production of monoclonal IgM. The incidence is 5-10% of multiple myeloma with approximately 1,500 new cases per year in the USA and 300 in the UK. The median age of presentation is 65 years. Median survival is 5 years but over 20% of patients live for more than 10 years and many die from unrelated causes. Typically, patients present with high concentrations of IgM and infiltration of the bone marrow, spleen and lymph nodes with plasmacytoid lymphocytes and mast cells. Patients may have suppression of bone marrow function, enlarged spleen, liver and lymph nodes, hyperviscosity syndrome, cryoglobulinaemia, neuropathy or AL amyloidosis. All aspects of WM have been reviewed in the April 2003 edition of Seminars in Oncology *(Owen 2)*. The diagnostic criteria for WM are shown below.

Proposed criteria for the diagnosis of WM

- IgM monoclonal gammopathy of any concentration
- Bone marrow infiltration by small lymphocytes showing plasmacytoid/plasma cell differentiation
- Inter-trabecular pattern of bone marrow infiltration
- Surface IgM+, CD10-, CD19+, CD20+, CD22+, CD23-, CD25+, CD27+, FMC7+, C D103-, CD138- immunophenotype (variations from this immunophenotypic profile can occur)

Serum IgM quantification is important for both diagnosis and monitoring. Unfortunately, nephelometric determinations may be unreliable because polymerisation of the IgM molecules distorts the results. At high concentrations in particular, accurate measurements require the use of serum protein electrophoresis (SPE) and scanning densitometry. At low concentrations no method is accurate because the inclusion of normal IgM leads to overestimation of the monoclonal IgM concentrations. IFE is more sensitive than SPE for detecting low concentrations of IgM but is non-quantitative. In addition, the presence of cryoglobulins or cold agglutinins affects IgM measurements by all methods, so serum samples may need to be assessed under warm conditions *(Owen 2)*.

Another laboratory assessment criterion is the presence of FLC proteinuria. This occurs in approximately 50% of patients and may exceed 1g/day. However, the amounts excreted are usually low and do not relate particularly well to changes in tumour burden *(Weber D, Bladé 4)*.

Since FLC proteinuria occurs in many patients it is likely that sensitive sFLC assays show abnormal results more frequently. Figure 18.5 shows sFLC concentrations in 37 patients (21 IgMκ, 15 IgMλ and one biclonal) at the time of plasma exchange for hyperviscosity syndrome. All but one had abnormal FLC concentrations and/or abnormal κ/λ ratios. The non-tumour FLCs were not elevated in any of the patients, indicating no significant renal impairment, but occasionally renal failure does occur *(Bradwell 5)*.

Since sFLCs are elevated in nearly all patients this may be clinically useful. Their short half-life and the large clinical range should provide a sensitive marker for treatment responses. Also, FLCs do not cryoprecipitate and are not affected by other factors that can make IgM measurements difficult.

Several studies have investigated sFLCs in WM. The largest cohort comprised 98 WM patients and 68 IgM monoclonal gammopathy of undetermined significance (MGUS) patients in studies by Leleu *(1,3)* et al. They found the following:

1. sFLCs were higher in WM (36mg/L; range 16-140), compared with IgM MGUS (20mg/L; range 16-33): p<0.0003. For sFLC ratios, 76.5% of WM patients were abnormal compared with 23.5% of IgM MGUS: p<0.001.

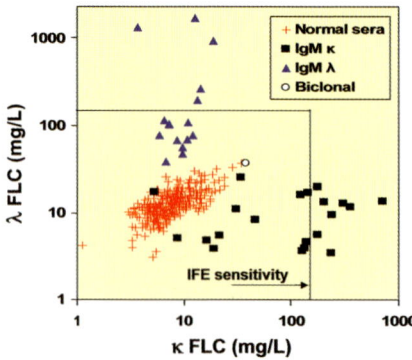

Figure 18.5. sFLC concentrations in normal sera and 37 patients with WM at the time of plasma exchange for hyperviscosity syndrome.

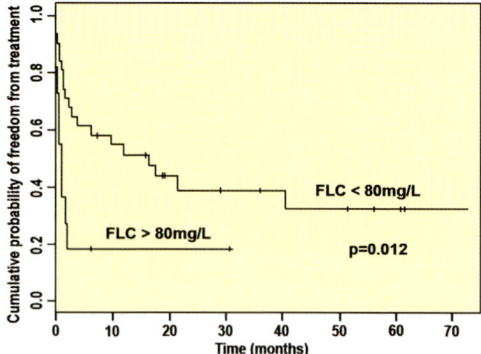

Figure 18.6. sFLC concentrations in WM at baseline compared with the time to treatment requirements by other clinical criteria. (Reproduced with permission from Haematologica *(Itzykson)* and R Itzykson).

2. sFLCs correlated with serum IgM (p<0.008) and viscosity (p<0.008) but not bone marrow involvement.

3. sFLCs were higher in symptomatic patients (p<0.001), and correlated with other poor prognostic markers of disease activity such as β_2-microglobulin, thrombocytopenia and leukopenia.

4. sFLCs >60mg/L separated WM from IgM MGUS with >95% specificity.

These authors also investigated the utility of sFLCs in monitoring patients *(Leleu 1)*. A prospective study of 32 patients with WM showed that using weekly sFLC measurements, response rates could be detected within a month, and earlier than using IgM measurements. They concluded that sFLCs were a sensitive and useful marker for WM management. Itzykson et al. studied 42 patients and showed that sFLCs >80mg/L were associated with a progressive disease and a shorter time to requirement for treatment *(Figure 18.6)*.

These studies now need to be reviewed in relationship to the current clinical response criteria of WM to determine whether improved patient management can be achieved.

In Waldenström's macroglobulinaemia, sFLCs may be helpful:

1. As prognostic markers.

2. To distinguish WM from IgM MGUS.

3. As an additional criteria for treatment responses or disease relapse.

18.6. B-cell, non-Hodgkin lymphomas

Non-Hodgkin lymphomas represent about 2.6% of all cancer deaths in the UK (approximately twice that of MM) and the incidence is rising by 3-4% per year in all age groups and both sexes. This is largely unexplained but immunosuppression is a well-defined causative factor, leading to a high excess risk. Approximately 80% of lymphoid malignancies are derived from B-lymphocytes at various stages of differentiation *(Table 18.1 and Figures 18.7 and 18.8)*.

Precursor B-lymphoblastic leukaemias/lymphomas	<1%
Chronic lymphocytic leukaemia (CLL)/B-cell small lymphocytic leukaemia	7%
B-cell prolymphocytic leukaemia	<1%
Lymphoblastic lymphoma	1%
Splenic marginal zone B-cell lymphoma	<1%
Hairy cell leukaemia	<1%
Plasma cell myeloma/plasmacytoma	<1%
Extranodal marginal zone B-cell lymphoma (MALT lymphoma)	8%
Nodal marginal zone B-cell lymphoma	2%
Follicular lymphoma	22%
Mantle cell lymphoma	6%
Diffuse large B-cell lymphomas (DLC)	33%
Burkitt's lymphoma/leukaemia	2%

Table 18.1. The REAL/WHO classification of B-cell, non-Hodgkin lymphomas and their frequency in relation to all non-Hodgkin lymphomas *(Evans)*.

Monoclonal immunoglobulins can be identified in the serum of 10-15% of patients using standard electrophoretic methods. The proteins may be IgG, IgA or IgM and are occasionally biclonal. Reports have indicated that monoclonal FLCs can be detected in the urine of 60-70% of patients with B-CLL if the urine is highly concentrated *(Deegan; Pezzoli; Pascali 1)*, but interpretation may be difficult if there is co-existing proteinuria.

In order to determine the frequency of abnormal sFLC concentrations in B-cell non-Hodgkin lymphomas, frozen sera from the Lymphoma SPORE serum bank at The Mayo

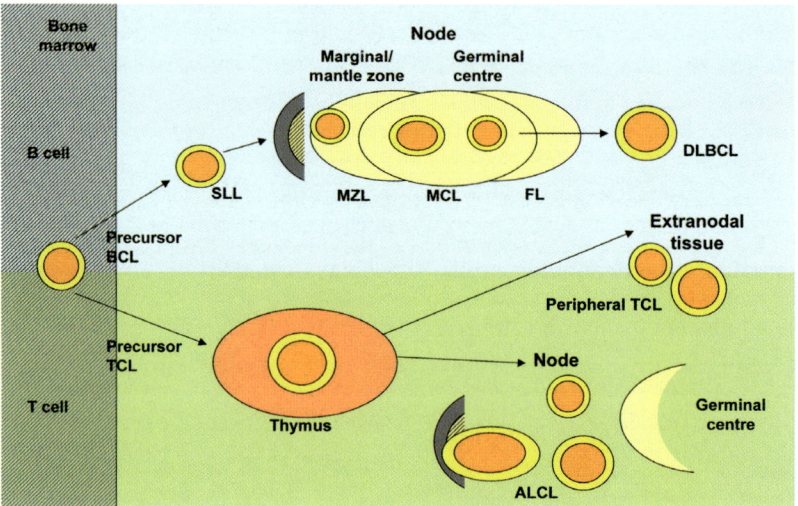

Figure 18.7. Origins of representative non-Hodgkin lymphomas *(Evans)*. ALCL: anaplastic large-cell lymphoma. BCL: B-cell lymphoma. DLBCL: diffuse large B-cell lymphoma. FL: follicular lymphoma. MCL: mantle-cell lymphoma (pre-germinal centre). MZL: marginal zone (MALT) lymphoma (post-germinal centre). SLL: small lymphocytic lymphoma. TCL: T-cell lymphoma.

Figure 18.8. Details of the origins of tumours of lymph node follicles in B-cell non-Hodgkin lymphoma. Most CLL originate from activated, antigen stimulated B-cells that have not undergone somatic hypermutation (unmutated). Some CLL may arise from a subset of mutated, memory B-cells that have transited through germinal centres. (Courtesy of J Hobbs).

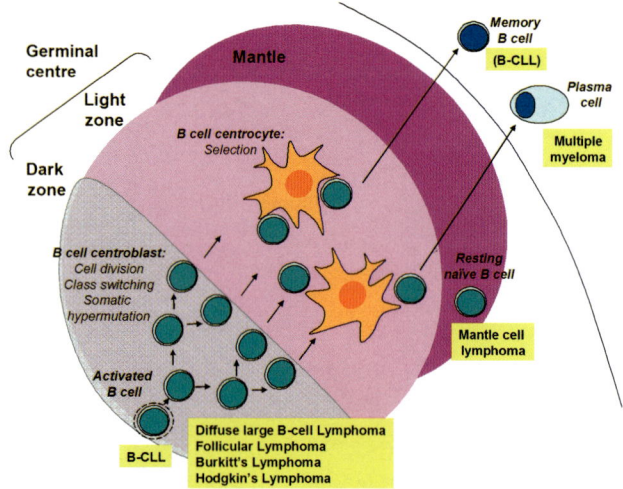

B-cell neoplasm	Number studied	FLC +ve	FLC +ve only	SPE/IFE +ve	SPE/IFE +ve only	Total +ve
Small lymphocytic	25	5 (20%)	3 (12%)	4 (16%)	2 (8%)	7 (28%)
Lymphoblastic	8	0	0	0	0	0
Lymphoplasmacytic	14	2 (14%)	0	4 (29%)	2 (14%)	4 (29%)
MALT lymphoma	19	3 (16%)	1 (5%)	7 (37%)	5 (26%)	8 (42%)
Follicular, stage I	25	0	0	4 (16%)	4 (16%)	4 (16%)
Follicular, stage II	25	2 (8%)	1 (4%)	5 (20%)	4 (16%)	6 (24%)
Follicular, stage III	25	1 (4%)	1 (4%)	3 (12%)	3 (12%)	4 (16%)
Mantle cell lymphoma	25	9 (36%)	5 (20%)	6 (24%)	2 (8%)	11(44%)
Diffuse large B-cell	25	2 (8%)	1 (4%)	2 (8%)	1 (4%)	3 (12%)
Burkitt's lymphoma	17	2 (12%)	1 (6%)	2 (12%)	1 (6%)	3 (18%)
Total	**208**	**26 (13%)**	**13 (6%)**	**37 (18%)**	**24 (12%)**	**50 (25%)**

Table 18.2. sFLC concentrations in B-cell non-Hodgkin lymphoma.

Clinic were studied by Martin et al. For comparison, samples were also tested for monoclonal immunoglobulins by SPE and IFE. Of 208 patients with non-Hodgkin lymphoma, a total of 13% (26/208) had abnormal sFLC concentrations (*Table 18.2 and Figures 18.9 and 18.10*). The highest incidences were in patients with B-cell small lymphocytic leukaemia (20%) and mantle cell lymphoma (36%). The concentrations of the FLCs were typically much lower than those found in patients with MM. Using SPE and IFE, 18% (37/208) of the patients had detectable monoclonal proteins. In 13 patients (6%), monoclonal proteins were detected only by sFLC immunoassays. The authors commented that these results highlighted the potential importance of sFLC tests in monitoring these patients and assessing complete responses to treatment.

Preliminary analysis of a larger serum bank of Italian patients (n=354) indicated a

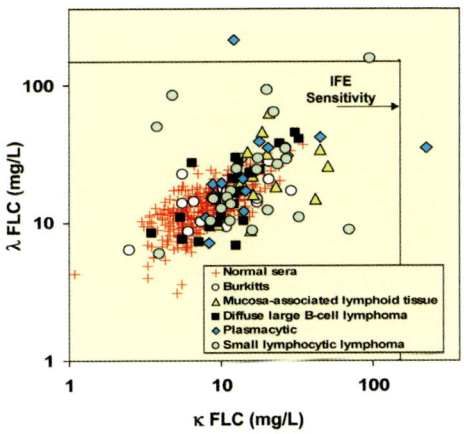

Figure 18.9. sFLC concentrations in non-Hodgkin B-cell lymphomas.

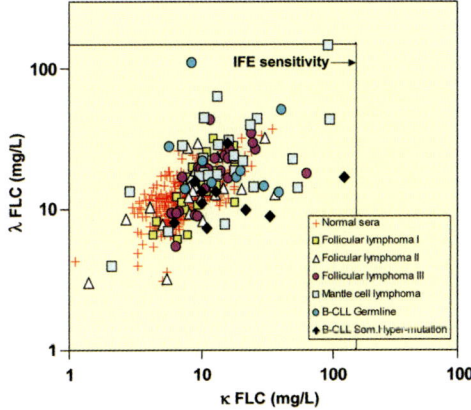

Figure 18.10. sFLC concentrations in non-Hodgkin lymphoma and B-CLL.

higher frequency of abnormal FLC ratios (e.g. 57% for mantle cell lymphoma, 33% for Burkitt's lymphoma and 25% for diffuse large B-cell lymphoma). Concentrations were also measured after treatment and were found to change in accordance with clinical assessments of response and relapse *(Pinto)*. A smaller study of patients from the UK (n=85) also reported a similar pattern of abnormalities *(Mead 5)*.

A more detailed study of FLC concentrations in 80 patients with diffuse large B-cell lymphoma, found that 29% had elevated (pre-treatment) concentrations of free kappa and 11% had abnormal FLC ratios. Elevated FLC concentrations were associated with worse event-free survival (log rank p=0.006), overall survival (log rank p=0.009) and concentrations were seen to decrease in response to treatment *(Maurer)*.

In a case series of 3 patients with primary effusion lymphoma (PEL) or PEL-like lymphoma it was reported that abnormal FLC concentrations changed in accordance with treatments and clinical assessments and the authors suggested FLC measurement could be useful for patient monitoring *(De Filippi 1)*.

HIV-infected patients are at increased risk of developing non-Hodgkin lymphoma (NHL) and Landgren *(2)* and colleagues compared FLC measurements in archived sera from patients who went on to develop NHL and others, who did not. The presence of an abnormal FLC ratio was not found to be prognostic but polyclonal FLC elevations were prognostic *(Chapter 21)*.

B-cell, non-Hodgkin lymphoma complicated by AL amyloidosis

Rarely, AL amyloidosis is associated with non-Hodgkin lymphoma. Six patients with this pattern of disease were studied by AD Cohen *(2)* et al. and comprised five patients with lymphoplasmacytic lymphoma and one with small lymphocytic lymphoma with plasmacytic features. Organ involvement with amyloid was characterised by bulky lymphadenopathy and visceral deposits but no cardiac disease. Measurements of sFLC concentrations showed elevations at diagnosis and responses to successful treatment. It was concluded that sFLCs were useful for monitoring these patients.

18.7. B-cell, chronic lymphocytic leukaemia

Several reports have suggested that a high percentage of patients with B-cell chronic lymphocytic leukaemia (B-CLL) may have urine monoclonal proteins *(Pezzoli; Pascali 2)*. These results are supported by the finding of raised sFLCs in many patients with B-CLL *(Martin)*. Of 18 sera studied, 6 patients (33%) had abnormal sFLCs alone *(Table 18.3)*. Using SPE and IFE, only 1 additional patient (6%) had an intact immunoglobulin monoclonal protein. Monoclonal proteins were more commonly found in patients with

B-cell CLL type	Number studied	FLC +ve	FLC +ve only	SPE/IFE +ve	SPE/IFE +ve only	Total +ve
Germline	9	4 (44%)	3 (33%)	2 (22%)	1 (11%)	5 (56%)
Somatic Hypermutation	9	3 (33%)	3 (33%)	0	0	3 (33%)
Total	18	7 (39%)	6 (33%)	2 (11%)	1 (6%)	8 (44%)

Table 18.3. sFLC concentrations in B-CLL.

germline B-CLL (56%) than those with somatic hypermutation (33%). As in patients with B-cell lymphomas, the concentrations of sFLCs were typically much lower than those found in MM (*Figure 18.10*).

A prospective study that screened 1,003 serum samples from symptomatic patients identified five new patients with B-CLL/ lymphoma *(Bakshi)*. This surprisingly high number of positive samples perhaps reflects the relative frequency of lymphomas compared with MM and has been observed in other studies *(Chapter 23)*. However, the concentrations of monoclonal FLCs were low, supporting the observations shown in Figures 18.9 and 18.10.

Pratt *(1)* et al. made a retrospective study of sFLC concentrations in samples collected at various time points in 226 CLL patients (183 Stage A, 18 Stage B, 16 Stage C, 9 unknown, mean age 74, male:female ratio 2.2) treated at 3 separate hospitals in the UK. sFLC concentrations were similar to those previously observed (*Figure 18.11*). Of greater interest was the observation that abnormal sFLC ratios were associated with poor outcome. Using Kaplan-Meier survival hazards, abnormal sFLC ratios were a significant indicator of poor survival (n=226, Log rank Mantel-Cox p=0.001) and also time to first treatment, particularly for patients who died from their disease (*Figure 18.12*). Using Cox regression analysis, in 142 patients with complete data sets, disease stage, CD38, Zap-70, *IGHV* mutation status, sFLC ratio, β_2-microglobulin and age were analysed in a forward stepwise analysis. 4 independent prognostic variables were identified: Zap-70 (p<0.001), β_2-microglobulin (p=0.002), IGHV mutation status (p=0.003) and sFLC ratio (p=0.009) (*Table 18.4,*).

Thus, abnormal sFLC ratios contributed significantly and independently to the prediction of a worse outcome. Further analysis of patients from the same cohort showed that those with abnormal sFLC ratios and sFLC concentrations >50mg/L had more progressive disease and a shorter time to treatment (median 83 months versus 241

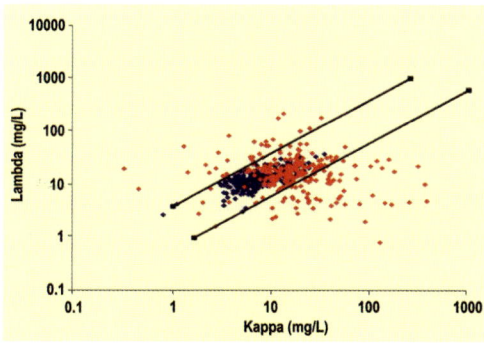

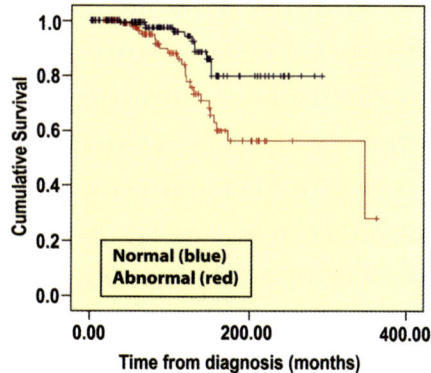

Figure 18.11. sFLC concentrations in 226 patients with B-cell chronic lymphocytic leukaemia. (Courtesy of G Pratt).

Figure 18.12. Kaplan Meier plot of cumulative survival for normal or abnormal sFLC ratios in 226 patients with B-CLL. (Courtesy of G Pratt).

	All CLL patients	Normal FLC	Abnormal Kappa	Abnormal Lambda	P value
No. of cases	259	159	66	34	N/A
Binet Stage A/B/C	209/23/21	135/10/10	50/9/5	24/4/6	N/A
Zap-70 (pos/neg)	89/146	46/97	35/24	18/14	0.034
CD38 (pos/neg)	83/164	46/105	22/41	15/18	0.259
Median TTFT (months)	84 (0-266)	117 (0-241)	52 (0-362)	33 (0-192)	0.001
Median OS (months)	209 (0-326)	254 (0-292)	201 (0-362)	124 (0-223)	0.002
Median Kappa (mg/L)	15.9 (0.32-382)	15(1.83-55.9)	39.05 (2.73-382)	6.52 (0.32-32.1)	<0.001
Median Lambda (mg/L)	16.45 (0.82-216)	17.3 (2.8-57.1)	7.53 (0.82-29.6)	38.4 (8.13-216)	<0.001
Median β_2M (mg/L)	3.4 (0.13-72.4)	2.95 (1.4-12.5)	4.35 (0.13-17.7)	4.53 (1.74-72.4)	<0.001

Table 18.4. Relationship of disease markers at presentation to outcome in patients with B-CLL. TTFT: time to first treatment. OS: Overall survival. (Courtesy of G Pratt).

months). As an indicator of time to first treatment, sFLC>50mg/L was independent of stage, Zap-70 and mutational status.

Results supporting the association of an abnormal sFLC ratio with a worse prognosis have also been reported by Ruchlemer et al., Yegin et al. and Matschke et al. Shustik et al. found less significant associations with outcome but the patients in that study had more advanced disease. It appears probable that sFLC measurements have more prognostic value in B-CLL patients with Rai stage I / Binet stage A disease. While many other prognostic factors have been identified in CLL, the ready availability and relatively low cost of sFLC measurement makes it an attractive option. Further studies to investigate the biological rationale for its prognostic associations would be valuable.

In a similar study to that already completed, looking for preceding MGUS in subjects who subsequently developed myeloma *(Landgren)*, archived sera from subjects who subsequently developed B-CLL were examined *(Tsai)*. Elevated sFLC levels and abnormal sFLC ratios were observed many years prior to the diagnosis of CLL and the authors suggested that chronic immune stimulation might play a role in CLL pathogenesis.

In B-cell, non-Hodgkin lymphoma and B-CLL:
1. Abnormal sFLC concentrations can be detected in a substantial fraction of patients
2. sFLC analysis identifies additional patients to those detected by SPE and IFE
3. Studies indicate sFLCs are markers of B-CLL disease activity and are independent prognostic factors for response and survival

18.8. POEMS syndrome
POEMS syndrome is an acronym for a rare paraneoplastic syndrome that includes Polyneuropathy, Organomegaly, Endocrinopathy, Monoclonal gammopathy, and Skin changes, among other manifestations. The disease is usually monoclonal λ restricted.

Drengler et al. studied 50 patients with newly diagnosed POEMS syndrome using sFLC analysis to determine its role in the disease management. 45 patients (90%) had an elevated λ sFLC but only 9 had abnormal sFLC ratios. The elevated sFLCs with normal ratios were due to a degree of renal impairment and/or polyclonal activation of the bone marrow but the underlying mechanisms causing these abnormalities were not apparent.

18.9. Cryoglobulinaemia

It is likely that elevated FLC concentrations are present in many patients with monoclonal cryoglobulinaemia and they should provide a useful tool because of the technical difficulties in measuring cryoglobulins. To date, there has been one significant report in patients with hepatitis C virus related lymphoproliferative disorders *(Terrier)*. This showed that abnormal sFLC ratios were related to both mixed cryoglobulin concentrations and lymphoma development *(Chapter 21)*.

Test questions
1. What is the frequency of abnormal sFLCs in solitary plasmacytomas of bone?
2. Are sFLC measurements helpful in patients with WM?
3. Are patients with B-CLL likely to be detected when screening for monoclonal gammopathies using sFLC assays?

Answers
1. About 80% of patients (Section 18.1, page 154).
2. Yes. They help distinguish WM from IgM MGUS and high levels predict shorter event free survival (Section 18.5, page 157).
3. Yes, quite frequently, because B-CLL is more common than WM (Section 18.7, page 162).

Chapter

19

Monoclonal gammopathies of undetermined significance (MGUS)

19.1. MGUS: Definition and frequency   *166*
19.2. MGUS and monoclonal FLCs   *167*
19.3. Risk stratification of MGUS using sFLC concentrations   *168*
19.4. Serum free light chain MGUS   *170*
19.5. MGUS as the precursor condition for MM   *173*

Summary: Elevated monoclonal serum free light chains (sFLCs):

1. Are found in 0.5-1% of elderly individuals and 30-60% of monoclonal gammopathy of undetermined significance (MGUS) individuals.
2. May identify individuals with MGUS missed by immunofixation electrophoresis (IFE).
3. Indicate poor outcome in MGUS.
4. Have been incorporated into a risk-stratification model for progression in association with the monoclonal immunoglobulin class and concentration.
5. FLC-only MGUS may progress to light chain only plasma cell dyscrasias.

19.1. MGUS: Definition and frequency

MGUS originally denoted the unexpected presence of an intact immunoglobulin monoclonal protein in individuals who have no evidence of MM, AL amyloidosis, Waldenström's macroglobulinaemia (WM), lymphoproliferative disorders, plasmacytoma or related conditions.

MGUS is defined as follows:

1. Monoclonal (M)-protein in serum <30 g/L
2. Bone marrow clonal plasma cells <10% and low level of plasma cell infiltration in a trephine biopsy (if done)
3. No evidence of other B-cell proliferative disorders
4. No related organ or tissue impairment

MGUS may be found in 1% of the population over 50 years, 3% over 70 years and up to 10% over 80 years of age *(Kyle 9,11,13)*, is 2-fold higher in African-Americans *(Landgren 1)*, and is associated with inflammatory and infectious disorders *(Brown)*. Because of the frequency of MGUS, between 50-65% of all monoclonal proteins detected fall into this category and vast numbers go undetected. In a Mayo Clinic study, 73% were IgG, 14% IgM, 11% IgA and 2% biclonal *(Kyle 11)*. Rarely, high concentrations of FLC MGUS are found in the urine *(Kyle 8)*.

19.2. MGUS and monoclonal FLCs

Although most people with MGUS die from unrelated illnesses, MGUS may transform into malignant monoclonal gammopathies. Patients should, therefore, be monitored on a regular basis to identify early signs of progression *(for recent guidelines see Chapter 25)*. In order to minimise therapeutic harm, treatment is given only when disease develops. During this monitoring phase, symptoms, signs and markers of malignancy are carefully observed, paying particular attention to serum and urine M-proteins.

In a long-term study of outcome in MGUS patients at the Mayo Clinic, 1,384 patients with MGUS have been continually monitored *(Kyle 13)*. Since enrolment between the years 1960 and 1994, 115 had progressed, a rate of approximately 1% per year. The most important prognostic factor for progression was the initial size of the serum monoclonal spike. Immunoglobulin class was also important; individuals with IgM and IgA, but not IgG, monoclonal proteins were 5 times more likely to progress. In a recent study by Kyle *(14)* et al. looking specifically at 213 patients with IgM MGUS, there was a very high relative risk of progression to WM (262 fold) or lymphoma (15 fold). Neither study showed any increased relative risk associated with the various immunoglobulin subclasses or urine FLC excretion.

In contrast, other studies have indicated that urine FLC excretion may be an important prognostic marker *(Dimopoulos 1; Baldini)*. In an Italian study of 1,231 patients, Bence Jones proteinuria was an independent risk factor for malignant transformation *(Cesana)*.

Since the amounts of FLC in the urine are restricted by renal catabolism, serum concentrations might be a more reliable predictor of disease progression. Initial studies indicated that FLC concentrations are raised in the serum of many patients with MGUS *(Marien)*. Examples from 31 patients are shown in Figure 19.1. 50% of the sera contained monoclonal FLCs as indicated by abnormal κ/λ ratios. Several others had raised concentrations of both FLCs because of renal impairment *(Chapter 20)*.

In a study by Tate *(1)* et al. sFLC concentrations and/or κ/λ ratios were abnormal in 26 of 32 MGUS patients. Serum intact immunoglobulin M-protein concentrations ranged from 1.0 to 22g/L. 9 patients had abnormal serum κ/λ ratios but a further 5 with

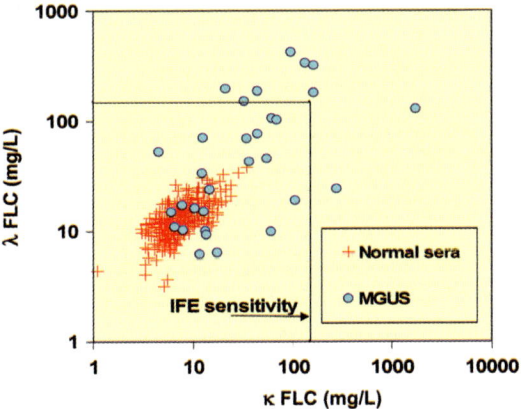

Figure 19.1. sFLCs in patients with MGUS. (Courtesy of H Lachmann).

normal sFLCs had small amounts of urine monoclonal FLCs identified by electrophoretic tests. Presumably, the levels of monoclonal FLCs were insufficient to cause serum abnormalities but accompanying renal leakage allowed detectable amounts to enter the urine. In contrast, when patients with MM have good renal function, urine tests may be negative for monoclonal FLCs, while serum monoclonal FLC concentrations are relatively high. The same applies to patients with MGUS. It should be born in mind that in patients with MM, sFLC abnormalities rather than urine FLCs are a more reliable measure of outcome (*Chapters 8 and 11*) and presumably the same applies to FLCs in MGUS. In addition, the urine IFE tests may be falsely identifying intact monoclonal immunoglobulins or ladder-banding as monoclonal FLCs *(Chapter 6.6)*.

19.3. Risk stratification of MGUS using sFLC concentrations

Rajkumar *(1,5)* et al. reported a large series of MGUS patients (1,148) in 2005. Results showed that the risk of progression in patients with abnormal sFLC κ/λ ratios was significantly higher (hazard ratio 2.6) than in patients with normal ratios and was independent of the quantity and type of MGUS *(Figure 19.2)*. Furthermore, the risk of progression increased as κ/λ ratios became more extreme *(Figure 19.3)*.

The data was used to produce a risk-stratification model based upon immunoglobulin MGUS class, its quantity above or below 15g/L and the presence or absence of an abnormal FLC κ/λ ratio *(Table 19.1 and Figure 19.4)*. The risk of progression, with time, after MGUS identification is shown in Figure 19.4. Another smaller study has reported similar results *(Giarin)*. Indeed, these observations were included in the 2010 IMWG MGUS guidelines *(Kyle 16)*. The recommendations indicate that patients with MGUS should be risk stratified and their follow up is to be determined by the 'Risk Stratification Model', outlined in Table 25.5. The 2009 IMWG guidelines on the use of sFLC *(Dispenzieri 7)* recommend sFLC determination at baseline for MGUS prognosis *(Chapter 25)*.

Figure 19.2. Risk of progression based on the presence or absence of an abnormal FLC κ/λ ratio. (This research was originally published in Blood *(Rajkumar 5)* © the American Society of Hematology).

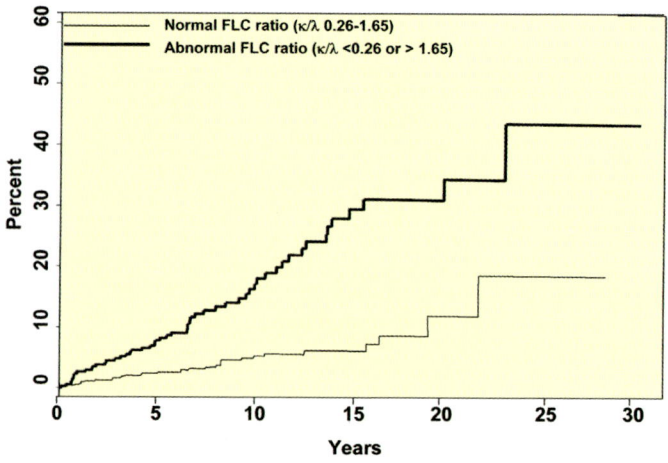

Figure 19.3. Effect of increasingly abnormal FLC κ/λ ratio on the relative risk of progression of MGUS. (This research was originally published in Blood *(Rajkumar 5)* © the American Society of Hematology).

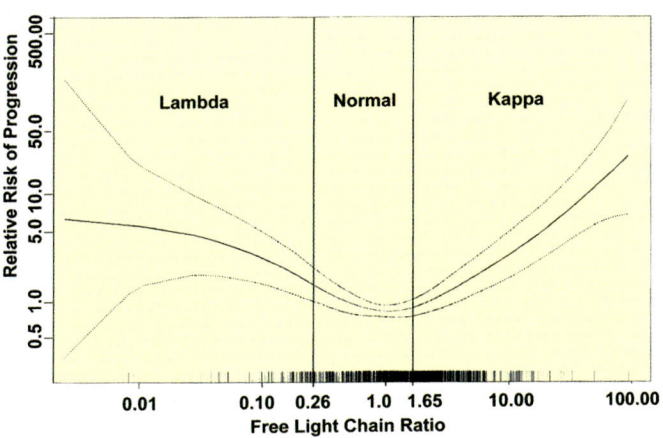

Figure 19.4. Risk of progression to myeloma or related condition in 1148 patients with MGUS. (This research was originally published in Blood *(Rajkumar 5)* © the American Society of Hematology).

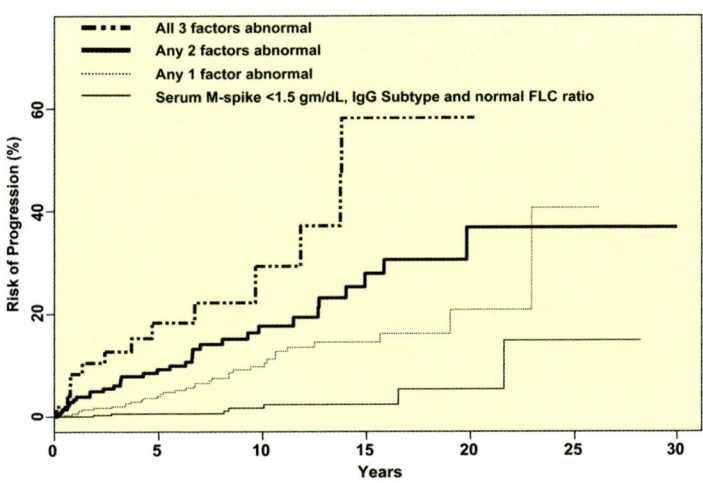

Table 19.1. Risk stratification model to predict progression of MGUS.

Risk of Progression	No. of abnormal risk factors	No. patients	Absolute risk of progression at 20 years*
Low	0	449	2%
Low intermediate	1	420	10%
High intermediate	2	226	18%
High	3	53	27%

* Accounting for death as a competing risk

The explanation for the increased risk from abnormalities of sFLC ratios may relate to the clonal evolution of the plasma cells. Genetic and molecular events involved in the transformation of MGUS to MM presumably lead to disordered heavy and light chain immunoglobulin synthesis and abnormal monoclonal FLC production *(Cavallo)*.

These important data have been widely reported and have led to considerable debate regarding the management guidelines for MGUS *(Bladé 2; Jagannath; Rajkumar 6)*. Typical historical practice has been to monitor all individuals on an annual basis in order to anticipate and prevent debilitating disease progression. The more recent consensus statement on MGUS *(Berenson)* states that all patients should be followed initially at 3-6 months, however, the frequency over time may vary depending upon the size of the M-protein and other risk factors for progression *(Chapter 25, Table 25.5)*. Low risk patients (~40%) could be reassured about their test results and followed up on a more infrequent basis (every 2-3 years) or when they attend for other illnesses. In contrast those at high risk might enter drug trials to prevent disease progression.

The UK Myeloma Forum and Nordic Myeloma Study Group published guidelines in 2009 *(Bird 2)* on the investigation of newly detected M-proteins and management of MGUS that are further detailed in Chapter 25.

In spite of these significant predictive factors for progression, there is insufficient data for preventative treatment of high-risk patients. In order to increase predictive value further studies have been performed. Rowstron et al. recently reported, in a small study, the addition of plasma cell phenotype (CD138/38/45 expression) to sFLCs and noted they were independent and complementary. Results of a larger study are awaited.

19.4. Serum free light chain MGUS

Monoclonal proteins in MGUS were historically considered to be intact immunoglobulins, as these patients progress to intact immunoglobulin plasma cell dyscrasias. So, what is the precursor protein for light chain multiple myeloma (LCMM) and AL amyloidosis? Occasional reports have described individuals with *"idiopathic"* Bence Jones proteinuria who progress to MM *(Kyle 8)*. However, it is probable that FLC MGUS exists as serum FLC κ/λ ratio abnormalities that are undetected by serum and urine electrophoretic tests. An example of such an individual is shown in Figure 19.5. Isolated, minimal urine FLC excretion is usually considered insignificant, but analysis of the corresponding serum indicated an abnormality of the κ/λ ratio in this patient.

The existence of FLC MGUS is also apparent from screening studies for monoclonal gammopathies that have incorporated serum FLC measurements. Some individuals have grossly abnormal serum FLC κ/λ ratios but completely normal serum and urine electrophoretic tests *(Chapter 23)*.

The frequency of serum FLC MGUS in a general population was initially addressed in a pilot survey at the Mayo Clinic *(Katzmann 7)*. 901 sera from the Olmsted County MGUS epidemiological study were investigated. All sera selected were defined as negative for serum or urine monoclonal immunoglobulins by IFE. However, 18 of the samples were abnormal when assessed for serum FLCs. 12 of the sera, with the most abnormal FLC κ/λ ratios (<0.2 or >2.0) were carefully re-assessed by IFE.

Figure 19.5. Serum and urine IFE on a patient with isolated urine FLC excretion. Serum analysis for FLCs showed an abnormal κ/λ ratio. TP: urine protein excretion. (Courtesy of RA Kyle and JA Katzmann).

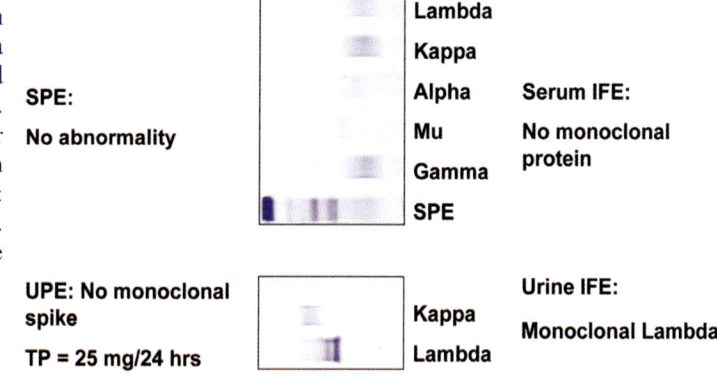

SPE: No abnormality

Lambda
Kappa
Alpha   Serum IFE:
Mu   No monoclonal
Gamma   protein
SPE

UPE: No monoclonal spike

TP = 25 mg/24 hrs

Kappa   Urine IFE:
Lambda   Monoclonal Lambda

Serum κ/λ ratio = 0.24* (ref range: 0.26-1.65)

One sample had a small IgA band and another a small IgM band hidden in the β region of the gel; 4 had monoclonal λ FLC bands; 3 were equivocal for monoclonal FLCs and 3 were negative. Thus, a total of 7 samples (0.78%) probably had only monoclonal FLCs. These FLC-only MGUS might be the *"missing"* individuals with preclinical LCMM or AL amyloidosis. Interestingly, this 0.78% is approximately 20% of the total MGUS incidence in such an age group - a similar percentage to the number of MM patients that are FLC-only (LCMM). This further supports the hypothesis that FLC MGUS are preclinical FLC plasma cell dyscrasias. Of importance is the observation that the FLC assays identified 2 patients with intact immunoglobulin monoclonal proteins that had been missed in the initial IFE tests. This is an additional reason to use FLC assays in a screening mode.

This study has been extended to a much larger proportion of the Olmsted County MGUS epidemiology survey *(Rajkumar 4),* originally consisting of 21,463 residents. More recent analysis of the data led to the proposal for a light chain equivalent of conventional MGUS as a new clinical entity *(Dispenzieri 8).* The authors defined this as an abnormal sFLC ratio with an increased concentration of the involved light chain and no expression of the intact immunoglobulin.

The study included 18,357 individuals, 610 of these had an abnormal ratio; 213 of these had an intact immunoglobulin MGUS. Of these, 57 had not been detected with SPE as the screening test and so the prevalence of conventional MGUS in this population was revised from 3.2% to 3.4% (95% CI 3.2-3.7). Of the remaining 397 individuals, 146 had an increase of at least one FLC, resulting in a calculated prevalence of light chain MGUS of 0.8% (95% CI 0.7-0.9) and led to an overall MGUS prevalence of 4.25% (95% CI 3.9-4.5). Of the light chain MGUS individuals 108 were κ and 38 λ. Figure 19.6 illustrates the observed ranges of FLC levels in FLC MGUS. Both κ and λ FLCs are raised in renal diseases *(Chapter 20)* and so the authors examined the application of the FLC renal reference ranges *(Figure 19.6).* This led to the exclusion of 69 individuals with apparent κ light chain MGUS and 57 with λ. This included one patient who subsequently progressed to IgG λ MM. An association of renal diagnosis with FLC MGUS was noted, however, as medical information was not available for all patients the prevalence data was not calculated.

Interestingly, the authors showed a 0.3% (95% CI 0.1-0.8) risk per 100 person years of progression for light chain MGUS to MM and related disorders. This is lower than the overall progression rate of MGUS (1.0%), although this did not differ from patients with low-risk conventional MGUS. The authors speculate that the transformation events resulting in either FLC MGUS or conventional MGUS may be the same and their presence in combination may indicate a step closer to full transformation to MM or a related disorder.

Further evidence of FLC MGUS may come from studies of chronic inflammatory conditions *(Brown)* and chronic renal failure populations *(Chapter 20)*. Notably, 23% of FLC MGUS individuals identified by Dispenzieri *(8)* et al. had an incidental renal diagnosis either at the time of sample acquisition or subsequently. Hutchison *(3,4)* et al. showed a high prevalence of monoclonal FLCs in patients with severe renal failure and speculated that there might be causal link between more rapid deterioration of renal function and FLC toxicity.

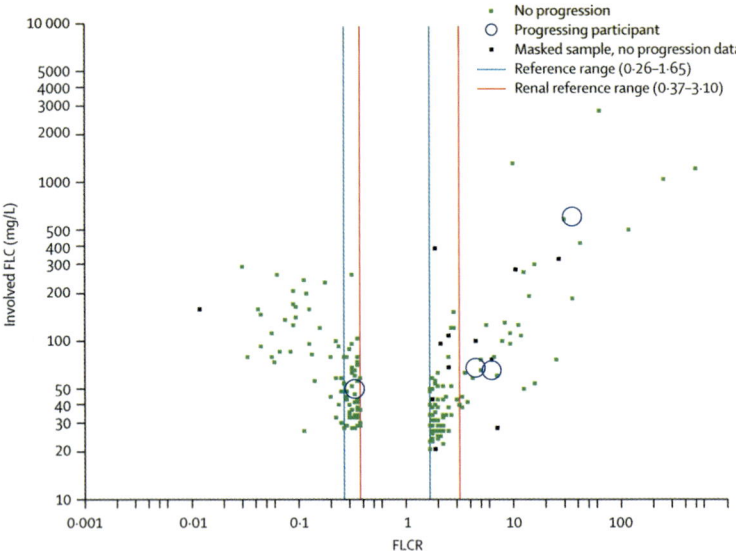

Figure 19.6. Light chain MGUS as defined by the normal sFLC ratio reference range (0.26-1.65) compared with the renal reference range ratio (0.37-3.10). (Reproduced with permission from the Lancet *(Dispenzieri 8)*).

19.5 MGUS as the precursor condition for MM

MGUS patients have an increased risk of developing MM and related disorders. Patients may be risk stratified based on the number of risk factors present *(Table 25.5)*. Until recently there has been little evidence supporting the supposition that MM always arises from an MGUS disorder or alternatively that MM more typically arises *de novo*. Such available evidence would provide important information for the study of MGUS and in developing new approaches to management.

Recently, Landgren *(3)* et al. examined the records from 77,469 healthy adults enrolled in the US PLCO (Prostate, Lung, Colorectal, and Ovarian) cancer screening trial to identify individuals who developed MM. Serum samples were available from 2 to 10 years prior to MM diagnosis for 71 individuals. Analysis by SPE, IFE and sFLC assay was carried out to define the prevalence of MGUS prior to MM diagnosis (MGUS was defined as having an M-protein visible by SPE or IFE, an abnormal sFLC ratio or both). The prevalence of MGUS was 100.0%, 98.3%, 97.9%, 94.6%, 100%, 93.3% and 82.4% at 2, 3, 4, 5, 6, 7, and 8+ years prior to MM diagnosis *(Table 19.2)*.

In a similar, but smaller study Weiss et al. demonstrated 27/30 MM diagnosis were preceded by MGUS in a group of autologous stem cell transplant patients. Monoclonal proteins were identified by SPE/IFE in 21 individuals (77.8%) and by sFLC ratio alone in 6 (22%). Of the 3 individuals where no pre-existing plasma cell disorder was shown, one had a single prediagnostic sample from 9.5 years prior to diagnosis and the other 2 were IgD MM patients with available samples 3.5 and 5 years prior to MM diagnosis. The true incidence of MGUS prior to MM diagnosis may therefore have been higher than the 27/30 indicated.

Blood draw prior to MM diagnosis	M-Spike		Abnormal κ/λ FLC Ratio		MGUS	
By year	n	%	n	%	n	%
2	25/27	92.6	23/27	85.2	27/27	100.0
3	54/58	93.1	46/58	79.3	57/58	98.3
4	45/48	93.8	29/46	63.0	47/48	97.9
5	34/37	91.9	25/37	67.6	35/37	94.6
6	25/25	100.0	19/25	76.0	25/25	100.0
7	14/15	93.3	11/15	73.3	14/15	93.3
8 or more	13/17	76.5	8/17	47.1	14/17	82.4

Table 19.2. Prevalence of serum protein abnormalities prior to MM diagnosis. (Adapted from *Landgren 3*). This data is based on the actual available serum sample at a given time point as opposed to inferred data from earlier time points. M-Spike is defined as a monoclonal immunoglobulin being detected by SPE, IFE, or both. The normal sFLC ratio reference range (0.26-1.65) was used. For this study MGUS was defined as evidence of an M-spike (as defined above), an abnormal sFLC ratio, or both.

Test Questions

1. How should patients with a sFLC MGUS be managed?
2. What clinical decisions should be made about MGUS patients who have associated monoclonal sFLCs?
3. Why are FLC MGUS rarely seen with serum protein electrophoresis (SPE) and IFE testing of serum and urine?
4. What are the clinical benefits of MGUS risk stratification?

Answers

1. *Unknown at present, but annual follow-up is prudent to determine if concentrations are rising (Section 19.3, page 168).*
2. *The risk of progression should be assessed and long-term monitoring discussed with the patient (Section 19.3, page 168).*
3. *Because the amounts of FLCs produced are below the sensitivity of electrophoretic tests and are insufficient to exceed the proximal tubular reabsorption mechanisms and overflow into urine.*
4. *Patients at high risk of progression can be more closely monitored while low-risk patients can be discharged from follow-up clinics (Section 19.3, page 168).*

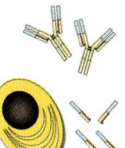

SECTION 3 - Diseases with increased polyclonal free light chains

Chapter 20. *Renal diseases and free light chains*   *175*
Chapter 21. *Immune stimulation and elevated polyclonal free light chains*   *184*
Chapter 22. *Cerebrospinal fluid and free light chains*   *194*

Chapter	Section 3 - Diseases with increased polyclonal free light chains
20	**Renal diseases and free light chains**

AR Bradwell, P Cockwell and CA Hutchison

20.1. *Introduction*   *175*
20.2. *Effect of renal impairment on sFLC concentrations*   *176*
20.3. *Effect of renal impairment on serum κ/λ ratios*   *178*
20.4. *Removal of FLCs by dialysis in chronic renal failure (CRF)*   *179*
20.5. *Monoclonal FLCs in CRF*   *181*
          Clinical case history No 10   *182*

Summary: In patients with renal impairment:

1. Polyclonal sFLC concentrations rise as glomerular filtration rate falls.
2. sFLC concentrations may increase 30-40 fold.
3. κ/λ ratios increase slightly with decreasing renal function as κ filtration slows.
4. Monoclonal FLCs are frequently found in patients with CRF.
5. sFLCs can be removed by haemodialysis.

20.1. Introduction

Elevated polyclonal FLCs in serum (associated with normal κ/λ ratios) result from increased polyclonal production, reduced renal clearance or a combination of both mechanisms. Increased production is due to proliferation of plasma cells and/or their progenitors and is frequently associated with polyclonal hypergammaglobulinaemia. This is a common finding in patients with liver diseases, connective tissue disorders, chronic infections, etc. A complete list of the relevant diseases is given in Chapter 29.

Reduced clearance of sFLCs results from impaired renal glomerular filtration rate (GFR) and is a frequent finding. This can be seen even in apparently healthy, elderly individuals who may have normal serum creatinine concentrations but elevated polyclonal sFLCs from slight reduction in GFR.

Frequently, acute and chronic inflammatory diseases are associated with some degree of renal impairment. The combination of increased production and reduced renal clearance leads to particularly high sFLC concentrations, classically observed in patients with systemic lupus erythematosus (SLE). Although not documented, it is likely that most patients with chronic inflammatory diseases and associated renal impairment have high concentrations of polyclonal sFLCs, but with substantially normal κ/λ ratios.

20.2. Effect of renal impairment on sFLC concentrations

The normally rapid renal clearance of sFLCs of 2-6 hours is increased to 2-3 days in complete renal failure *(Chapter 3) (Abraham GN)*. Removal from the circulation then occurs through pinocytosis in all tissues but particularly by capillary endothelial cells *(Chapters 10 and 32)*. As GFR falls, sFLC concentrations rise and may be 20-30 times normal in end-stage renal failure *(Figures 20.1 and 20.2) (Hutchison 4 ,7; Scherberich)*.

Figure 20.1. Elevated sFLC concentrations in 107 patients with varying degrees of CRF. Numbers 1-8 refer to patients with monoclonal gammopathies shown in Table 20.3. Line A is at a κ/λ ratio of 0.6 and line B is the mean κ/λ ratio in patients with renal failure.

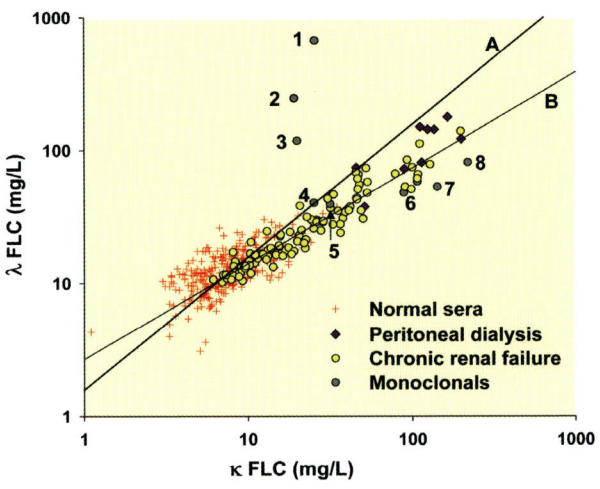

Figure 20.2. Serum κ (white) and λ (grey) FLC concentrations in CKD stages 1 – 5, in 688 patients attending a renal disease clinic plus patients on peritoneal dialysis (PD), haemodialysis (HD) and controls (Con). Data presented as box plots (1st – 3rd inter-quartile ranges, central line is median value) with whiskers (5th – 95th percentile values). Courtesy of Colin Hutchison.

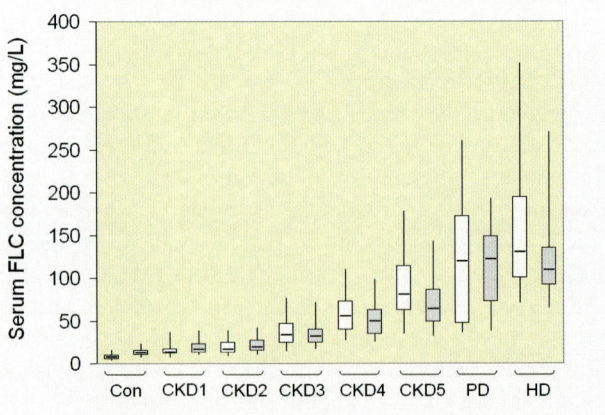

The relationship between sFLC concentrations and GFR is shown in Table 20.1. The MDRD index (Modification of Diet in Renal Disease) of GFR produces a worse correlation with FLC levels than serum creatinine levels but better than the Cockcroft-Gault formula *(Levey; Cockcroft)*. Cystatin C concentrations have the highest correlation with FLC concentrations and arguably are the most accurate simple measure of GFR. The rising concentrations of sFLC with progressive chronic kidney disease (CKD) stage (from MDRD) is shown in Figure 20.2.

	Kappa	**Lambda**
Serum creatinine	0.697	0.701
Cockcroft-Gault formula	0.521	0.490
MDRD	0.631	0.613
Cystatin C	0.778	0.727

Table 20.1. Relationship between sFLC concentrations and different markers of GFR in 107 patients with CRF (correlation coefficient - R^2). MDRD: Modification of Diet in Renal Disease.

Preliminary analysis of follow-up data from a prospective cohort of 1,394 patients with CKD indicated that the summation of free kappa plus free lambda concentrations was prognostic for all-cause mortality and change in GFR *(Figure 20.3 and 20.4) (Hutchison (7) et al.* – presented at the Renal Association meeting, 2010). In univariate analysis, the combined FLC concentrations had greater prognostic value than the creatinine-based CKD staging system. The principal causes of death associated with high FLC concentrations were cardiovascular disease, infections and cancer.

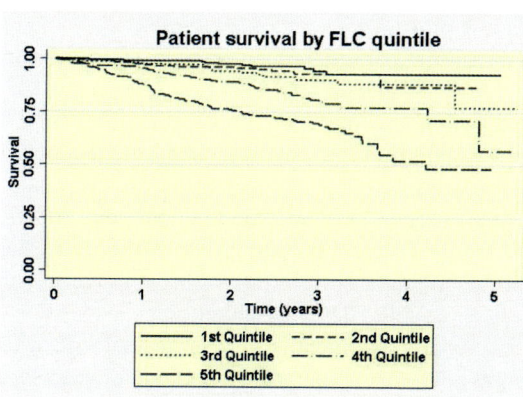

Figure 20.3. Relationship of patient survival with serum concentration of polyclonal FLCs (P<0.001). (Courtesy of Colin Hutchison).

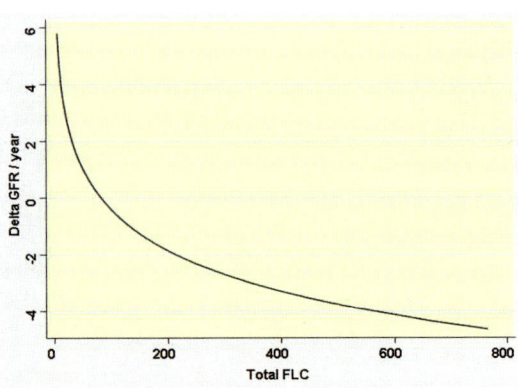

Figure 20.4. Relationship of loss of renal function with sFLC concentrations (Estimated GFR was based on the MDRD equation). (Courtesy of Colin Hutchison).

20.3. Effect of renal impairment on serum κ/λ ratios

As both serum κ and λ FLC concentrations increase with deteriorating renal function their relative amounts change slightly. While glomeruli clear monomeric κ molecules approximately 3 times faster than dimeric λ molecules *(Chapter 3)*, pinocytosis removes both proteins at the same rate. With deteriorating renal function κ/λ ratios gradually increase and eventually equal the κ:λ production rates of approximately 1.8:1 in end-stage renal failure *(Hutchison 4, 7; Scherberich; Wells)*.

The effect of renal impairment on κ/λ ratios is seen in several patient groups:

1. κ and λ concentrations increase with age in apparently normal individuals because of minor degrees of renal impairment *(Figure 5.1)*. This is associated with an increase in the median κ/λ ratio from 0.49 to 0.70 in older people *(Table 5.3)*. This is shown as a deviation in the κ/λ ratios from 0.6 on a κ/λ plot *(Figure 5.2)*.

2. Patients in CRF, but without FLC monoclonal gammopathies, have higher κ/λ ratios than normal individuals *(Figure 20.5)*.

3. Patients with AL amyloidosis caused by λ FLC monoclonal gammopathies tend to have κ/λ ratios closer to the normal range population than κ-producing patients *(Figure 20.6)*. There is also asymmetry of κ compared with λ FLC concentrations in patients with light chain multiple myeloma (LCMM) *(Figure 8.2)* and monoclonal gammopathy of undetermined significance (MGUS) *(Figure 19.4)*. The patients with these diseases frequently have some degree of renal failure so that κ concentrations are relatively high. Individuals with normal renal function clear κ molecules more quickly.

The change in κ/λ ratios with increasing renal impairment is of clinical relevance when interpreting borderline results. Patients might be misclassified as having minor κ monoclonal gammopathies because of slightly increased κ/λ ratios when, in fact, they have renal impairment. Equally, patients with minor λ monoclonal gammopathies could have normal κ/λ ratios because of the relative increase in κ concentrations *(Figure 20.6)*. A higher cut-off should be used when assessing patients with acute renal failure *(Figure 13.7)*.

Figure 20.5. Increase in sFLC κ/λ ratio with increasing CKD stage plus patients on peritoneal dialysis (PD) and haemodialysis (HD) and controls (Con). Same patient data as used in Figure 20.2. Data presented as box plots (1st – 3rd inter-quartile ranges, central line is median value) with whiskers (5th – 95th percentile values). Courtesy of Colin Hutchison.

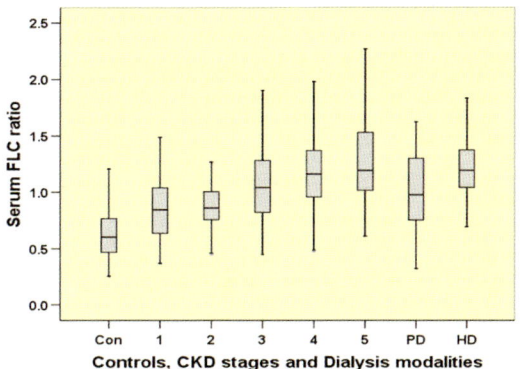

Figure 20.6. sFLCs in blood donors (red) and patients with AL amyloidosis (yellow), many of whom have renal impairment. (The λ–producing patients tend to cluster against the normal samples because of increased κ levels from reduced renal clearances. The arrows A and B show potential changes in FLC concentrations with recovery of normal renal function).

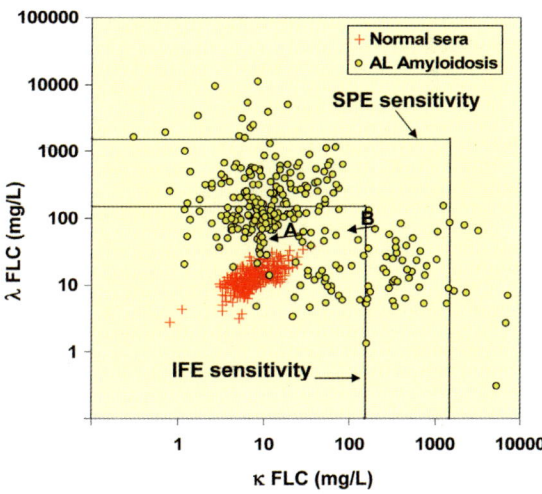

20.4. Removal of FLCs by dialysis in chronic renal failure (CRF)

The pore size of haemodialysis membranes (dialysers) allows removal of small to medium-sized protein molecules from the blood. Membranes typically have a molecular weight cut-off of 10-15kDa so the filtration efficiency for FLCs is very low. However, some dialysers are more permeable and others adsorb FLCs on to their surfaces. Figure 20.7 shows the reduction in sFLC concentrations in 40 patients with CRF over a 4-hour dialysis period using high-flux polysulphonate membranes. 60% of κ and 37% of dimeric λ molecules were removed and median κ/λ ratios changed from 1.23 to 0.7. The quantities of FLCs in the filtrate fluid accounted for approximately 60% of the observed reductions in serum concentrations while the remainder was adsorbed onto the filtration membranes *(Cohen 2)*. Multiple sampling of the dialysate fluid from a patient during the 4-hour haemodialysis period showed slowly reducing FLC concentrations as serum levels fell *(Figure 20.8)*. Dialysate fluid κ/λ ratios changed from 1.6 to 1.4 during the dialysis period as greater amounts of the smaller monomeric κ molecules were preferentially cleared. Some dialysers are more efficient at FLC removal because of larger pore sizes and different suface charges. However, the amounts present in CRF are hundreds of times lower than those found in LCMM. In the latter patients *"high cut-off"* dialysers must be used for effective FLC removal (*Chapter 13*). Whether such large pore dialysers are clinically useful in preventing dialysis associated diseases in CRF is currently being assessed.

Peritoneal dialysis may be less efficient at removing FLCs *(Figure 20.1)*. The fluid volumes exchanged during this process are much less than in haemodialysis, which presumably accounts for the poor removal.

The clinical consequences of elevated sFLCs in renal impairment are unclear. Reports have suggested that the elevated FLCs lead to reductions in immune function and should therefore be classified as uraemic toxins *(Cohen 1)*. Indeed, studies have shown that FLCs are inflammatory proteins *(van der Heijden)*. It is thought that their toxicity may partly account for the relatively poor survival of many patients undergoing chronic haemodialysis.

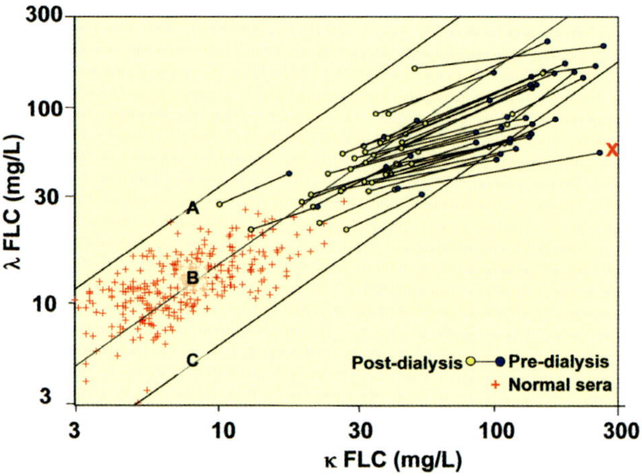

Figure 20.7. Reduction of sFLCs in 40 patients with CRF undergoing four hours of haemodialysis. (Patient X had an IgAκ monoclonal gammopathy with monoclonal κ FLCs. A, B and C show axes for κ/λ ratios of 0.2, 0.6 and 2.0 respectively).

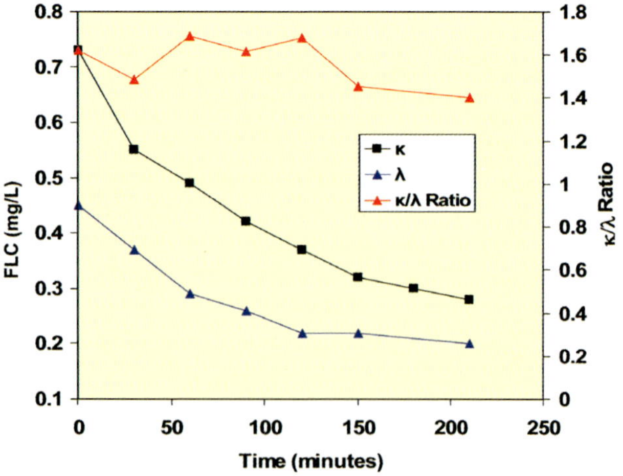

Figure 20.8. Concentrations of FLCs in the dialysate fluid of a patient with CRF, sampled every 30 minutes, during 4 hours of haemodialysis. (κ black, λ blue, κ/λ ratios red).

There is an additional concern about the toxicity of the high FLC load on the kidneys in patients during the progression to end-stage renal failure. As the nephron mass declines with increasing renal impairment the remaining nephrons, which are hyperfiltrating, are exposed to increasing levels of potentially toxic FLCs. This may contribute to an accelerating decline in renal function *(Kriz)*.

20.5. Monoclonal FLCs in CRF

It is of interest that 8 of 107 (7%) patients in CRF in Figure 20.1 and Table 20.3 had monoclonal gammopathies. None were associated with clinically apparent plasma cell dyscrasias and, indeed, only one had been identified prior to the study. 3 of the patients had FLC-only MGUS. 5 other patients had intact immunoglobulin monoclonal proteins identified by serum IFE of which 2 had associated abnormal FLC ratios.

Hutchison *(3,4)* et al. assessed the prevalence of MGUS in patients with CKD and renal transplant recipients using sFLC analysis and immunofixation electrophoresis (IFE) for intact monoclonal immunoglobulins. Of 595 patients studied in the CKD group, 10.6% had MGUS which was 3 times higher than an aged-matched population *(Figure 20.9)*. In the transplant patients, 8.3% had MGUS *(Figure 20.10)*.

The reason for the high MGUS prevalence is unclear. However, monoclonal FLCs may contribute to an increased rate of renal deterioration in patients with or without underlying renal damage. Dingli *(2)* et al. showed that there was a relationship between focal and segmental glomerulosclerosis and plasma cell disorders in a study of 13 patients. It was considered likely that some of the pathological damage was due to monoclonal FLCs. Studies of renal biopsies in patients with chronic renal disease have shown that AL-amyloidosis is under-recognised and approximately 3% of biopsies have monoclonal light chains deposited in renal tissues without systemic features *(Sanders 3; Novak)*.

Many of these patients with light chain monoclonal diseases require chemotherapy, but monitoring treatment from repeated biopsies is impractical. sFLC analysis should allow these patients to be identifed early and monitored for the appropriate chemotherapy in order to eliminate their clonal disease *(Montseny)*. It may be that these patients with renal failure should not be transplanted until their clonal plasma cell disorder is properly treated as the donated kidney may not survive long *(Leung; Tanenbaum)*.

As regards monoclonal FLCs in the renal transplant recipients, MGUS may result from the effects of immunomodulatory drugs. The time course of MGUS development and its clinical and pathological consequences are unknown. Further studies are needed.

Table 20.3. Analysis of 8 samples with monoclonal gammopathies (5 with monoclonal light chains only) in 107 patients with CRF. (From Figure 20.1).

Patient	Kappa	Lambda	κ/λ ratio	IFE
1	25.42	671.77	0.04	Normal
2	19.1	246.77	0.08	Normal
3	19.88	118.13	0.17	IgGλ
4	25.15	40.31	0.62	IgGκ
5	33.06	46.58	0.71	IgGλ
6	31.52	39.41	0.80	IgGκ
7	141.32	53.15	2.66	IgGκ
8	216.91	81.31	2.67	Normal

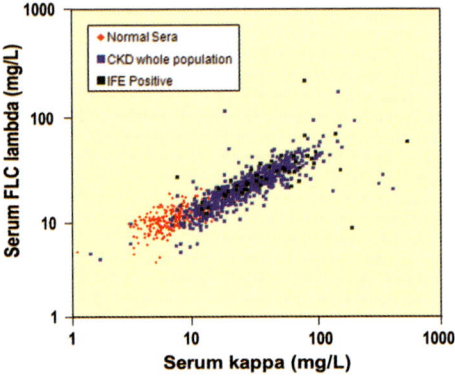

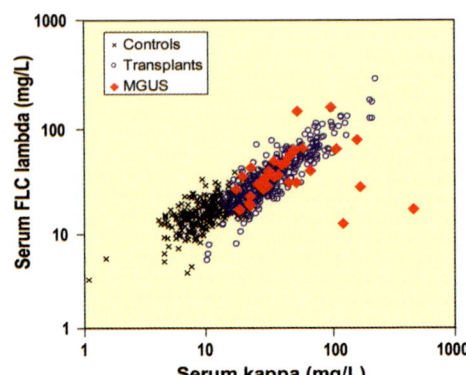

Figure 20.9. sFLC concentrations in 595 patients with CKD (blue squares) showing patients with MGUS (black squares) and the normal population (red diamonds).

Figure 20.10. sFLC concentrations in 465 renal transplant recipients (circles) showing patients with MGUS (red diamonds) and the normal population (crosses).

Clinical case history No 10. Influence of renal function on sFLC concentrations.

A 55-year-old woman presented with uncomplicated IgGκ MM and was given three courses of vincristine, adriamycin (doxorubicin), dexamethasone (VAD) followed by high dose melphalan and a peripheral blood stem cell transplant (PBSCT). She was well immediately following the transplant but then developed a degree of renal impairment. Careful attendance to fluid and electrolyte balance restored normal renal function.

While in hospital, she was monitored daily with sFLCs in an on-going investigation into their role as markers of treatment responsiveness (*Figure 20.11 and Chapter 13*). The effect of high dose melphalan (HDM) was to reduce both κ and λ FLCs to below normal concentrations with a favourable relative reduction in the tumour FLC levels. Subsequently, from day 15, as bone marrow engraftment took place, both FLCs increased and the κ/λ ratio normalised. Then, the FLCs increased above normal with a rising κ/λ ratio indicating some renal impairment. This observation was supported by an increase in serum creatinine concentrations. After day 22, FLC concentrations normalised and the κ/λ ratio fell, alongside serum creatinine levels.

In this patient, the combination of serial sFLC and serum creatinine measurements allowed assessment of renal dysfunction and bone marrow recovery. Changes in the κ/λ ratio may need to be corrected for renal impairment if there is confusion about interpreting the results. In addition, since patients with plasma cell dyscrasias are frequently elderly and are treated with nephrotoxic drugs, repeated measurements of sFLCs might assist in assessing renal function.

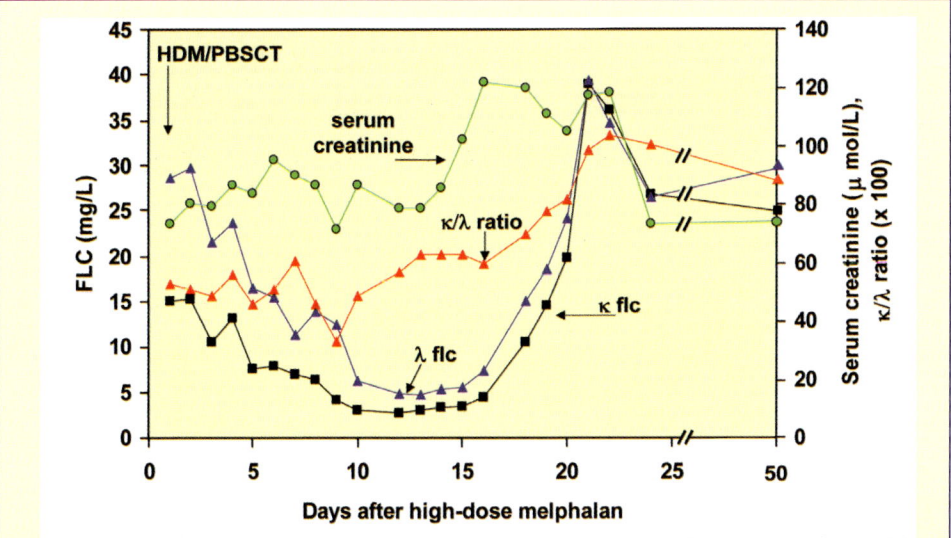

Figure 20.11. Changes in FLCs and creatinine while monitoring a patient with MM after HDM and PBSCT (Courtesy of G Pratt).

Test Questions

1. What is the mechanism of increased sFLC levels in renal impairment?
2. How are increases in polyclonal and monoclonal FLCs differentiated?
3. Why are κ/λ ratios slightly increased in patients with renal failure compared to patients having normal renal function?
4. Which is more efficient for FLC removal: plasma exchange or haemodialysis?
5. Why may patients with CKD have unexpected monoclonal FLCs?

Answers.

1. *Increase in serum half-life from reduced glomerular filtration (Section 20.2, page 176).*
2. *By use of a κ/λ log plot and κ/λ ratios (Section 20.2, page 168 and Section 20.3, 178).*
3. *Faster removal of κ molecules with increasing renal failure (Section 20.3, page 178).*
4. *Haemodialysis (Section 20.4, page 179 and Chapter 13).*
5. *Light chain deposition is not easily identified in renal biopsies so the clone of plasma cells remains unrecognised, and IFE fails to identify low concentrations of monoclonal sFLCs (Section 20.5, page 181).*

Chapter

21

Immune stimulation and elevated polyclonal free light chains

21.1. *Introduction* **184**
21.2. *Rheumatic diseases* **184**
    Systemic Lupus Erythematosus (SLE) **185**
    Primary Sjögren's Syndrome (pSS) **187**
    Rheumatoid Arthritis (RA) **188**
    Dermatitis **188**
21.3. *Diabetes mellitus* **189**
21.4. *Infectious diseases with elevated polyclonal serum free light chains (sFLCs)* **190**
21.5. *Lymphoma* **192**
21.6. *Free light chains as bioactive molecules in inflammatory diseases* **192**

Summary: In patients with immune stimulation:

1. There is an increase in polyclonal FLC production.
2. sFLC concentrations may increase 10-20 fold.
3. There is no increase in serum κ/λ ratios unless there is associated renal impairment.
4. There is frequently some increase in renal FLC excretion.
5. Increased polyclonal sFLCs are a sensitive marker of B-cell stimulation.

21.1. Introduction

Diseases associated with generalised increased B-cell activation frequently have high concentrations of polyclonal immunoglobulins and coexisting high polyclonal sFLCs *(Chapter 29)*. This relationship was demonstrated in 25 patients with polyclonal hyper-gammaglobulinemia studied at the Mayo Clinic *(Katzmann 6)*. In some patients, concentrations of both FLCs were highly elevated, although κ/λ ratios were always within the normal range *(Figure 21.1)*. Total immunoglobulin concentrations were as high as 54g/L while maximum FLC concentrations were 273mg/L for κ and 307mg/L for λ. The correlation between immunoglobulin and sFLC levels was modest. This may have been due to impaired renal function elevating the FLC concentrations in some patients. However, data on glomerular filtration rate (GFR) were not available.

There have been a number of studies published recently, exploring the pathological associations of high, polyclonal FLC concentrations in both renal *(Chapter 20)* and non-renal disease *(below)*. And it has been suggested that FLC measurements could form a useful early investigation in a general health assessment *(Bradwell 6)*.

21.2. Rheumatic diseases

Many rheumatic diseases feature polyclonal B-cell activation, high concentrations of autoimmune antibodies and polyclonal elevations of serum immunoglobulins. Excess polyclonal FLCs have been detected in the urine of these patients and indeed, their

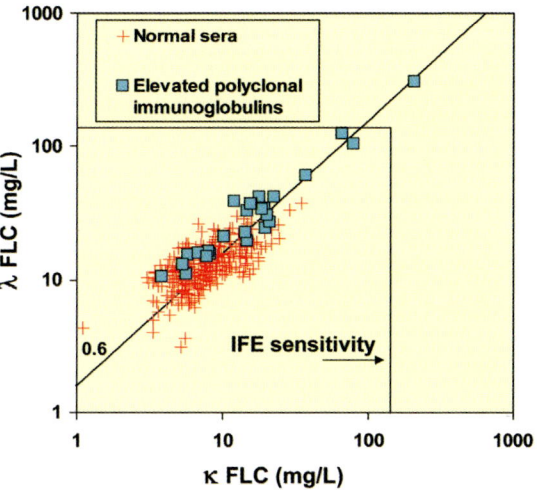

Figure 21.1. sFLCs in 25 patients with polyclonal hypergammaglobulinaemia (blue squares), compared with normal individuals (red crosses). (Courtesy of RA Kyle and JA Katzmann).

measurement may be useful for assessing disease activity. Presumably, serum analysis of FLCs in such patients would be more reliable. This is particularly applicable to patients with systemic lupus erythematosus (SLE), many of whom have high levels of urine polyclonal FLCs *(Epstein; Hopper 1, 2)*. Since these patients frequently have renal impairment, sFLC concentrations may also be highly elevated.

Hoffman et al. investigated the relationship between sFLCs and other markers of disease activity in patients with rheumatic diseases. Figure 21.2 shows the concentrations of sFLCs in the different disease groups. Patients with intercurrent illnesses were excluded from analysis to ensure that the changes were due exclusively to the disease under study.

High FLC concentrations were found in rheumatoid arthritis, SLE, Sjögren's syndrome, vasculitis and systemic sclerosis compared with control groups of 28 patients with fibromyalgia and 19 blood donors ($p<0.05$). Furthermore, sFLC concentrations were more frequently elevated than intact immunoglobulins. In all individuals, κ/λ ratios were normal, indicating polyclonal synthesis. It was also found that sFLCs were more frequently elevated than C-reactive protein (CRP) in patients with SLE, Sjögren's syndrome and systemic sclerosis. However, the numbers of patients in some groups were insufficient for statistical analysis. As might be expected, there was a positive correlation between concentrations of sFLCs and creatinine in all patient groups.

Systemic Lupus Erythematosus (SLE)

Hoffmann et al. studied 45 patients with SLE and showed that sFLCs were elevated approximately 3-fold *(Figure 21.2)*. Predictably, FLC concentrations were higher in SLE patients who had renal involvement compared with those having normal renal function *(Figure 21.3)*.

Clinical scores of SLE correlated with sFLC levels, particularly when the disease was active. In a subsequent prospective study, the clinical scores (European Consensus

Lupus Activity Measurement (ECLAM) *(Vitali)*) in 8 patients were compared with a variety of laboratory parameters *(Urban)*. sFLC concentrations showed a strong correlation with disease activity that was not observed for CRP or erythrocyte sedimentation rate (ESR). A larger study with 75 SLE patients *(Aggarwal)* also showed a strong association between total FLC concentrations and disease activity.

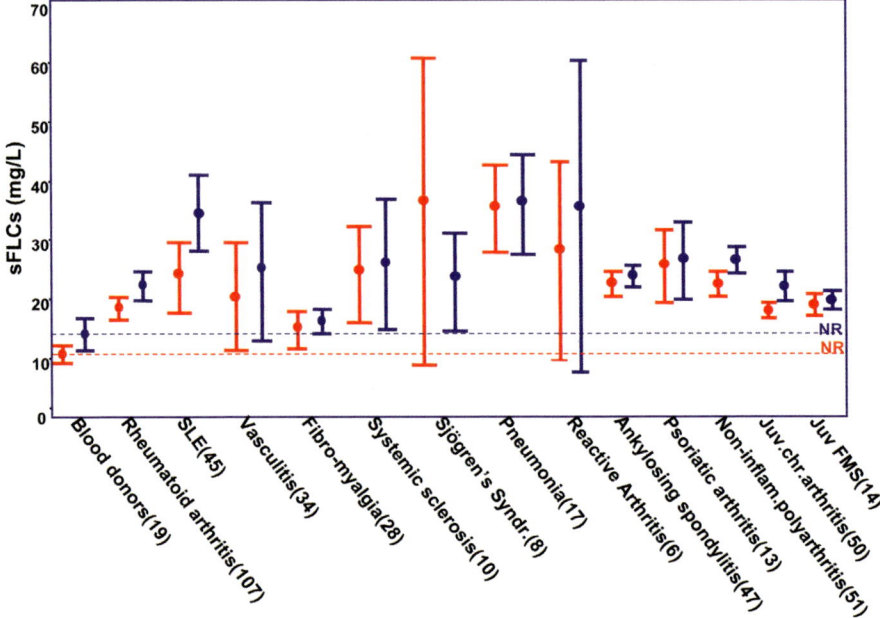

Figure 21.2. sFLCs in different diseases compared with blood donors. Bars show range and mean concentrations of κ (red) and λ (blue). Juv chr arthritis: juvenile chronic arthritis. Juv FMS: juvenile fibromyalgia syndrome. Patient Nos. in brackets. (Courtesy of U Hoffmann).

Figure 21.3. sFLC in SLE with (A) and without (B) renal involvement. The numbers of individuals studied are shown in brackets. Bars show range and mean concentrations of κ (red) and λ (blue). (Courtesy of U Hoffmann).

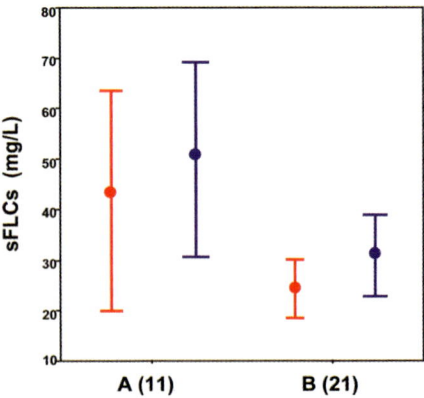

Primary Sjögren's Syndrome (pSS)

Gottenberg et al. studied 139 patients with pSS. 22% had raised sFLC, and mean levels were significantly higher than controls: p<0.001 *(Figure 21.4)*, while κ/λ FLC ratios were normal in all but one patient. sFLC concentrations were significantly correlated with IgG (p<0.001), rheumatoid factor (p<0.005), β_2-microglobulin (p<0.001) and B-cell activating factor (p<0.01).

Mean sFLCs were higher in patients with autoantibodies, particularly when both anti-SSA and anti-SSB were co-occurring *(Figure 21.5)*. Also, patients with extra-glandular involvement had higher levels than those with only glandular involvement. Interestingly, 15 patients had monoclonal FLCs, a much higher proportion than might be expected by chance. These results indicate that extra-glandular involvement in pSS is associated with intense stimulation of B-cells.

Some of these patients progress to non-Hodgkin lymphomas, [particularly mucosa-associated lymphoid tissue lymphoma (MALToma) with an odds ratio of 12.9 *(Smedby)*] but currently no biological marker is available to evaluate the individual risk for lymphoma. However, the above results show that abnormal κ/λ ratios are associated with loss of control over the proportion of heavy and light chains synthesised. [Also, it was recently shown to be a relevant clinical marker of malignant evolution in B-cell chronic lymphocytic leukaemia (CLL) *(Chapter 18)* and monoclonal gammopathy of undetermined significance (MGUS) *(Chapter 19)*]. Among 5 patients with pSS without MGUS who had abnormal κ/λ ratios, one had purpura and two had decreased complement C4 levels, both of which are risk factors for lymphoma.

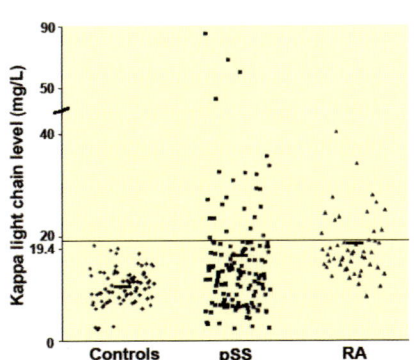

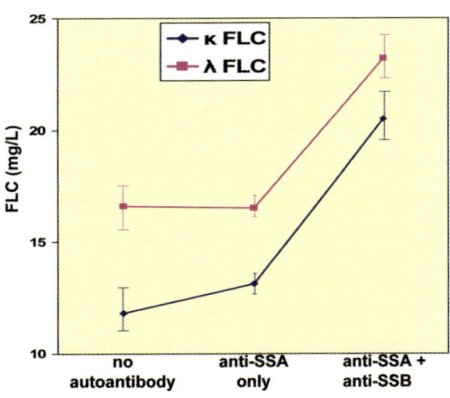

Figure 21.4. Serum κ FLC concentrations (and mean values) in patients with Primary Sjögren's Syndrome (pSS), rheumatoid arthritis (RA) and healthy controls. NR: Normal range. (Reproduced with permission from Ann Rheum Dis (Gottenburg) and J-E Gottenburg).

Figure 21.5. sFLCs in relation to the presence of SSA and SSB antibodies in patients with Sjögren's Syndrome. (Courtesy of X Mariette).

No clonal B-cell populations could be detected in the blood of these patients, which suggests that an abnormal κ/λ ratio could be a more sensitive marker of clonality, possibly restricted initially to the site of autoimmunity. Hence, Gottenberg et al. suggested that the predictive value of abnormal κ/λ ratios regarding the occurrence of lymphoma should be investigated in a longitudinal study of this disease.

Rheumatoid Arthritis (RA)

Gottenberg et al. studied 50 patients with RA. 36% had raised sFLCs with mean values significantly higher than controls: p<0.001 *(Figure 21.4)*, while sFLC κ/λ ratios were normal in all but 3 patients. sFLC concentrations were significantly correlated with IgG (p<0.04), CRP (p<0.04), and rheumatoid factor (for κ only: p<0.03), but not with anti-cyclic citrullinated peptide (CCP) antibodies. Significant correlations were observed between disease activity assessed by the Disease Activity Score 28 (DAS28) *(Prevoo)* and both κ (p=0.0004 - *Figure 21.6*) and λ concentrations (p=0.05 - data not shown). This supports the functional relationship between B-cells and disease activity. Interestingly, no correlation was observed between DAS28 and IgG, another marker of B-cell activation, but which has a much longer half-life (20–25 days) than FLCs (2–6 h). The faster turnover of sFLCs might account for their observed disease activity correlation, and suggests that they might be a good early surrogate marker for responses to treatments. Studies linking FLC concentrations to the use of drugs such as Rituximab that deplete B-cell numbers, and the development of clonal disease are underway. A further study including 710 patients with arthritis *(Gottenberg 2)* found that polyclonal FLC (and other markers of B-cell activation) were higher in early RA than in undifferentiated arthritis. The authors concluded that B-cell activation is an early pathogenic event in the disease.

Dermatitis

A recent study of children with atopic dermatitis *(Kayserova)* reported significantly higher FLC concentrations in patients compared with controls. Levels were also higher in those with severe forms of the disease but there were no significant associations between FLC levels and IgE or age.

Figure 21.6. Correlation between serum κ FLC concentrations and Disease Activity (DAS28) in patients with RA. (Reproduced with permission from Ann Rheum Dis (Gottenberg) and J-E Gottenberg).

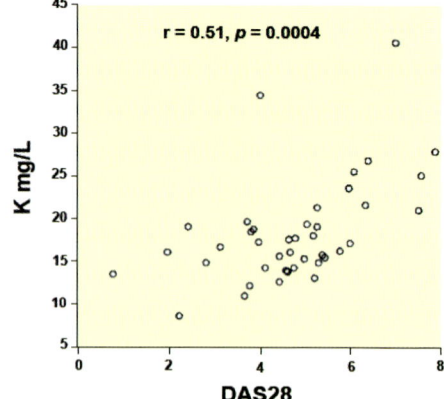

21.3. Diabetes mellitus

Two early studies identified a relationship between urine FLC concentrations and rapidly progressive diabetes mellitus. Thus, 20 years ago, it was noted that patients with proliferative retinopathy had both higher urine κ FLC excretion (λ FLC assays were not available) than those without retinopathy and higher than in patients with proteinuria from other causes *(Teppo)*. There was an associated elevated κ FLC/albumin excretion ratio. Subsequently, the same authors suggested that elevated FLC/albumin excretion ratios were an early indication of diabetic nephropathy and they directly implicated a renal cause of the FLC leakage rather than excess production *(Groop)*. This was supported by their finding of normal serum FLC concentrations. However, in the absence of sensitive serum assays this interpretation was, perhaps, premature.

A recent study by our group analysed FLC levels in both serum and urine of Type 2 diabetic patients to determine if they were an early marker of diabetic kidney disease *(Hutchison 1)*. It was clear that diabetic patients had significantly raised serum and urine concentrations of polyclonal FLCs *(Figure 21.7)* before overt renal impairment developed (P<0.001). κ concentrations were higher than λ concentrations and 1.9% of patients had MGUS (confirmed by immunofixation electrophoresis (IFE)). There was a

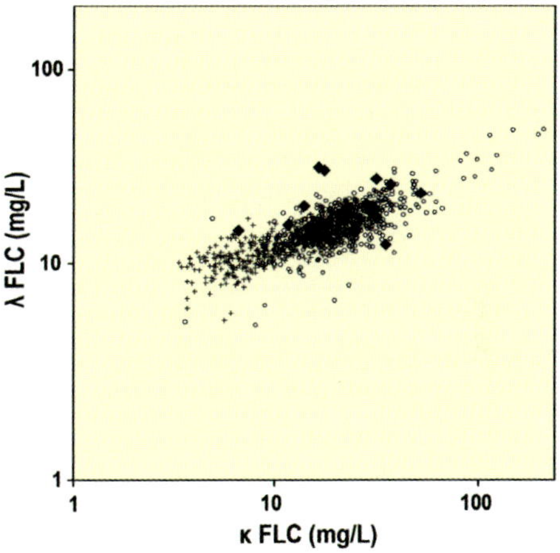

Figure 21.7. sFLC concentrations in 745 patients with early diabetes mellitus (circles) compared with the normal population (crosses) and showing patients with MGUS (diamonds) (1.9%).

Figure 21.8. κ sFLCs in 745 patients with early Type 2 diabetes correlated with the estimated GFR - MDRD (p<0.001).

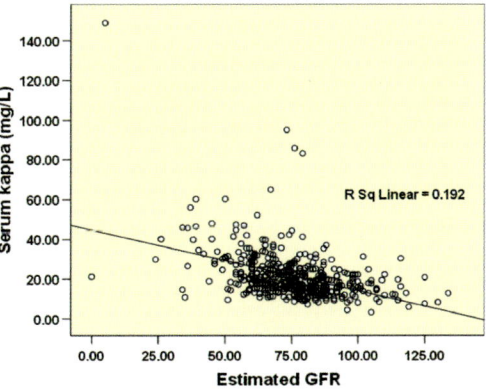

good correlation between sFLC concentrations and various markers of GFR including serum creatinine, cystatin-C [κ; R=0.55 (P<0.01), λ; R=0.56 (P<0.01)] and estimated GFR *(Figure 21.8 shown for κ only)*. South-Asian diabetic patients had higher sFLCs than Caucasian diabetic patients and this was independent of renal function suggesting more underlying inflammation.

Urinary FLC concentrations were raised in diabetic patients (P<0.001). 68% of patients with normal urinary albumin/creatinine ratios (ACRs) had abnormal urinary FLC/creatinine ratios. Urine FLC concentrations correlated with urinary ACR: κ, R=0.32, P<0.01 and λ, R=0.25, P<0.01 respectively. However, some patients had normal GFR (estimated using the Modification of Diet in Renal Disease (MDRD) study equation) with high concentrations of sFLCs indicating increased production. This is suggestive of generalised inflammation/vasculopathy. Perhaps retinopathy and nephropathy are the most readily observed clinical signs of a generalised inflammatory process that is apparent from raised FLC production.

Since polyclonal FLCs are potentially nephrotoxic, increased concentrations may contribute to progressive nephropathy. It has also been suggested that monoclonal FLCs may play a role in some patients' renal diseases *(Dillon)*. Indeed, mesangial monoclonal FLC deposits observed in renal biopsies of patients with renal impairment are sometimes similar in appearance to those found in diabetic glomerulosclerosis *(Sanders 3)*. Furthermore, FLC MGUS is observed in patients with renal impairment *(Chapter 20)*, so it may be an additional risk factor for progressive nephropathy.

Thus, Type 2 diabetic patients have significantly raised concentrations of serum and urinary polyclonal FLCs before overt renal disease occurs. Possibly, measurement of polyclonal FLCs could provide a useful tool in early diagnosis of diabetic kidney disease.

21.4. Infectious diseases with elevated polyclonal sFLCs

There was one early report by Sölling *(7)*, and a more recent study of patients with acute pneumonia *(Figure 21.9)* by Hoffman et al. As expected, both publications showed that polyclonal sFLCs were elevated. In the latter study, the median κ/λ ratio was higher than in other disease groups suggesting modest impairment of renal function *(Chapter 20)*.

One potent cause of elevated polyclonal immunoglobulins and sFLCs is chronic viral infection. Terrier et al. studied 59 patients with chronic hepatitis C virus infections (HCV) and mixed cryoglobulinaemia (MC) at different stages of evolution to non-Hodgkin lymphoma (NHL). The MC comprised type II cryoglobulins with immune complexes of monoclonal IgM directed against polyclonal IgG. 17 patients had no MC, 7 had asymptomatic MC, and 35 had MC vasculitis, 9 of whom had B-cell NHL.

The results showed elevated sFLCs in nearly 50% of patients. Furthermore, mean polyclonal sFLC concentrations and the frequency of abnormal sFLC κ/λ ratios progressively increased with worsening disease category *(Figure 21.10)* (p<0.001 and p=0.002 respectively), increasing cryoglobulin concentrations (P<0.0001 and P=0.0016 respectively) and the severity of the B-cell disorder (P=0.045 and P=0.0012 respectively). Among patients with an abnormal sFLC ratio at baseline, FLC ratios

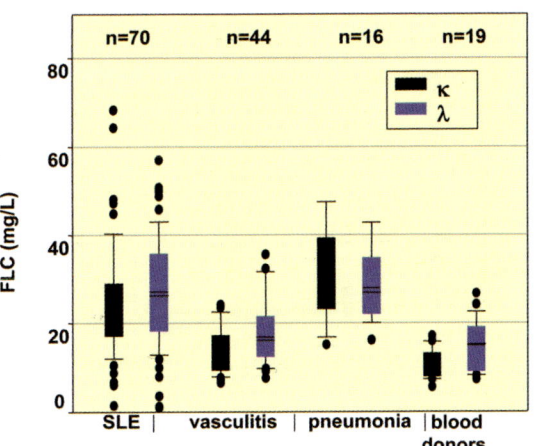

Figure 21.9. sFLCs in patients with pneumonia compared with other diseases. The slight increase in κ (black) compared with λ concentrations (blue) may be related to renal impairment. (Reproduced with permission from Zeitschrift für Rheumatologie (Hoffman) and U Hoffman).

correlated with the virological response to HCV treatment *(Figure 21.11)*. The authors concluded that in HCV-infected patients, abnormal sFLC ratios were very interesting markers, and were consistently associated with the presence of MC vasculitis and/or B-cell NHL. After anti-viral therapy, the sFLC ratio could be used as a surrogate marker for the control of the HCV-related lymphoproliferation.

In 2010, the Journal of Clinical Oncology published a study describing FLC and immunoglobulin concentrations in HIV positive patients *(Landgren 4)*. FLC concentrations were clearly higher than those found in the general population and higher levels were associated with significantly increased risk of progression to non-Hodgkin lymphoma. Neither abnormal FLC ratio nor immunoglobulin concentration showed a similar association. The authors speculated that FLC measurement may have a clinical utility for assessing risks of NHL in HIV patients and/or may have more general applications as a marker of polyclonal B-cell activation.

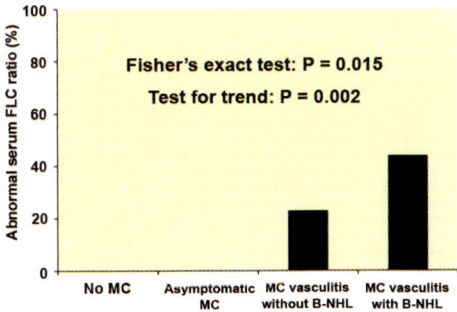

Figure 21.10. In virus C hepatitis, abnormal sFLC ratios are progressively more frequent with worsening disease category. (MC: mixed cryoglobulinaemia. B-NHL: B cell non-Hodgkin lymphoma). Reproduced with permission from Ann Rheum Dis (Terrier) and B Terrier).

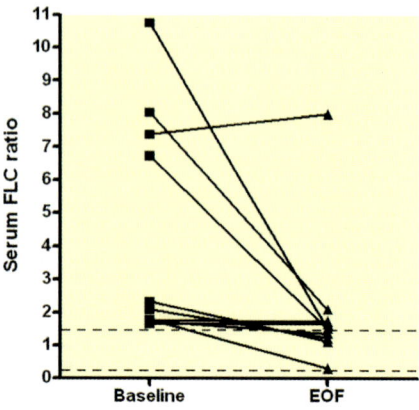

Figure 21.11. Abnormal sFLCs ratios in virus hepatitis C infections may normalise with successful anti-viral treatment. (EOF: end of follow-up). Reproduced with permission from Ann Rheum Dis (Terrier) and B Terrier).

21.5 Lymphoma

The poor prognosis associated with high FLC concentrations in diffuse large B-cell lymphoma has already been described *(Chapter 18)*. The normal FLC ratio present in many of these patients indicates that the elevations were due to polyclonal B-cell activation and were not tumour-derived. This is certainly the case for Hodgkin's lymphoma where the tumour cells are incapable of producing immunoglobulin proteins. De Filippi and colleagues reported elevated FLC concentrations in 47% of 119 patients with Hodgkin's lymphoma. The FLC concentration predicted event-free survival in patients with early-stage but not late-stage disease (a situation also seen in CLL, *Chapter 18)*. In a separate case series of 3 patients with refractory Hodgkin's lymphoma *(Corazzelli)*, it was noted that serum FLC concentrations fell after lenalidomide treatment and rose transiently during disease "flares" suggesting an association with the disease process.

21.6. Free light chains as bioactive molecules in inflammatory diseases

Since FLCs are part of the antigen binding site of intact immunoglobulin molecules they have bioactivity. This has been observed for both polyclonal and monoclonal FLCs and has recently been reviewed *(Figure 21.12) (van der Heijden, Thio)*.

For instance, biological activities of FLCs have been shown in patients with immediate hypersensitivity-like responses and contact sensitivity dermatitis. Since FLCs can activate mast cells (which contain a range of biologically active molecules), their potential for causing or contributing to inflammatory and other diseases such as asthma is high. Possibly, FLCs are one of the components in the active inflammatory processes that are apparent in chronic renal failure. Furthermore, different monoclonal FLCs have been shown to bind specific target molecules that can stimulate anti-angiogenic activity or have proteolytic potential. The complementarity determining regions of FLCs have sufficient variability and flexibility that they could mimic almost any biological molecule.

In this context, removal of FLCs using *"high cut-off"* dialysers may be helpful in reducing inflammation in chronic kidney disease *(Chapter 13)*. There has also been a suggestion that their inflammatory actions might be usefully blocked by novel FLC binding peptides *(Thio)*. To help resolve these issues, studies on purified polyclonal FLCs from patients with different diseases are required.

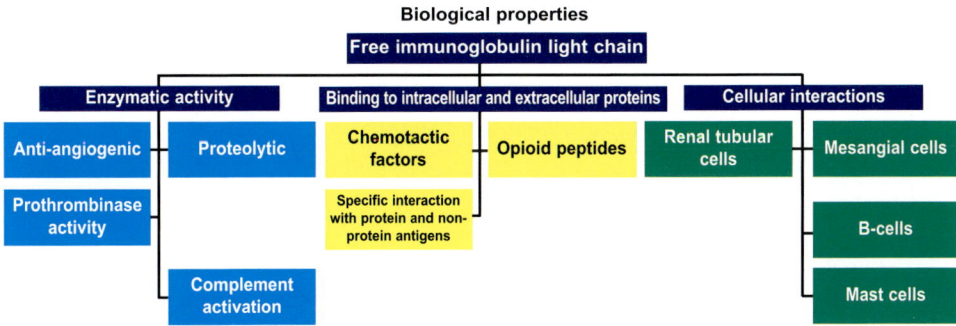

Figure 21.12. Biological actions of sFLCs include enzymatic acitivity, binding to tissue substrates and mast cells. (Reproduced with permission from Trends Pharmacol Sci (Thio) and M Thio).

Test questions

1. Do conditions causing hypergammaglobulinaemia produce increases in sFLCs?

2. How are unexpected increases in polyclonal FLCs evaluated?

3. Why are sFLCs highly elevated in patients with SLE?

4. Why might urine FLCs be an early marker of renal disease in diabetes mellitus?

5. What percentage of hepatitis C patients have abnormal sFLCs?

Answers

1. FLC production normally increases alongside increased production of the intact immunoglobulin molecules (Section 21.1, page 184).

2. Assess renal status first and if elevated FLC concentrations remain unexplained, there may be occult inflammatory disease processes.

3. Because of increased production and reduced renal clearance (SLE section, page 185).

4. Because hyperfiltering glomeruli leak albumin. This competes with normal FLC removal in the proximal tubules thereby displacing it into the urine (Section 21.3, page 189).

5. Nearly 50% if there is associated cryoglobulinaemic vasculitis and NHL (Section 21.4, page 190).

Cerebrospinal fluid and free light chains

Summary:

1. Abnormal intrathecal production causes free light chains (FLCs) to accumulate in the cerebrospinal fluid (CSF).
2. CSF FLCs can be readily measured by immunoassays.
3. CSF κFLCs are a sensitive marker of intrathecal inflammation and lymphomatous meningitis.

During inflammation of the central nervous system there is usually synthesis of intrathecal immunoglobulins. Since the blood-brain barrier largely prevents their escape into blood, they gradually accumulate in the CSF. They are then detectable as oligoclonal bands on electrophoretic gels or can be quantitated by protein assays.

When determining clinical relevance, abnormal immunoglobulins need to be assessed in the context of small amounts of serum immunoglobulins that may have diffused into the CSF from the blood. In particular, if the patient's serum contains monoclonal immunoglobulins produced in the bone marrow, some will cross the blood-brain barrier making interpretation of intrathecal production difficult. Similarly, if there is inflammation of the meninges, serum proteins will enter the CSF more readily and not only be of intrathecal origin. Hence, CSF measurements are always made in relation to serum proteins. When assessing immunoglobulin production, CSF/blood immunoglobulin concentration ratios are compared with CSF/blood albumin ratios because albumin is never synthesised in the brain. Similarly, CSF IgG oligoclonal bands are assessed on electrophoretic gels run alongside corresponding serum samples. However, commonly used techniques such as iso-electric focussing are only qualitative, rather time consuming and interpretation may be difficult *(Luxton)*.

Consequently, alternative markers of intrathecal inflammation have been studied, particularly CSF FLCs *(DiCarli; Khoury; Lamers; Krakauer; Jenkins)*. These are produced alongside intact immunoglobulins and accumulate in the CSF. Moreover, they do not enter CSF from the blood in significant amounts because of their large size and low serum concentrations. The detection methods for CSF FLCs have included iso-electric focusing, quantitation by enzyme/radio-immunoassays and nephelometry, but results have not been of general interest. With the development of simple FLC nephelometric assays, there has been renewed interest in their measurement in CSF.

Using these new FLC assays, Fischer et al. studied CSF/serum pairs from 95 patients who had been investigated for intrathecal immunoglobulin synthesis. Of these, 24 were negative for oligoclonal immunoglobulin synthesis and 71 were positive, comprising 49 with multiple sclerosis and 22 with other neurological diseases *(Figure 22.1)*. The median κ concentrations in the patients with neurological diseases were high. (Curiously, the λ FLC concentrations were unreliable). A κ FLC index was then constructed: **κFLC index** = κ FLC in CSF/κ FLC in serum divided by albumin in CSF/albumin in serum.

When samples with increased albumin leakage were excluded, there was no overlap between normal and disease groups (cut-off level of κ concentrations: 0.5mg/L). This indicated that determination of κ FLC concentrations in CSF provided information similar to that of oligoclonal band measurements, providing samples with raised albumin levels were excluded. Furthermore, when using the κFLC index *(Figure 22.2)* only two

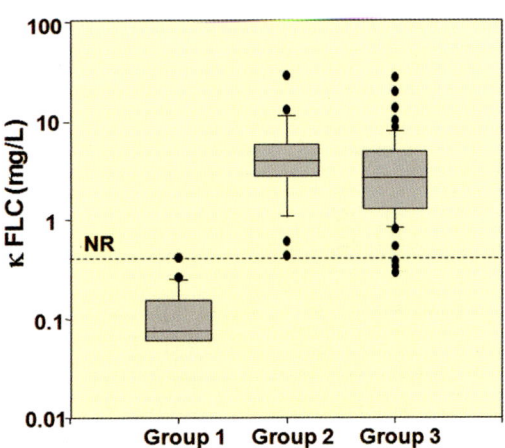

Figure 22.1. Box plot of κ FLC concentration (mg/L) in CSF from (1) normal individuals, (2) patients with multiple sclerosis and (3) other neurological diseases. NR: upper limit of normal range for κ in the CSF was 0.5mg/L. (Courtesy of KJ Lackner).

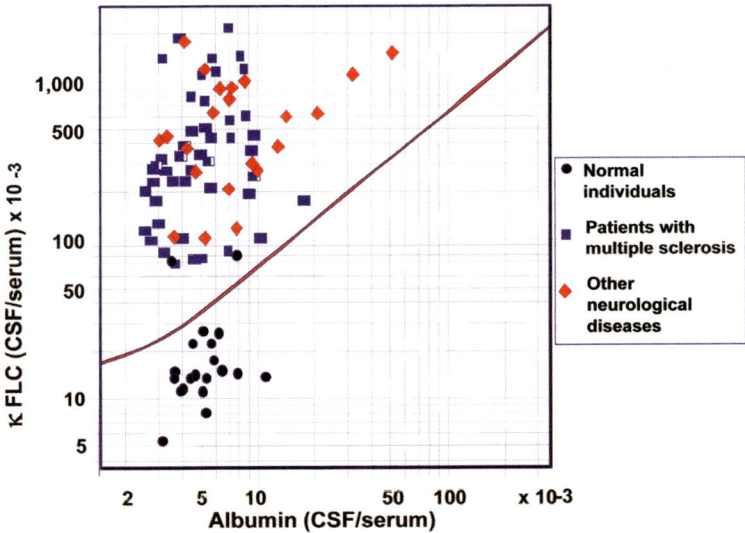

Figure 22.2. Quotients of κ FLC concentrations in CSF and serum plotted against the respective albumin quotients. Samples are from patients in Figure 22.1. (Courtesy of KJ Lackner).

normal samples were misclassified. They concluded that CSF κ FLC measurements may be a useful diagnostic procedure for detecting and potentially monitoring intrathecal immunoglobulin synthesis.

Others have published similar findings. Desplat-Jego et al. showed that of 89 patients studied, the κFLC index was more sensitive but less specific than the comparable IgG CSF index or CSF oligoclonal bands. 2 patients were only positive by the κFLC index, so it was a useful complementary test for the diagnosis of multiple sclerosis. It was considered that because the CSF FLC assay was easy, rapid and automated, it could be included in CSF studies in patients with suspected intrathecal inflammation. Further studies by Lewis et al. with 27 patients, Presslauer et al. with 367 patients and Arneth et al. with 110 patients supported these findings.

Additional support for the clinical value of CSF κ FLC measurements has come from its relationship with disability outcome. Rinker et al. reviewed the clinical records of 57 patients with multiple sclerosis over a 15 year median follow-up period. Raised levels of CSF κ FLCs (by radioimmunoassay) predicted progression to the need for ambulatory assistance during the study with a specificity of 87.5% and a positive predictive value of 89%.

Monoclonal FLC production in the brain has been assessed in patients with lymphomatous meningitis (LM). This is a serious complication of malignant lymphoma and can be difficult to diagnose. Hildebrandt et al. studied 17 patients, 5 of whom were positive for LM by CSF cytology, immunocytochemistry and imaging procedures. All patients with LM had elevated CSF κ/λ ratios as did 4 of the control population of patients with cerebral infections. There is some doubt about the sensitivity of the λ FLC assay for CSF samples but a larger study is underway.

Test question
1. Are raised FLCs in the CSF of clinical relevance?

Answer
1. Yes, they are indicative of intrathecal immunoglobulin synthesis and are clinically useful.

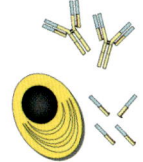

SECTION 4 - General applications of free light chain assays

Chapter 23. *Screening studies using serum free light chain analysis*   *197*
Chapter 24. *Serum versus urine tests for free light chains*   *208*
Chapter 25. *Guidelines for use of serum free light chain assays*   *221*

Chapter

23

Section 4 - General applications of free light chain assays

Screening studies using serum free light chain analysis

23.1. *Introduction*   *197*
23.2. *Diagnostic protocols for monoclonal gammopathies*   *198*
23.3. *Screening studies using sFLC analysis*   *199*
23.4. *Audit of sFLC usage*   *207*

Summary:

1. Screening symptomatic patients using SPE/sIFE and sFLCs is a clinically sensitive strategy for identifying patients with monoclonal gammopathies.
2. Extra patients with monoclonal FLCs are identified that require further investigations including renal function studies.
3. Adding urine tests to the initial screen is of no extra clinical benefit.
4. The combination of capillary zone electrophoresis and FLC immunoassays allows automated screening of symptomatic patients for monoclonal gammopathies.

23.1. Introduction

Whether or not to measure serum free light chains (sFLCs) at the time of the initial diagnostic request for *"possible myeloma - please investigate"* is an important issue. By tradition, serum protein electrophoresis (SPE) and/or serum immunofixation electrophoresis (sIFE) tests have been performed first, sometimes alongside immunoglobulin (Ig) G, IgA and IgM measurements. SPE can detect intact immunoglobulin monoclonal proteins in the 1-5g/L range *(Chapter 6)* which is more than adequate to identify all intact immunoglobulin multiple myeloma (IIMM) patients. Some centres screen with IFE because it is ten times more sensitive. However, occasional monoclonal FLCs are missed and IFE is non-quantitative. Furthermore, many low-level monoclonal gammopathy of undetermined significance (MGUS) proteins are detected that are unlikely to progress to malignancy, provided sFLC levels are also normal *(Chapter 19) (Rajkumar 5)*. A few laboratories measure total serum κ and λ in the initial screen but this is clinically inadequate *(Chapter 6)*.

Because of these deficiencies, urine protein electrophoresis (UPE) has been performed alongside serum tests, but many patients with nonsecretory multiple myeloma (NSMM), AL amyloidosis and other FLC-associated disorders are still missed. Furthermore, only a minority of patients have accompanying urine samples (typically 5-25% in the UK). It is, therefore, logical to test for sFLCs on receipt of the first blood sample. The results will also provide baseline values for subsequent disease monitoring.

Hence, a strategy of using SPE/sIFE combined with sFLC analysis *"up-front"* allows identification of all clinically significant monoclonal gammopathies. It also allows risk assessment for disease progression in MM, MGUS, asymptomatic (smouldering) multiple myeloma (ASMM), plasmacytomas etc., and appropriate clinical decisions for monitoring patients.

This chapter discusses the current and potential screening options for identifying monoclonal proteins. As a general rule, intact monoclonal immunoglobulins can be identified using serum electrophoretic tests while monoclonal light chain diseases should be identified using sFLC assays. The combination of the 2 procedures produces good diagnostic accuracy and *urine FLC analysis is not required* at initial clinical presentation *(Chapter 24)*.

23.2. Diagnostic protocols for monoclonal gammopathies

The accuracy of different diagnostic protocols used for identifying monoclonal gammopathies are shown in Table 23.1. The basis of the numerical analysis is as follows:

1. SPE identifies all patients with IIMM. By definition, they have at least 10g/L of monoclonal protein which is much greater than the sensitivity of the assays (1-5g/L). The test fails to identify approximately 30% of light chain multiple myeloma (LCMM) patients (the other 70% have sFLC bands or hypogammaglobulinaemia) and all NSMM *(Chapter 9)*. Approximately 50% of AL amyloidosis patients are abnormal by SPE *(Chapter 15)*. Many intact immunoglobulin MGUS individuals are also identified.

2. The combination of SPE and sIFE is more sensitive but fails to identify a few patients with LCMM and all patients with NSMM. 70% of patients with AL amyloidosis are

Accuracy of diagnostic tests at clinical presentation			
Protocols	Myeloma	AL amyloidosis	MGUS
1 SPE alone	90	50	45
2 SPE and sIFE	95	70	80
3 SPE and UPE	95	75	70
4 SPE, UPE, serum and uIFE	97	90	80
5 FLC alone	96	95	30-65
6 SPE and FLC	99	98	85
7 SPE, FLC, sIFE	99	99	100

Table 23.1. Approximate diagnostic sensitivity of tests for monoclonal gammopathies. (See relevant clinical chapters for details).

identified but those producing FLCs only are frequently missed. Many more MGUS individuals are identified but monoclonal proteins of less than 1-5g/L are of little clinical consequence if sFLCs are normal *(Chapter 19)*.

3+4. The addition of UPE/urine IFE (uIFE) to either of the above protocols identifies the remaining 30% of patients with LCMM and many patients with AL amyloidosis. Few additional MGUS patients are identified because they rarely produce sufficient monoclonal FLCs to exceed the renal threshold and enter the urine *(Chapter 19)*. Also, since serum tests for FLCs are clinically more useful than UPE tests, these protocols are illogical *(Chapter 24)*.

5. sFLC analysis identifies FLC monoclonal gammopathies but cannot be used alone. It is, by definition, a test for FLCs and not for intact monoclonal immunoglobulins. 30-60% of patients with MGUS, 10% with ASMM and 5% with IIMM do not have excess FLC production *(see Chapters 11, 14 and 19)*.

6. A sensitive protocol is to use SPE to identify all the intact monoclonal immunoglobulins and sFLC immunoassays to identify all the monoclonal FLCs. Approximately 20% of NSMM patients will be missed with this strategy but are not detected by any other serum or urine tests *(Chapter 9)*. Such patients are non-producers or non-excretors and may have mutations in the light chain DNA causing protein shape distortion. Nearly all patients with AL amyloidosis and light chain deposition disease (LCDD) are identified *(Chapters 15 and 17)*. Some additional MGUS individuals are identified who produce only monoclonal sFLCs *(Chapter 19)*. Patients with renal impairment are also identified *(Chapter 20)*.

7. The addition of sIFE to protocol 6 is not likely to identify any more patients with MM. It identifies approximately 2-10% additional AL amyloidosis patients *(Chapter 15)*. It also identifies some MGUS individuals with minor monoclonal intact immunoglobulins who are unlikely to progress to overt disease if there is no accompanying abnormal FLC κ/λ ratio *(Chapter 19)*. The extra cost, the inconvenience of performing the test and problems interpreting the clinical relevance of minor MGUS bands may not be justified in all clinical laboratories. A recent study by Katzmann *(8)* et al. *(see below, screening study 12)* concludes that the use of SPE plus FLC provides a simple and efficient initial diagnostic screen for the high tumour burden monoclonal gammopathies and that urine studies and urine IFE can be ordered more selectively.

In the unlikely event that SPE/sIFE and serum FLC tests are normal but the clinical picture still suggests a plasma cell disorder, urinalysis for FLCs is required. However, caution should be exercised when considering the clinical importance of minor urine FLC bands when serum FLC tests are normal *(see below and Chapter 24)*. After a monoclonal protein has been identified, IFE is required to characterise the heavy chain type. The FLC type can be identified by serum κ/λ ratios or from the IFE gels.

23.3. Screening studies using sFLC analysis

Twelve screening studies evaluating the use of sFLC assays, alongside other tests, as part of the initial diagnostic screen for monoclonal gammopathies, are described below:

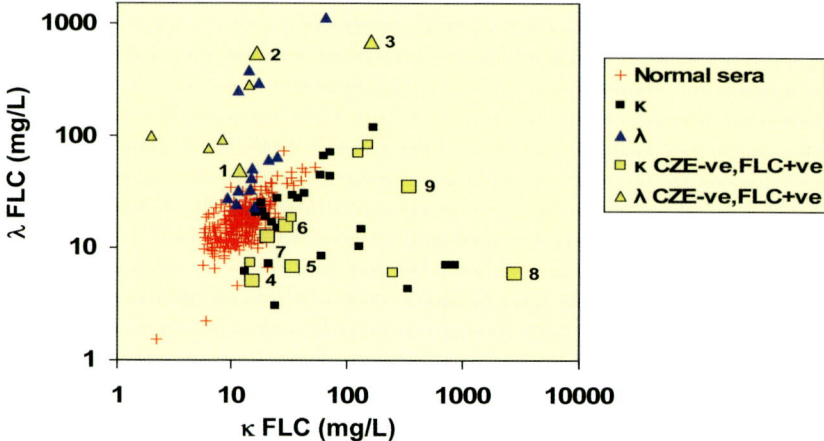

Figure 23.1. κ/λ ratios in 55 abnormal sera from a CZE and FLC screening study of 1,003 patients *(Bakshi)*. Sera with intact monoclonal immunoglobulins by CZE are shown in black or blue and samples containing only abnormal FLC κ/λ ratios are shown in yellow. Numbers against large symbols refer to additional patients with confirmed monoclonal gammopathy identified by FLC tests *(Table 23.2)*.

1. Bakshi and colleagues used a combination of CZE and sFLC analysis. 1,003 consecutive unknown samples were studied. 39 contained monoclonal proteins by CZE and 33 had abnormal κ/λ ratios to give a total of 55 abnormal sera.

16 samples (11 κ and 5 λ) were abnormal for monoclonal proteins only by sFLC analysis *(Figure 23.1 and Table 23.2)*. Subsequent IFE showed that 5 of the 16 samples were abnormal, although some had barely visible bands, but 11 samples were clearly only abnormal by FLC assays. Of the 39 samples that were positive by CZE, 17 were associated with abnormal FLC κ/λ ratios. In every case, the FLC test results concurred with the light chain type of the intact monoclonal immunoglobulin.

9 of the extra monoclonal gammopathy patients identified from abnormal κ/λ ratios were eventually shown to have plasma cell dyscrasias: 3 with MM, 4 with B cell chronic lymphocytic leukaemia (B-CLL)/small cell lymphoma, 1 aplastic anaemia and 1 lymphoma. 7 samples were not associated with significant diseases and were classified as FLC MGUS *(Chapter 19)*.

Addition of the FLC analysis to the CZE test in this screening trial increased the yield of lymphocyte and plasma cell proliferative diseases by 56%. The authors noted that the study emphasised the utility of sFLC testing for patients with monoclonal proteins when CZE was essentially non-diagnostic.

2. Hill et al. studied 923 consecutive serum samples referred to a UK District General Hospital. 71 had abnormal sFLC ratios *(Figure 23.2)*. 8 new B-cell dyscrasias were detected among 43 patients with negative SPE but abnormal sFLC κ/λ ratios. 2 patients with κ/λ ratios of 50 and 193 had κ FLC LCMM; 5 patients had FLC MGUS and one a malignant lymphoma (but with no monoclonal band). 35 patients with negative SPE had

5 sera : CZE negative but lambda FLC positive

n°	κ	λ	κ/λ Ratio	CZE	IFE	Diagnosis
1	11.8	48.5	0.24	Hypoalbuminaemia	Normal	B-CLL/SLL
2	163.0	681.0	0.24	Polyclonal with β-γ bridging	Tiny IgMλ	Aplastic anaemia
3	16.9	537.0	0.03	Normal pattern	λ light chain	λ LCMM
	8.5	91.5	0.09	Normal pattern	IgAλ(too small to measure)	MGUS
	14.3	279.0	0.05	Normal pattern	Normal	MGUS

11 sera : CZE negative but kappa FLC positive

n°	κ	λ	κ/λ Ratio	CZE	IFE	Diagnosis
4	15.3	5.1	3.04	Normal pattern	Normal	B-CLL/SLL
5	33.8	6.8	4.94	Hypogammaglobulinaemia	Normal	B-CLL/SLL
6	29.7	15.6	1.90	Normal pattern	Normal	Possible early B-CLL/SLL
7	21.0	12.4	1.69	Mild protein loss pattern	NI	B-NHL, possible MZL or LPL
8	2830.0	5.9	482.94	Mild protein loss pattern/reactive	Free κ in serum & urine	κ LCMM
9	350.0	35.0	10.00	Oligoclonal bands	Normal	NSMM
	155.0	83.1	1.87	Polyclonal-chronic inflammation	Normal	Borderline ?MGUS
	127.0	68.9	1.84	Polyclonal-chronic inflammation	Broad IgGκ	Borderline ?MGUS
	33.4	18.5	1.81	Normal	Normal	Borderline ?MGUS
	14.6	7.3	1.99	Normal	Normal	Borderline ?MGUS
	255.0	6.1	42.08	Hypogammaglobulinemia	Normal	FLC MGUS

Table 23.2. Clinical and laboratory data in 16 patients (confirmed monoclonal lymphoproliferative diseases) that were normal by CZE but abnormal by sFLCs (Bakshi). NI: Not indicated, B-CLL: B-cell chronic lymphocytic leukaemia, SLL: B-cell small lymphocytic lymphoma, LPL: lymphoplasmacytic lymphoma, MZL: marginal zone lymphoma. Case numbers refer to the patients also shown in Figure 23.1.

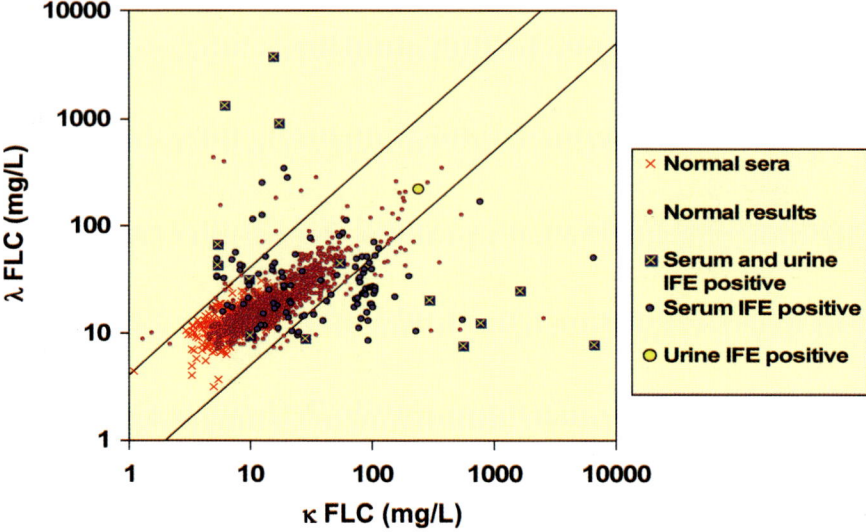

Figure 23.2. sFLCs in 925 sera screened at Derby, UK. Normal results are the FLC levels in samples with normal SPE. (Courtesy of P Hill).

abnormal κ/λ ratios but no monoclonal diseases. This false positive rate for a ratio of >1.65 was associated with polyclonal increases in immunoglobulins and renal impairment and was higher than previously noted. Such minor abnormalities need to be considered in the context of reference ranges for hospitalised patients and renal diseases *(Chapter 5 and Chapter 20)*.

In this study, comparison was made between the utility of serum and urine tests for monoclonal FLCs. 370 (40%) of the serum samples were accompanied by urine samples. Of these, 141 (38%) had suspicious UPE tests and were further analysed by uIFE. Only 15 of these urine samples (4% of the total tested) had confirmed monoclonal FLC indicating that 126 tests were performed unnecessarily *(Figure 23.3)*. In 11 of the samples there was a corresponding sFLC abnormality but 4 were normal. One of the 4 samples was associated with a serum IgGλ of 7g/L that was identified by IFE but had been missed on the initial SPE. The remaining 3 samples had urine FLC concentrations of approximately 50mg/L. One normalised as the patient recovered from illness, another was subsequently interpreted as polyclonal urine FLCs and the 3rd patient had persisting FLC proteinuria but was asymptomatic and had no evidence of a B-cell disorder from bone marrow biopsy. Hence, none of the *"urine only"* monoclonal FLCs had clinically active disease.

On the basis of these results the hospital practice was changed to a policy of no urine tests when screening for monoclonal gammopathies. This allowed all patients to have proper assessments for monoclonal FLCs, improved operational efficiencies and increased costs by less than £5 per patient. The cost increase is largely due to testing all patients for FLCs whereas previously urines were available in only 40% of cases.

3. Augustson *(1)* et al. studied 217 consecutive samples referred to a UK district general hospital. They found an extra 8 monoclonal gammopathies by abnormal sFLC κ/λ ratios. 3 of these patients had LCMM that were missed by the routine SPE tests and resulted in serious diagnostic delays. Subsequently, the study was increased to 971 patients *(Reid 3)* and results were similar to other screening studies *(Figure 23.4)*.

4. Abadie et al. reported results on 312 consecutive samples from Seattle hospitals in the USA. sFLC tests identified a higher percentage of true positive and true negative samples and a lower percentage of false positive and false negative samples than SPE.

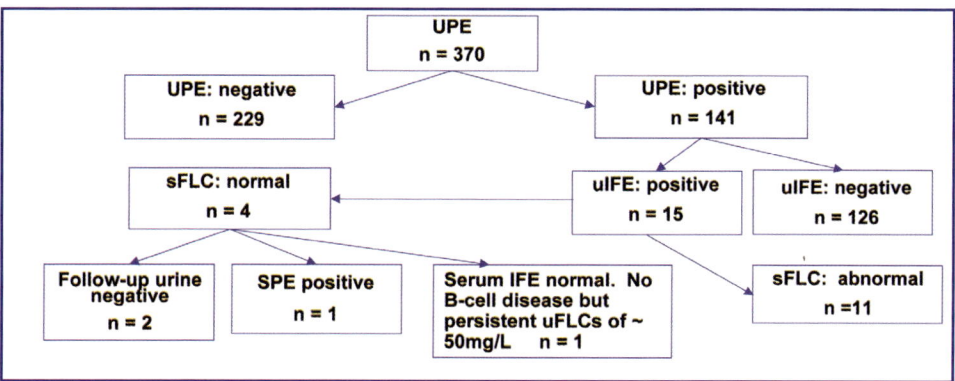

Figure 23.3. Test results on 370 urines screened for monoclonal proteins at a UK Hospital. Of 925 sera sent for screening, 370 (40%) had accompanying urine samples of which 126 were suspicious by UPE but were negative by uIFE. (Courtesy of P Hill).

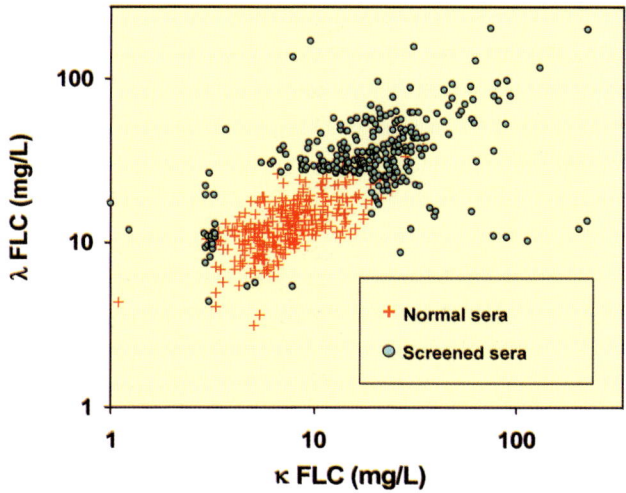

Figure 23.4. Test results on 971 sera screened for monoclonal proteins at Nuneaton, UK *(Reid)*. Patient samples in the normal range have been excluded.

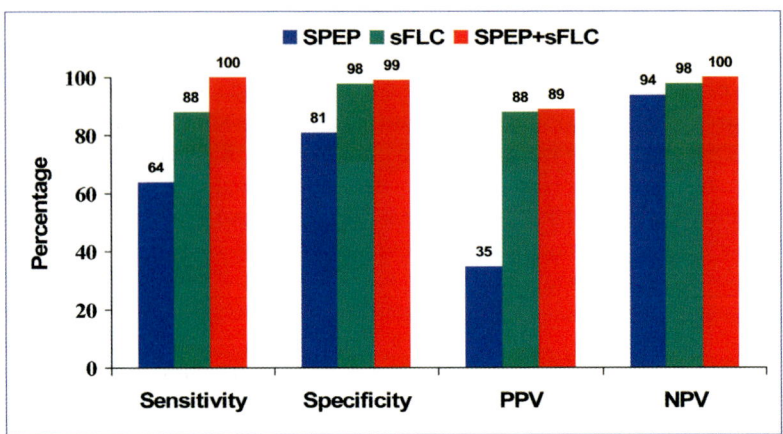

Figure 23.5. Combined assay performance of SPE and sFLC assays for identifying monoclonal gammopathies in a screening study (Abadie). PPV: Positive predictive value, NPV: Negative predictive value. (Courtesy of DE Smith).

The two assays in combination offered the most reliable results *(Figure 23.5)*.

5. Foray and Chapuis-Cellier studied 75 patients with FLC monoclonal gammopathies. They observed 6 patients who were normal by urine tests but had abnormal sFLC results. They concluded that the assays should be used whenever monoclonal FLC diseases were suspected.

6. Katzmann *(4)* et al. in a large retrospective study, made a direct comparison of the relationship between the diagnostic sensitivity of sFLC analysis and urine studies. 428 patients with monoclonal urine proteins were studied, a 10 times larger cohort than any other study *(for details see Chapter 24)*. The clinical diagnoses had been established in all patients. The authors concluded that by adding sFLC analysis to sIFE, urine screening tests were no longer necessary. These results addressed the uncertainties that had been discussed by Katzmann *(1)* in an earlier editorial in the Journal of Clinical Chemistry. Furthermore, the data supported the diagnostic sensitivity of sFLC assays that Katzmann *(3)* had previously reported for assessing patients with known monoclonal gammopathies *(see audit below)*.

7. Beetham et al., in the UK, prospectively investigated 932 consecutive patients by SPE and sFLCs plus serum or urine IFE where appropriate. 449 had serum studies only (because no urine samples were available) of which 53 had monoclonal proteins but importantly, no significant diseases were missed. 3 new cases of serum monoclonal FLCs only were detected but unfortunately no information was provided regarding their clinical relevance. The authors indicated that the results supported other published studies in that SPE, sIFE and sFLC analysis could replace urine studies for detecting monoclonal gammopathies. The other 483 samples were investigated in the context of a paired urine study *(Chapter 24.6)*.

8. Holding et al. assessed impact on disease detection from 753 patient sera tested by SPE and FLC and 128 patient matched associated urine samples. Sensitivity of FLC for

BJP was 98%. Use of FLC in routine testing increased the number of monoclonal gammopathies detected by 7%.

9. Piehler et al. evaluated, in a general hospital population, the combination of SPE and sFLC analysis where additional clinical/biochemical information suggested a myeloma related disorder. Of the 332 patients investigated, 85 were diagnosed with a monoclonal gammopathy (MGUS, plasmacytoma, MM, WM and AL amyloidosis). The inclusion of the sFLC measurement detected 8 additional monoclonal gammopathies: 5 LCMM, 1 NSMM, 1 AL amyloidosis patient and 1 MGUS. One NSMM, one plasma cell leukemia and two plasmacytomas were not detected by the combined approach. The authors concluded that the combined approach is capable of detecting nearly all patients with clinically relevant monoclonal gammopathies. They also highlighted the number of borderline κ/λ ratios and thus the importance of considering additional clinical and laboratory parameters when interpreting sFLC results.

10. Robson *(4)* et al. examined the additional value of sFLC analysis on all 653 SPE requests over a 2 month period. The sFLC test was positive in 43 individuals, 17 of which were also positive by SPE/IFE. 5 of the remaining 26 sFLC positive patients had previously been identified as CLL and 21 were being screened for M-proteins; one was an IgD myeloma patient, 1 an IgA MGUS and 1 had CLL. A further 14 had no other indication of haematological malignancy and ratios close to normal (0.2 – 0.25 or 1.7-2.0) and were not investigated further. The remaining 4 patients had more extreme ratios and were classified as FLC MGUS and annual follow up was recommended. Notably, the poor provision of urine samples (4.6%) meant that sFLC analysis provided the most effective means of determining monoclonal FLC production.

11. Vermeersch *(2)* et al. examined combinations of SPE, sFLC, sIFE and uIFE analysis on 833 consecutive patients being investigated for monoclonal gammopathy. 28 were diagnosed with malignant plasma cell disorders, 25 with B cell NHL and 156 with MGUS. sFLC was abnormal in 24 of the patients with malignant plasma cell disorders (15 IIMM, 2 LCMM, 3 AL amyloidosis, 2 WM, 1 plasma cell leukaemia, 1 osteosclerotic myeloma). Nine of the 25 B-NHL patients and 44 of the 156 MGUS patients had abnormal sFLC results. SPE and FLC (with follow up IFE) had a slightly higher sensitivity than SPE and uIFE (with follow up IFE) (81.8% vs 82.3%). The authors noted their observed specificity for sFLC (96.8%) to be similar to that reported by Hill et al. The authors highlighted the lower sensitivity of sFLC analysis for MGUS. This is of limited concern as risk stratification would suggest a lower risk of progression to a malignant plasma cell disorder for these individuals *(Chapter 19, Rajkumar 5)*.

12. To date, the most extensive study aimed at evaluating screening strategies for the detection of monoclonal gammopathies was carried out at the Mayo Clinic with 1877 samples tested across 5 assay formats (SPE, UPE, sIFE, uIFE and sFLC), all within 30 days of initial diagnosis *(Katzmann 8)*. The cohort comprised 467 MM, 191 ASMM, 524 MGUS, 29 plasmacytoma, 26 WM, 581 AL amyloidosis, 18 LCDD and 31 POEMS patients. The sensitivities for each testing combination across the separate plasma cell disorders is shown in Table 23.3. When comparing the combination of SPE and sFLC with SPE, sFLC and sIFE, 58 patients were missed (44 MGUS, 7 POEMS, 5 AL amyloidosis, 1 plasmacytoma and 1 ASMM). However, no MM, WM or LCDD patients were missed. Furthermore the addition of the sFLC assay identified 23 AL amyloidosis patients, 6 MM and 1 LCDD that were not detected by the traditional panel of serum and urine tests. The omission of urine analysis from the testing panel missed 23 patients (15 MGUS, 1 extramedullary myeloma, 1 LCDD and 6 AL amyloidosis patients). As part of their conclusion, Katzmann et al. stated that due to the small incremental sensitivity provided by urine studies and sIFE, the use of SPE with sFLC analysis provides a simple and efficient initial diagnostic screen for the high tumour burden monoclonal gammopathies such as MM, WM and ASMM. They observed that urine studies and sIFE may be ordered more selectively.

The results from the 12 studies described above are consistent. Adding sFLC assays to the SPE/IFE tests for routine screening (in the absence of urine studies), represents a sensitive combination and increased the tumour pick-up rate by approximately one patient per 100 samples tested. Some of the additional patients identified had significant changes to therapy resulting from their earlier diagnoses. It is of interest that many of the patients identified with this strategy had B-cell non-Hodgkin lymphoma or CLL *(Chapter 18)*.

The diagnostic value of sFLC testing at the intial screening stage has recently been reviewed *(Mayo; Jagannath; Pratt 3; Katzmann 9)*. Furthermore, the international Myeloma Working Group (IMWG) has recently published specific guidelines on the use of sFLC analysis in diagnosis/screening, monitoring and prognosis *(Chapter 25)*. These state that in the context of screening for the presence of MM or related disorders, the sFLC assay in combination with SPE and sIFE yields high sensitivity and negates the requirement for 24h-urine studies for diagnosis other than AL amyloidosis *(Dispenzieri 7)*.

Diagnosis, %	n	All 5 tests	SPE & IFE; uIFE	SPE, IFE & sFLC	SPE& sFLC	sIFE	SPE	sFLC
All	1877	98.6	97.0	97.4	94.3	87.0	79.0	74.3
MM	467	100.0	98.7	100.0	100.0	94.4	87.6	96.8
Macroglobulinemia	26	100.0	100.0	100.0	100.0	100.0	100.0	73.1
ASMM	191	100.0	100.0	100.0	99.5	98.4	94.2	81.2
MGUS	524	100.0	100.0	97.1	88.7	92.8	81.9	42.4
Plasmacytoma	29	89.7	89.7	89.7	86.2	72.4	72.4	55.2
POEMS	31	96.8	96.8	96.8	74.2	96.8	74.2	9.7
Extramedullary plasmacytoma	10	20.0	20.0	10.0	10.0	10.0	10.0	10.0
Primary AL	581	98.1	94.2	97.1	96.2	73.8	65.9	88.3
LCDD	18	83.3	77.8	77.8	77.8	55.6	55.6	77.8

Table 23.3. Sensitivity of monoclonal gammopathy screening panels *(as shown by Katzmann (8) et al. (screening study 12, above).*

23.4. Audit of sFLC usage

Katzmann *(3)* et al. performed an audit of sFLC analysis for the year 2003, on patients attending the Mayo Clinic *(Table 23.4)*. Of the 1,020 patients, 88% had plasma cell disorders. All 120 patients who had no plasma cell disorder had normal κ/λ ratios, in spite of a complex variety of different diseases. Thus, there were no false positive results in this study. The diagnostic sensitivity for AL amyloidosis by serum κ/λ ratios was 91% compared with 69% for sIFE and 83% for uIFE. Serum κ/λ ratios plus sIFE produced a sensitivity of 99% which was not improved by adding uIFE *(Table 15.1)*. For 5 NSMM patients at clinical presentation, all had abnormal κ/λ ratios (100%). 5 of the 6 NSMM patients who had normal sFLCs had achieved complete clinical remission after peripheral blood stem cell transplant (PBSCT) and had normal bone marrow plasma cell content. The conclusion of the study was that the performance of the FLC assays, in a prospective analysis, matched the results from published retrospective validation studies.

1,020 sample requests	
120	normal individuals had normal κ/λ ratios
900	**with plasma cell disorders comprising:**
330	MM
269	AL amyloidosis (91% sensitivity at diagnosis)
115	MGUS (44% sensitivity)
72	ASMM (88% sensitivity)
22	Plasmacytoma
20	NSMM (all of 5 at diagnosis were abnormal)
9	Waldenström's macroglobulinaemia
7	LCDD (100% sensitivity)
56	Miscellaneous

Table 23.4. Audit of sFLC requests at the Mayo Clinic for the year 2003. Sensitivity for detection by FLC analysis was available for some of the diseases.

Test Questions

1. *How many extra patients with monoclonal immunoglobulins are detected when screening symptomatic patients with sFLC assays?*
2. *How do sFLC assays fit into routine testing for monoclonal proteins?*

Answers
1. *20%-50% more than by SPE and 1% extra in the population (Section 23.3, page 199).*
2. *FLC tests should be added to SPE/IFE tests (Section 23.2, page 198).*

24.1. *Introduction*   208
24.2. *Renal threshold for FLC excretion*   208
24.3. *Problems collecting satisfactory urine samples*   210
24.4. *Problems measuring urine samples*   211
24.5. *Clinical benefits of sFLC analysis*   212
24.6. *Elimination of urine studies when screening for monoclonal gammopathies*   213
24.7. *Comparison of sFLCs and urinalysis for monitoring patients*   217
24.8. *Organisational, cost and other benefits of sFLC analysis*   218
       Clinical case history No 11   219

Summary:

1. Normal renal tubular metabolism prevents urine excretion of small amounts of monoclonal FLCs.

2. This mechanism ensures that some patients have abnormal sFLCs but normal urine.

3. Urine studies are more difficult and more expensive than sFLC tests.

4. Urine is usually unavailable at initial patient screening.

5. Screening symptomatic patients for monoclonal gammopathies using serum protein electrophoresis (SPE) and sFLCs is a clinically sensitive strategy and eliminates the need for urine studies except for AL amyloidosis.

24.1. Introduction

The purpose of this chapter is to bring together all the arguments for the use of serum rather than urine for FLC measurements. The preceding chapters have covered most of the issues, and in some detail, but they are scattered throughout the book rather than being focussed into a single coherent discussion. Since there is a residue of informed opinion that continues to favour urine over serum measurements, it is time to persuade them otherwise. An analogy with diabetes mellitus is helpful. 40 years ago, all patients were monitored using urine glucose tests. Now they are monitored using blood glucose because of its overwhelming clinical advantages. Because glucose and FLCs are handled in a similar manner by the kidneys, similar benefits accrue from serum over urine, for FLC analysis.

"If free light chains are in the urine they are always in the serum first."

24.2. Renal threshold for FLC excretion

As described in Chapter 3, sFLCs are primarily cleared through the renal glomeruli and then metabolised in the proximal tubules of the nephrons. Only when the tubular absorptive capacity is exceeded are significant amounts of FLCs seen in the urine as *"overflow proteinuria"*. Since normal production is about 500mg/day and the renal absorptive capacity is 10-30g/day, production must increase many times before urine contains significant amounts of FLCs *(Bradwell 1)*.

The clinical effect of renal tubular absorption on urine FLC (uFLC) concentrations is shown in Figure 24.1. Serum and urine FLC concentrations are compared in 4 patients undergoing treatment *(Alyanakian)*. Patients 1 and 2 had large amounts of serum and urine FLCs with good correlations between changes in concentrations. In patients 3 and 4, urine excretion was minimal and unchanging over many months while serum levels could be used to monitor the changing tumour burden. In spite of similar sFLC concentrations, there was no uFLCs in these latter patients because there was no renal impairment and, therefore, no overflow proteinuria.

The concentrations of monoclonal sFLCs necessary to cause overflow proteinuria was studied by Nowrousian et al. in a group of patients attending a myeloma clinic. In 131 samples from patients with elevated monoclonal serum κ FLC concentrations, 82 had uFLCs by IFE while 49 had FLC-negative urine *(Figure 24.2)*. The median serum κ FLC concentrations associated with monoclonal FLCs in urine was 113mg/L (range 7-39,500) and for normal urines 40mg/L (range 6-710). Monoclonal λ FLC producing patients had median serum values of 278mg/L (range 5-7,060) for FLC positive urines and 44mg/L (range 3-561) for FLC negative urines. The wide range of renal thresholds observed presumably reflected different degrees of renal damage.

Thus, for κ-producing patients, median serum levels associated with abnormal urine FLCs were 5-fold above normal (upper limit of normal range: 19.4mg/L). For λ patients median sFLCs were 10-fold above normal (upper limit of normal range: 26.3mg/L)

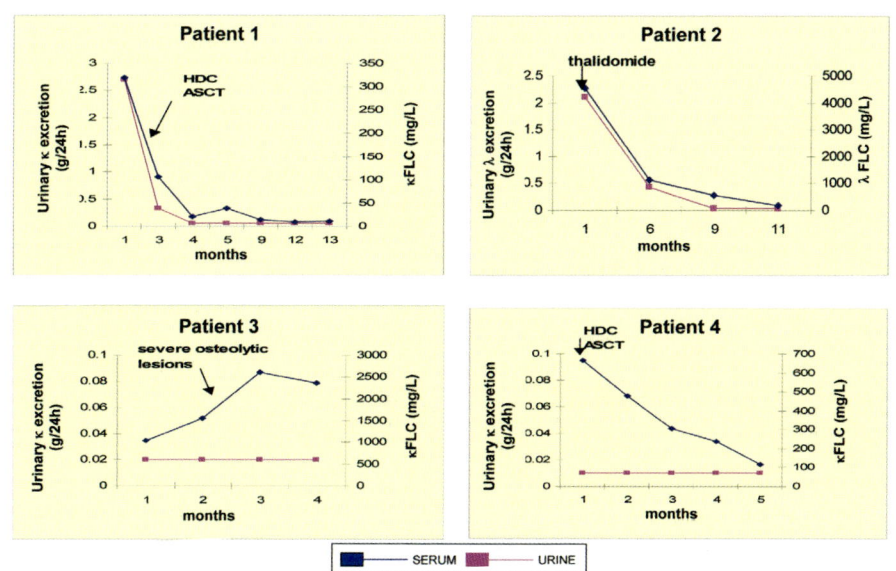

Figure 24.1. Serum and urine FLCs in 4 patients with light chain multiple myeloma (LCMM). Serum tests (blue) were useful even when urine FLC excretion (pink) was minimal in patients 3 and 4. (Courtesy of M-A Alyanakian).

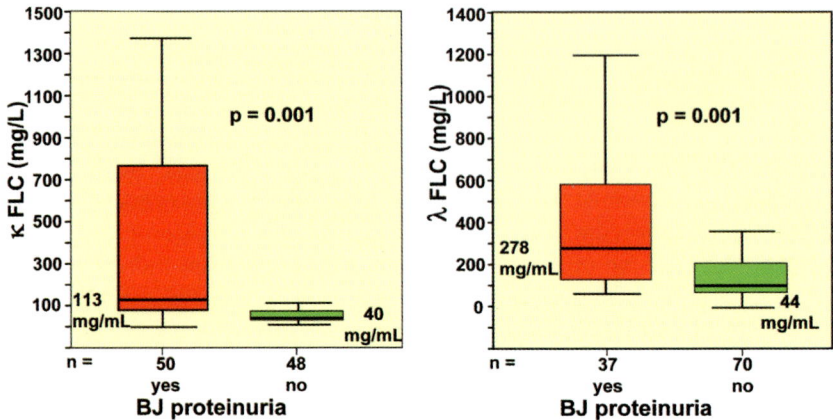

Figure 24.2. Amounts of sFLCs required to produce light chain proteinuria for κ and λ myelomas. Median, 95% and ranges are shown. (Courtesy of MR Nowrousian).

when the urine contained monoclonal FLCs. The higher serum levels necessary for λ overflow proteinuria can be explained by the dimerisation of λ molecules. This reduces glomerular filtration compared with monomeric κ molecules *(Chapter 3)*.

Thus, when FLC production is below the renal clearance threshold, serum tests are more reliable than urinalysis. This study by Nowrousian et al. showed that the extra sensitivity of the serum tests translated directly into greater clinical sensitivity for evaluating disease stage *(Figure 11.9)*. This is particularly relevant for identifying patients with residual disease when urine assessments indicate complete remission *(Chapter 12),* and has been incorporated into international guidelines *(Chapter 25)*.

24.3. Problems collecting satisfactory urine samples

Even if there is significant urine excretion of FLCs, accurate quantification requires a proper 24-hour urine collection. This may be particularly difficult because:

- Accurate timing of the collection is hard for ill patients.
- Large volumes are produced in polyuric patients - perhaps larger than the bottle volume.
- Night-time collections are difficult for patients with painful or fractured bones.
- Problems occur sending voluminous urines to the laboratory by post.
- Collections may be demeaning in front of friends or work colleagues.

Hence, even if there is significant renal leakage of FLCs, urine measurements may not be as reliable as those in serum. Figure 24.3 compares serum and urine results in a patient with relapsing LCMM. The concentrations of the FLCs in both fluids are considerably elevated indicating that the renal threshold is exceeded (compare with patients 3 and 4 in Figure 24.1). However, the urine measurements are highly variable and do not show a definitive rise until day 160. In contrast, the steady rise in sFLC concentrations from day 40 indicates relapse of the tumour 3-4 months earlier. Presumably, the 24-hour urine

Figure 24.3.
Serum and urine FLCs in a patient with LCMM. Disease relapse can be identified 3 months earlier using serum rather than urine samples. (Courtesy of G Wieringa).

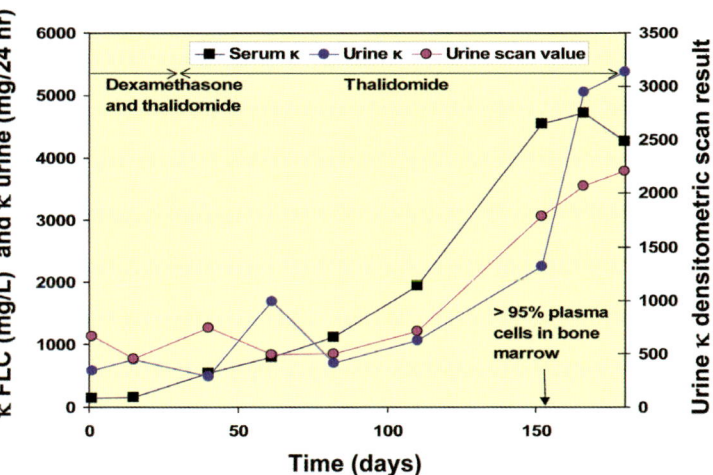

collections were inaccurate, but there may have been additional inaccuracies in the measurements of the monoclonal FLC by urine protein electrophoresis (UPE) (*see below*).

24.4. Problems measuring urine samples

uFLC measurements are normally based upon electrophoretic tests (*Table 4.1 and Chapter 6*). These may require samples to be concentrated prior to analysis by up to 200-fold. They are then analysed by UPE and scanning densitometry or by IFE with visual interpretation.
Problems interpreting FLC bands in urine samples include (*Chapter 6.6*):

- High background staining in the presence of heavy proteinuria.
- Ladder banding - false bands that may hide monoclonal FLCs.
- Difficulties identifying the correct band amongst other protein bands.
- Poor precision compared with immunoassays.
- Non-specificity of antibodies used on IFE.

Consistent with the various technical issues that may affect urine results, Siegel (2) et al. reported that urine electrophoresis results can be highly susceptible to error. Analysis of 623 24h-urine collections showed fluctuating urine electrophoresis results. 19% of samples demonstrated spuriously increased monoclonal levels. In contrast, the sFLC values did not show these fluctuations.

As an alternative to urine electrophoretic tests, FLC immunoassays can be used on urine samples. Nowrousian et al. compared the sensitivity of urine FLC immunoassays with urine IFE (uIFE) in patients with multiple myeloma (MM). 98 κ and 107 λ samples that had abnormal serum κ/λ ratios were studied. Urines positive by IFE contained a median of 448 mg/L of κ (range: 5-70,800) and 313 mg/L of λ (range: 17-11,100) by FLC immunoassays (*Figure 24.4*). Urines negative by IFE contained a median of 23 mg/L of κ (range 0-251) and 9mg/L of λ (range 1-196) by FLC immunoassays. Similar findings have been reported by others who concluded that uIFE was more reliable for

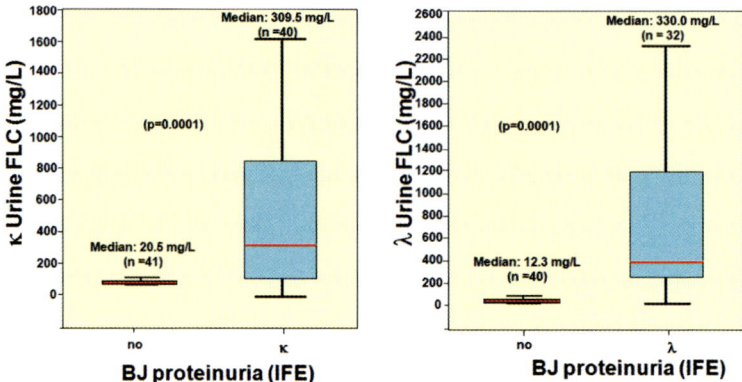

Figure 24.4. Comparison of FLC immunoassays and IFE for detecting uFLCs. Median, 95% and ranges of FLCs are shown. (Courtesy of MR Nowrousian).

detecting monoclonal diseases than urine FLC analysis by immunoassay *(Viedma)*. Alternatively, one could conclude that urinalysis by FLC immunoassays and IFE are complementary.

Because the ranges of FLC concentrations and κ/λ ratios in normal sera are far narrower than in urine, they are clinically more reliable *(Chapter 5.3)*. Furthermore, uFLC immunoassays do not solve any of the renal threshold, urine collection and urine measurement problems indicated above.

There are, of course, many other problems with urinalysis. Urine is less easily handled than serum; samples may be unpleasant; they need to be stored in large volumes if further analysis is required, and FLCs are more prone to precipitation in urine compared with serum.

24.5. Clinical benefits of sFLC analysis

The improved sensitivity of serum over urine FLC measurements has had a major impact on the ease of diagnosis, monitoring and assessing risk of progression for many patients with the following diseases:

1. Light chain multiple myeloma (LCMM) *(Chapter 8)*.
2. Nonsecretory multiple myeloma (NSMM) *(Chapter 9)*.
3. Intact immunoglobulin multiple myeloma (IIMM) *(Chapters 11 and 12)*.
4. Asymptomatic multiple myeloma (ASMM) *(Chapter 14)*.
5. Myeloma kidney *(Chapter 13)*.
6. Plasmacytoma *(Chapter 18)*.
7. AL amyloidosis *(Chapter 15)*.
8. Light chain deposition disease (LCDD) *(Chapter 17)*.
9. Monoclonal gammopathy of undetermined significance (MGUS) *(Chapter 19)*.

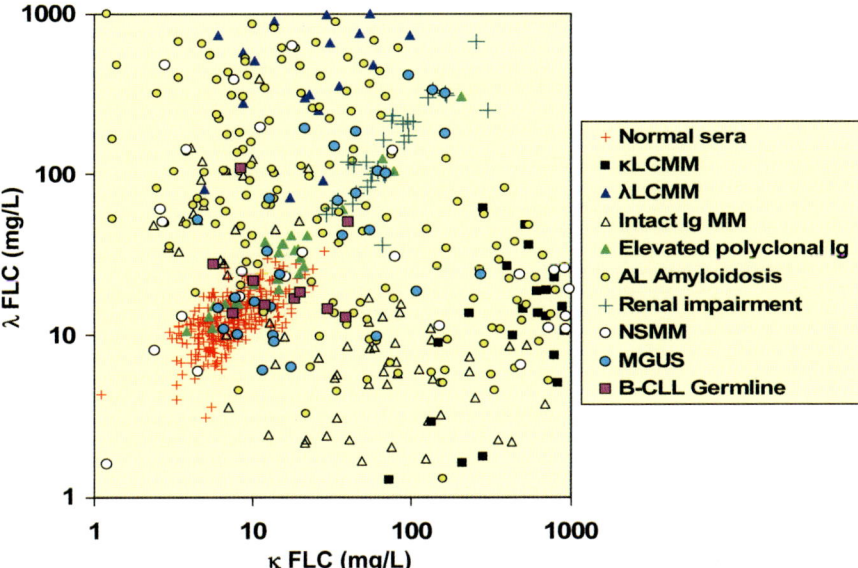

Figure 24.5. sFLCs in patients with low monoclonal immunoglobulin production rates. By electrophoretic tests, sFLCs are usually undetectable or unquantifiable in most of these patients. (See relevant chapters for details of the patient data).

Figure 24.5 shows a set of sFLC concentrations in patients with low production rates at the time of clinical diagnosis. Samples from patients with NSMM are shown as white circles that, by definition, have no detectable monoclonal proteins by both serum and urine electrophoretic tests. Hence, other patients with monoclonal sFLCs at or below these concentrations but with other types of plasma cell dyscrasias are difficult to identify by conventional tests. The figure also includes samples from many patients with AL amyloidosis and IIMM who were in remission by IFE.

24.6. Elimination of urine studies when screening for monoclonal gammopathies

An extensive comparison of the relative diagnostic contributions of serum and urine studies for the detection of plasma cell proliferative disorders has recently been reported *(Katzmann 8)*. Samples from 1,877 untreated patients with various diseases *(Table 24.1)* received a full panel of screening tests (SPE, UPE, sIFE, uIFE and sFLCs). This facilitated the determination of the more sensitive combinations of screening tests and addressed the question of whether sFLC analyses could replace urine studies.

The combined results from all five tests identified 1,851 (98.6%) samples as abnormal, of those not detected 11 were AL amyloidosis (1.9% of total), 8 extramedullary plasmacytoma (80%), 3 plasmacytoma (10.3%), 3 LCDD (16.7%) and 1 POEMS syndrome (3%). The combination of serum and urine IFE in the absence of sFLC tests resulted in a further 6 MM, 23 AL amyloidosis and 1 LCDD being missed.

Diagnosis	No. of Patients
Multiple myeloma	467
AL amyloidosis	581
LCDD	18
Waldenströms macroglobulinemia	26
Plasmacytoma	29
Extramedullary plasmacytoma	10
POEMS	31
Asymptomatic smouldering myeloma	191
MGUS	524

Table 24.1. Clinical diagnosis of the 1,877 patients studied by Katzmann *(8)* et al.

In contrast, the combination of sFLC and serum IFE in the absence of urine IFE resulted in 6 AL amyloidosis, 15 MGUS, 1 extramedullary MM and 1 LCDD being missed (in addition to those not detected by all five tests combined). It can be concluded from this that sFLC analysis can play an essential part in the primary screen through contributing to the detection of a significant number of malignant diseases, particularly MM and AL amyloidosis which would not be detected by the less sensitive urine IFE. The urine IFE allowed the detection of 6 AL amyloidosis and 1 LCDD patient that were normal by the serum tests; similar results have been reported by Palladini *(5)* et al.

It may not always be practical for laboratories to perform serum and urine IFE on all samples and so the authors also considered the possibility of avoiding urine analysis by combination of SPE and sFLC analyses with follow up sIFE where indicated *(Figure 24.6)*. This simplified algorithm showed the same sensitivity as sIFE plus sFLC for MM and WM but missed a further 44 MGUS patients, 7 POEMS, 5 AL amyloidosis, 1 plasmacytoma, 1 ASMM and 1 LCDD. The 44 MGUS patients, by risk stratification *(Chapter19)*, were considered low risk for progression to malignant disease. The impact of savings from reduced follow-up of low-risk MGUS and reduced patient anxiety was considered a reasonable outcome of this screening algorithm. The overall findings of the study provided significant support for the IMWG guidelines *(Dispenzieri 7)* that recommended screening using a combination of serum electrophoresis plus sFLC tests and that uIFE needed only be additionally carried out to maximise the sensitivity when AL amyloidosis was suspected.

The guidelines are supported by numerous smaller screening studies. One *(Katzmann 4)* used a historical cohort of 428 unselected patient samples consisting of a variety of plasma cell proliferative diseases. These were analysed with all 5 screening tests. Pre-selection criteria of the cohort included the requirement to have a monoclonal protein detected by uIFE. This facilitated the study aim of determining if serum tests could identify all samples found positive by uIFE.

Performance of the various tests and combinations are shown in Table 24.2. sIFE was the most sensitive serum test at 93.5% followed by sFLC κ/λ ratios at 85.7%. Of the 61

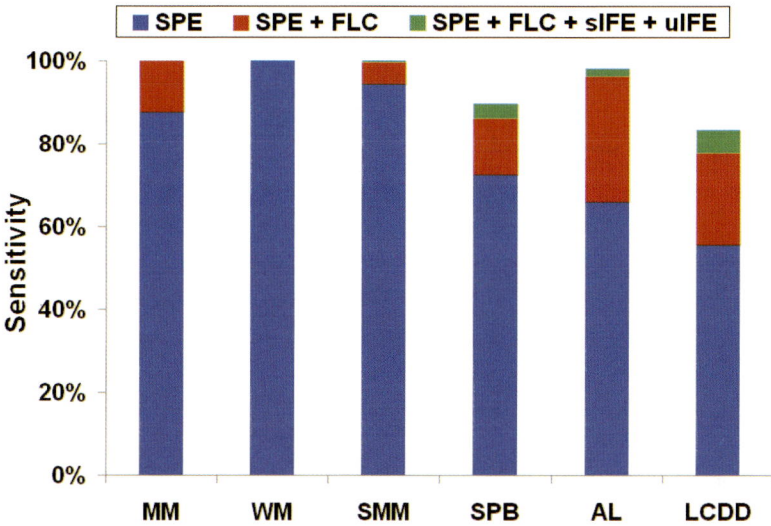

Figure 24.6. Sensitivity of different testing panels for a range of lymphoproliferative disorders *(Katzmann 8)*. Urine IFE provides no additional sensitivity for MM and WM in comparison to SPE and FLCs.

uIFE positive patients with normal sFLCs, 30 had intact monoclonal immunoglobulins in the urine but no urine FLCs. The sFLC assays, therefore, missed 31 of the 61 urine FLC positive monoclonal samples, so that the diagnostic sensitivity for sFLC detection was 93%. All but 2 of these patients were identified by sIFE. One missed sample was from a patient with a uFLC only MGUS. The other was a monoclonal urine IgAκ that was considered to be a contamination as it was not found in subsequent samples from the patient and was absent in the serum. Clinically, neither required medical attention.

28 patients in the study had negative sIFE results of which 19 had AL amyloidosis, 3 solitary plasmacytoma, 3 MGUS, 2 MM and 1 ASMM. All these were identified both by sFLC analysis and by urine studies. The authors concluded that by adding sFLC analysis to sIFE, urine screening tests were no longer necessary. Furthermore, since urine samples are frequently not included with initial diagnostic serum samples, sFLC testing has considerable diagnostic utility. For example, Robson *(4)* et al. reported a urine compliance of <5% in a study carried out at New Cross Hospital, Wolverhampton *(Table 24.3)*. The extremely poor provision of matched urine samples led the authors to comment that "the debate over the relative merits of the sFLC assay versus urine BJP analysis borders on the irrelevant". This is in addition to its value as a prognostic marker in patients with MM, ASMM, MGUS and AL amyloidosis etc.

Fulton et al. retrospectively analysed 219 sFLC requests that had matching SPE, sIFE, UPE and uIFE test results. They showed that the sFLC ratio was abnormal in 12% more samples than using UPE plus uIFE. Furthermore, the combination of an SPE and sFLC analysis allowed the detection of 6% more abnormal samples than the combined use of SPE, sIFE, UPE and uIFE. The authors, therefore, proposed the incorporation of sFLC

Laboratory test	No. (%) abnormal
uIFE	428 (100)
sIFE	400 (93.5)
SPE	346 (80.8)
sFLC κ/λ ratio	367 (85.7)
sIFE or κ/λ ratio	426 (99.5)

Table 24.2. Diagnostic sensitivity of various tests for monoclonal proteins in patients with positive uIFE

Study	Number of sera	Urine compliance
Hill et al. 2006	923	40%
Holding et al. 2007	753	17%
Beetham et al. 2007	932	52%
Abadie et al. 2009	-	35%
Robson et al. 2009	653	<5%

Table 24.3. Published urine compliance data at five centres.

tests into a primary screening algorithm, thus reducing the amount of labour intensive sIFE and uIFE.

A further screening investigation on 3,818 sera received for SPE analysis over a 1 year period has also been reported *(Piehler)*. To minimise the requirement of sFLC testing the authors used a series of criteria based on clinical presentation and other laboratory tests to determine the need for a sFLC evaluation; thus, 1,067 sFLC tests were carried out. By combining SPE, sFLC and clinical details their algorithm resulted in the detection of 95% of all new patients with MM, WM or AL amyloidosis. Overall, 4 patients with clinical disease were missed (1 NSMM, 1 plasma cell leukemia and 2 plasmacytomas), however all 4 were negative by urine electrophoresis.

A prospective screening study of 370 patients by Hill et al. directly compared the clinical sensitivity of sFLC tests and UPE. 15 samples were Bence Jones protein positive of which 11 were also sFLC positive. The 4 discordant samples (1%) were of no clinical consequence *(Chapter 23.3)*, so no significant disease was missed.

A similar, prospective, serum versus urine study of 483 patients was recently published by Beetham et al. Monoclonal proteins were detected in 105 (22%) patients of whom 34 had Bence Jones proteins at greater then 5mg/L. 8 of these 34 samples had normal sFLC κ/λ ratios. However, 7 were positive by SPE/sIFE for intact monoclonal immunoglobulins, so only one sample of 105 positive urines was not identified by serum studies (0.2%). This patient was considered to have a urine only MGUS of no apparent clinical consequence (<50mg/L). These results supported other published studies in that

SPE/sIFE and sFLC analysis can, in practice, replace urine studies.

However, while the serum tests gave the same practical results as urine tests, Beetham et al. expressed some disquiet as to why monoclonal proteins were present in urine when sFLC κ/λ ratios were normal. There are a variety of mechanisms whereby this may occur, which are discussed below and elsewhere *(see FcRn transport mechanism, Chapter 10)*. Also, it might have been useful if the authors had assessed the urines of the discordant samples by quantitative FLC immunoassays rather than assume that the electrophoretic techniques gave the correct results. Furthermore, 5 samples had abnormal sFLC κ/λ ratios as the only significant anomaly, but no clinical details were provided. The data of Katzmann *(3)* et al. suggests that AL amyloidosis or another subtle B-cell dyscrasia should have been considered.

24.7. Comparison of sFLCs and urinalysis for monitoring patients

There are huge clinical benefits derived from the improved sensitivity and specificity of sFLC analysis versus urine studies for monitoring patients with light chain diseases. The main studies are reported in detail in the appropriate clinical sections. However, as an example, Mazurkiewicz et al. in a small UK study, considered that 11 of 16 patients might have had better treatment decisions based upon sFLCs rather than the urine studies that had been performed.

Nevertheless, sFLC κ/λ ratios are occasionally normal when urine tests for FLCs are abnormal. Some of the discrepancies may be explained on technical grounds or on sampling errors *(Chapter 6.6)*. For instance, when monitoring patients, 24-hour urine collections may be taken earlier than the corresponding serum samples. Since FLC concentrations can fall rapidly following treatment *(Chapter 13)*, urine samples collected a few days before attending the clinic for sFLC analysis could produce quite different results. The sensitivity of the urine measurements is also highly dependent upon technique. Obviously, highly sensitive uIFE gels identify more abnormal samples than UPE used in routine hospital practice.

The frequency of discordant results has been analysed in patients with AL amyloidosis *(Stubbs)*. 260 sera from patients with AL amyloidosis were studied alongside their corresponding urine samples. 5 samples had normal sFLC κ/λ FLC ratios but more than 200mg/day of uFLCs by UPE *(Table 24.4)*. The serum samples were taken during the period of the 24-hour urine collection to avoid errors of timing and all gels were carefully analysed retrospectively.

One of the sFLC results in the 5 patients was borderline abnormal. Since the normal serum reference range for FLCs was based on a screening study that included all samples tested, it may be that when monitoring patients, narrower normal range criteria can be used *(Chapter 5)*. Also, these patients usually have a degree of renal failure that needs to be considered in all patients with borderline results *(Chapter 20)*. Slower clearance of monomeric κ molecules occurs when glomerular filtration is impaired so that serum κ/λ ratios increase slightly. In contrast, urine polyclonal FLCs increase, with a greater proportion of λ FLCs because of the impaired κ clearance. This may explain the predominance of λ discordant patient samples in Table 24.4.

	uIFE	UPE FLC mg/day	Urine κ (mg/L)	Urine λ (mg/L)	Serum κ/λ ratio
1	Lambda	530	70.2	53.5	0.31
2	Lambda	700	211	152	0.45
3	Lambda	210	96.5	69.8	0.62
4	Lambda	380	130	98	0.70
5	Lambda	410	117	42.6	0.72

Table 24.4. Discordant results in 5 patients with AL amyloidosis who had abnormal uFLCs by IFE and UPE but normal sFLC κ/λ ratios *(see Chapter 6)*.

Quantification of the urine using FLC immunoassays showed lower concentrations than suggested by UPE and all 5 had normal κ/λ ratios *(Table 24.4)*. It may be that UPE testing is inaccurate because polyclonal FLCs are included in the analysis. Moreover, low concentration urine monoclonal proteins in patients with normal sFLCs and κ/λ ratios are of doubtful clinical importance. International guidelines indicate that monoclonal uFLCs below 200mg/24 hours are of little clinical relevance *(Chapter 25)*.

These results suggest that it may be unwise to treat patients with chemotherapy based only upon minor urine monoclonal FLC bands detected by IFE when sFLC tests are normal. In spite of these observations, there is support for the continuing use of 24-hour urinalysis for monitoring patients. This is based upon the clear evidence that urine tests can be positive when serum is negative in some patients *(Singhal; Dispenzieri 6)*. Further studies are required.

24.8. Organisational, cost and other benefits of sFLC analysis

As well as improved clinical diagnosis, there are organisational, cost and other benefits of introducing sFLC assays. The laboratory issues were analysed at the Christie Hospital in Manchester, UK *(Carr-Smith 2)*. Superior analytical performance of the serum assays, faster reporting times and reduced laboratory costs were clearly identified *(Table 24.5)*. Cost benefits in relation to clinical outcomes were not analysed in this study but they may accrue from earlier diagnosis and treatments that reduce morbidity.

Hill et al. compared the costs of urine electrophoretic tests with sFLC immunoassays in routine screening of patients for monoclonal gammopathies *(Chapter 23)*. They considered that on a patient-by-patient basis the costs were increased by less than £5. But, in their study only 40% of serum samples were accompanied by urines so the overall costs increased considerably. However, better clinical governance was achieved and more clinical diagnoses were made.

Katzmann *(3)* et al. compared the costs of sFLC screening with urine testing. The 2006 Medicare reimbursement for sFLCs was $38 compared with urine assays at $71 (total protein, UPE and uIFE). With sFLC tests costing approximately half the urine tests per patient, considerable laboratory savings could be made.

There have been no direct assessments of clinical cost benefits. However, one relatively simple clinical situation that may be improved by measuring sFLCs is when determining the underlying pathology of patients presenting with acute renal failure *(Chapter 13)*. If MM is suspected, the normal procedure is to perform SPE/IFE and UPE. A simpler and better approach would be to measure sFLCs. This would identify all patients with FLC as a cause of their renal impairment.

If a choice has to be made between serum or urine tests then serum is clearly preferable for the many reasons given above *(Table 24.6) (Carr-Smith 4)*. When both serum and urine tests are available, it is clinically reassuring to have 2 separate tests giving the same results. Clearly, samples do occasionally get incorrectly analysed, mislabelled or misplaced, so supporting evidence for making a diagnosis or changing treatment is always helpful. In the context of a stem cell transplant in MM patients, for example, the additional cost of performing both serum and urine tests is inconsequential. And, there are some patients with positive uFLCs and negative sFLCs although the clinical importance is unclear.

New national and international guidelines for patients with monoclonal gammopathies include recommendations on sFLC analysis in relationship to urine tests *(Chapter 25)*.

Table 24.5. Analytical and cost/benefit study of sFLC and urine electrophoresis tests.

	sFLCs	Urine electrophoresis
Sensitivity	1.5mg/L	50mg/L
Precision	5%	>15%
Analysis time	15 minutes	1 hour
Reporting turnaround	1 hour	1 week
Cost per year (700 requests)	£6,500	£4,500
Extra staff costs per year	£0	£1,000
24hr urine bottle usage	Not relevant	£1,000
Storage needs	30cm^3	10m^3

Serum versus urine measurements	
sFLC	**Urine electrophoresis**
Easy to collect	Difficult to collect
κ/λ ratio little affected by renal function	Renal function affects levels
Easily analysed	Samples may need concentrating
Easily stored	More difficult to store
More frequently abnormal in NSMM and AL amyloidosis	Less frequently abnormal
More sensitive for monitoring patients	Less sensitive for monitoring patients
Of prognostic importance in MGUS	Of no importance in MGUS

Table 24.6. Summary of clinical and analytical comparisons of sFLC and urine electrophoresis tests.

Clinical case history No 11. Is urine examination a mandatory procedure?

An 86-year-old woman presented to her general practitioner (GP) with a short history of malaise and weight loss. Initial investigations included erythrocyte sedimentation rate (ESR), which was raised at 92mm/hr and haemoglobin at 11.5g/dL. The GP also requested a serum immunoglobulin profile and SPE, the results of which were interpreted as an acute phase response. No urine investigations were requested. Three months later, still suffering from malaise, her ESR was 103mm/hr.

Two months later, on admission to the Medical Assessment Unit with increasing malaise, her ESR remained at 103mm/hr. A repeat immunoglobulin profile was requested, which showed very similar results with an SPE again showing an acute phase response. On this occasion, however, the blood was accompanied by a request for Bence Jones protein testing on a urine sample. Urinary total protein was 0.49g/L, and UPE and IFE, followed by scanning densitometry, showed the presence of Bence Jones protein at 360mg/L. Following this result, sFLCs were measured and were highly abnormal *(Table 24.7)*. Also, IgD and IgE monoclonal proteins were excluded by IFE. Bone marrow studies were performed which showed 12% plasma cells, i.e., a plasma cell dyscrasia. Subsequent sFLC results are shown in Table 24.7.

In the light of the equivocal percentage of plasma cells in the bone marrow, the patient was diagnosed as having ASMM. During the follow up period, sFLCs were used in preference to the collection of 24-hour-urines for estimation of Bence Jones proteinuria.

Date (nr)	κ FLC (3.3-19.4)	λ FLC (5.71-26.3)	κ/λ ratios (0.26-1.65)	IgG (6.0-16.0)	IgA (0.8-4.0)	IgM (0.33-1.94)
1st sample	104.33	14.39	7.25	ND	ND	ND
13/01/05	98.14	17.12	5.73	10.6	1.81	1.08
16/03/05	105.48	16.77	6.29	9.4	1.87	0.96
18/05/05	109.21	16.31	6.70	9.2	1.80	1.00
20/07/05	114.50	16.56	6.91	ND	ND	ND
26/09/06	110.54	16.69	6.89	11.5	2.08	1.37

Table 24.7. sFLCs and polyclonal immunoglobulin concentrations in a patient with ASMM. ND: not determined; nr: normal range. (Reproduced with permission from Clin Lab *(Sinclair 2) and* D Sinclair).

Test Questions

1. *Is there any remaining role for urine tests for FLCs?*
2. *Why are patients with excess monoclonal serum κ FLCs more likely to have positive urine results?*
3. *How do the costs of sFLC immunoassays compare with urine tests?*

Answers

1. *Not when screening for monoclonal gammopathies and it remains unclear whether there is a remaining role for monitoring patients (Sections 24.6 and 24.7, page 213-218).*
2. *Monomeric κ molecules filter through the glomeruli more readily than λ dimeric molecules (Section 24.2, page 208).*
3. *Serum tests cost less than full urine analysis on a patient-by-patient basis (Section 24.8, page 218).*

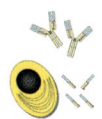

25.1. *International guidelines for the classification of MM and MGUS*   *221*

25.2. *International Myeloma Working Group guidelines for serum free light chain analysis in multiple myeloma and related disorders (2009)*   *222*

25.3. *UK Myeloma Forum, Nordic Myeloma Study Group and the British Committee for Standards in Haematology (2006)*   *223*

25.4. *UK Myeloma Forum and Nordic Myeloma Study Group: Guidelines for the investigation of newly detected M-proteins and management of MGUS (2009)*   *223*

25.5. *International Staging System for Multiple Myeloma (2005)*   *224*

25.6. *"Uniform Response Criteria for MM" incorporating FLCs (2009)*   *224*

25.7. *National Comprehensive Cancer Network. Clinical Practice Guidelines in Oncology: Multiple myeloma 2007*   *226*

25.8. *European Society of Medical Oncology (ESMO)*   *226*

25.9. *USA National Academy of Clinical Biochemistry guidelines*   *226*

25.10. *Guidelines for AL amyloidosis (2005)*   *228*

25.11. *International Myeloma Working Group guidelines for MGUS and smoldering (asymptomatic) multiple myeloma (2010)*   *229*

25.12. *Consensus statement for the screening, evaluation and management of MGUS (2010)*   *230*

International and national guidelines for identifying and managing patients with plasma cell dyscrasias are published and updated on a regular basis. These are widely adopted for assessing new patients and for patient entry into clinical trials. This chapter provides an overview of key available guidelines with an emphasis on-recommendations relating to the serum FLC (sFLC) component (identified in blue). Classification criteria for the less common plasma cell dyscrasias (some of which do not yet include the use of (sFLCs) are described in Chapter 18.

25.1. International guidelines for the classification of MM and MGUS

The most comprehensive guidelines were published in 2003 *(Kyle 10)*, these have subsequently been updated *(Durie 1)*, further clarified in 2009 *(Kyle 15)* and comprise the following:

Symptomatic multiple myeloma (MM)
- Monoclonal-(M-) protein in serum and/or urine.
- Bone marrow (clonal) plasma cells ≥10% or plasmacytoma.
- Related organ or tissue impairment (end organ damage, including bone lesions*).

Nonsecretory multiple myeloma (NSMM)
- No M-protein in serum and urine with immunofixation.
- Bone marrow clonal plasmacytosis ≥10% or plasmacytoma.
- Related organ or tissue impairment (end organ damage, including bone lesions*).

Asymptomatic (smouldering) multiple myeloma (ASMM)
- M-protein in serum (IgG or IgA) ≥30g/L: and/or
- Bone marrow clonal plasma cells ≥10%.
- No symptoms, related organ or tissue impairment (no end organ damage, including bone lesions*).

Monoclonal gammopathy of undetermined significance (MGUS)
- M-protein in serum <30g/L.
- Bone marrow clonal plasma cells <10% and low level of plasma cell infiltration in a trephine biopsy (if done).
- No evidence of other B-cell proliferative disorders.
- No related organ or tissue impairment (no end organ damage, including bone lesions).

*Criteria for end organ damage in MM "CRAB" criteria (**C**alcium increased, **R**enal insufficiency, **A**naemia or **B**one lesions).
Calcium levels increased: serum calcium >0.25mmol/L above the upper limit of normal or >2.75mmol/L.
Renal insufficiency: creatinine >173µmol/L.
Anaemia: haemoglobin 2g/dL below the lower limit of normal or haemoglobin <10g/dL.
Bone lesions: lytic lesions or osteoporosis with compression fractures.
Others: symptomatic hyperviscosity, AL amyloidosis, recurrent bacterial infections (>2 episodes in 12 months).

25.2. International Myeloma Working Group guidelines for serum free light chain analysis in multiple myeloma and related disorders (2009)

These guidelines, published in 2009 *(Dispenzieri 7)*, discuss published evidence for the utility and application of the sFLC assay for most plasma cell disorders and include symptomatic MM, NSMM, light chain multiple myeloma (LCMM), ASMM, MGUS, solitary plasmacytoma and AL Amyloidosis. Furthermore, these guidelines highlight key recommendations for the use of the sFLC assay in screening, prognosis and in the assessment of patient response to treatment. Specific emphasis is placed on distinguishing between proven utility and those potential utilities which remain under investigation.

Screening – The sFLC assay is recommended to be used in combination with SPE and serum IFE and deemed highly sensitive and sufficient to screen for pathological monoclonal plasmaproliferative disorders other than AL amyloidosis, which additionally requires a 24 hour urine IFE.

Prognosis – It is recommended that baseline sFLC assay results are obtained at diagnosis for all patients with MGUS, ASMM or active MM, solitary plasmacytoma and AL amyloidosis. Highly abnormal results have been shown to be prognostic in solitary plasmacytoma and MM. Notably, in MGUS, ASMM and plasmacytoma, a highly abnormal sFLC result indicates a substantial risk of progression to systemic disease.

In monitoring and response assessment – The sFLC assay is recommended for the quantitatitve monitoring of patients with oligosecretory plasma cells disorders, including patients with AL amyloidosis, oligosecretory myeloma and nearly two thirds of patients who had previously been classified as NSMM. Furthermore, in the absence of urinary evaluations or FLC measurements light chain escape can be missed and so these tests should be performed periodically. Baseline results of sFLC testing are required prior to initiating new chemotherapy regimens for all patients with MM to allow for the assessment of stringent complete response where a complete response has been shown.

25.3. UK Myeloma Forum, Nordic Myeloma Study Group and the British Committee for Standards in Haematology (2006) *(A Smith)*

The 2006 guidelines include the following use of sFLC measurements:

Investigation and diagnosis.

Quantification of sFLCs and κ/λ ratio can be used as an alternative to measuring urine FLCs. sFLCs are particularly useful for the diagnosis of LCMM and patients in whom the serum and urine is negative on IFE (NSMM) *(Chapters 8 and 9)*.

Measuring responses to therapy.

sFLCs are useful for monitoring LCMM and NSMM *(Chapters 8 and 9)*. sFLCs may also be helpful in monitoring responses in the many patients with IIMM *(Chapter 10)*. Because of the short half-life of sFLCs this can give an earlier indication of response to therapy than changes in intact monoclonal immunoglobulins.

25.4. UK Myeloma Forum and Nordic Myeloma Study Group: Guidelines for the investigation of newly detected M-proteins and management of MGUS (2009)

These 2009 guidelines *(Bird 2)* recommend SPE and UPE analysis in patients with a clinical presentation suspicious for monoclonal gammopathy: elevated erythrocyte sedimentation rate, anemia, hypercalcemia, renal failure and high or low serum concentrations of total immunoglobulins or the individual classes. IFE analysis is advised, in the absence of a detectable M-protein, where there is a strong clinical suspicion of underlying plasma cell dyscrasia.

Where urine samples are unavailable, sFLC analysis is required for the detection of NSMM and certain cases of AL amyloidosis and LCMM. Serum FLC analysis is also advised where serum immunoglobulin levels are low and no serum M-protein is identified, alternatively a urine sample may be requested for IFE.

The guidelines also discuss the current evidence for differential diagnosis of monoclonal gammopathies including the definitions for diagnosis, stated in Section 25.1 *(Kyle 10)*. Added to this is emphasis on low level monoclonal protein concentrations usually being observed with MGUS, although commenting that some patients within this group will have AL amyloidosis, light chain myeloma and solitary plasmacytoma. The guidelines further discuss prognosis and risk factors for MGUS and its malignant transformation. The value of kappa/lambda light chain ratio, along with M-protein level and immunoglobulin type, in differentiating high and low risk of malignant transformation is discussed *(Rajkumar 1)*.

According to these guidelines *(Bird 2)*, the frequency of monitoring MGUS patients considered at low risk, particularly those with low paraprotein concentrations, could be reduced if actual life expectancy is low and all lymphoproliferative diseases other than MGUS have been excluded. This is in agreement with the IMWG guidelines Section 25.9 *(Kyle16)*. For those patients with longer life expectancies, higher M-protein levels, and non-IgG subtypes, monitoring every 3-4 months within the first year was considered advisable and thereafter once or twice yearly in the absence of symptoms of progression. For patients with an abnormal baseline sFLC ratio or significant Bence Jones proteinuria, the risk of renal failure is increased, and more frequent monitoring along with ample hydration was recommended.

25.5. International Staging System for Multiple Myeloma (2005) *(Greipp)*

In May 2005, the International Staging System (ISS) for Multiple Myeloma was published. This is based upon serum albumin and serum β_2M alone *(Table 25.1)*. Because of its detailed analysis, wide international agreement and simplicity, it has been adopted quickly.

Earlier staging systems included concentrations of monoclonal immunoglobulins. It has now been realised that they have little relevance to MM outcome. For IgG MM this may be due to variable recycling by FcRn receptors *(Chapters 10 and 32)*. Monoclonal sFLC concentrations and abnormal κ/λ ratios do relate to disease stage and outcome *(Chapters 11 and 12)*. This may be because their clearance rate is more constant and because they can cause renal damage which influences mortality *(Chapter 13)*.

Stag	Criteria	Median survival
I	Serum β_2M < 3.5mg/L. Serum albumin $\geq$35g/L	62 months
II*	Not stage I or III	44 months
III	Serum β_2M $\geq$5.5mg/L	29 months

Table 25.1. International Staging System. *There are 2 categories for stage II: serum β_2M <3.5mg/L but serum albumin <35g/L; or serum β_2M 3.5 to <5.5mg/L, irrespective of the serum albumin level.

25.6. "Uniform Response Criteria for MM" incorporating FLCs (2009)

There have been extensive international discussions on guidelines that utilise the high sensitivity of sFLC measurements *(Rajkumar 3; Kumar 2; Durie 1)*, and more recently summarised in 2009 *(Kyle 15)*. One important new recommendation is a change to the way that the κ/λ ratios are reported when monitoring for partial or very good partial response to treatment. Subtraction of the tumour FLC from the non-tumour FLC provides a single value that is easy to report and understand. This also provides an interpretable result when the non-tumour FLC is either below the detection limit or is fluctuating widely. It is similarly helpful when interpreting high concentrations of the alternate FLC that are seen in patients with impaired renal function (*Chapter 20.2*). Limited clinical studies indicate no loss of clinical utility. The International Uniform Response Criteria include the following FLC guidelines, including those shown in Tables 25.2, 25.3 and 25.4:

Rationale for the development of uniform response criteria
1. Facilitate precise comparisons of efficacy between new treatment and strategies in trials
2. *Incorporation of the sFLC assays*
3. *Stricter definitions of complete response (CR)*
4. Incorporate standard definition of near complete response
5. Stricter definition of disease progression
6. *Enable greater inclusion of patients with oligo-secretory and nonsecretory disease into clinical trials*
7. Provide clarifications, improve detail and correct inconsistencies in prior response criteria

Diagnostic criteria for MM requiring systemic therapy
1. Presence of an M-protein* in serum and/or urine plus clonal plasma cells in the bone marrow and/or a documented clonal plasmacytoma

   Plus one or more of the following:**
2. **C**alcium elevation (>11.5 mg/dL)
3. **R**enal insufficiency (creatinine>2mg /dL)
4. **A**nemia (Hemoglobin <10g/dL or 2g/dL <normal)
5. **B**one disease (lytic or osteopenic)

* *In patients with no detectable M-protein, serum Freelite chain assays can substitute and satisfy this criterion. For patients with no serum or urine M-protein, the baseline bone marrow must have ≥30% plasma cells. These patients are referred to as "nonsecretory myeloma".*
** Must be attributable to the underlying plasma cell disorder.

Practical Details of Response Evaluation
1. Laboratory tests for measurement of M-proteins.

Serum M-protein level is quantified using densitometry on serum protein electrophoresis (SPE) except in cases where the SPE is felt to be unreliable such as in patients with IgA monoclonal proteins migrating in the β-region. If SPE is not available for routine M-protein quantitation during therapy, then nephelometry or turbidimetry can be accepted. However, this must be explicitly reported; nephelometry can be used only for individual patients to assess response; SPE and nephelometric values cannot be used interchangeably.

Urine M-protein measurement is estimated using 24-hour urine protein electrophoresis (UPE) only. *Random or 24 hour urine tests measuring κ and λ FLC measurements are not reliable and are not recommended.*

2. Definitions of measurable disease for MM

Response criteria for all categories and subcategories of response except CR are applicable only to patients who have "measurable" disease defined by at least one of the following 3 measurements:-

Serum M-protein ≥1 g/dL (10 g/L)

Urine M-protein ≥200 mg/24 hours

Serum FLC assay: Involved FLC ≥10 mg/dL (≥100 mg/L), provided the FLC ratio is abnormal

3. Response criteria for complete response (CR)

These are applicable to patients who have no abnormalities on one of the three measurements listed above. Note that patients who do not meet any of the criteria for measurable disease as listed can only be assessed for stringent CR and cannot be assessed for any of the other response categories.

4. Follow-up to meet criteria for partial response (PR) or stable disease (SD)

It is recommended that patients undergoing therapy be tracked monthly for the first year of new therapy and every alternate month thereafter.

Patients with "measurable disease" as defined above by SPE and UPE, need to be followed by both SPE and UPE for response assessment and categorization.

Except for assessment of CR, patients with measurable disease restricted to the SPE will need to be followed only by SPE; correspondingly patients with measurable disease restricted to UPE will need to be followed only by UPE.

Patients with measurable disease in either SPE or UPE, or both, will be assessed for response based only on these two tests and not by FLC assays. FLC response criteria are only applicable to patients without measurable disease in the serum or urine, and to fulfill the requirements in the category of stringent CR.

To be considered CR, both serum and urine IFE must be done and be negative regardless of the size of baseline M protein in the serum or urine; patients with negative UPE values pre-treatment still require UPE testing to confirm CR and exclude light chain escape.

25.7. National Comprehensive Cancer Network. Clinical Practice Guidelines in Oncology: Multiple myeloma 2007 *(KC Anderson)*

sFLCs are included in these guidelines based upon the *"Uniform Response Criteria for MM"* described in Section 25.6 above.

25.8. European Society of Medical Oncology (ESMO): Multiple myeloma; Recommendations for diagnosis, treatment and follow-up *(Harousseau)*

sFLC measurements are useful for identifying and monitoring patients with NSMM. Furthermore, free light chain determination can be used as well as or instead of urine electrophoresis in follow up.

25.9. USA National Academy of Clinical Biochemistry. Guidelines for the use of tumor markers in monoclonal gammopathies (2005) *(Gupta)*.

sFLC measurements are useful for the diagnosis and follow up of NSMM, MGUS and AL amyloidosis. They are more sensitive than urine tests in patients with LCMM and in other patients with MM who may have coexistent systemic lupus erythematosus (SLE) or renal impairment.

Response Subcategory	Response Criteria
Complete Response (CR)	· Negative IFE in the serum and urine *and:* · Disappearance of any soft tissue plasmacytomas *and:* · ≤5% plasma cells in bone marrow · *If the serum and urine M-protein are unmeasurable*, CR is defined as a normal FLC ratio of 0.26-1.65 in addition to the CR criteria listed above*
Stringent CR (sCR)	CR as defined above plus: · *Normal FLC ratio and:* · Absence of clonal cells in the bone marrow (BM) by immuno-histochemistry or immunofluorescence
Very Good Partial Response (VGPR)	· Serum and urine M-protein detectable by IFE but SPE-ve or: ≥ 90% reduction in serum M-protein and urine M-protein <100 mg per 24 hours · *If the serum and urine M-protein are unmeasurable*, VGPR is defined as a >90% decrease in the difference between the involved and uninvolved sFLC levels*
Partial Response	≥50% reduction of serum M-protein *and:* · Reduction in 24-hour urinary M-protein by≥90% or to ≤200 mg per 24 hours and · If present at baseline, a ≥50% reduction in size of soft tissue plasmacytomas · *If the serum and urine M-protein are unmeasurable*, a ≥50% decrease in the difference between involved and uninvolved FLC levels is required in place of the M-protein criteria* · *If serum and urine M-protein are unmeasurable, and sFLCs are also normal, ≥50% reduction in plasma cells is required in place of M-protein, provided baseline BM plasma cell percentage is ≥30%*
Stable Disease	Not meeting criteria for complete response, near complete response, partial response, or progressive disease

Table 25.2. International Myeloma Working Group Uniform Response Criteria (Section 25.6): Complete Response and Other Response Categories

Note that all response categories require two consecutive assessments done at anytime before the institution of any new therapy. **Refer to page 225 for definitions of measurable disease.*

Timing	Recommended testing of M-protein, sFLCs and CRAB features.
Post-Induction	At least one assessment pre-harvest. If harvest occurs during induction then assessment should be immediately pre-transplant
Post-Harvest	At least one assessment post-harvest if cytoreductive therapy used for harvest and/or if patient proceeding directly to maintenance only
Post-Transplant	One measurement required at ≤100 days. This re-staging requires bone marrow biopsy

Table 25.3. Response Evaluation for Patients proceeding to Stem Cell Harvesting and Transplantation

Relapse Subcategory	Relapse Criteria
Progressive disease.* (Includes primary progressive disease and disease progression on or off therapy; to be used for calculation of time to progression and progression-free survival endpoints.	**Any one or more of the following:** · Increase of ≥25% from baseline in - Serum M-protein and/or (the absolute increase must be ≥ 0.5g/dL) - Urine M-protein and/or (the absolute increase must be ≥200mg/24 hours - *However, in patients without measurable serum and urine M-protein levels: Increase of ≥ 25% from baseline in the difference between the involved and uninvolved sFLC levels and the absolute increase must be >10 mg/dL.* - Bone marrow plasma cell %. The absolute % must be ≥10%. **PLUS any one or more of:** **Direct indicators of increasing disease and/or end organ dysfunction (CRAB features)**** 1. Development of new soft tissue plasmacytomas or bone lesions 2. Definite increase in size of existing plasmacytomas or bone lesions. A definite increase is defined as a 50% (and at least 1 cm) increase as measured serially by the sum of the products of the cross diameters of the measurable lesion 3. Hypercalcemia (>11mg/dL) 4. Decrease in haemoglobin of 20g/L or more 5. Rise in serum creatinine by 2mg/dL or more
Relapse from CR.* (To be used only if the endpoint studied is "Disease free survival"; otherwise CR patients should also be evaluated using criteria for progressive disease).	**Any one or more of the following:** · Reappearance of serum and urine M-protein by IFE, SPE or UPE · Development of ≥5% plasma cells in the bone marrow · Appearance of any other sign of progression (i.e. new plasmacytoma, lytic bone lesion, or hypercalcemia)

Table 25.4. International Myeloma Working Group Uniform Response Criteria (Section 25.6): Disease Progression and Relapse from Complete response*

* All relapse categories require two consecutive assessments done at anytime before the institution of any new therapy. **Patients developing two or more of the CRAB features without increases in M-protein or FLC levels are also considered to have progressive disease.

25.10. Guidelines for AL amyloidosis (2005) *(Gertz 1)*

A Consensus Opinion from the 10th International Symposium on Amyloid and Amyloidosis. Definition of Organ Involvement and Treatment Response in Immunoglobulin Light Chain Amyloidosis (AL):

Complete response (CR)

Serum and urine negative for a monoclonal-protein by IFE.

sFLC κ/λ ratio normal.

Marrow contains < 5% plasma cells.

Partial response (PR)

If serum M-protein > 0.5g/dL, a 50% reduction.

If FLCs in the urine with a visible peak and >100 mg/day, a 50% reduction.
If sFLC >10 mg/dL (100mg/L): reduction by >50 %.

Progression

From CR, any detectable monoclonal-protein or abnormal FLC κ/λ ratio (light chain must double).

From PR or stable response, 50% increase in serum M-protein to > 0.5g/dL or 50% increase in urine M-protein to > 200mg/day; a visible peak must be present.
sFLC increase of 50% to >10mg/dL.

Stable disease

No CR, no PR, no progression

UK Guidelines for AL Amyloidosis (2004) *(Bird).* (Similar guidelines have been adopted in France *[Jaccard]*)

The report includes comments on sFLC measurements as follows:

1. sFLCs are abnormal in 98% of patients with systemic AL amyloidosis including those that cannot be identified by conventional methods. Patients should be assessed for sFLCs during the diagnostic procedure.
2. sFLC measurements appear to be the most effective current method of monitoring patients.
3. Changes in amyloid load correlate with changes in FLC concentrations and with survival.
4. Treatment should be continued, when feasible, until sFLC concentrations have fallen by at least 50-75%.
5. Treatment strategies are best guided by their early effect on concentrations of sFLCs.

25.11. International Myeloma Working Group guidelines for MGUS and smoldering (asymptomatic) multiple myeloma (2010)

Until recently there have been no useful markers for predicting outcome in MGUS because they lacked statistical power. In 2005, Rajkumar *(5)* et al. showed that the combination of immunoglobulin MGUS class, its quantity above or below 15g/L and the presence or absence of an abnormal FLC κ/λ ratio can be used for risk stratification *(Table 19.2)*. Following this, the IMWG has published consensus perspectives of risk factors for progression and provided guidelines for monitoring and management *(Kyle 16)*. These new recommendations are outlined below, where the relative risk factors are as defined by Rajkumar *(5)* and summarized in Table 25.5.

Risk Group	Criteria	Absolute Risk* (%)	Recommended Follow-up
Low	Serum M-protein <1.5 g/dL) IgG subtype Normal FLC ratio (0.26-1.65)	2	6 months initially, and if stable every 2-3 years
Low-intermediate	Any one risk factor present	10	6 months initially and then annually
High-intermediate	Any two risk factors present	18	
High	All three risk factors present	27	

Table 25.5. Summary of MGUS risk groups and recommended follow-up.
*of progression at 20 years accounting for death as a competing risk

25.12 Consensus statement for the screening, evaluation and management of MGUS (2010)

Following a meeting of international experts on plasma cell dyscrasia and skeletal disease, February 2009, a consensus statement for the screening and treatment of MGUS was published *(Berenson)*. Table 25.6 summarises these recommendations.

Who to screen for MGUS	1. Age inappropriate osteoporosis/osteopenia 2. All African Americans with osteoporosis 3. Unexplained proteinuria or elevated total protein levels in the blood 4. Unexplained peripheral neuropathy
Which tests to use	1. Quantification of the M-protein in the serum and urine by electrophoresis 2. Typing of the monoclonal-protein 3. Blood counts and routine chemistries 4. sFLC assay may have prognostic value 5. Bone marrow aspirate and biopsy may be performed for individuals with high-risk features
Follow up testing and frequency	1. Every 3-6 months initially 2. Subsequent frequency may vary depending on the size of the M-protein and other risk factors for progression to a more serious B-cell disorder 3. Patients should have tests performed that are similar to those done at the time of initial diagnosis
Further discussion with respect to bone surveys, considerations following the ruling out of MM in MGUS and considerations in MGUS where patients develop fractures is also found in these guidelines *(Berenson)*.	

Table 25.6. Summary of the recommendations for evaluation and management of MGUS (adapted from *Berenson*).

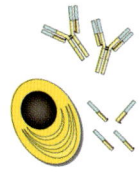

Chapter 26. *Implementation and interpretation of free light chain assays*   *231*
Chapter 27. *Instrumentation for free light chain immunoassays*   *239*
Chapter 28. *Quality assurance for serum free light chain analysis*   *252*

Chapter

26

Section 5 - Practical aspects of serum free light chain testing

Implementation and interpretation of free light chain assays

26.1. *Why measure FLCs in serum?*   *231*
26.2. *Getting started*   *232*
26.3. *Use and interpretation of sFLC results*   *233*
26.4. *Limitations of sFLC analysis*   *236*

Summary: The full benefits of sFLC analysis require:

1. Knowledge of where the clinical and laboratory benefits occur.
2. Consideration of close links between clinical and laboratory staff.
3. An understanding of when to use FLC concentrations or κ/λ ratios.
4. Interpretation of differences between sFLC, immunoglobulin and urine results.

26.1. Why measure FLCs in serum?

It is logical to measure sFLCs in the situations listed below. Justification can be found in the relevant chapters. Most of the indications are well established while others are still under evaluation. A selection of uses include:

^ Diagnostic test when suspecting a monoclonal gammopathy *(Chapters 6 and 23)*.
^ Replacement of urine tests for Bence Jones protein *(Chapters 8 and 24)*.
^ Monitoring of patients that cannot be assessed by electrophoretic tests *(Section 2)*.
^ Rapid assessment of treatment responses in multiple myeloma (MM) *(Chapter 12)*.
^ Assessment of residual disease and complete responses in MM *(Chapter 12)*.
^ Risk stratification for progression in monoclonal gammopathies *(Many chapters)*.
^ Monitoring MM patients in renal failure undergoing haemodialysis *(Chapter 13)*.

FLCs are preferably measured in serum rather than urine. Urine samples have a wider normal range, are more difficult to collect and process and are less sensitive when FLC production is low *(Chapter 24)*. Both FLCs should be measured and κ/λ ratios calculated. Results should be reported on log/log graphs that include normal range data and results from a variety of clinical conditions *(Figure 26.1)*.

26.2. Getting started

There are many implementation issues to be considered: clinical, technical, educational, political, etc. Laboratories are familiar with the introduction of new tests so it is not appropriate for these issues to be discussed here. Implementation of FLC tests normally come under the category of "*service development*". Because FLCs are widely measured in urine they can usually be introduced without seeking new test approval.

In analytical terms, FLC molecules are stable in serum, kits are available for many routine laboratory instruments and clinical interpretation of the results is well established. Practical issues directly related to FLC analysis such as the most appropriate instrumentation and QA schemes are discussed below and in Chapters 27 and 28.

Pre-analytical

Serum or plasma samples can be used and FLCs are stable for many weeks when stored at 4°C. Longer-term storage should be at -20°C with preservatives *(Chapter 4)*. There is minimal variation in FLC concentrations in samples taken from patients at different times of the day.

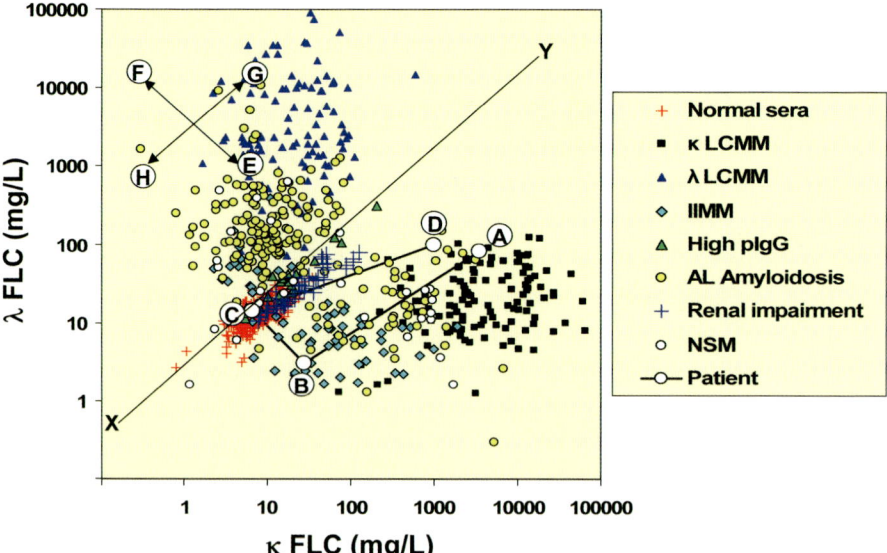

Figure 26.1. sFLCs in several clinical conditions and changes seen during chemotherapy. Results at presentation (A), after high-dose melphalan showing treatment response and bone marrow suppression (B), in complete remission (C) and in relapse with renal impairment from FLC toxicity (D). For explanations of E-F: G-H see text in Section 26.3. X---- Y: mid point of normal κ/λ ratios, LCMM: light chain multiple myeloma, IIMM: intact immunoglobulin multiple myeloma, High pIgG: polyclonal hypergammaglobulinaemia, NSM: nonsecretory multiple myeloma).

Analytical

FLC kits are available for use on many instruments. Results are generally more precise on the large clinical chemistry analysers. Details can be found in Chapter 27. External quality control schemes are available and should be used (*Chapter 28*).

Post-analytical

Reporting of results for diagnosis should be on κ/λ log plots that include normal range data and results from a variety of clinical situations *(Figure 26.1)*. Serial monitoring should include both FLCs and κ/λ ratios. It may be useful to include a renal function marker such as serum creatinine or cystatin C.

26.3. Use and interpretation of sFLC results

A. Screening symptomatic patients for monoclonal gammopathies

sFLC analysis should be used alongside serum protein electrophoresis (SPE) and serum immunofixation electrophoresis (sIFE) tests and then >99% of patients with monoclonal gammopathies are identified. Results are considered abnormal when they are outside the following normal ranges *(Figure 26.2)*:

Serum κ concentrations: 3.3-19.4mg/L
Serum λ concentrations: 5.7-26.3mg/L
Serum κ/λ ratio: 0.26-1.65

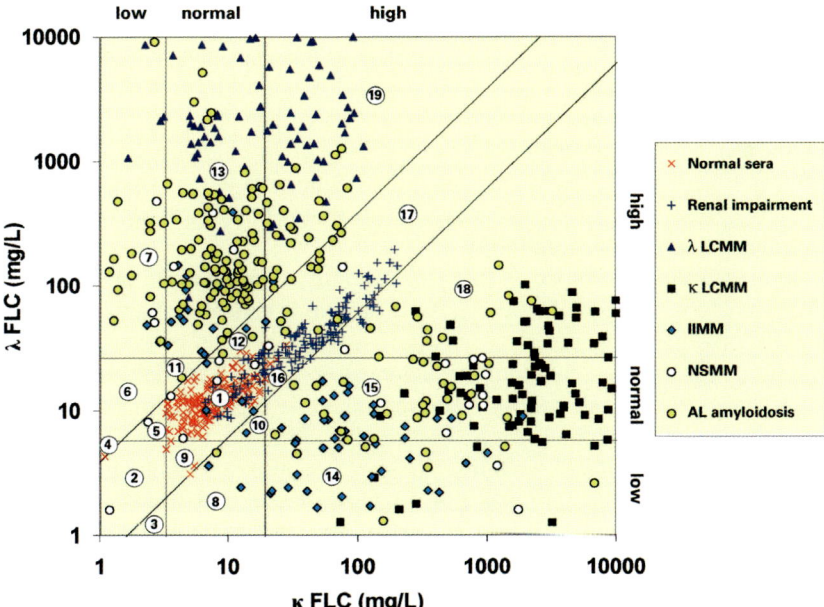

Figure 26.2. Serum κ and λ concentrations in a selection of clinical conditions. Patients are categorised according to FLC concentrations and κ/λ ratios. *See Table 26.1 for interpretation.* The axes are truncated for clarity compared with Figure 26.1.

Sector	Kappa	Lambda	κ/λ Ratio	Interpretation
1	Normal	Normal	Normal	Normal serum
2	Low		Normal	BM suppression without MG
3		Low	High	MG with BM suppression
4			Low	
5		Normal	Normal	Normal serum or BM suppression
6			Low	
7		High	Low	MG with BM suppression
8	Normal	Low	High	MG with BM suppression
9			Normal	Normal serum or BM suppression
10		Normal	High	MG with BM suppression
11			Low	
12		High	Normal	pIg or renal impairment
13			Low	MG without BM suppression
14	High	Low	High	MG with BM suppression
15		Normal	High	MG without BM suppression
16			Normal	pIg or renal impairment
17		High	Normal	
18			High	MG with renal impairment
19			Low	

Table 26.1. Classification of monoclonal gammopathies according to serum FLC concentrations (*also see Figure 26.2*). Borderline FLC ratios may be observed due to renal impairment or a polyclonal inflammatory response. In patients with renal impairment it may be preferable to use the renal reference range (κ/λ ratio; 0.37-3.1) *(Hutchison 5)*. BM: bone marrow, MG: monoclonal gammopathy, pIg: raised polyclonal immunoglobulins.

Patients' results separate into different categories depending upon several factors: whether the clone is κ or λ, the presence of renal failure or polyclonal hypergammaglobulinaemia and the degree of bone marrow impairment from the growing tumour or from drug therapy (*Figure 26.2*). An accompanying table for this figure provides a simplistic guide to interpretation of results (*Table 26.1*).

1. Normal samples. Serum κ, λ and κ/λ ratio are all within the normal ranges. If accompanying serum electrophoretic tests are normal it is most unlikely that the patient has a monoclonal gammopathy.

2. Abnormal κ/λ ratios. Support the diagnosis of a monoclonal gammopathy and require an appropriate tissue biopsy. Borderline elevated κ/λ ratios (up to 3.1) occur with renal impairment and may require appropriate renal function tests *(Hutchison 7)*.

3. Low concentrations of κ, λ or both. Indicate bone marrow function impairment.

4. Elevated concentrations of both κ and λ with a normal κ/λ ratio. May be due to the following:-

    renal impairment (common).

    over-production of polyclonal FLCs from inflammatory conditions (common).

    biclonal gammopathies of different FLC types (rare).

5. Elevated concentrations of both κ and λ with an abnormal κ/λ ratio. Suggest a combination of monoclonal gammopathy and renal impairment.

B. Replacement of urine electrophoretic tests for FLCs

Many clinical studies have shown that urine FLC tests offer little or no additional benefits over sFLC tests in assessing patients with MM, AL amyloidosis and monoclonal gammopathy of undetermined significance (MGUS). Laboratory comparison of the sensitivity of sFLC tests with urine tests indicates greater sensitivity for serum tests. Occasional patients have normal sFLC results but minor monoclonal FLCs in the urine. However, their clinical relevance is doubtful *(Chapter 6 and 24) (Nowrousian; Hill; Katzmann 4, 8; Palladini 5).*

C. Monitoring patients using sFLC assays

Patients with monoclonal gammopathies can be monitored serially using the tumour FLC and κ/λ ratios. When sFLC concentrations are very high, the level of the individual monoclonal light chain can be used for monitoring changes in disease. This is particularly important when the non-tumour FLC concentrations are very low and measurement precision is poor. As levels trend to normal or there is renal impairment, κ/λ ratios become better monitoring tools. As an alternative, the numerical value of the concentration of the tumour (involved) FLC, minus the non-tumour (uninvolved) FLC appears satisfactory. This technique of subtracting the polyclonal FLC component is included in the new response criteria for monitoring MM *(Chapter 25, Table 25.4).* *Also, it is important to ensure that sFLC concentrations are given in consistent units. All results outside the USA are in mg/L whereas within the USA results may be either in mg/L or mg/dL (Figure 28.7).*

Results from sFLC measurements can be presented in different ways. Figures 26.1 and 26.3 show changes in sFLC concentrations in a patient with MM from the time of presentation to disease relapse. In Figure 26.1, at presentation (A), κ concentrations were highly elevated at 3,500mg/L and the non-tumour FLC was mildly elevated because of FLC deposition in the kidneys. While under treatment (B), both sFLC concentrations fell because of bone marrow suppression but there was selective tumour cell killing reflected in a reduction of the κ/λ ratio. Successful treatment is shown by

Figure 26.3. Data from the patient shown in Figure 26.1. Results at clinical presentation (A), after HDM showing treatment response and bone marrow suppression (B), in complete remission (C) and in relapse with renal impairment from FLC damage (D).

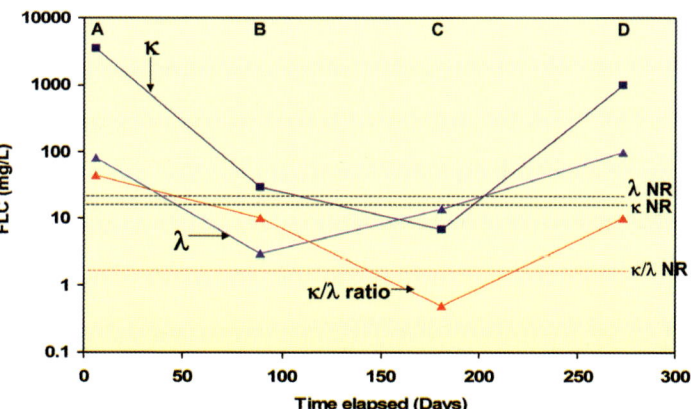

return of the FLC concentrations and κ/λ ratio to normal (C). At subsequent relapse (D), the tumour FLC and the κ/λ ratio increased, as might be expected, but the alternate FLC also increased as a result of renal impairment from chemotherapy and FLC deposition in the nephrons.

In Figure 26.1 the mid point of normal κ/λ ratios is marked as **X---Y**. Changes away from this axis indicate decreasing (E) or increasing λ tumour FLCs (F). Changes parallel to the κ/λ ratio axis and both FLC concentrations increasing (G) indicate renal impairment. Similarly, decreasing concentrations indicate renal function recovery, with the addition of bone marrow suppression if below the normal range (H).

Another way to present the results is against time. Data should include κ, λ and κ/λ ratios *(Figure 26.3)*, perhaps with the addition of other markers such as intact immunoglobulins. Examples of patients being monitored using sFLCs can be found in the relevant chapters.

26.4. Limitations of sFLC analysis

While sFLC immunoassays have many advantages over electrophoretic tests, there are limitations. When results give doubt, monoclonal FLCs should be assessed by SPE and immunofixation electrophoresis (IFE). Difficulties that may occur are described below. Details can be found in the relevant chapters of the book, particularly Chapter 4.

1. sFLC assays do not measure intact monoclonal immunoglobulins.

This may seem a rather obvious statement, yet there are several publications in which the authors criticise the sFLC assays for missing patients with monoclonal gammopathies *(Tate 1; Jaskowski; Mehta)*. FLC antisera do not, by definition, react with intact monoclonal immunoglobulins and will fail to identify these patients if monoclonal FLCs are absent.

2. Clonality assessment

Using FLC immunoassays, monoclonal FLCs are determined by abnormal κ/λ ratios. This is clearly different from the assessment of bands on electrophoretic gels. Arguably, κ/λ ratios are preferable because numerical limits are easily established while visual impressions of bands on gels are more difficult to assess and quantitate. However, some find this difficult to accept because of its novelty.

In some situations analysis of κ/λ ratios is clearly preferable. For example, patients with NSMM have highly abnormal κ/λ ratios yet electrophoretic tests are normal. In AL amyloidosis *(Chapter 15)* and light chain deposition disease (LCDD) *(Chapter 17)*, sFLCs are the actual molecules causing the disease so quantification is preferable.

Rare patients have monoclonal FLC bands when tested by urine IFE that are not abnormal by sFLC immunoassays *(Chapter 24)*.

3. Inaccuracy of sFLC measurements

Each monoclonal protein is structurally unique, so results depend upon how well the antibodies recognise molecular variants and polymeric configurations. Some of these variants give rise to inaccuracies in measurement. Historically, this line of argument has been used to discredit measurements of IgG, IgA and IgM by nephelometry. However, experience has shown that this has not invalidated their utility and widespread acceptance in a clinical setting. The same can now be said for sFLC immunoassays. Nevertheless, in some patients huge discrepancies are seen between monoclonal FLCs measured by immunoassays and other techniques, particularly in urine samples. In the latter fluid, FLC immunoassays may give higher concentration values than other methods of FLC measurement (*Chapters 4 and 6*).

4. Standardisation

While current standards may well be accurate, there are no international standards or traceable international materials and none will be available in the short term. Although every effort might be made by the manufacturer to ensure good quality, the assays may drift with time and in an unpredictable manner. Local standards and reference materials should, therefore, be used with the assays, when possible. Initially, a set of local serum samples should be compared with published reference ranges and this is usually satisfactory. If not, the manufacturer should be contacted to determine if the cause of excessively frequent abnormal results is due to local inexperience or an instrument related problem. When results are still unsatisfactory, a local reference range can be established *(Pattenden)*.

5. Non-linearity of monoclonal FLC proteins

This refers to variations in quantification that occur when a sample is diluted and the result is different from the starting value. A number of factors may be involved. Details can be found in Chapter 4 and include the following:-
- Monoclonal FLCs of unusual conformation, partially missed by the antisera (see below).
- Antibody bias to one form of the FLC proteins when polymerisation or fragments are present.
- Antisera cross-reactivity with intact immunoglobulins.
- Non-specific assay interference (lipids, haemoglobin etc).
- Use of unsuitable materials for assay calibrators.

6. Different batches of antisera

Consideration should be given to apparent changes in sFLC concentrations that might occur when changing to a different batch of antisera. Batches are made to react in a characterised manner with a variety of standards and control sera. These are manufactured to defined limits that are achieved during kit production. Within these limits there is some small but quantifiable variation.

In addition, each monoclonal FLC is unique and will react in its own particular manner in the assays. While great effort is made during manufacture to maintain lot-to-lot consistency, antisera cannot be made to recognise every individual monoclonal FLC equally. Some structurally abnormal molecules may not, therefore, be reproducibly measured. In such circumstances, the previous sample should be re-run alongside the new sample, using the new batch of antiserum, and the results compared *(Tate 3; Robson 2; Mazurkiewicz)*.

7. Presence of polyclonal FLCs

Polyclonal FLCs in all sera lead to overestimations of monoclonal FLCs. This is particularly apparent in patients with renal failure in whom polyclonal, non-tumour FLCs are greatly elevated (*Chapter 20*). However, κ/λ ratios usually remain abnormal if monoclonal FLCs are present. Borderline results may need to be corrected against serum creatinine (or a better marker of impaired glomerular filtration such as cystatin C) in patients with renal failure.

8. Biclonal gammopathies of different light chain types

Approximately 1-2% of patients with MM have bi-clonal gammopathies. When the FLC types differ (~50%), so that the patient has both types of FLC, κ/λ ratios can be normal. Since it is likely that both FLC concentrations would be elevated, and in different amounts, the clinician would usually be alerted to an abnormality *(Figure 26.4)* *(Ramasamy)*. The issue can be resolved by testing the sample using IFE and identifying two monoclonal bands of different FLC types. Renal function should also be determined to assess the degree of polyclonal elevation of sFLCs from reduced glomerular clearance.

Figure 26.4. sFLC concentrations in 5 patients with biclonal gammopathies. (Courtesy of I Ramasamy)

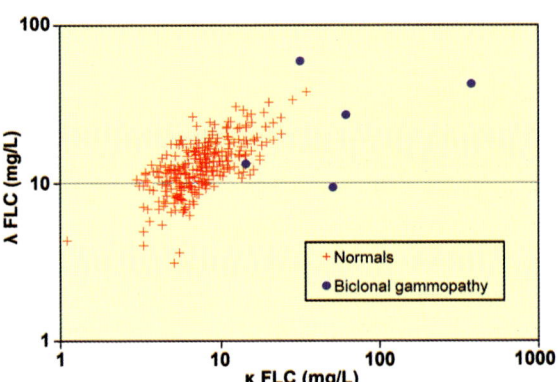

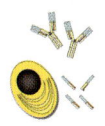

Introduction   *239*

27.1. *Beckman Coulter AU® (400, 640, 2700 and 5400)*   *241*

27.2. *Beckman Coulter IMMAGE® and IMMAGE 800®*   *242*

27.3. *Binding Site MININEPHPLUS™*   *243*

27.4. *Binding Site SPAPLUS bench-top analyser*   *244*

27.5. *Radim Delta™*   *245*

27.6. *Roche Cobas® c501 automated analyser*   *246*

27.7. *Roche Cobas Integra® 400, 400plus and 800 automated analyser*   *246*

27.8. *Roche Hitachi 911/912/917 and Modular P*   *247*

27.9. *Siemens Advia® 1650, 1800, 2400*   *248*

27.10. *Siemens BN™II*   *249*

27.11. *Siemens BN ProSpec®*   *250*

Introduction

Immunoassays for serum free light chains (sFLCs) (**Freelite**™) are available for the majority of nephelometric and turbidimetric laboratory instruments. They all utilise latex-enhancement to allow detection of FLCs at low concentrations. Nephelometers and turbidimeters have similar levels of sensitivity and precision but instruments vary in their ability to handle samples, clean reaction cuvettes, identify antigen excess, etc. Smaller laboratories tend to use nephelometers that are dedicated to protein measurements but sample handling is relatively slow. Such instruments may not have *"open"* channels, so technical assistance is frequently required to implement the FLC tests.

Large laboratories tend to use multifunctional chemistry instruments that analyse proteins turbidimetrically. These instruments are usually more precise and stable than dedicated protein analysers and may be more *"user-friendly"* for new assay implementation. However, some external assistance is usually required to ensure proper running of the assays and interpretation of results.

A comparison of the various attributes of 8 different instruments is shown in Table 27.1 *(Showell 9)*. These assessments were made at the Binding Site production laboratory and because of their extensive experience the results are equivalent to those obtained in an expert laboratory. Throughput was tested by measuring 20 normal serum samples (requiring no re-measurement) and 20 multiple myeloma (MM) samples (requiring multiple remeasurements). These were run separately so that each instrument's ability to make automatic redilutions could be isolated from its basic throughput. The hands-on time, e.g. manual dilutions, was also recorded. Total precision (n=100) was assessed, and the availability of antigen excess protection was also documented. Instrument reliability was assessed by calculating the average number of breakdowns per unit, per year requiring engineer intervention. Finally, overall performance was assessed by awarding marks from 0 – 3 for each category, apart from 0 or 1 for antigen excess detection.

	Beckman IMMAGE®	Binding Site SPAPLUS	Cobas Integra® 400*	Roche Hitachi Modular P*	Beckman Coulter AU®400	Siemens ADVIA® 1650*	Siemens BN™II	Siemens BN ProSpec®
Sample redilution**	None	1	3	1	1	1	2	2
20 Normal samples	40 mins	37 mins	33 mins	29 mins	36 mins	18 mins	52 mins	52 mins
20 MM samples	84 mins	68 mins	75 mins	51 mins	71 mins	106 mins	127 mins	172 mins
Hands-on time	21 mins	10 mins	5 mins	5 mins	15 mins	14 mins	0 mins	10 mins
Kappa precision	2.7%	4.1%	1.8%	6.3 %	2.5 %	4.5 %	4.8%	1.6%
Lambda precision	2.7%	3.9%	1.2%	5.4 %	2.2 %	3.1 %	4.7%	1.3%
Antigen excess	No	Yes	Yes	No	No	No	No	No
Breakdown unit/year	5	0.2	1	6	0	6	1.4	6
Overall performance	12/22	16/22	20/22	13/22	16/22	12/22	14/22	14/22

Table 27.1. Comparison of sFLC assays (**Freelite**™) on eight different laboratory instruments. *Other compatible instrument models are available but there is no data. ** 1: single auto redilution, 2: multiple auto redilutions, 3: multiple manually ordered redilutions. Marks from 0-3 were awarded for each category apart from 0 for 1 for antigen excess capability. 20 samples were tested for each parameter.

Basic sample throughput was similar for all instruments but the BNII and ProSpec were affected by longer assay times. Instruments offering multiple automatic redilutions showed faster throughput for MM sera with the Modular P and SPAPLUS showing the best processing speed, with only a few minutes hands-on required. Precision was good overall, although the best precision was seen with instruments using disposable cuvettes (Integra 400/800 and BN ProSpec). Antigen excess protection was only available on the SPAPLUS, MININEPHPLUS and Integra 400/800, giving reassurance that high samples are not mis-reported. Where reliability was concerned, the Beckman Coulter AU 400 and SPAPLUS performed best, with less than a single breakdown per unit, per year. Overall performance was best for the Integra, SPAPLUS and Beckman Coulter AU 400.

Inevitably, there are occasional assay related *"issues"* that lead to customer questions or complaints. Analysis of The Binding Site database has showed that approximately 50% of problems were related to inadequate maintenance of the analysers or incorrect parameter programming *(Figure 4.17)*. About 20% were due to problems with calibrators and 15% were related to samples behaving in a non-linear manner. The precision values shown in the tables for each instrument are those found at Binding Site laboratories. If customer attained precision figures are significantly worse, then the instrument or the programming are probably at fault. It should be noted that the majority of complaints have been restricted to two instruments. This is largely because they have been more widely used. Exact information on the performance of the assays and their implementation is contained in the package inserts *(also see Chapter 26)*. Measuring ranges may change slightly with different batches. In addition, for several of the clinical chemistry analysers the FLC assay comes with batch specific technical range limits which must be applied with each new batch of reagents.

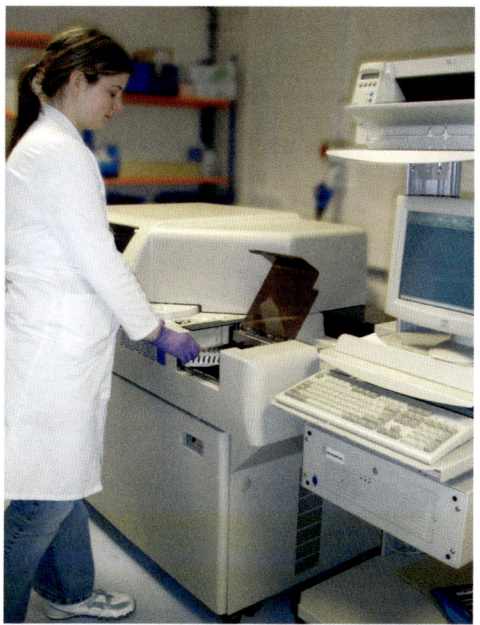

27.1. Beckman Coulter AU® (400, 640, 2700 and 5400)

(Showell 6; Ramasamy).

The Beckman Coulter chemistry analysers use turbidimetry for the measurement of plasma proteins. All have similar hardware and software but vary in their sample handling capacity. The instruments are open and can be programmed by the user. For the FLC assays, 6 pre-diluted standards are provided. The standard sample dilution is 1/10, sample results outside this range are automatically re-measured at 1/100 or 1/5, while off-line dilutions are required for higher concentrations. Results correlate well with other FLC immunoassays and precision is good *(Table 27.2)*.

At present, no specific technical issues have been encountered and comparison with other instruments shows impressive reliability *(Table 27.1)*. Fogging of the glass cuvettes with the latex reagent is a potential problem. This can be prevented by adding a W2 wash with 1% Decon 90 before the normal weekly W2 wash programme.

	Serum κ FLC	Serum λ FLC
Range at 1/10	6.0-150.0mg/L	6.0-150.0mg/L
Sensitivity at 1/5	3.0 mg/L	3.0 mg/L
Assay time	8 min	8 min
Precision: within-run	2.0% at 22.1mg/L	1.6% at 31.4mg/L
	1.2% at 40.8mg/L	0.8% at 66.1mg/L
	1.9% at 153.3mg/L	6.8% at 183.1mg/L
Precision: between-run	5.8% at 22.2mg/L	5.4% at 29.6mg/L
	5.2% at 41.4mg/L	4.9% at 62.9mg/L
	6.0% at 135.4mg/L	4.5% at 177.8mg/L

Table 27.2. Assay performance on the Beckamn Coulter AU® 400 analysers.

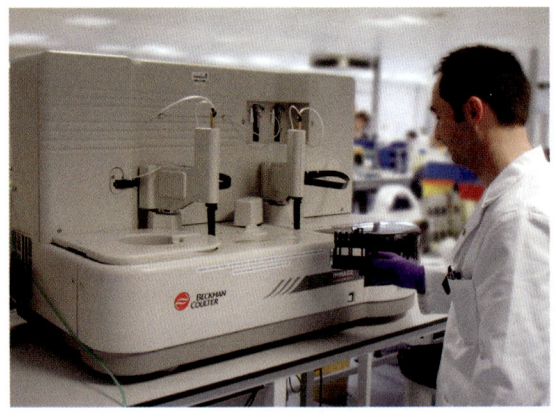

27.2. Beckman Coulter IMMAGE® and IMMAGE® 800 *(Tate 1; Harris; Ong)*

The IMMAGE can be used as either a rate nephelometer or as a rate turbidimeter that measure plasma proteins by homogeneous immunoassay. In the case of the FLC assays the latter function is used. The instrument has two basic modes of operation; a fully automated *"Beckman"* mode, controlled by the manufacturer and a *"user-defined"* reagent mode. The latter allows assays to be developed by individual laboratories.

The Binding Site provides FLC kits for use in 'user-defined' mode. Each contains a pre-prepared, six level calibrator set, controls, κ or λ latex reagents and product information. Support is usually required for setting up the assays and some disposable components and buffers are required from Beckman Coulter.

The assays use serum samples diluted to 1/10 and the instrument can be programmed to analyse samples at 1/5 if the result is lower than the initial measuring range. Offline dilutions of 1/25 and 1/500 may be required for higher concentrations. The 1/10 dilutions cover the FLC concentrations found in most samples *(Table 27.3)*.

Assay trouble shooting

1. **Cuvette fogging:** Latex particles accumulate on the cuvettes so they must be changed regularly (every 1,200 Freelite tests).

2. **Washer probe:** This needs particular attention. It must be clean, free from corrosion and correctly aligned.
3. **Syringes and buffer lines:** Check there is no leakage or air bubbles.
4. **Buffers:** The No 1 system diluent must be used.

	Serum κ FLC	Serum λ FLC
Range at 1/10	6.0-180.0mg/L	4.8-162.0mg/L
Sensitivity at 1/5	3.0mg/L	2.4mg/L
Assay time	10 min	10 min
Precision: within-run	8.1% at 10.7mg/L	2.2% at 13.1mg/L
	5.4% at 17.7mg/L	2.0% at 25.5mg/L
	4% at 147.2mg/L	6.8% at 138.2mg/L
Precision: between-run	15% at 9.3mg/L	7.2% at 11.9mg/L
	5.8% at 27.8mg/L	7.2% at 28.0mg/L
	7.2% at 111.9mg/L	11.7% at 190.9mg/L

Table 27.3. Assay performance on the Beckman Coulter IMMAGE®.

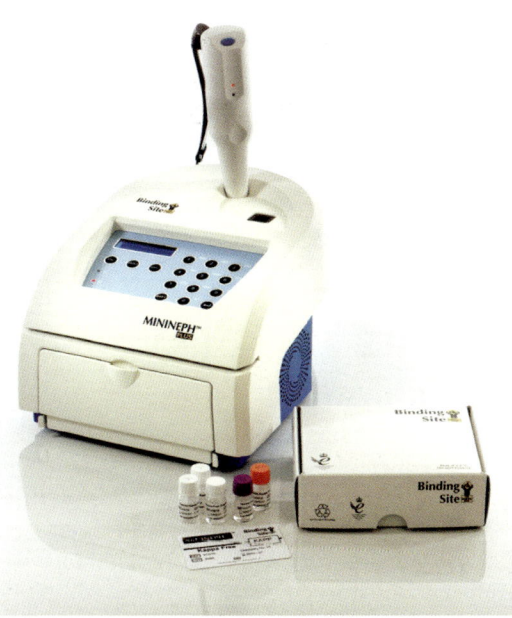

27.3. Binding Site MININEPHPLUS™

This is a small, manual nephelometer designed for sFLC analysis. The calibration curves are stored on magnetic swipe-cards. Once loaded into the instrument's memory, curve validity is confirmed by assaying control samples. Patient samples are analysed individually by semi-automated addition of reagents into cuvettes within the instrument. The assays show good linearity, identify antigen excess, have good agreement with the Siemens BNII and are fast (*Table 27.4)*. The instrument may find use in laboratories with low work loads (<10 samples per day) or when the cost of a larger instrument cannot be justified.

	Serum κ FLC	Serum λ FLC
Range at 1/20	3.0-72.4mg/L	4.9-98.3mg/L
Sensitivity at 1/20	3.0mg/L	4.9mg/L
Assay time	5 min	5 min
Precision: within-run	7.3% at 55.82mg/L	5.1% at 80.54mg/L
	5.7% at 18.67mg/L	3.9% at 29.42mg/L
	4.9% at 4.79mg/L	3.2% at 7.62mg/L
Precision: between-run	6.8% at 55.82mg/L	6.7% at 80.54mg/L
	5.7% at 18.67mg/L	3.8% at 29.42mg/L
	5.0% at 4.79mg/L	7.4% at 7.62mg/L

Table 27.4. Assay performance on the Binding Site MININEPHPLUS.

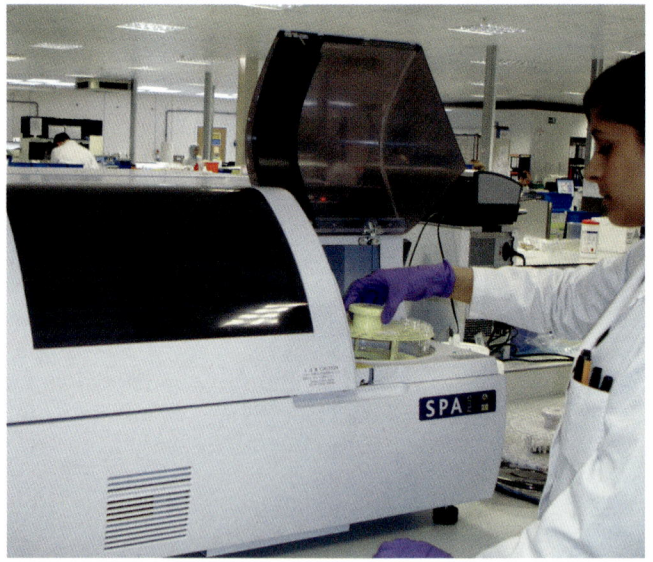

27.4. Binding Site SPAPLUS bench-top analyser *(Marionneaux; Showell 1,8)*

The instrument is an automated, random access turbidimeter with host interface capability, barcoded sample identification and reagent management systems. Precision is maintained through a combination of acid/alkali cuvette washing and an innovative reaction cuvette mixing system. Air pressure is used in place of stirrers to mix the reaction mixture in a U-shaped cuvette. No physical contact is made with the reaction mixture, thereby removing any possibility of carry-over on a stirrer. Calibration curves are made from calibrator sets and validated by assay of control fluids supplied with the kits. Samples are initially measured at the standard programmed sample dilution and if out of range, the instrument automatically re-measures the samples at the appropriate alternative dilutions. All dilutions are made with the instrument's pipetting system which is capable of dilutions between 1/10 and 1/100. An antigen excess protection function is available. The instrument has a good overall performance compared with other analysers *(Tables 27.1 and 27.5)* and provides good throughput for myeloma samples in spite of its relatively small size.

	Serum κ FLC	**Serum λ FLC**
Range at 1/10	4.0-180.0mg/L	4.5-165.0mg/L
Sensitivity at 1/1	0.4mg/L	0.5mg/L
Assay time	15 min	15 min
Precision: within-run	3.3% at 7.2mg/L	3.4% at 10.4mg/L
	1.6% at 35.7mg/L	2.4% at 35.1mg/L
	1.8% at 123.8mg/L	2.0% at 142.1mg/L
Precision: between-run	4.2% at 7.2mg/L	2.2% at 10.4mg/L
	1.9% at 35.7mg/L	0.0% at 35.1mg/L
	2.3% at 123.8mg/L	2.4% at 142.1mg/L

Table 27.5. Assay performance on The Binding Site SPAPLUS.

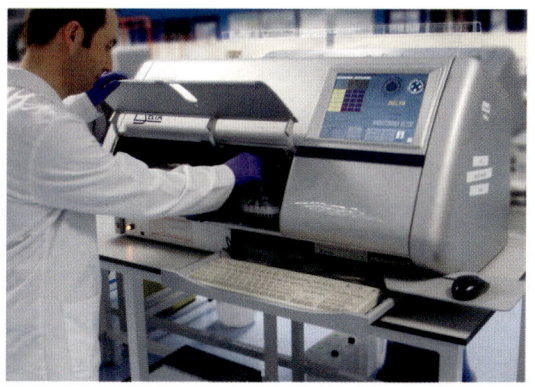

27.5. Radim Delta™ *(Showell 3)*

This is a medium-sized, fully automated bench-top nephelometer, with similar characteristics and performance to the Siemens BNII *(Table 27.6).* The instrument produces calibration curves from a single calibration fluid. Samples with concentrations of FLCs outside the initial measuring range are automatically re-diluted by the instrument.

	Serum κ FLC	**Serum λ FLC**
Range at 1/100	5.9-190.0mg/L	5.0-160.0mg/L
Sensitivity at 1/5	0.3 mg/L	0.25 mg/L
Assay time	18 min	18 min
Precision: within-run	7.1% at 8.8mg/L	4.3% at 15.5mg/L
	2.3% at 41.0mg/L	2.0% at 75.2mg/L
	7.7% at 143.6mg/L	4.1% at 202.8mg/L
Precision: between-run	8.2% at 7.2mg/L	5.3% at 13.1mg/L
	7.8% at 40.05mg/L	5.2% at 47.6mg/L
	7.0% at 144.8mg/L	6.4% at 143.5mg/L

Table 27.6. Assay performance on the Radim Delta.

27.6. Roche Cobas® c501 automated analyser *(Lynch 4)*

The c501 is part of the cobas 6000 range of instruments intended to replace Roche's Modular and Integra analysers. This is a random access, turbidimetric analyser using end-point measurements. The instrument features ultrasonic, contact-free stirring. Most parameter channels are dedicated to Roche assays, however parameters for Binding Site FLC channels can be downloaded by the user from the Roche database. Single-vial calibrators are used that are diluted automatically to generate the calibration curves. Starting sample dilutions are 1/5 for κ assays and 1/8 for λ assays. The instrument can be programmed to carry out a single, automatic re-dilution when samples are out of range. Further dilutions for very high samples must be made off-line. Results show a good correlation with the Siemens BNII *(Table 27.7)*.

	Serum κ FLC	Serum λ FLC
Range 1/5 (κ); 1/8 (λ)	3.7-56.2mg/L	5.6-74.8mg/L
Sensitivity (neat)	0.8 mg/L	0.7 mg/L
Assay time	10 min	10 min
Precision: within-run	1.4% at 5.6mg/L	4.1% at 7.7mg/L
	2.5% at 18.5mg/L	5.5% at 27.3mg/L
	1.7% at 41.1mg/L	2.8% at 60.3mg/L
Precision: between-run	7.2% at 5.8mg/L	10% at 8.0mg/L
	7.0% at 19.8mg/L	3.6% at 28.6mg/L
	2.6% at 41.4mg/L	2.3% at 63.3mg/L

Table 27.7. Assay performance on the Roche Cobas® c501.

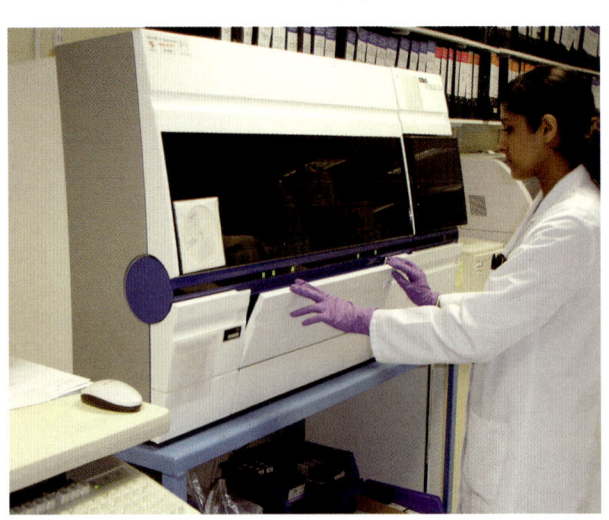

27.7. Roche Cobas Integra® (400, 400 *plus* & 800) automated analyser *(Showell 10)*

The Integra 400 is a random access, bench-top turbidimetric analyser using end-point measurement. Although it is a closed system, Binding Site FLC channels are provided as part of the standard menu and can be loaded from a TAS/TASU disc provided by Roche. Single-vial calibrators are used that are diluted automatically to generate the calibration curves. Starting sample dilutions are 1/10 for κ assays and 1/8 for λ assays.

Samples outside of the measuring range can be manually selected for on-board redilution and re-measurement. The range of on-board dilution factors is broader on the integra 400 and 400 *plus)* than most biochemistry instruments, but very high samples may still require off-line dilution. Use of disposable cuvettes results in precision that is superior to many analysers and it gives the best overall performance *(Tables 27.1 and 27.8).* FLC assays are required to be run in batch mode to avoid carry-over from Roche chemistries.

	Serum κ FLC	**Serum λ FLC**
Range 1/10 (κ); 1/8 (λ)	2.9-127.0mg/L	5.2-139.0mg/L
Sensitivity (neat)	0.6mg/L	1.3mg/L
Assay time	10 min	10 min
Precision: within-run	5.8% at 6.0mg/L	2.3% at 7.7mg/L
	2.1% at 18.7mg/L	0.7% at 27.0mg/L
	1.4% at 95.6mg/L	0.7% at 99.2mg/L
Precision: between-run	2.7% at 6.0mg/L	2.5% at 7.7mg/L
	2.7% at 18.7mg/L	0.8% at 27.0mg/L
	1.8% at 95.6mg/L	0.7% at 99.2mg/L

Table 27.8. Assay performance on the Roche Cobas Integra 400.

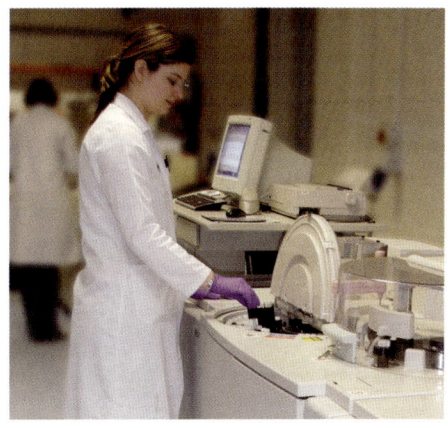

27.8. Roche Hitachi 911/912/917 and Modular P *(Showell 2; Overton)*

These instruments measure serum proteins by turbidimetry. Most parameter channels are dedicated to Roche assays but a few can be used with other manufacturers' products. Single-vial calibrator fluids are used that are diluted automatically to generate the calibration curves. The starting sample dilutions are 1/5 for κ assays and 1/8 for λ assays. The instruments can be programmed to automatically re-dilute samples when results are outside the range of the calibration curve *(Table 27.9)*. Very high samples must be diluted off-line. Results show a good correlation with the Siemens BNII.

Assay trouble-shooting

The following issues need to be considered with the Modular P.

1. Ensure correct settings for the parameters and technical limits. These change with each batch of kits.
2. Add an additional *"joker-labelled"* open channel reagent bottle with saline for diluent.
3. If the sodium hydroxide wash buffer bottle is empty, cuvettes become turbid and an "OVER" error message appears.

4. If tolerance limits have been exceeded after the water wash, a "CELL" error message appears.
5. Wait until the calibrator rack has cleared the pipetting area before loading new controls. This ensures that the new curve information is allocated to the correct samples.

	Serum k FLC	Serum λ FLC
Range at 1/5 (κ), 1/8 (λ)	3.7-56.2mg/L	5.6-74.8 mg/L
Sensitivity (neat)	0.8 mg/L	0.7 mg/L
Assay time	10 min	10 min
Precision: within-run	8.0% at 8.2mg/L	3.7% at 10.7mg/L
	3.7% at 30.0mg/L	2.6% at 42.8 mg/L
	3.0% at 48.9mg/L	3.5% at 57.4mg/L
Precision: between-run	7.3% at 6.7mg/L	9.5% at 11.9mg/L
	4.1% at 18.2mg/L	6.8% at 29.1mg/L
	5.7% at 35.3mg/L	6.3% at 55.4mg/L

Table 27.9. Assay performance on the Roche Modular P.

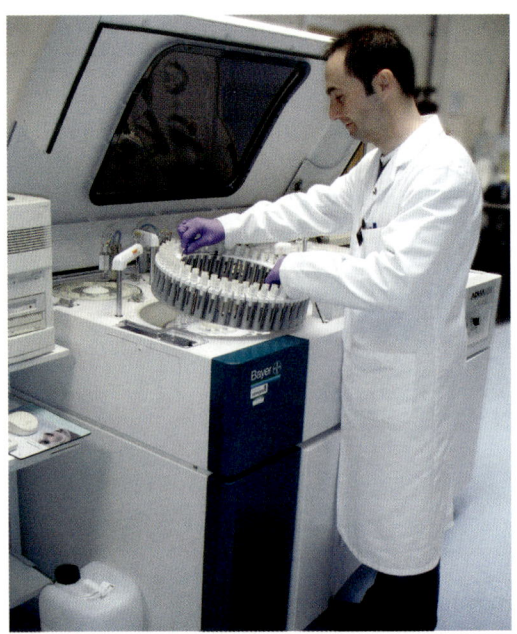

27.9. Siemens ADVIA® 1650, 1800 and 2400 *(Wands; Carr-Smith 3; Higgins)*

The Advia series are turbidimetric, floor standing instruments of varying size and throughput, with the old 1650 being replaced by the 1800 model. The software system is open and parameter set-up can be carried out by the user. However, due to its complexity, assistance from the Binding Site may be required. All the instruments feature an optional rack handling system with separate dilution and reaction cuvettes to improve throughput. The starting sample dilutions are 1/5 for κ assays and 1/8 for λ assays and if out of range, the instrument automatically re-measures the samples at a higher or lower dilution. Because repeat values are only checked and flagged using the range for the standard dilution, all repeat samples should also be checked manually (against the measuring range for the repeated dilution) to ensure valid results. The instrument is fast for normal samples *(Tables 27.1)*. It is linear across the measuring range, has good sensitivity and curve stability and shows good intra-run and inter-run precision *(Table 27.10)*.

	Serum κ FLC	**Serum λ FLC**
Range at 1/5 (κ); 1/8 (λ)	3.7-56.2mg/L	5.6-74.8mg/L
Sensitivity neat	0.75mg/L	0.70mg/L
Assay time	15 min	15 min
Precision: within-run	6.5% at 9.4mg/L	6.3% at 9.7mg/L
	2.7% at 17.7mg/L	6.4% at 23.0mg/L
	2.3% at 40.7mg/L	1.8% at 73.3mg/L
Precision: between-run	5.0% at 9.6mg/L	7.1% at 11.2mg/L
	1.9% at 18.2mg/L	2.9% at 24.0mg/L
	2.6% at 40.7mg/L	3.0% at 77.1mg/L

Table 27.10. Assay performance on the Siemens (Bayer) Advia 1650.

27.10. Siemens BN™II *(Bradwell 2; Carr-Smith 6; Katzmann 3; Showell 5)*

The BNII is a nephelometer designed for measuring plasma proteins by homogeneous immunoassays. The instrument can be programmed for non-Siemens assays using a software key from the company's engineers. Once programmed, the FLC assays run in a similar manner to other protein tests. A single calibrator fluid is automatically diluted to form a calibration curve. The initial sample dilution is 1/100 with automatic dilution of samples that are outside the calibration range *(Table 27.11)*.

Assay trouble-shooting

If the assay is performing poorly the following issues should be considered and an engineer may be needed for assistance:

1. **Wash system** (wash shoe, alignment and tubing). Problems with these components may lead to liquid remaining in the bottom of the cuvettes after cleaning. This may lead to carry-over problems in subsequent assays causing dilution errors and variable results. If fluid remains in the cuvettes after laundering, a service engineer should assess the wash system.

2. **Syringes.** Fluid leakage or excessive air in the syringes and buffer lines.
3. **Programming**. The assays require software programming with a service key (dongle). Check carefully for correct parameter programming before the engineer departs.
4. **Leaking three-way valves.** N-Reaction buffer used in many of the assays may leak across the three-way valve into the FLC assays. This leads to loss of curve reproducibility and poor assay precision (particularly at the low end of the curve). If this occurs the syringes should be primed with N-Diluent prior to running the assays. Avoid running assays that use N-Reaction buffer alongside the FLC assays until after the valve has been replaced.

	Serum κ FLC	**Serum λ FLC**
Range at 1/100	5.9-190.0mg/L	5.0-160.0mg/L
Sensitivity at 1/5	0.30mg/L	0.25mg/L
Assay time	18 min	18 min
Precision: within-run	3.1% at 13.6mg/L	8.4% at 15.1mg/L
	4.8% at 32.8mg/L	5.2% at 21.9mg/L
	4.2% at 51.1mg/L	4.8% at 71.8mg/L
Precision: between-run	6.3% at 12.0mg/L	8.1% at 18.4mg/L
	8.4% at 32.2mg/L	4.7% at 24.2mg/L
	7.4% at 54.7mg/L	7.5% at 71.7mg/L

Table 27.11. Assay performance on the Siemens BNII.

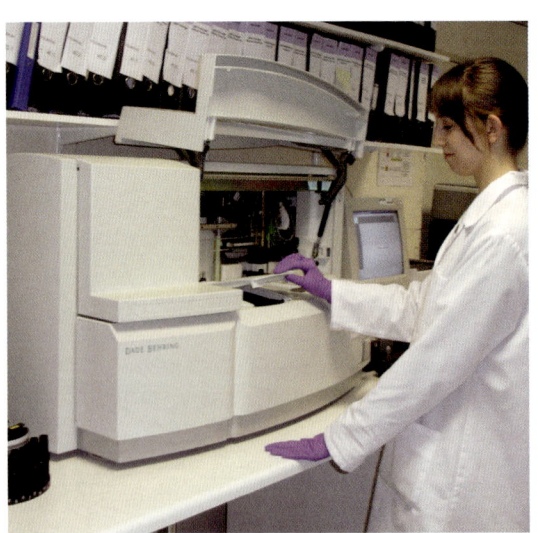

27.11. BN ProSpec® *(Showell 5)*

This instrument is available in some markets in an 'open-mode', or can be opened for serum FLC assays by the Siemens engineers with a software programme called *"Assay Builder"*. Precision of the instrument is good, in part, because of the use of disposable cuvettes *(Table 27.12)*.

	Serum κ FLC	**Serum λ FLC**
Range at 1/100	6.0-190.0mg/L	8.0-260.0mg/L
Sensitivity at 1/5	0.30mg/L	0.40mg/L
Assay time	18 min	18 min
Precision: within-run	4.0% at 16.0mg/L	2.0% at 25.0mg/L
	6.0% at 30.0mg/L	3.0% at 56.0mg/L
	4.2% at 51.0mg/L	4.8% at 72.0mg/L
Precision: between-run	4.0% at 25.0mg/L	2.0% at 22.0mg/L
	2.0% at 36.0mg/L	1.0% at 44.0mg/L
	4.0% at 80.0mg/L	1.0% at 141.0mg/L

Table 27.12. Assay performance on the Siemens BN ProSpec.

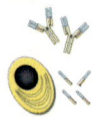

28.1.	*Introduction*	*252*
28.2.	*The Binding Site QA scheme (QA003)*	*252*
28.3.	*College of American Pathologists (CAP) QA scheme*	*254*
28.4.	*AFSSAPS QA Scheme (Agencé Française de Securité Sanitaire des Produits de Santé)*	*254*
28.5.	*UK NEQAS Monoclonal Protein Identification scheme*	*255*
28.6.	*Randox Laboratories Ltd. UK*	*255*
28.7.	*Instand e.V. Laboratories (Germany)*	*255*
28.8.	*Practical aspects of The Binding Site QA scheme QA003*	*255*

28.1. Introduction

It is essential that all laboratories participate in external quality assurance schemes when performing monoclonal protein analysis. Guidelines are described in many publications and are summarised in Chapter 25. There are several national quality assurance (QA) schemes for assessing monoclonal immunoglobulins. Typically, distributed samples comprise serum or urine and contain combinations of intact monoclonal immunoglobulins and monoclonal free light chains (FLCs). Results are generally satisfactory for serum samples but less good for urine, particularly when FLC concentrations are low. The explanations for poor urine results are: variations in electrophoretic techniques, the requirement to concentrate urine samples and difficulties with interpretation of the gels *(Chapter 25)*.

Many laboratories measuring monoclonal-proteins could improve their results by the addition of FLC immunoassays. If the QA schemes encouraged the use of serum rather than urine tests, this might further improve clinical diagnostic accuracy. Wide availability of QA schemes for serum FLC (sFLC) analysis should be encouraged.

28.2. The Binding Site QA scheme (QA003)

This was the first sFLC QA programme and was initiated to fulfill customer demands in the absence of any national schemes at that time. To date, there have been over 55 distributions in 9 years. There are currently over 300 participants for the sFLC component and approximately 100 for the urine tests. Figure 28.1 shows the results of analysis of one of these samples, from a patient with intact immunoglobulin multiple myeloma (IIMM), that had a κ sFLC concentration of 420 mg/L. The monoclonal band was undetectable by serum protein electrophoresis (SPE) and immunofixation electrophoresis (IFE).

Another of the distributed samples was from a patient with nonsecretory multiple myeloma (NSMM) and contained no detectable monoclonal protein by all serum and urine electrophoretic tests *(Figure 28.2)*. This included urine IFE (uIFE) on highly concentrated samples. sFLC immunoassays showed 250 mg/L of κ FLC and 11 mg/L of λ FLC. This was identified correctly only by those laboratories using sFLC immunoassays. It should be noted that weak, false-positive bands for FLCs may sometimes be seen by IFE. So-called *"free"* light chain antisera used in IFE usually have some cross-reactivity with bound light chains in order to improve IFE sensitivity (see information leaflet in IFE kits).

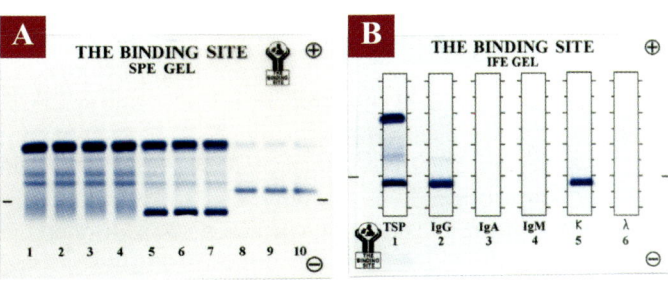

Figure 28.1. (A) SPE and **(B)** serum IFE (sIFE) in a patient with IIMM. SPE comprised normal sera (lanes 1-4) a patient sample (IgGκ) applied 3 times (lanes 5-7) and a κ FLC positive urine (lanes 8-10). The κ sFLC concentration was 420 mg/L.

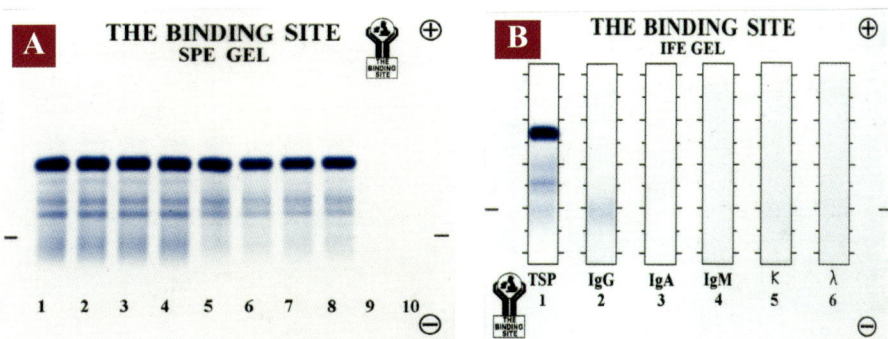

Figure 28.2. (A) SPE, **(B)** sIFE and **(C)** sFLC concentrations in a patient with NSMM. SPE comprised a normal sera (lanes 1-4) and the patient sample (lanes 5-8).

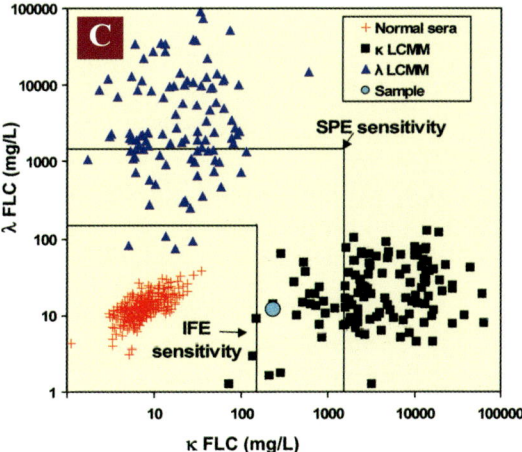

28.3. College of American Pathologists (CAP) QA scheme

The College of American Pathologists (CAP) produces a serum paraprotein QA scheme with over 700 participants. 2 samples are distributed twice a year. Reporting methods include SPE, IFE and monoclonal protein quantification and *"Binding Site Freelite"* participants. Unfortunately, the scheme does not report the FLC results. However, a further scheme has recently been launched (CAP-SFLC) with 3 sample distributions per year where both FLC results and FLC ratios can be reported.

One of the earlier distributed samples, in the former screen, was particularly difficult to characterise for monoclonal immunoglobulins, yet contained a monoclonal IgAλ of 6g/L (*Figure 28.3*). This was missed by 65% (593/916) of laboratories using SPE and 6% of laboratories using IFE but only 43% (398/916) of laboratories used the latter method. However, sFLC measurements showed an elevated λ of 39mg/L (normal range 5.7-26.3mg/L) and an abnormal κ/λ ratio of 0.24 (normal range 0.26-1.65) indicating monoclonality (*Figure 28.3*). This abnormal FLC result would have alerted many laboratories to the presence of a monoclonal plasma cell disease.

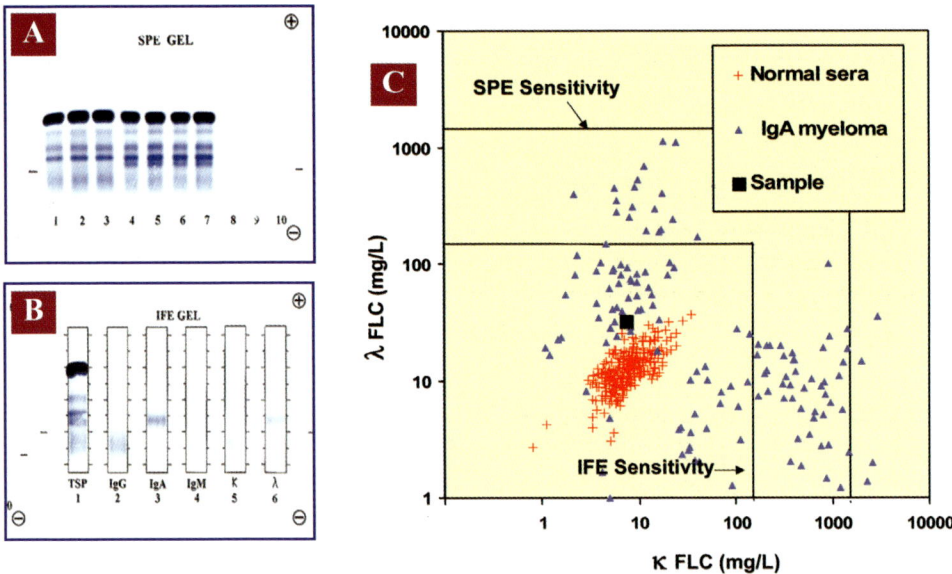

Figure 28.3. (A) SPE, **(B)** IFE and **(C)** sFLC concentrations in a patient with IgAλ MM in relation to other patients with IgA MM (*Figure 11.2*). SPE comprised 3 normal sera (lanes 1-3) and the patient sample applied 4 times (lanes 4-7).

28.4. AFSSAPS QA Scheme (Agencé Française de Securité Sanitaire des Produits de Santé) (143-147 Boulevard Anatole, 93285 St Denis cedex, France).

France has a national quality control scheme for monoclonal immunoglobulins. One distributed serum sample was from a patient with IgD multiple myeloma (MM). Almost 95% of laboratories identified the IgDλ monoclonal immunoglobulin correctly, but only 17% of 1,118 laboratories reported the associated λ FLC. However, sFLC immunoassays showed a λ FLC concentration of 254mg/L and a κ/λ ratio of 0.0096 -

the latter result being 30-fold outside the normal range *(Figure 28.4).*

Figure 28.4. sFLC concentrations in a patient with IgDλ MM (black circle) and results from other patients with IgD MM (blue squares) *(see Figure 11.2).*

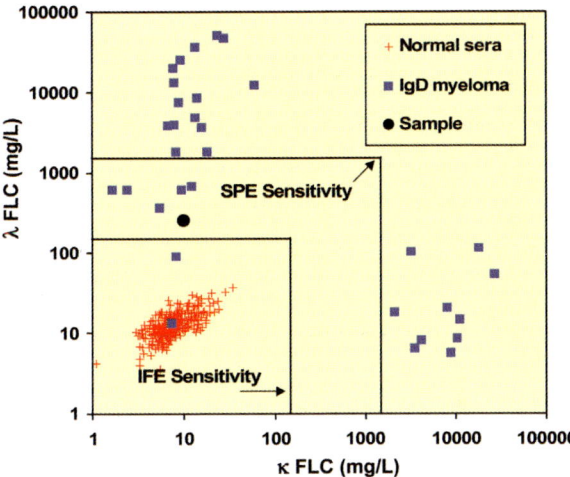

28.5. UK NEQAS Monoclonal Protein Identification Scheme (PO Box 401, Sheffield, S5 7XY, UK).

The UK National External Quality Assessment Service for Immunology and Immunochemistry (Sheffield, UK) provides a scheme for monoclonal proteins and FLCs. Approximately 100 laboratories currently participate in the FLC component and the usual quality assessment parameters are analysed. There are 6 distributions per year containing unmatched serum and urine samples. Antigen excess has been identified by many laboratories as an issue. This is discussed in Chapter 4, Section 4.2 I.

28.6. Randox Laboratories Ltd. UK

This is a recent, international FLC scheme and the first clinical external quality control scheme to gain UKAS accreditation. 12 serum samples are distributed per 6-monthly cycle. There have been over 40 samples distributed, to date, and there are currently around 40 participants.

28.7. Instand e.V. Laboratories (Germany)

This company provides a monoclonal protein scheme with over 180 participating laboratories reporting IFE. Since 2007, sFLCs have been added and there are currently over 70 users. 2 plasma samples are distributed 4 times per year.

28.8. Practical aspects of The Binding Site QA scheme QA003

There are 6 sets of samples issued per year and each contains unmatched serum and urine samples. The results that are returned to the laboratories include photocopies of a summary of the results, the electrophoretic gels, comparisons with other laboratories' results and an *"expert opinion"* regarding the sample and results. Typical reports are shown in Figures 28.5 and 28.6. Results can now also be returned electronically via the internet. Laboratories are given a lab number and password when joining the scheme which enables result entry. Full instructions are provided in the welcome pack.

The left hand side of Figure 28.5 shows an SPE gel with normal sera for reference on the 4 left-hand lanes, then 3 lanes loaded with the unknown serum sample and on the right, 3 lanes loaded with the urine sample. In these samples the serum is negative but the urine positive by protein electrophoresis, and a λ band is visible by uIFE. Figure 28.6 shows the λ results reported from all laboratories with the mean, standard deviation and the result from the reporting laboratory. It should be noted that an aberrant sample result from one laboratory was 1000 fold lower from the majority view. Results 10-fold or 100-fold different may be due to mathematical errors in calculating the sample concentration. In the USA, mg/dL is commonly used, while elsewhere mg/L is the rule. *Care should be taken when calculating the results.*

Laboratory and clinical reports accompanying results from The Binding Site QA003 Scheme *(Figures 28.5-28.7)*.

Serum sample: IgG 7.9g/L, IgA 1.6g/L, IgM 0.77g/L, κFLC 13.7mg/L (NR 3.3-19.4), λ FLC 70mg/L (NR 5.7-26.3), κ/λ ratio 0.096 (NR 0.26-1.65) and β_2-microglobulin 1.8mg/L. The sample appeared normal by SPE but a λ FLC band was detected by IFE.

Urine sample: κ FLC 4.2mg/L (NR 0.36-20.3mg/L), λ FLC 62mg/L (NR 0.81-17.3mg/L), κ/λ ratio 0.07. UPE showed a minor band on concentrated urine and IFE confirmed an abnormal λ FLC band.

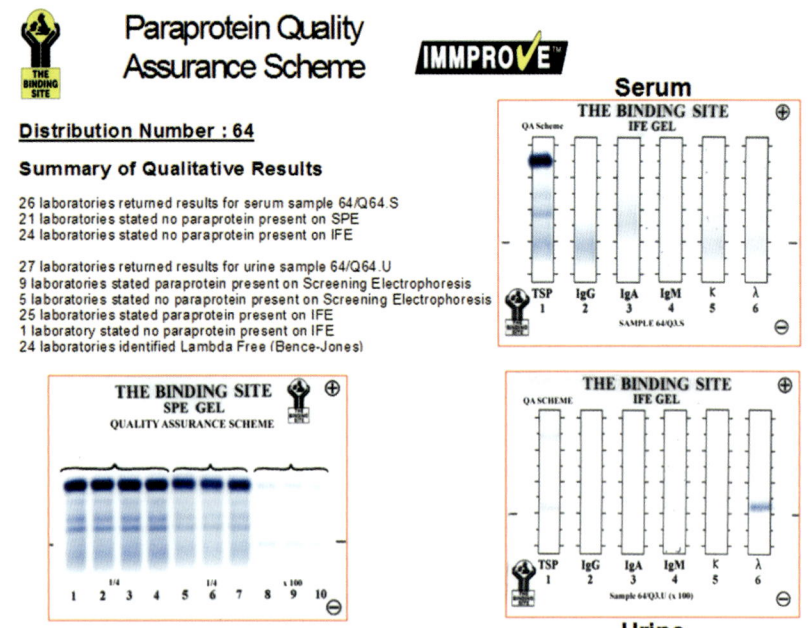

Figure 28.5. Typical report accompanying results from The Binding Site QA003 scheme showing electrophoretic gels.

Comment: The serum sample contained elevated λ FLCs of 70mg/L with no evidence of immunosuppression. The FLC band was not visible by SPE since the concentration was below the detection limit. By IFE, a λ monoclonal band was detected with no coincident heavy chains for IgG, A, M, D or E. sFLC assays showed an elevated λ concentration and a reduced κ/λ ratio. Tests on the urine sample showed similar results by electrophoretic tests and uFLC immunoassays. The latter can often be used to decypher urine electrophoretic test results that are difficult to interpret.

Figure 28.6. Frequency distribution of reported λ sFLC concentrations from QA003 97. The reporting laboratory's result is indicated by an arrow.

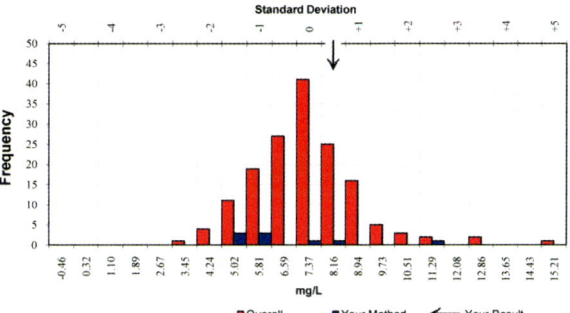

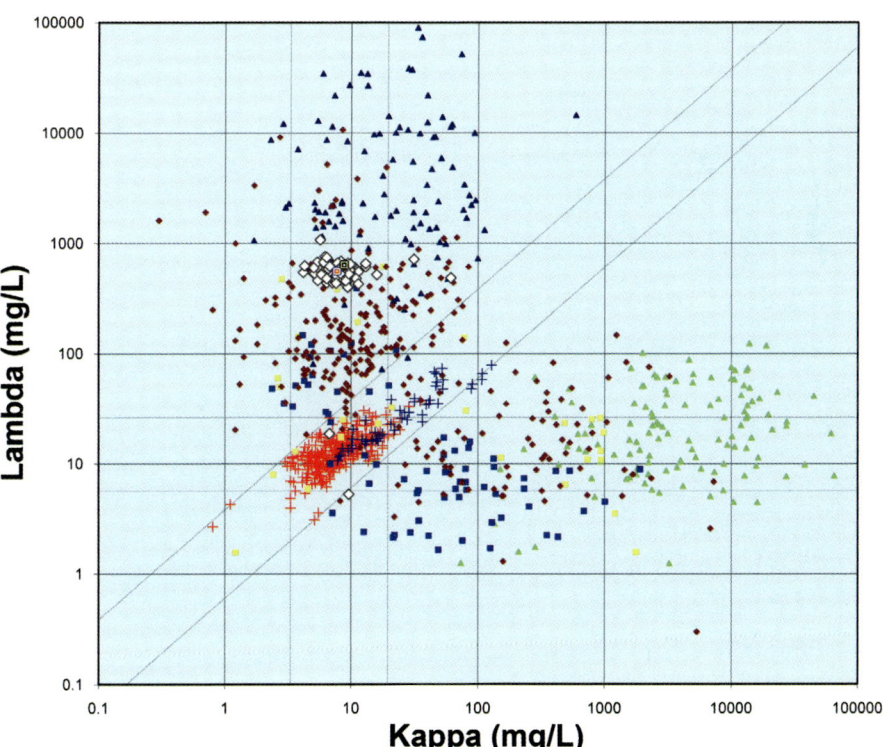

Figure 28.7. Combined results of The Binding Site Paraprotein Quality Assurance Scheme (QA003) distribution 97.

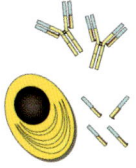

SECTION 6 - Appendices

Chapter 29. *Classification of diseases with increased immunoglobulins*   *259*
Chapter 30. *Questions and answers about FLCs*   *263*
Chapter 31. *Serum free light chain publications*   *272*

Chapter	Section 6 - Appendices
29	**Classification of diseases with increased immunoglobulins**

29.1. *Monoclonal gammopathies*   *259*
29.2. *Polyclonal gammopathies*   *260*

29.1. Monoclonal gammopathies

A. New classification

In May 2003, The International Myeloma Working Group published a review of the criteria for the diagnosis and classification of monoclonal gammopathies, multiple myeloma (MM) and related disorders *(Kyle 5)*. The aim was to standardise and simplify previous classification systems and provide easy definitions based on routinely available investigations. A uniform approach should facilitate comparisons of therapeutic trial data. The disease groups are as follows and their definitions are found in the respective chapters of this book.

1. Monoclonal gammopathy of undetermined significance (MGUS). (*Chapter 19*)
2. Asymptomatic (smouldering) multiple myeloma (ASMM)or Durie and Salmon Stage I). (*Chapter 14*)
3. Symptomatic MM. (*Chapters 8-13*)
4. Nonsecretory multiple myeloma (NSMM). (*Chapter 9*)
5. Solitary plasmacytoma of bone. (*Chapter 18.1*)
6. Extramedullary plasmacytoma. (*Chapter 18.2*)
7. Multiple solitary plasmacytoma. (*Chapter 18.3*)
8. Plasma cell leukaemia. (*Chapter 18.4*)

B. Old classification

This classification includes the diseases indicated above and other types of monoclonal gammopathies. The distribution of clinical diagnoses in patients with monoclonal gammopathies is discussed further in Section 2. The co-occurrence of monoclonal immunoglobulins with non-malignant conditions is uncommon but a proportion of such patients will contain chance associations with MGUS.

I. MGUS
    1. Benign (IgG, IgA, IgD, IgM and FLCs
    2. Biclonal and triclonal gammopathies
II. Malignant Monoclonal Gammopathies
    1. MM (IgG, IgA, IgD, IgE and free k or λ light chains)
        a. Overt MM
        b. ASMM
        c. Plasma cell leukaemia
        d. NSMM
        e. POEMS (polyneuropathy, organomegaly, endocrinopathy,
        monoclonal protein, skin changes; osteosclerotic myeloma)
    2. Plasmacytoma
        a. Solitary plasmacytoma of bone
        b. Extramedullary plasmacytoma
    3. Malignant lymphoproliferative disorders
    4. Waldenström's macroglobulinemia (primary macroglobulinemia)
    5. Malignant Lymphoma
    6. Chronic lymphocytic leukemia or lymphoproliferative disorders
    7. Heavy chain diseases
        a. γ heavy chain disease
        b. α heavy chain disease
        c. μ heavy chain disease
    8. Amyloidosis
        a. Primary amyloidosis
        b. With MM (secondary, localised, and familial amyloidosis have no
        M-protein)

Non-malignant disorders associated rarely with monoclonal proteins (usually MGUS)
1. Dermatological diseases
    Lichen myxoedematosus (IgGl), scleroderma, pyoderma gangrenosum, necrobiotic xanthogranuloma, discoid lupus erythematosus, psoriasis, cutaneous lymphoma
2. Immunosuppression
    AIDS and HIV infection, renal transplantation, bone marrow transplantation
3. Liver diseases
    Chronic hepatitis, cirrhosis, primary biliary cirrhosis
4. Miscellaneous
    Rheumatoid arthritis, inflammatory seronegative polyarthritis, polymyositis (IgGκ), polymyalgia rheumatica, myasthenia gravis, angioneurotic oedema (C1 inactivator deficiency)

29.2. Polyclonal gammopathies
    Chronic infections, autoimmune diseases and many tumours cause increases in

polyclonal immunoglobulins and presumably polyclonal FLC concentrations *(Chapter 21)*. Skin, pulmonary and gut diseases are more likely to cause increases in IgA concentrations while systemic infections will increase all immunoglobulins but particularly IgG. The percentages indicate the frequency of the various disorders from a study at The Mayo Clinic *(Dispenzieri 2)*. Clearly, results will vary in different parts of the world.

1. Connective tissue diseases (22%)
Sjögren's syndrome, rheumatoid arthritis, systemic lupus erythematosus, mixed connective tissue disease, overlap syndrome, juvenile rheumatoid arthritis, progressive systemic sclerosis, ankylosing spondylitis, fibrosing alveolitis, CREST syndrome, temporal arteritis, Raynaud's phenomenon, cutaneous vasculitis, familial mediterranean fever, eosinophilic fasciitis and inclusion body myositis.

2. Liver diseases (61%)
Autoimmune hepatitis, viral hepatitis, primary biliary cirrhosis, primary sclerosing cholangitis, cryptogenic cirrhosis, primary hemochromatosis, ethanol-induced liver injury and α1-antitrypsin deficiency.

3. Chronic Infections (6%)
Subacute bacterial endocarditis, renal abscess, cystic fibrosis, Whipple's disease, brucellosis, Lyme disease, malaria, worm infestations, tropical splenomegaly syndrome, mycobacterium tuberculosis, mycobacterium leprae, Leishmania organisms, Trypanosoma cruzi, Toxocara canis, HIV-1, varicella and vaccinia.

4. Lymphoproliferative disorders (5%)
Pseudolymphoma, Kikuchi disease, malignant lymphoma, Castleman disease, angioimmunoblastic lymphadenopathy with dysproteinemia, large granular lymphocytic leukemia, chronic lymphocytic leukemia, hairy cell leukemia, plasma cell leukaemia, histiocytosis X, sinus histiocytosis with massive lymphadenopathy, cutaneous eruptive histiocytoma, intracranial plasma cell granulomata, systemic cutaneous plasmacytosis, proteinaceous lymphadenopathy with hypergammaglobulinemia, chronic active EBV infection syndrome and severe autoimmune lymphoproliferative syndrome.

5. Other haematological conditions
Myelodysplastic syndromes, idiopathic neutropenia, idiopathic thrombocytopenic purpura, severe hemophilia A, Thalassemia major, Sickle cell anemia, benign hyper-gammaglobulinemic purpura of Waldenström, cryoglobulinemia and Fanconi anemia.

6. Non-hematological malignancies (3%)
Gastric carcinoma, lung cancer, hepatocellular carcinoma, renal cell carcinoma, ovarian cancer and chondrosarcoma.

7. Neurological conditions
Acquired chronic dysimmune demyelinating polyneuropathy, HTLV-1-associated myelopathy. chronic progressive sensory ataxic neuropathy, pure motor neuron disease and plasma cell dyscrasia and microangiopathy of vasa nervorum in dysglobulinemic neuropathy.

8. Diseases with associated immune system abnormalities
Graves' disease, chronic ulcerative colitis, chronic autoimmune pancreatitis, sarcoidosis, syndrome of IgG2 subclass deficiency, hyper IgE syndrome, hyperimmunoglobulinemia D and periodic fever syndrome.

9. Drugs
Aminophenazone, asparaginase, ethotoin, hydralazine hydrochloride, mephenytoin, methadone, oral contraceptives, phenylbutazone and phenytoin.

10. Miscellaneous conditions
Gaucher's disease, Meniére's disease, cardiac myxoma, asbestos exposure, cryptogenic organising pneumonitis, lymphoid interstitial pneumonia, distal renal tubular acidosis and hyperimmunisation.

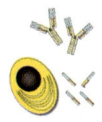

30.1   *Questions about urine testing for free light chains (FLCs)*   *263*
30.2   *Clinical questions about serum testing for FLCs*   *264*
30.3   *Laboratory questions about serum testing for FLCs*   *267*

30.1 Questions about urine testing for FLCs

1. Why are existing assays for uFLCs poor?

There are several reasons. Urine protein elctrophoresis (UPE) with scanning of the bands is inaccurate if there is accompanying proteinuria. Urine immunofixation electrophoresis (uIFE) is non-quantitative but is more sensitive than UPE. Dye uptake tests, dipsticks and many other widely used tests for proteinuria are unreliable as they largely fail to measure small, cationic proteins such as free κ and λ. 24-hour urine collections are frequently not collected reliably so measurements of FLC excretion may be inaccurate.

2. Can I replace existing UPE and uIFE tests with uFLC assays?

No. FLC assays will produce direct, quantitative results for uFLCs but are not always more sensitive than UPE tests. It is preferable to measure serum rather than uFLCs because the removal of FLCs by the renal tubules has a huge effect on the amount of FLCs entering the urine. Also, FLC κ/λ ratios in urine are much more variable than in serum because of renal tubular metabolism.

3. Why might patients show a poor correlation between uFLC M-spike concentrations measured by UPE and uFLCs measured by immunoassays?

UPE involves scanning gels for FLC bands but these can be obscured or confused by other proteins present in urine. In contrast, FLC assays are specific but will also measure polyclonal FLCs in urine. Polymerisation or fragmentation of FLC molecules will affect measurements by either assay, but to different extents.

4. How does proteinuria affect FLC measurements?

Moderate to heavy proteinuria makes interpretation of UPE difficult particularly when obscuring immunoglobulins are present. FLC assays are affected only by increased monoclonal and polyclonal FLC excretion and not by other proteins.

5. How are assays for monoclonal uFLCs affected by polyclonal uFLCs?

FLC assays measure both polyclonal and monoclonal FLCs. Renal impairment will lead to increases in both polyclonal κ and λ urine FLC concentrations and may produce abnormal urine κ/λ ratios. Large amounts of polyclonal FLCs may be excreted in diseases such as systemic lupus erythematosus (SLE) and may prevent low concentrations of monoclonal FLCs from being detected as abnormal κ/λ ratios.

6. In urine, how do FLC assays and IFE compare?

Urinary FLCs immunoassays are more sensitive than urine IFE and are quantitative. Clonality is judged by κ/λ ratios for FLCs and visually by IFE. There is a poor correlation between FLC concentrations and the intensity of staining on IFE because of the IFE antibody. However, clonality is better judged by uIFE than by immunoassays because of variable levels of polyclonal FLCs.

7. Will ladder banding be a problem using FLC tests?

Ladder banding is the term used to describe repeating bands of κ and/or λ seen in concentrated urine samples on IFE gels. It is due to minor differences in charge on polyclonal FLC molecules. The pattern may be confused with, and potentially obscure, genuine monoclonal FLC bands. Ladder banding does not indicate the presence of monoclonal FLCs. uFLC immunoassays are not affected by ladder banding but the presence of polyclonal FLCs will affect measurements of monoclonal FLCs.

8. How do urine and sFLC tests compare?

There is a poor correlation between serum and uFLC concentrations in individual patients. This is because the kidney metabolises large amounts of FLC molecules, preventing them from entering the urine. There is a much better correlation between changes in serum and uFLC concentrations.

9. Can sFLC tests replace diagnostic tests for urine FLCs?

Yes. There is considerable evidence that sFLC tests alongside serum protein electrophoresis (SPE) and serum immunofixation electrophoresis (sIFE) identify all significant patients with monoclonal proteins. Urine tests do identify extra monoclonal light chains but there is no evidence that they are clinically significant except for AL amyloidosis which requires an additional 24 hour uIFE *(Chapter 25)*. As regards sFLC tests replacing urine tests for monitoring multiple myeloma (MM), the jury is still out. Serum and urine tests are complementary in terms of sensitivity but it is unclear if positive urine results are clinically significant when serum tests are negative.

30.2 Clinical questions about serum testing for FLCs

10. What diseases cause elevations of monoclonal FLC levels in serum?

Monoclonal elevations of free κ and λ occur in the same diseases that produce monoclonal gammopathies with intact immunoglobulins. The number of diseases is extensive but clearly MM and AL amyloidosis are important. A list of diseases associated with monoclonal gammopathies is given in Chapter 29.

11. Why are serum tests preferable to urine tests?

Since the kidney can metabolise between 10 and 30g of FLCs per day in the proximal tubules, urinary FLC results do not accurately reflect FLC production by the tumour. This results in a poor correlation between serum and uFLC concentrations and makes urine testing unreliable.

12. How many more MM patients will be detected if sFLC assays are run alongside SPE?

Light chain multiple myeloma (LCMM) comprises between 15-18% of all MM. Approximately, half to two thirds of these patients will have abnormalities by SPE. This especially applies to patients who are producing large quantities of FLCs, are in renal failure or have associated hypogammaglobulinaemia. The remaining LCMM patients with normal SPE (5-7%) and 70% of those with nonsecretory multiple myeloma (NSMM) (2%) will be detected by sFLC assays. Most patients with AL amyloidosis and other rare monoclonal gammopathies will also be detected. In addition, normal individuals with FLC monoclonal gammopathy of undetermined significance (MGUS) will be identified. Intact immunoglobulin MGUS occurs in approximately 3% of people aged over 70 and more frequently with increasing age. Therefore, at least 10-15% additional patients with monoclonal gammopathies will be detected using sFLCs.

The study by Bakshi et al. *(Chapter 23)* detected an additional 50% of patients and included B-cell chronic lymphocytic leukaemia and other plasma cell dyscrasias *(Chapter 23)*.

13. I currently screen for MM using IFE. Will sFLC assays be of additional benefit?

Yes, for two reasons. sFLC analysis will detect monoclonal gammopathies that are missed by IFE. Also, quantification of sFLCs, at the time of clinical presentation provides a base line result for subsequent disease monitoring *(Chapter 23)*.

14. How many more MM patients will be detected if I perform sFLC assays together with SPE and IFE?

sFLC assays are more sensitive for FLC detection than both SPE and IFE so approximately 5% extra patients with low concentration monoclonal FLCs will be detected. Some of these patients will have LCMM, NSMM or AL amyloidosis. Many of these patients also have FLCs in the urine but some do not. If monoclonal FLCs are detected in the serum then there is little value in performing additional urine tests. However, there are rare patients that have normal sFLC concentrations but have detectable FLCs in concentrated urine by IFE. These include patients with AL amyloidosis but not patients with LCMM.

15. How many more MGUS patients will be detected using sFLC assays?

MGUS have historically comprised intact immunoglobulins and these are detected using SPE and sIFE. FLC only MGUS have been observed in the urine but only rarely. FLC MGUS have recently been detected in serum using FLC immunoassays at a frequency of approximately 20% of intact immunoglobulin MGUS (3% of samples from individuals over 70 years of age). Since MGUS comprises 60-80% of monoclonal gammopathies, many new FLC MGUS patients will be detected. These may be the precursor of LCMM and AL amyloidosis and are a focus of considerable clinical interest.

16. Can sFLC assays be used to follow-up patients with LCMM?

Yes, serum is preferable to urine for monitoring patients for several reasons. Serum samples are easier to collect than 24-hour urine samples and serum is a more reliable fluid for assessing changes in production of FLCs. Also, normal serum concentrations of FLCs are less variable than urine concentrations so abnormal results are more easily assessed. Since serum is more sensitive for detection of FLCs than urine, sFLCs are more effective for assessing minimal residual disease. The new international guidelines for MM monitoring include the use of sFLCs for meeting stringent remission criteria.

17. How will sFLC tests benefit patients with NSMM?

By definition, these patients have no detectable monoclonal proteins in their serum and urine by conventional electrophoresis tests and have to be monitored by marrow biopsies or bone scans. sFLC tests are clearly useful in these patients and more accurate than bone marrow biopsies that can miss patchy tumour deposits. sFLC concentrations assess FLC production by the whole bone marrow (and extramedullary sites) so are probably a better reflection of overall tumour activity than isolated bone marrow aspirations. sFLC tests have led to a reduced number of bone marrow biopsies in these patients. However, as for all tumours, a tissue diagnosis, obtained by biopsy, is essential to establish the initial diagnosis, even if the FLC test results are grossly abnormal *(Chapter 9)*.

18. Are sFLC measurements helpful in patients with AL amyloidosis?

AL amyloidosis is a difficult disease to diagnose because existing serum and urine tests are not sufficiently sensitive to identify all patients. sFLC measurements not only identify more patients than electrophoresis tests of serum and urine but also they are important for monitoring disease progress. The short serum half-life of FLCs makes them a mandatory test for evaluating responses to treatment and identifying relapse *(Chapter 15)*.

19. What are the normal values for serum and urine FLCs?

Publications from Birmingham University and The Mayo Clinic have shown similar results. The data from The Mayo Clinic contains results from older people and indicates that sFLC concentrations increase in normal individuals aged >70 years due to deteriorating glomerular filtration. Results from patients should be compared with the age-matched, normal range data. However, all laboratories should make some assessment of normal ranges in their own laboratories since there will be minor variations resulting from differences in race, age, exposure to infections, the use of different instruments, etc.

Normal ranges are usually established by reference to standards (comprising purified proteins) that have been agreed by international committees. Since there is no international agreement, at present, and because of the novelty of the FLC tests, no definitive statement on the accuracy of the normal ranges can be made. In time, agreed reference materials will be manufactured and international agreement established. Existing normal ranges will then be adjusted to take into account any recommendations *(Chapter 5)*.

20. Is the absolute value of the FLCs or the free κ/λ ratio more important?

sFLC abnormalities should be assessed from the concentrations of the clonal FLC, the alternate FLC and the κ/λ ratio. This is optimally performed using a κ/λ log plot for each patient's result (see Figure on page iv) which is compared with normal range and disease group data. This elegantly distinguishes monoclonal FLC diseases from polyclonal FLC abnormalities and individuals that are in the normal range. Since renal function is frequently affected in MM patients, the alternate FLC concentrations are often elevated but the κ/λ ratio remains abnormal.

If the patient has bone marrow suppression, the individual FLC concentrations might be low, in which case the κ/λ ratio may be more helpful. This is typical of patients undergoing chemotherapy and in those patients with NSMM and AL amyloidosis with bone marrow suppression. Clinical judgement should be used in these cases. If either the FLC concentrations or the κ/λ ratios are outside the normal range then the cause should be investigated. Consideration should be given to assessment of renal function and causes of increased immunoglobulin production such as autoimmune diseases, chronic infections and some malignant tumours.

21. If MM patients are in complete remission, how useful is the sFLC result?

FLC results are more useful than existing SPE or IFE because they are more sensitive and, therefore, more likely to detect residual disease. Also, sFLC concentrations are more likely to be elevated than uFLC levels when patients are in remission. Thus, some patients apparently in full remission by existing tests might have abnormal FLC concentrations and their clinical status will need to be revised *(Chapter 25, Section 25.6)*. Some patients who relapse between monitoring periods may be more easily identified using sFLC assays rather than SPE or IFE.

22. What happens to monoclonal FLC concentrations when patients develop renal impairment?

As patients develop renal impairment the concentrations of κ and λ polyclonal sFLCs increase. This is associated with an increase in the monoclonal FLC but the κ/λ ratio will also increase slightly because of the relative reduction in clearance of κ molecules *(Chapter 11.2)*. However, changes in the κ/λ ratio are a better guide to changes in clinical status when glomerular filtration rates (GFRs) are changing than the concentrations of the monoclonal FLC. Changes in the concentrations of creatinine or cystatin C should be assessed, under these circumstances, in order to provide an independent assessment of renal function and a correction factor for the κ/λ ratios.

23. Does bone marrow suppression affect FLC concentrations?

Bone marrow suppression, either because of bone marrow replacement by tumour or resulting from chemotherapy, leads to a reduction in the concentrations of polyclonal FLCs. Typically, at the time of diagnosis, the alternate FLC is suppressed and there is a grossly distorted κ/λ ratio. κ/λ ratios play an important role in assessing changing FLC concentrations in these patients and may even be useful when the concentrations of the monoclonal FLCs are below normal levels.

24. Will sFLC concentrations help in understanding tumour kinetics?

This is an important issue. At present, intact monoclonal immunoglobulins in serum or FLCs in urine are used to monitor the progress of patients. The half-life of IgG is 3-4 weeks, so reductions in tumour mass with chemotherapy may not be reflected in serum monoclonal protein changes for several weeks.

The half-life of FLCs is only a few hours - extending to 2-3 days when renal function is impaired. Thus, reductions in tumour mass with chemotherapy can be identified earlier when the patients are being monitored using sFLC tests. Indeed, changes in tumour mass can be assessed between each cycle of chemotherapy allowing subsequent treatments to be specifically tailored to individual patients. Changes in urine and serum levels of FLCs broadly correspond but renal tubular reabsorption prevents accurate assessment of tumour responses from urine measurements.

Approximately 95% of patients with MM, excreting intact immunoglobulins have abnormal sFLCs. It is likely, because of the short half-life of FLCs, discussed above, that disease monitoring in many of these patients will be better using sFLC levels rather than intact immunoglobulin concentrations.

25. Can sFLC assays be used for transplantation monitoring?

Yes, sFLC assays are useful for monitoring post transplantation progress in AL amyloidosis and MM and may indicate relapses and responses earlier.

26. What diseases cause elevated serum polyclonal FLC concentrations?

Serum polyclonal FLC concentrations increase if there is increased production or reduced glomerular filtration of κ and λ proteins. Increased production results from any disease that stimulates B-cell proliferation such as infections, autoimmune diseases, various tumours etc. For example, in active SLE, total immunoglobulin production increases 3-4 fold with a corresponding increase in sFLC concentrations *(Chapter 21)*.

Reduced glomerular filtration of FLCs occurs in renal damage because almost all FLC removal is via the glomerular pores and the proximal renal tubules. In renal failure, sFLC concentrations may rise 10-20 fold but in all cases both free κ and λ are elevated, so the free κ/λ ratio remains normal. In many diseases, particularly SLE, increases in polyclonal FLC concentrations are due to a combination of increased FLC production and reduced renal filtration.

In situations of increased polyclonal FLC production or reduced filtration, there may be moderate distortions of κ/λ ratios (e.g., 3-4 standard deviations from the normal range). These patients are borderline abnormal and should be investigated appropriately *(Chapter 20)*.

30.3 Laboratory questions about serum testing for FLCs

27. Should sFLC tests be used as a screen for monoclonal proteins instead of SPE?

No. SPE detects intact immunoglobulin monoclonal proteins and some FLC monoclonal proteins. In contrast, sFLC assays detect FLC monoclonal gammopathies, either alone, or in association with intact monoclonal immunoglobulins. Approximately 95% of patients with intact monoclonal immunoglobulins have abnormal sFLCs, but not all, particularly those patients with low concentration MGUS.

28. Will any monoclonal proteins detected by SPE and/or IFE be missed by sFLC tests?

Yes. monoclonal proteins can be intact immunoglobulins or FLCs. Since sFLC assays are >100 times more sensitive than electrophoretic tests it is most unlikely that FLCs will be detected in serum by SPE or IFE yet be normal by FLC assays. However, intact monoclonal immunoglobulins by tradition (especially MGUS) are detected by IFE and SPE, but may have normal FLC concentrations. Studies indicate that all LCMM and ~95% of AL amyloidosis patients are correctly identified by sFLC assays.

29. If the sFLC concentration is many thousands of mg/L why is there no band on SPE?

The sensitivity of SPE for monoclonal proteins depends upon the width of the band and its position in the gel in relation to other plasma proteins. Narrow monoclonal proteins in the gamma region bands in association with hypogammaglobulinaemia will be visible at 200-400mg/L. The same band in a beta position, perhaps superimposed on transferrin, will be invisible. The monoclonal protein may need to be over 2,000mg/L to be visible in this area of the gel. In addition, monoclonal FLCs may be polymerised to different extents and then they migrate on electrophoresis gels as diffuse bands. This is frequently found in NSMM and is well-documented in LCMM. In these patients, even 5,000mg/L of monoclonal FLC may be difficult to detect above the background of the other plasma proteins.

30. Many sera tested by SPE have bands that are barely visible. I worry that I might be missing MM patients. At present I ask for IFE on these samples. Can sFLC testing help?

Yes, the sFLC assays will detect all patients with LCMM and most patients with NSMM and AL amyloidosis. IFE will not detect many of these patients. Since sFLC immunoassays are 100-fold more sensitive than serum IFE, sFLC abnormalities, visible by IFE, will be exceptionally rare if the sFLC tests are normal. uIFE or uFLC measurements may be helpful in these rare cases.

31. Serum albumin levels are reduced in patients with nephrotic syndrome and gross proteinuria. Are sFLC levels also reduced in these patients with proteinuria and will FLC monoclonal gammopathies be missed?

No. Renal damage never increases the GFR of small molecules such as FLCs or creatinine since they normally pass relatively unhindered through the glomerular pores. Molecules as large as albumin are not normally filtered by the kidney but they are cleared in nephrotic syndrome as the glomerular pores become damaged. The extra protein leakage overwhelms the proximal tubular reabsorption mechanisms allowing many different proteins to appear in the urine. The protein leakage damages the tubules in the process which become non-functional with glomerular death. Renal clearance of all small proteins is then reduced. This leads to an increase in sFLC levels (and creatinine). In the early stages of the process, FLCs are increased in the urine because of increased competition with albumin for reabsorption by the proximal tubules.

Renal impairment leads to increases in both κ and λ FLCs in the serum. Therefore, when both are elevated the likely cause is a reduction in glomerular filtration. There is a correlation between changes in the concentrations of serum creatinine, cystatin C and FLCs during changes in renal function *(Chapter 20)*.

32. How can clonality be judged using FLC assays?

Using electrophoresis methods, clonality is judged by the appearance of a narrow protein band. Using FLC assays, clonality is judged by the numerical ratio of free κ to free λ concentrations. In a similar manner, B-cell clonality in leukaemia is assessed by cellular κ/λ ratios using flow cytometry. Arguably, numerical FLC ratios are more accurate than visual assessments of stained bands on electrophoresis gels. Furthermore, in NSMM, clonality may not be apparent by any electrophoretic procedure but is usually identified by serum κ/λ ratios. In the situation of biclonal gammopathies, with increased synthesis of both free κ and λ molecules, free κ/λ ratios may be normal but the concentrations of both FLCs will be raised.

33. How do we report borderline results?

All tests have borderline results. For FLCs, results should be judged against normal and disease state sera from the laboratory, from national and international reference ranges and in clinical context alongside other laboratory results. The normal range recommended for the free κ/λ ratios is greater than that used for most tests in order to provide a large safety margin for normal individuals.

Since FLC results are quantitative, less experience is required compared with protein electrophoresis. This leads to less subjective interpretation of results.

34. What is the frequency of false positive and false negative results?

All tests produce false positive and false negative results and these need to be assessed for clinical significance. For FLCs, reference ranges have been developed in collaboration with The Mayo Clinic and include individuals up to 90 years of age. Some of these individuals have minor degrees of renal impairment. This increases the concentrations of the FLCs, and the κ/λ ratios, and is apparent on a κ/λ log plot. The difference between the normal and abnormal samples is then selected using standard deviations from the mean. If all of the 282 normal samples in the Mayo Clinic study are used, this represents four standard deviations from the mean and is greater than normally chosen cut off levels. Therefore, test samples outside this range will most likely indicate patients with monoclonal gammopathies.

Negative sFLC results occur in a few patients with NSMM, AL amyloidosis and LCDD. Also, rare patients have monoclonal proteins in the urine detected only by IFE. There is no evidence that they are abnormal in shape or size because they react with FLC antibodies. Their origin is unclear but it could be from minor tubular reabsorption failure or from breakdown of monoclonal immunoglobulins in the tubules. They appear to be of little clinical consequence.

35. Why is the free κ/λ ratio different from the total κ/λ ratio in normal individuals?

Approximately twice as many κ molecules are produced as λ. Since free λ is mostly in dimeric form it has a half-life (determined by glomerular filtration) that is approximately three times that of monomeric κ FLCs. This causes free λ molecules to accumulate in the serum more than free κ molecules and alters the free κ/λ ratio from 1.8 to 0.6. When the light chains are bound to immunoglobulins they are metabolised as the whole immunoglobulin, which is independent of light chain type, so total κ/λ ratio is 1.8:1.

36. Why should I change from using a total light chain measurement to a more expensive FLC assay?

Normal sFLC concentrations are <30mg/L, which is much lower than total light chain concentrations of 1,000-3,000 mg/L. Patients who have levels of monoclonal FLCs between these two ranges cannot be assessed using total light chain assays. Since this applies to most patients with

AL amyloidosis, NSMM and many LCMMs, sFLC assays have considerable benefit in these diseases.

37. How does the sensitivity of sFLC tests compare with capillary zone electrophoresis (CZE)?

CZE of serum is more sensitive than SPE but less sensitive than IFE for detecting monoclonal proteins *(Chapter 6)*. In a recent study, it was shown that sFLC assays detected all monoclonal FLCs from patients with LCMM that were missed by CZE but detected by IFE. If CZE is used for initial detection of monoclonal proteins, sFLC assays will detect additional patients.

38. Since FLC assays use polyclonal antibodies, how is batch-to-batch variation minimised?

The FLC antisera are produced by immunisation with many different monoclonal FLC proteins. These are not representative of all monoclonal FLCs but the antibody target is the constant region of the molecule that has little structural variation. However, tumour produced monoclonal FLCs may be truncated, have amino acid substitutions or additions and may be abnormally polymerised. Therefore, occasional patients' monoclonal FLCs may not be detected reliably by the immunoassays or may be detected differently with different antiserum batches. It is, therefore, ideal laboratory practice to assay current and previous samples alongside each other when there is a change in reagent batch. This is no different from the situation when quantifying IgG with different antiserum batches.

To minimise batch-to-batch variation, antisera pools are large, are prepared from multiple immunisations and are carefully controlled to maintain consistency. Many monoclonal proteins are tested when new batches are prepared but there is a limit to the number of different monoclonal proteins that can be used.

Monoclonal antibodies have been assessed in some studies to measure FLCs but they have proved to be unreliable. Polyclonal antisera are superior since they detect more monoclonal FLC molecules and they detect them more reliably.

39. If I am going to use sFLC tests how do they fit into my laboratory protocols?

The preferred option is to measure FLCs alongside SPE/IFE at the time of the presentation blood sample. SPE will identify all intact immunoglobulin multiple myeloma patients while FLC assays will identify LCMMs, most NSMMs and other FLC diseases such as AL amyloidosis. Low concentration, intact immunoglobulin MGUS sera (less than 2-5g/L) will not be detected using these two procedures. A strategy of performing SPE/IFE as a screen for FLC monoclonal proteins and not FLC immunoassays will result in some patients with LCMM, NSMM and AL amyloidosis being missed. FLC assays, performed on patients' presentation samples, are also important for providing a baseline for subsequent disease monitoring. For easy interpretation, results should be reported using a logarithmic κ/λ plot, alongside existing clinical data.

When monitoring patients with FLC diseases, results should be reported alongside other analyses. IFE may add little to the combined use of SPE/FLC tests apart from identifying some low level intact immunoglobulin MGUS samples and rare patients with AL amyloidosis.

40. Does antigen excess occur with FLC assays?

Yes, for two reasons. The range of monoclonal sFLCs is huge, from a few mg/L to many g/L. Hence, assay conditions causing antigen excess occur on a regular basis. In addition, the small size of FLC molecules and the variety of different shapes and sizes may produce antigen excess conditions at relatively low concentrations for some samples. For accurate results, care must be taken to dilute samples for which antigen excess is suspected (e.g. based on clinical information or other laboratory test results) into the appropriate assay range. If in doubt, samples should be re-analysed at higher dilutions according to the product insert.

41. Which instrument should I use for measuring FLCs?

Nephelometers and turbidimeters have a similar level of sensitivity and precision for measuring sFLCs. Instruments vary in their ability for sample handling, in providing clean cuvettes for each test, for evaluating antigen excess, etc. Generally, the large clinical chemistry analysers are the best platforms for measuring sFLCs. The assays are available for most analytical platforms and others will be available in due course *(Chapter 27)*. Since the FLC kits are specifically prepared for each instrument, cross-usage will produce unreliable results.

42. How accurate are the quantitative FLC results?

Quantification of monoclonal FLCs by immunoassay is less accurate than scanning bands on SPE gels. This is the same situation as using nephelometry for measuring intact immunoglobulin monoclonal proteins. Studies have shown that purified monoclonal FLCs assessed by accurate quantitative protein tests may give quite different results compared with immunoassays for FLCs. The explanation is that the antibody assays cannot be expected to produce consistent results for all molecular shapes and polymeric forms of FLCs.

Since the exact amount of serum or uFLCs, at the time of diagnosis, bears little relationship to disease outcome, accurate quantification is relatively unimportant. Of greater concern is the reproducibility of the assay results in individual patients during treatment. It is apparent that FLC measurements produce consistent results during chemotherapy and this provides the basis for their value in managing patients with the various monoclonal diseases. Indeed, sFLC tests are much more reproducible than electrophoresis tests.

43. The patient has high serum and urine polyclonal FLCs with an abnormal urine κ/λ ratio. What does this mean?

High concentrations of both FLCs in the urine and serum indicate a degree of renal impairment *(Chapter 20)*. If the kidney becomes further damaged then both serum and urine concentrations may rise further. This is because with increasing renal damage, glomerular filtration falls and sFLC concentrations increase. This leads to the circulating FLCs being filtered by the remaining nephrons. As their proximal tubules become overwhelmed by the increase in filtered FLCs there is more leakage into the urine. Improving renal function is characterised by reductions in both serum and uFLC concentrations.

The mechanism of abnormal uFLC κ/λ ratios seen in some of these patients can be explained by the renal handling of the molecules. In patients with renal damage, the glomerular pores become altered in size so that monomeric and dimeric FLCs may be filtered differentially. In addition, the proximal tubular reabsorption mechanism is partly dependent upon molecular charge, which is different for each FLC type. Thus, there may be differential clearance and adsorption of κ or λ molecules. This may lead to small distortions of serum and urine κ/λ ratios in patients with renal impairment.

44. Why don't my in-house results compare with results from other laboratories?

There are a number of reasons why results would differ to a moderate extent. Instruments vary in their optical capabilities, different dilution capabilities, normal range usage, etc.. For example the Siemens BNII uses nephelometric, end-point reactions, the Beckman Immage assays use a turbidimetric rate reaction and the Roche Hitachi and Beckman Coulter AU400 analysers both use turbidimetric end-point reactions. Since the normal range for Freelite was established on the Siemens BNII, results on other instruments may show minor differences. While every effort is made to ensure cross-comparability of results, precise matching of results is not always possible.

Abstracts in black, papers and book chapters in blue.

1999

1. **Bradwell AR, Drew R, Showell PJ, Carr-Smith HD, Mead GP.** Clinical potential of free immunoglobulin light-chain measurements. Clin Chem 1999; 45: 181a

2. **Tang LX, Showell PJ, Mead GP, Carr-Smith HD, Drew R, Bradwell AR.** An automated nephelometric immunoassay for quantification of free light-chains in human serum and urine. Clinical Chemistry 1999; 45: 179a

2000

1. **Bradwell AR, Tang LX, Drayson MT, Drew RL.** Immunoassays for detection of free light chains in sera of patients with nonsecretory myeloma. Blood 2000; 96: 4906a

2. **Carr Smith HD, Edwards J, Showell P, Drew R, Tang LX, Bradwell AR.** Preparation of an immunoglobulin free light-chain reference material. Clin Chem 2000; 46: 699a

3. **Harris J, Tang LX, Showell PJ, Carr-Smith HD, Drew R, Bradwell AR.** Assays for immunoglobulin free light chains in serum on the Beckman IMMAGE™. Clin Chem 2000; 46: A180a

4. **Tang LX, Showell P, Carr-Smith HD, Mead GP, Drew R, Bradwell AR.** Evaluation of F(ab)2-based latex enhanced nephelometric reagents for free immunoglobulin light-chains on the Behring Nephelometer II. Clin Chem 2000; 46: 705a

2001

1. **Abraham RS, A. KJ, Clark RJ LJF, Dispenzieri A, Lust JA, Bradwell AR.** Light chain myeloma: Correlation of serum nephelometric analysis for the quantitation of immunoglobulin free light chain with urine Bence Jones protein. Clin Chem 2001; 47: 108a

2. **Abraham RS, Bergen RH, Naylor S, Katzmann JA, Bradwell AR, Kyle RA, Fonseca R.** Characterisation of free immunoglobulin light chains (LC) by mass spectroscopy in light chain-associated (AL) amyloidosis. Blood 2001; 98: 3772a

3. **Abraham RS, Katzmann JA, Clark RJ, Dispenzieri A, Lust JA, Bradwell AR, Kyle RA.** Detection of serum immunoglobulin free light chains in primary amyloidosis and light chain deposition disease by nephelometry. Clin Chem 2001; 47: 107a

4. **Bradwell AR, Mead GP, Carr Smith HD, Drayson MT.** Detection of Bence Jones myeloma and monitoring of myeloma chemotherapy using serum immunoassays specific for free immunoglobulin light chains. Blood 2001; 98: 633a

5. **Bradwell AR, Carr-Smith HD, Mead GP, Tang LX, Showell PJ, Drayson MT, Drew R.** Highly sensitive, automated immunoassay for immunoglobulin free light chains in serum and urine. Clin Chem 2001; 47: 673 – 80

6. **Bradwell AR, Drayson MT, Mead GP.** Measurement of free light chains in urine (letter - reply). Clin Chem 2001; 47: 2069 – 70

7. **Bradwell AR, Smith L, Drayson MT, Mead GP, Carr Smith HD.** Serum free light chain measurements for identifying and monitoring patients with nonsecretory myeloma. Clin Chem Lab Med 2001; 39: PO-F038a

8. **Bradwell AR, D. C-SH, L. S, P. MG.** Detection and monitoring of nonsecretory myeloma using an assay for free light chains in serum. Clin Chem 2001; 47: 142a

9. **Carr-Smith HD, Smith L, Showell P, Mead GP, Drayson MT, Bradwell AR.** Development of serum free immunoassays for the detection and monitoring of patients with Bence Jones myeloma. Hematol J 2001; 1: 70a

10. **Carr-Smith HD, Smith L, Showell P, Mead GP, Drayson MT, Bradwell AR.** Detection and montioring of nonsecretory myeloma using assay for free light chains in serum. Hematol J 2001; 1: 71a

11. **Clark RJ, Katzmann JA, Abraham RS, Lymp JF, Kyle RA, Bradwell AR.** Detection of monoclonal free light chains by nephelometry: Normal ranges and relative sensitivity. Clin Chem 2001; 47: 90a

12. **Drayson M, Tang LX, Drew R, Mead GP, Carr-Smith H, Bradwell AR.** Serum free light-chain measurements for identifying and monitoring patients with nonsecretory multiple myeloma. Blood 2001; 97: 2900 – 2 Chem 2001; 47: 673 – 80

13. **Graziani MS, Merlini G.** Measurement of free light chains in urine. Clin Chem 2001; 47: 2069 - 70

14. **Hansson L-O, Dijlai-Merzog R, Danielsson O, Bradwell AR.** Quantitation of the free light chain kappa and lambda serum, urine and cerebrospinal fluid (CSF) using The Binding Site (F)ab2-based nephelometric method on the IMMAGE® system. Clin Chem 2001; 47: 106a

15. **Katzmann JA, Clarke RJ, Abraham RS, Lymp JF, Carr-Smith HD, Kyle RA, Bradwell AR.** Detection of monoclonal free light chains in serum by nephelometry: Normal ranges and relative sensitivity. Proceedings of the Monoclonal Gammopathies & the Kidney 2001: OC1a

16. **Mead GP, Carr-Smith HD, Drayson MT, Hawkins PN, Bradwell AR.** Serum free light chain immunoassays as an aid in the diagnosis and monitoring of light chain monoclonal gammopathies J Bone & Mineral Metab 2001; 19: 44a

17. **Smith L, Mead GP, Carr-Smith HD, Drayson MT, Bradwell AR.** Detection and monitoring of Bence Jones myeloma using serum free light chain measurements. Clin Chem Lab Med 2001; 39: F039a

18. **Smith L, Mead GP, Carr-Smith HD, Drayson MT, Bradwell AR.** Development of serum free light chain immunoassays for the detection and monitoring of patients with Bence Jones myeloma. Clin Chem 2001; 47: 141a

2002

1. **Abraham RS, Clark RJ, Bryant SC, Lymp JF, Larson T, Kyle RA, Katzmann JA.** Correlation of serum immunoglobulin free light chain quantification with urinary Bence Jones protein in light chain myeloma. Clin Chem 2002; 48: 655 – 7

2. **Bradwell AR, Carr-Smith HD, Mead GP, Drayson MT.** Serum free light chain immunoassays and their clinical application. Clin Appl Imm Rev 2002; 3: 17 – 33

3. **Bradwell AR, D. C-SH, P. MG, C. HT.** Management of patients with light chain myeloma using serum free light chain immunoassays. Clin Chem 2002; 48: A51a

4. Bradwell AR, Mead GP, Drayson MT, Carr-Smith HD. Serum immunoglobulin free light chain measurement in intact immunoglobulin myeloma. Blood 2002; 100: 5054a

5. Bradwell AR, Drayson MT, Mead GP, Galvin G, Gasparetto C. Serum free light chain immunoassay for monitoring patients with light chain producing and nonsecretory myeloma. Blood 2002; 100: 5091a

6. Carr-Smith HD, Showell P, Bradwell AR. Antigen excess assessment of free light chain assays on the Dade-Behring BNII nephelometer. Clin Chem 2002; 48: 23a

7. Carr-Smith HD, Mead GP, Drayson MT, Hawkins PN, Bradwell AR. Detection and monitoring of light chain monoclonal gammopathies using serum free light chain immunoassays. Clin Chem Lab Med 2002; s: 42a

8. Carr-Smith HD, Showell P, Bradwell AR. Development and evaluation of turbidimetric reagents for measuring free immunoglobulin light-chains on the Hitachi 911/912. Clin Chem Lab Med 2002; s: 43a

9. Carr-Smith HD, Showell PJ, Matters DJ, Long JM, Bradwell AR. Development and evaluation of nephelometric reagents for measuring free immunoglobulin light-chains on a modified Minineph™. Clin Chem Lab Med 2002; s: 44a

10. Hermsen D, Bradwell AR, Reinaauer H. Evaluation of an automated nephelometric immunoassay for quantification of free light chains in serum. J Lab Med 2002; 26: Q18a

11. Hofmann W, Guder WG, Bradwell AR, Garbrecht M. Kappa and lambda light chain in serum and urine in patients with monoclonal gammopathy. J Lab Med 2002; 26: 19a

12. Hofmann W, Guder WG, Bradwell AR, Garbrecht M. Detection of free kappa and lambda light chain in serum and urine in patients with monoclonal gammopathy. Onkologie: Int J Canc Res & Treatment 2002; 25: 752a

13. Katzmann JA, Clark RJ, Abraham RS, Bryant S, Lymp JF, Bradwell AR, Kyle RA. Serum reference intervals and diagnostic ranges for free kappa and free lambda immunoglobulin light chains: relative sensitivity for detection of monoclonal light chains. Clin Chem 2002; 48: 1437 – 44

14. Lachmann HJ, Gallimore R, Gillmore JD, Smith L, Bradwell AR, Hawkins PN. Detection of monoclonal free light chains by nephelometry in systemic AL amyloidosis. Clin Chem 2002; 48: E45a

15. Lachmann HJ, Gallimore R, Gillmore JD, Smith L, Bradwell AR, Hawkins PN. Correlation of changes in nephelometric quantification of serum monoclonal free light chains following chemotherapy and outcome in systemic AL amyloidosis. Clin Chem 2002; 48: E46a

16. Lachmann HJ, Gallimore R, Gillmore JD, Carr-Smith HD, Bradwell AR, Hawkins PN. Changes in the concentration of circulating free immunoglobulin chains and outcome in systemic AL Amyloidosis. Blood 2002; 100: 5090a

17. Le Bricon T, Bengoufa D, Benlakehal M, Bousquet B, Erlich D. Urinary free light chain analysis by the Freelite immunoassay: a preliminary study in multiple myeloma. Clin Biochem 2002; 35: 565 – 7

18. Marien G, Bradwell AR, Blanckaert N, Bossuyt X. Detection of monoclonal proteins in sera by capillary zone electrophoresis and free light-chain measurments. Clin Chem 2002;48:E52a

19. Marien G, Oris E, Bradwell AR, Blanckaert N, Bossuyt X. Detection of monoclonal proteins in sera by capillary zone electrophoresis and free light chain measurements. Clin Chem 2002; 48: 1600 – 1

20. Mead GP, Carr-Smith HD, Drayson MT, Bradwell AR. Detection of Bence Jones myeloma and monitoring of myeloma chemotherapy using immunoassays specific for free immunoglobulin light chains. Br J Haematol 2002; 117: 195a

21. Mead GP, Stubbs PD, Carr-Smith HD, Drew R, Drayson MT, Bradwell AR. Nephelometric measurement of serum free light chains in nonsecretory myeloma. Clin Chem 2002; 48: 70a

22. Showell PJ, Long JM, Carr-Smith HC, Bradwell AR. Evaluation of latex-enhanced turbidimetric reagents for measuring free immunoglobulin light-chains on the Hitachi 911/912. Clin Chem 2002; 48: A66a

23. Showell PJ, Matters DJ, Long JM, Carr-Smith HD, Bradwell AR. Evaluation of latex-enhanced nephelometric reagents for measuring free immunoglobulin light-chains on a modified Minineph™. Clin Chem 2002; 48: A67a

24. Sirohi B, Powles R, Kulkarni S, Carr-Smith HD, Sankpal S, Patel G, et al. Serum free light chain assessment in myeloma patients who are in complete remission by immunofixation. Blood 2002; 100: 5095a

25. Tate JR, Grimmett K, Mead GP, Cobcroft R, Gill D. Free light chain ratios in the serum of myeloma patients in complete remission following autologous peripheral blood stem cell transplantation. Clin Chem 2002; 48: E50a

2003

1. Abraham RS, Katzmann JA, Clark RJ, Bradwell AR, Kyle RA, Gertz MA. Quantitative analysis of serum free light chains. A new marker for the diagnostic evaluation of primary systemic amyloidosis. Am J Clin Pathol 2003; 119: 274 – 8

2. Alyanakian M-A, Abbas A, Delarue R, Arnulf B, Bradwell AR, Aucouturier P. Free immunoglobulin light chains serum levels in the follow-up of patients with monoclonal gammopathies: correlation with the 24H urinary light-chain excretion. Clin Chem 2003; 49: D54a

3. Arneth B, Fischer C, Birklein F, Lackner KJ. The suitability of kappa free light chains in cerebrospinal fluid. Clin Chem Lab Med 2003; 41: 107a

4. Bradwell AR. Clinical applications of serum free light chain immunoassays. Clin Lab Invest 2003; November

5. Bradwell AR, Carr-Smith HD, Mead GP. Clinical utility of serum free light chain assays. Clin Chem Lab Med 2003; 41

6. Bradwell AR, Carr-Smith HD, Mead GP, Harvey TC, Drayson MT. Serum test for assessment of patients with Bence Jones myeloma. Lancet 2003; 361: 489 – 91

7. Bradwell AR, Galvin GP, Mead GP, Carr-Smith HD. Practical use of serial measurements of serum free light chains to monitor response to treatment of multiple myeloma. Blood 2003; 102: 5234a

8. Bradwell AR, Mead G, Carr-Smith H, Galvin G, Pratt G. Efficacy of high dose myeloma treatment and response to individual chemotherapy agents in myeloma is indicated by changes in serum free light chain concentrations. Blood 2003; 102: 2547a

9. Bradwell AR, Mead GP, Drayson MT, Kyle RA, Katzmann JA. Role of serum free light chain measurements in disease diagnosis and monitoring. Hematol J 2003; 4

10. Carr-Smith HD, Smith LJ, Bradwell AR. Serum free light chain assay development - specificity, sensitivity, standardisation and protocols. Clin Chem Lab Med 2003; 41: S72a

11. Carr-Smith HD, Mead GP, Smith L, Drayson MT, Bradwell AR. Serum free light chain levels in patients with intact immunoglobulin myeloma. Br J Haematol 2003; 121: 198a

12. Cohen AD, Zhou P, Xiao Q, Fleisher M, Kalakonda N, Akhurst T, et al. An unusual clinicopathologic association: Systemic AL amyloidosis due to Non-Hodgkin's Lymphoma. Blood 2003; 102: 4838a

13. **Graziani M, Merlini G, Petrini C.** Guidelines for the analysis of Bence Jones protein. Clin Chem Lab Med 2003; 41: 338 – 46

14. **Helmke KH, Oppermann MO, Teuber WT, Michels HM, Ventur YV, Welcker MW, Landenberg PL.** Freie immunoglobulin-leichtketten im serum bei rheumatischen erkrankungen. Z Rheumatol 2003; 62: Fr40a

15. **Hermsen D, Bradwell AR.** Nephelometric detection of free kappa and lambda light chains in serum in patients with suspected monoclonal gammopathy. Onkologie 2003; 26: 697a

16. **Herzog W, Hofmann W.** Detection of free kappa and lambda light chains in serum and urine in patients with monoclonal gammopathy. Blood 2003; 102: 5190a

17. **Hofmann W, Herzog W.** Detection of free kappa and lambda light chain in serum and urine in patients with monoclonal gammopathy. Onkologie 2003; 26: 694a

18. **Hoffman U, Opperman M, Kuchler S, Ventur Y, Teuber W, Michels H, et al.** Free immunoglobulin light chains in patients with rheumatic diseases. J Rheumatology 2003; 62: Fr40a

19. **Hunter HM, Peggs K, Powles R, Apperley J, Mahendra P, Canenagh J, et al.** A comparison of outcome for patients undergoing non-myeloablative stem cell transplantation compared to conventional conditioning for multiple myeloma. Blood 2003; 102: 2705a

20. **Jaccard A, Moreau P, Aucouturier P, Ronco P, Fermand J-P, Hermine O.** Amylose immunoglobulinique. Hématologie 2003; 9: 485 – 95

21. **Katzmann JA, Clark RJ, Rajkumar VS, Kyle RA.** Monoclonal free light chains in sera from healthy individuals: FLC MGUS. Clin Chem 2003; 49: A74a

22. **Keren DF.** Techniques to measure free kappa and free lambda light chains in serum and/or urine. Protein electrophoresis in Clinical Diagnosis, Arnold (Hodder Headline), 2003

23. **Lachmann HJ, Gallimore R, Gillmore JD, Carr-Smith HD, Bradwell AR, Pepys MB, Hawkins PN.** Outcome in systemic AL amyloidosis in relation to changes in concentration of circulating free immunoglobulin light chains following chemotherapy. Br J Haematol 2003; 122: 78 – 84

24. **Martin W, Clark RJ, Shanafelt T, Katzmann JA, Bradwell AR, Abraham R, et al.** Detection of serum free light chains in patients with B-cell non-Hodgkin lymphoma (NHL) and chronic lymphocytic leukemia (CLL). Blood 2003; 102: 4827a

25. **Mead GP, Carr-Smith H, Drayson M, Bradwell AR.** Serum free light chain concentrations and their use for disease monitoring in multiple myeloma patients with intact immunoglobulin monoclonal proteins. Hematol J 2003; 170

26. **Mead GP, Carr-Smith HD, Drayson MT, Bradwell AR.** Elevated serum free light chain concentrations in multiple myeloma patients with intact immunoglobulin monoclonal proteins; their use for disease monitoring. Hematol J 2003; 4: 407a

27. **Mead GP, Carr-Smith HD, Drayson MT, Bradwell AR.** Diagnosis and monitoring of multiple myeloma using serum free light chain concentrations. Onkologie 2003; 26: 695a

28. **Mead GP, Carr-Smith HD, Drayson MT, Bradwell AR.** Serum free light chain assays provide improved monitoring of myeloma therapy. Br J Haematol 2003; 121: 205a

29. **Mead GP, Carr-Smith HD, Drayson MT, Bradwell AR.** Serum free light chain levels in patients with intact immunoglobulin myeloma. Clin Chem 2003; 49: D61a

30. **Mead GP, Drayson MT, Carr-Smith HD, Bradwell AR.** Measurement of immunoglobulin free light chains in serum. Clin Chem 2003; 49: 1957 – 8; author reply

31. **Mead GP, Pratt G, Bradwell AR.** Frequent measurement of total immunoglobulin and serum free light chains in myeloma patients during peripheral blood stem cell transplantation. Hematol J 2003; 171

32. **Mead G, Bradwell AR, Lovell R, Pratt G.** Changes in serum free light chain concentrations as a marker of disease response in myeloma patients after autologous stem cell transplantation. Onkologie 2003; 26

33. **Myers B, Lachmann H, Russell NH.** Novel combination chemotherapy for primary (AL) amyloidosis myeloma: clinical, laboratory and serum amyloid-P protein scan improvement. Br J Haematol 2003; 121: 816 – 7

34. **Myers B, Russell NH, McMillan AK.** Use of a novel combination chemotherapy for AL-Amyloidosis: cyclophosphamide, thalidomide and dexamethasone - serum free light chain (SFLC) and serum amyloid protein P (SAP) scan results. Blood 2003; 102: 5249a

35. **Nowrousian MR, Brandhorst D, Daniels R, Sammet C, Schuett P, Ebeling P, et al.** Free light-chain measurement in serum compared with immunofixation of urine in patients with multiple myeloma. Blood 2003; 102: 5197a

36. **Nowrousian MR, Brandhorst D, Daniels R, Sammet C, Schuett P, Ebeling P, Seeber S.** Comparison between free light chain measurement in serum and immunofixation of urine in patients with multiple myeloma. Onkologie 2003; 26: 696a

37. **Ong S, Sethi S.** Assessment of free light chain assay in serum on the Beckman Immage™. Clin Chem 2003; 49: D59a.

38. **Overton J, Goodier D, Carr-Smith HD, Bradwell A.** Evaluation of latex-enhanced turbidimetric reagents for measuring free immunoglobulin light-chains on the Roche Modular P. Clin Chem 2003; 49: D60a

39. **Patten PE, Ahsan G, Kazmi M, Fields PA, Chick GW, Jones RR, et al.** The early use of the serum free light chain assay in patients with relapsed refractory myeloma receiving treatment with thalidomide analogue (CC-4047). Blood 2003; 102: 1640a

40. **Rajkumar SV, Kyle RA, Therneau TM, Bradwell AR, Melton J, Katzmann JA.** Presence of monoclonal free light chains in serum predicts risk of progression in monoclonal gammopathy of undetermined significance (MGUS). Blood 2003; 102: 3481a

41. **Reid SD, Mead GP, Drayson MT, Auguston B, Bradwell AR.** Serum free light chains as a sensitive marker of minimal residual disease in patients with multiple myeloma. Onkologie 2003; 26

42. **Showell P, Matters D, Bradwell AR.** Evaluation of latex-enhanced turbidimetric reagents for measuring free immunoglobulin light-chains on the Olympus AU400. Proceedings of ACB National Meeting 2003: 13a

43. **Showell PJ, Matters DJ, Long JM, Carr-Smith HD, Bradwell AR.** Evaluation of latex-enhanced turbidimetric reagents for measuring free immunoglobulin light-chains on the Olympus AU400. Clin Chem 2003; 49: D55a

44. **Sirohi B, Powles R, Kulkarni S, Carr-Smith HD, Patel G, Das M, et al.** Serum free light chain assessment in myeloma patients who are in complete remission (CR) by immunofixation predicts early relapse. Blood 2003; 102: 5195a

45. **Smith L, Showell P, Matters D, Carr-Smith HD, Bradwell AR.** Evaluation of latex-enhanced turbidimetric reagents for measuring free immunoglobulin light-chains on the Olympus AU400. Clin Chem Lab Med 2003: W497a

46. **Smith LJ, Long J, Matters DJ, Carr-Smith HD, Bradwell AR.** Sample storage and stability for free light chain assays. Proceedings of ACB National Meeting 2003:2a

47. **Smith LJ, Long H, Matters DJ, Carr-Smith HD, Bradwell AR.** Sample storage and stability for serum free light chain assays. Clin Chem 2003; 49: D57a

48. **Smith LJ, Long J, Carr-Smith HD, Bradwell AR.**

Measurement of immunoglobulin free light chains by automated homogeneous immunoassay in serum and plasma samples. Clin Chem 2003; 49: D58a

49. **Smith LJ, Mead GP, Bradwell AR.** Comparative sensitivity of serum and urine assays for free light chains. Clin Chem Lab Med 2003; 41: S72a

50. **Tate JR, Gill D, Cobcroft R, Hickman PE.** Practical considerations for the measurement of free light chains in serum. Clin Chem 2003; 49: 1252 – 7

51. **Tate JR, Mollee P, Gill D.** Measurement of immunoglobulin free light chains in serum: Response. Clin Chem 2003; 49: 1958

52. **Terpos F, Politou M, Szydio R, Nadal E, Avery S, Olavarria E, et al.** Autologous stem cell transplantation normalises abnormal bone resorption through the reduction of RANK/OPG ratio in multiple myeloma. Blood 2003; 102: 3658a

53. **Wands C, Powell M, Jupp R.** The development of serum free light chain immunoassay on Bayer Advia. Proceedings of ACB National Meeting 2003; 2: 3a

2004

1. **Alyanakian MA, Abbas A, Delarue R, Arnulf B, Aucouturier P.** Free immunoglobulin light-chain serum levels in the follow-up of patients with monoclonal gammopathies: correlation with 24-hr urinary light-chain excretion. Am J Hematol 2004; 75: 246 – 8

2. **Augustson BM, Reid SD, Mead GP, Drayson MT, Child JA, Bradwell AR.** Serum free light chain levels in asymptomatic myeloma. Blood 2004; 104: 4880a

3. **Berard A, Bouabdallah K.** Chaines legeres libres: interet pour le diagnostic et le suivi des gammapathies monoclonales. Infos Biologiste Tribune 2004; 8: 12 – 3

4. **Bergen HR, 3rd, Abraham RS, Johnson KL, Bradwell AR, Naylor S.** Characterization of amyloidogenic immunoglobulin light chains directly from serum by on-line immunoaffinity isolation. Biomed Chromatogr 2004; 18: 191 – 201

5. **Bird JM, Cavenagh J, Samson D, Mehta A, Hawkins P, Lachmann H.** Guidelines on the diagnosis and management of AL amyloidosis. Br J Haematol 2004; 125: 681 – 700

6. **Bouabdallah K.** Interet des chaines legeres libres dans la prise en charge des patients atteints de myelome. Biologiste infos 2004; 8: 12 – 3

7. **Bradwell AR, Garbincius J, Holmes EW.** Serum free kappa to free lambda ratios as an adjunct to serum protein electrophoresis for the detection of monoclonal proteins in the serum. Blood 2004; 104: 4856a

8. **Carr-Smith HD, Harland B, Anderson J, Overton J, Wieringa G, Bradwell AR.** The effect on laboratory organisation of introducing serum free light chain assays. Clin Chem 2004; 50: A76a

9. **Carr-Smith HD, Harland B, Anderson J, Overton J, Wieringa G, Bradwell AR.** Evaluation of latex-enhanced turbidimetric reagents for measuring free immunoglobulin light chains on the Bayer Advia 1650. Clin Chem 2004; 50: C44a

10. **Chou PP, Praither JD, Grainger A, Knight R.** Serum free light chain assays. Clin Chem 2004; 50: C24a

11. **Cohen AD, Zhou P, Reich L, Quinn A, Fircanis S, Drake L, et al.** Risk-adapted intravenous melphalan followed by adjuvant dexamethazone (D) and Thalidomide (T) for newly diagnosed patients with systemic AL amyloidosis (AL): Interim results of a phase II study. Blood 2004; 104: 542a

12. **Cohen AD, Zhou P, Xiao Q, Fleisher M, Kalakonda N, Akhurst T, et al.** Systemic AL amyloidosis due to non-Hodgkin's lymphoma: an unusual clinicopathologic association. Br J Haematol 2004; 124: 309 – 14

13. **Dispenzieri A, Gertz MA, Kyle RA.** Response: Determining appropriate treatment options for patients with primary systemic amyloidosis. Blood 2004; 104: 2992 – 3

14. **Engelhardt M, Rapple D, Weis A, Bisse E, Thorst G.** Serum free light chain (FLC) measurement in multiple myeloma (MM) patients (pts) correlate with known monoclonal paraprotein, disease stage and therapy response. Blood 2004; 104: 4907a

15. **Fischer CB, Arneth B, Koehler J, Lotz J, Lackner KJ.** Kappa free light-chains in cerebrospinal fluid as markers of intrathecal immunoglobulin synthesis. Clin Chem 2004;50:1809-13

16. **Fischer CB, Arneth B, Koehler J, Lackner K.** The suitability of kappa free light chains in cerebrospinal fluid diagnostics. Clin Chem 2004; 50: B47a

17. **Gertz MA.** Freelite immunoglobulin free light chain assay diagnostic: Serum test enables identification and monitoring of MM. Myeloma Today 2004; 6: 5 – 6

18. **Gertz M, Comenzo R, Falk RH, Fermand J-P, Hazenberg BP, Hawkins PN, et al.** Definition of organ involvement and treatment response in primary systemic amyloidosis (AL): A consensus opinion from the 10th International Symposium on amyloid and amyloidosis. Blood 2004; 104: 754a

19. **Gertz MA, Lacy MQ, Dispenzieri A.** Therapy for immunoglobulin light chain amyloidosis: the new and the old. Blood Rev 2004; 18: 17 – 37

20. **Gertz MA, Merlini G, Treon SP.** Amyloidosis and Waldenstrom's macroglobulinemia. Hematology Am Soc Hematol Educ Program 2004: 257 – 82

21. **Goodman HJB, Lachmann HJ, Bradwell AR, Hawkins PN.** Intermediate dose intravenous melphalan and dexamethasone treatment in 144 patients with systemic AL amyloidosis. Blood 2004; 104: 755a

22. **Guis L, Diemert MC, Ghillani P, Choquet S, Leblond V, Vernant JP, Musset L.** The quantitation of serum free light chains: Three case reports. Clin Chem 2004; 50: F38a

23. **Hermsen D, Herzog W.** Nephelometric detection of serum free kappa and lambda light chains in patients with suspected monoclonal free light chain gammopathies. Clin Chem 2004; 50: C19a

24. **Herzum I, Heinz R, Bruder-Burzlaff B, Renz H, Wahl HG.** Reliability of the new freelite assay for quantification of free light chains in urine. Clin Chem 2004; 50: C36a

25. **Hofmann W, Garbrecht M, Bradwell AR, Guder WG.** A new concept for detection of Bence Jones proteinuria in patients with monoclonal gammopathy. Clin Lab 2004; 50: 181 – 5

26. **Katzmann JA, Dispenzieri A, Abraham R, Kyle R.** Performance of free light chain assays in clinical practice Blood 2004;104:757a

27. **Mead GP, Bradwell AR, Lovell R, Pratt G.** Changes in serum free light chain concentrations after high-dose melphalan and autologous stem cell transplant in myeloma patients. Bone Marrow Transplant 2004; 33: 380a.

28. **Mead GP, Carr-Smith HD, Drayson MT, Morgan GJ, Child JA, Bradwell AR.** Serum free light chains for monitoring multiple myeloma. Br J Haematol 2004; 126: 348 – 54

29. **Mead GP, Carr-Smith HD, Galvin G, Pratt G, Bradwell AR.** Efficacy of high dose treatment and response to individual chemotherapy agents in myeloma is indicated by changes in serum free light chain concentrations. Clin Chem 2004; 50: C20a

30. **Mead GP, Reid SD, Augustson BM, Drayson MT, Bradwell AR, Child JA.** Correlation of serum free light chains and bone marrow plasma cell infiltration in multiple myeloma. Blood 2004; 104: 4865a

31. Merlini G. Editorial: Sharpening therapeutic strategy in AL amyloidosis. Blood 2004; 104: 1593 – 4

32. Merlini G, Palladini G, Bosoni T, Lavatelli F, D'Eril GM, Bradwell AR, Moratti R. Circulating free light chain concentration correlates with degree of cardiac dysfunction in AL amyloidosis. Clin Chem 2004; 50: B113a

33. Nakano T, Nagata A, Takahashi H. Ratio of urinary free immunoglobulin light chain kappa to lambda in the diagnosis of Bence Jones proteinuria. Clin Chem Lab Med 2004; 42: 429 – 34

34. Pratt G, Mead G, Lovell R, Bradwell AR. Changes in serum-free light chain concentrations as a marker of disease response in myeloma patients after autologous stem cell transplantation. Br J Haematol 2004; 125: 94a

35. Rajkumar SV, Kyle RA, Therneau TM, Clark RJ, Bradwell AR, Melton LJ, 3rd, et al. Presence of monoclonal free light chains in the serum predicts risk of progression in monoclonal gammopathy of undetermined significance. Br J Haematol 2004; 127: 308 – 10

36. Rajkumar SV, Kyle RA, Therneau TM, Melton LJI, Bradwell AR, Clark RJ, et al. Presence of an abnormal serum free light ratio is an independent risk factor for progression in monoclonal gammopathy of undetermined significance (MGUS). Blood 2004; 104: 3647a

37. Räpple D, Weis A, Deschler B, Bisse E, Ihorst G, Englehardt M. Lambda (λ) - and kappa (κ) - free light chains in multiple myeloma (MM) patients correlate with known monoclonal paraprotein, disease stage and therapy response. Onkologie 2004; 27: 771a

38. Reid SD, Drayson MT, Mead GP, Augustson B, Bradwell AR. Serum free light chains are a more sensitive marker of serological remission in multiple myeloma patients. Clin Chem 2004; 50: C34a

39. Reid SD, Mead GP, Drayson MT, Bradwell AR. Comparison of immunofixation electrophoresis with serum free light chain measurement for determining complete remission in light chain multiple myeloma. Bone Marrow Transplant 2004; 33: 632a

40. Sanchorawala V, Seldin DC, Wright DG, Skinner M, Finn KT, Falk RH. Pulsed low dose intravenous melphalan in patients with AL amyloidosis, ineligible for aggressive treatment with high-dose melphalan and stem cell transplantation. Blood 2004; 104: 2393a

41. Sanchorawala V, Wright DG, Magnani B, Skinner M, Seldin DC. Serum free light chain responses after high-dose intravenous melphalan and autologous stem cell transplantation for AL (primary) Amyloidosis. Blood 2004; 104: 942a

42. Showell PJ, Lynch EA, Carr-Smith H, Bradwell AR. Evaluation of latex-enhanced nephelometric reagents for measuring free immunoglobulin light-chains on the Radim Delta. Clin Chem 2004; 50: C40a

43. Sirohi B, Powles R, Kulkarni S, Carr Smith HD, Patel G, Das M, et al. Serum free light chain ratio is predictive of early relapse in patients who are in complete remission by immunofixation. Bone Marrow Transplant 2004;33: 649a

44. Tate JR, Mollee P, Gill D. Serum free light chain ratios do not detect relapse in some patients with intact immunoglobulin myeloma post autologous peripheral blood stem cell transplantation. Clin Chem 2004; 50: B54a

45. Thompson EJ. Editorial: Quality versus quantity: Which is better for cerebrospinal fluid IgG? Clin Chem 2004; 50: 1721-2

46. Urban S, Oppermann M, Reucher SW, Schmolke M, Hoffmann U, Hiefinger-Schindlbeck R, Helmke KH. Free light chains (FLC) of immunoglobulins as parameter resembling disease activity in autoimmune rheumatic diseases. Ann Rheum Dis 2004; 63: 141a

2005

1. Abdalla S, Goodman HJB, Hawkins PN. Thalidomide alone and in combination with other agents in the treatment of patients with AL amyloidosis. Haematologica 2005; 90: PO1411a

2. Abdalla S. The use of serum free light chain assay in clinical practice. Haematologica 2005; 90: PO405a

3. Akar H, Seldin DC, Magnani B, O'Hara C, Berk JL, Schoonmaker C, et al. Quantitative serum free light-chain assay in the diagnostic evaluation of AL amyloidosis. In: Grateau G, Kyle RA, Skinner M, eds. Amyloid and Amyloidosis, CRC Press, 2005

4. Akar H, Seldin DC, Magnani B, O'Hara C, Berk JL, Schoonmaker C, et al. Quantitative serum free light chain assay in the diagnostic evaluation of AL amyloidosis. Amyloid 2005; 12: 210 – 5

5. Augustson BM, Katsavara H, Reid SD, Mead GP, Shirfield M, Bradwell AR. Monoclonal gammopathy screening: Improved sensitivity using the serum free light chain assay. Haematologica 2005; 90: PO1302a

6. Augustson BM, Reid SD, Cohen D, Hawkins K, Mead G, Drayson M, et al. Normalisation of serum free light chains and negative immunofixation electrophoresis may be predictive of progression free survival and overall survival following high dose melphalan. Haematologica 2005; 90: PO403a

7. Augustson BM, Reid SD, Mead GP, Drayson MT, Child JA, Bradwell AR. Serum free light chain levels in asymptomatic myeloma. Haematologica 2005; 90: PO1303a

8. Bakshi NA, Gulbranson R, Garstka D, Bradwell AR, Keren DF. Serum free light chain (FLC) measurement can aid capillary zone electrophoresis in detecting subtle FLC-producing M proteins. Am J Clin Pathol 2005; 124: 214 – 8

9. Beetham R, Howie N, Soutar R. Can opportunistic case-finding of paraproteins be clinically justified? Ann Clin Biochem 2005; 42: 245 – 53

10. Berard A. Quelques notions de rappel sur le myelome? Biologists Infos 2005; 15: 1 – 3

11. Bergon E, Miravalles E, Miranda I, Bergon M. The predictive power of serum kappa/lambda ratios for discrimination between monoclonal gammopathy of undetermined significance and multiple myeloma. Clin Chem Lab Med 2005; 43: 32 – 7

12. Bradwell AR. Serum free light chain measurements move to center stage. Clin Chem 2005; 51: 805 - 7

13. Bradwell AR, Evans ND, Chappell MJ, Cockwell P, Reid SD, Harrison J, et al. Rapid removal of free light chains from serum by hemodialysis for patients with myeloma kidney. Blood 2005; 106: 3482a

14. Bradwell AR, Mead GP, Chappell MJ, Evans ND. Model for assessing free light chain kinetics when monitoring patients with multiple myeloma. Haematologica 2005; 90: PO411a

15. Brockhurst I, Harris KP, Chapman CS. Diagnosis and monitoring a case of light-chain deposition disease in the kidney using a new, sensitive immunoassay. Nephrol Dial Transplant 2005; 20: 1251 – 3

16. Carr-Smith HD, Abraham R, Mead GP, Goodman HJ, Hawkins PN, Bradwell AR. Measurement of serum free light chains in AL amyloidosis. In: Grateau G KR, Skinner M, ed. Amyloid and Amyloidosis, CRC Press, 2005: 154 – 6

17. Carr-Smith HD, Mead GP, Bradwell AR. Serum free light chain assays as a replacement for urine electrophoresis. Haematologica 2005; 90: PO404a

18. Cavallo F, Rasmussen E, Zangari M, Tricot G, Fender B, Fox M, et al. Serum free-lite chain (sFLC) assay in

Multiple Myeloma (MM): Clinical correlates and prognostic implications in newly diagnosed MM patients treated with Total Therapy 2 or 3 (TT2/3). Blood 2005; 106: 3490a

19.   **Chapuis-Cellier C, Foray V, Chazaud A, Troncy J.** Contribution of the quantification of free light chains in 273 patients presenting with a newly discovered monoclonal gammopathy. Haematologica 2005; 90: PO408a

20.   **Chapuis-Cellier C, Foray V, Chazaud A, Troncy J.** Apparent discrepancies in the quantitation of free light chains in serum of patients presenting with a monoclonal gammopathy. Haematologica 2005; 90: PO409a

21.   **Cohen AD, Zhou P, Reich L, Ford A, Hedvat C, Teruya-Feldstein J, et al.** Interim analysis of a Phase II study of risk-adapted intravenous melphalan followed by adjuvant dexamethasone (D) and thalidomide (T) for newly diagnosed patients with systemic AL Amyloidosis (AL). Haematologica 2005; 90: PO1407a

22.   **Comenzo RL.** Light chains ahoy: Pirating Thal/Dex for AL too. Blood 2005; 105: 2625

23.   **Comenzo RL, Zhou P, Reich L, Costello S, Quinn A, Fircanis S, et al.** Risk-adapted intravenous melphalan with adjuvant Thalidomide and Dexamethasone for newly diagnosed untreated patients with systemic AL amyloidosis: interim report of a phase II trial. In: Grateau G, Kyle RA, Skinner M, eds. Amyloid and Amyloidosis, CRC Press, 2005: 112 – 5

24.   **Comenzo RL, Zhou P, Reich L, Costello S, Quinn A, Fircanis S, et al.** Prospective evaluation of the utility of the serum free light chain assay (FLC), Clonal Ig VLR gene identification and troponin 1 levels in a phase II trial of risk-adapted intravenous melphalan with adjuvant Thalidomide and Dexamethasone for newly diagnosed untreated patients with systemic AL amyloidosis. In: Grateau G, Kyle RA, Skinner M, eds. Amyloid and Amyloidosis, CRC Press, 2005: 167 – 9

25.   **Comenzo RL, Zhou P, Wang L, Nimer SD, Olshen AB.** Plasma cell gene-expression profiles in patients with systemic AL amyloidosis: Responses to melphalan and stem cell transplant are associated with differential expression of genes involved in translation, protein degradation and detoxification. Blood 2005; 106: 3405a

26.   **Das M, Mead GP, Sreekanth V, Anderson J, Blair S, Howe T, et al.** Serum free light chain (sFLC) concentration kinetics in patients receiving Bortezomib: Temporary inhibition of protein synthesis and early biomarker for disease response. Blood 2005; 106: 5094a

27.   **Dellerba MP, James M, Butler SJ, Kelsey PR.** Serum free light chain measurement in patients with multiple myeloma and Amyloidosis AL. Clin Chim Acta 2005;355: MP4.35a

28.   **Desplat-Jego S, Feuillet L, Pelletier J, Bernard D, Cherif AA, Boucraut J.** Quantification of immunoglobulin free light chains in cerebrospinal fluid by nephelometry. J Clin Immunol 2005; 25: 338 – 45

29.   **Dingli D, Kyle RA, Rajkumar VS, Nowakowski GS, Larson DR, Bida JP, et al.** Immunoglobulin free light chains at diagnosis: Predictors of progression and survival in solitary plasmacytoma of bone. Blood 2005; 106: 5080a

30.   **Dispenzieri A, Lacy MQ, Katzmann JA, Rajkumar SV, Abraham RS, Hayman SR, et al.** Absolute values of serum immunoglobulin free light chains predict for survival in patients with primary systemic amyloidosis undergoing peripheral blood stem cell transplant. Blood 2005; 106: 422a

31.   **Foray V, Chapuis-Cellier C.** Contribution of serum free light chain immunoassays in diagnosis and monitoring of free light chain monoclonal gammopathies. Immuno-analyse et Biologie Spécialisée 2005; 20: 385 – 93

32.   **Forsyth JM, Hill PG, Rai BS, Mayne S, Mead GP.**

Serum free light chain measurement can replace urine electrophoresis in the detection of B cell proliferative disorders. Blood 2005; 106: 5081a

33.   **Forsyth JM, Hill PG, Rai BS, Mayne S, Mead GP.** Serum free light chains in screening for B cell proliferative disorders. Clin Chim Acta 2005; 355: WP18.30a

34.   **Gertz MA, Comenzo R, Falk RH, Fermand JP, Hazenberg BP, Hawkins PN, et al.** Definition of organ involvement and treatment response in primary systemic amyloidosis (AL): A concensus opinion from the 10th International Symposium on Amyloid and Amyloidosis. Haematologica 2005; 90: PO1405a

35.   **Gertz MA, Comenzo R, Falk RH, Fermand JP, Hazenberg BP, Hawkins PN, et al.** Definition of organ involvement and treatment response in immunoglobulin light chain amyloidosis (AL): a consensus opinion from the 10th International Symposium on Amyloid and Amyloidosis, Tours, France, 18-22 April 2004. Am J Hematol 2005; 79: 319 – 28

36.   **Gertz MA, Lacy MQ, Dispenzieri A, Hayman SR, Kumar SK, Ansell SM, et al.** Role of second stem cell transplant in patients with amyloidosis who are refractory or relapsing. Blood 2005; 106: 5469a

37.   **Gertz MA, Lacy MQ, Dispenzieri A, Hayman SR.** Amyloidosis: diagnosis and management. Clin Lymphoma Myeloma 2005; 6: 208 – 19

38.   **Giarin MM, Di Bello C, Battaglio S, Falco P, Giaccone L, Boccadoro M.** Serum free light chains: a new tool for diagnosis and management of multiple myeloma. Haematologica 2005; 90: 235a

39.   **Gillmore JD, Wechalekar AD, Goodman HJB, Lachmann HJ, Offer M, Joshi J, Hawkins PN.** Cardiac followed by autologous stem cell transplantation for systemic AL Amyloidosis. Blood 2005;106:1158a

40.   **Goodman HJB, Wechalekar AD, Lachmann HJ, Bradwell AR, Hawkins PN.** Clonal disease response and clinical outcome in 229 patients with AL amyloidosis treated with VAD-like chemotherapy. Haematologica 2005; 90: PO1408a

41.   **Goodman HJ, Bridoux F, Lachmann HJ, Gilbertson JA, Gallimore R, Joshi J, et al.** Localised amyloidosis: Clinical features and outcomes in 235 cases. Haematologica 2005; 90: PO1413a

42.   **Gupta S, Comenzo RL, Hoffman BR, Fleisher M.** National Academy of Clinical Biochemistry guidelines for the use of tumor markers in monoclonal gammopathies. http://www.aacc.org/sitecollectiondocuments /nacb/lmpg/tumor/chp3k_gammopathies.pdf

43.   **Hammer F, Rolinski B, Scherberich JE.** Impact of chronic renal failure on serum concentrations of free polyclonal immunoglobulin light chains. Nephro-News, Proceedings of Congress of Nephrology 2005; 10: 5a

44.   **Hassoun H, Reich L, Klimek VM, Dhodapkar M, Cohen A, Kewalramani T, et al.** The serum free light chain ratio after one or two cycles of treatment is highly predictive of the magnitude of final response in patients undergoing initial treatment for Multiple Myeloma. Blood 2005; 106: 972a

45.   **Hazenberg BPC, Bijzet J, de Wit H, van Steijn J, Vellenga E, van Rijswijk MH.** Diagnostic value of free kappa and lambda light chains in fat tissue of patients with systemic AL Amyloidosis. In: Grateau G, Kyle RA, Skinner M, eds. Amyloid and Amyloidosis, 2005: 107 – 8

46.   **Henon KT, Dispenzieri A, Katzmann JA, Lacy MQ, Ramirez-Alvarado M, Gertz MA, et al.** Circulating soluble light chain oligomers in sera of patients with light chain amyloidosis. Haematologica 2005; 90: PO1406a

47.   **Herzog W, Mead GP, Reid SD, Hewins P, Cockwell**

P, Bradwell AR. The effect of renal impairment and dialysis on serum free light chain measurement. Nephro-News, Proceedings of Congress of Nephrology 2005: P8.03a

48. **Herzog W, Mead GP, Drayson MT, Bradwell AR.** Serum free light chain immunoassays in the diagnosis of monoclonal gammopathies. Nephro-News, Proceedings of Congress of Nephrology 2005: P12.04a

49. **Hill PG, Forsyth JM, Mayne S, Mead GP.** Comparison of serum free light chain measurement and urine electrophoresis for detection of B cell proliferative disorders. Haematologica 2005; 90: PO406a

50. **Hill PG, Forsyth JM, Rai BS, Mayne S, Mead GP.** Serum free light chain measurement can replace urine electrophoresis for detecting B cell proliferative disorders. Clin Chem 2005; 51: B6a

51. **Ihenetu KU, Abudu N, Miller J, Elin RJ.** Free light chain assay shows greater clinical sensitivity than electrophoresis for detecting plasma cell dyscrasias. Clin Chem 2005; 51: B157a

52. **Jagannath S, Durie BG, Wolf J, Camacho E, Irwin D, Lutzky J, et al.** Bortezomib therapy alone and in combination with dexamethasone for previously untreated symptomatic multiple myeloma. Br J Haematol 2005; 129: 776 – 83

53. **Kang SY, Suh JT, Lee HJ, Yoon HJ, Lee WI.** Clinical usefulness of free light chain concentration as a tumor marker in multiple myeloma. Ann Hematol 2005; 84: 588– 93

54. **Katsavara H, Reid SD, Augustson BM, Mead GP, Shirfield M, Drayson MT, et al.** Screening for monoclonal gammopathy: Improved sensitivity using serum free light chain assays. Clin Chim Acta 2005; 355: TP4.06a

55. **Katzmann JA.** Quantitative free light chain assays for the diagnosis and monitoring of monoclonal gammopathies. J Ligand Chem 2005; 27: 246 – 55

56. **Katzmann JA, Abraham RS, Dispenzieri A, Lust JA, Kyle RA.** Diagnostic performance of quantitative kappa and lambda free light chain assays in clinical practice. Clin Chem 2005; 51: 878 – 81

57. **Katzmann JA, Dispenzieri A, Abraham RS, Lust JA, Kyle RA.** Diagnostic Performance of Free Light Chain Assays in Clinical Practice. Clin Chem 2005; 51: D16a

58. **Keren D.** Serum protein electrophoresis evaluation of monoclonal gammopathies (M-proteins). J Ligand Chem 2005; 27: 218 – 26

59. **Kühnemund A, Liebisch P, Bauchmüller K, Haas P, Kleber M, Bisse E, et al.** Secondary light chain multiple myeloma with decreasing IgA paraprotein levels correlating with renal insufficiency and progressive disease: Clinical course of two patients and review of the literature. Onkologie 2005; 28

60. **Kumar S, Gertz MA, Hayman SR, Lacy MQ, Dispenzieri A, Zeldenrust SR, et al.** Use of the serum free light chain assay in assessment of response to therapy in multiple myeloma: Validation of recently proposed response criteria in a prospective clinical trial of lenalidomide plus dexamethasone for newly diagnosed multiple myeloma. Blood 2005; 106: 3479a

61. **Kyrtsonis MC, Sachanas S, Vassilakopoulos TP, Kafassi N, Tzenou T, Papadogiannis A, et al.** Bortezomib in patients with relapsed-refractory multiple myeloma (MM). Clinical observations. Blood 2005; 106: 5193a

62. **Leleu X, Moreau A-S, Coiteux V, Guieze R, Hennache B, Facon T, et al.** Serum free light chain assays in solitary bone plasmacytoma. Clin Chim Acta 2005; 355: TP4.02a

63. **Leleu X, Moreau AS, Hennache B, Dupire S, Faucompret JL, Facon T, et al.** Serum free light chain immunoassays measurement for monitoring solitary bone plasmacytoma. Haematologica 2005; 90: PO410a

64. **Leleu X, Moreau A-S, Coiteux V, Guieze R,** Hennache B, Facon T, et al. Serum free light chain assays in solitary bone plasmacytoma. Br J Haematol 2005; 129: 186a

65. **Lorenz EC, Gertz MA, Fervenza FC, Leung N.** Long-term renal outcome of autologous stem cell transplantation in light chain deposition disease. Blood 2005; 106: 5518a

66. **Matsuda M, Yamada T, Gono T, Shimojima Y, Ishii W, Fushimi T, et al.** Serum levels of free light chain before and after chemotherapy in primary systemic AL amyloidosis. Intern Med 2005; 44: 428 – 33

67. **Mead GP, Carr-Smith HD, Drayson MT, Morgan GJ, Child AJ.** Response to: Serum free light chains for monitoring multiple myeloma. Br J Haematol 2005; 128: 406 – 7

68. **Mead GP, Reid S, Augustson B, Drayson MT, Bradwell AR, Child JA.** Comparison of serum and urine free light chain measurements with bone marrow assessments in multiple myeloma. Haematologica 2005; 90: PO407a

69. **Mead GP, Reid SD, Cockwell P, Hewins P, Bradwell AR.** The effect of renal impairment and dialysis on serum free light chain measurement. Haematologica 2005;90: PO237a

70. **Merkel S, Peest D, Witte T, Haller H, Schwarz A.** Rekurrenz einer Leichtkettennephropathie im Transplantat - Therapiemöglichkeiten. Nephro-News, Proceedings of Congress of Nephrology 2005: P3.08a

71. **Mollee P, Tate J, Dimeski G, Gill D.** Falsely low serum free light chain concentration in patients with monoclonal light chain diseases. Blood 2005; 106: 5077a

72. **Mösbauer U, Schieder H, Renges H, Ayuk F, Zander A, Kroger N.** Serum free light chain [FLC] assay in multiple myeloma patients who achieved negative immunofixation after allogeneic stem cell transplantation. Blood 2005; 106: 2023a

73. **Munshi NC.** Editorial: Determining the undetermined. Blood 2005; 106: 767 – 8

74. **Myers B.** To the editor: Cardiac amyloidosis. Clin Med 2005; 5: 2 – 3

75. **Nowrousian MR, Brandhorst D, Sammet C, Kellert M, Daniels R, Schuett P, et al.** Serum free light chain analysis and urine immunofixation electrophoresis in patients with multiple myeloma. Clin Cancer Res 2005; 11: 8706 – 14

76. **Nowrousian MR, Brandhorst D, Sammet C, Kellert M, Daniels R, Schuett P, et al.** Relationship between serum concentrations and urinary excretions of monoclonal free light chains (mFLC) detectable as Bence Jones proteins (BJP) by immunofixation electrophoresis (IFE) in patients with multiple myeloma (MM). Blood 2005; 106: 5060a

77. **Offer M, Wechalekar AD, Goodman HJB, Gillmore JD, Lachmann HJ, Bradwell AR, Hawkins PN.** Standard oral melphalan chemotherapy for AL amyloidosis revisited using the serum free light chain assay. Blood 2005; 106: 3495a

78. **Palladini G, Perfetti V, Perlini S, Obici L, Lavatelli F, Caccialanza R, et al.** The combination of thalidomide and intermediate-dose dexamethasone is an effective but toxic treatment for patients with primary amyloidosis (AL). Blood 2005; 105: 2949 – 51

79. **Palladini G, Perlini S, Vezzoli M, Perfetti V, Lavatelli F, Ferrero I, et al.** The reduction of the serum concentration of the amyloidogenic light-chain in cardiac AL results in prompt improvement of myocardial function and prolonged survival despite unaltered amount of myocardial amyloid deposits. In: Grateau G, Kyle RA, Skinner M, eds. Amyloid and Amyloidosis, CRC Press, 2005: 73 – 5

80. **Peterson MR, Sumabat F, Nesbet L, Mullaney S, Smith D, Herold DA.** Analysis of serum immunoglobulin free light chains in chronic hemodialysis patients. Am J Clin Path 2005; 124

81. **Pietrantuono A.** Que sont les tests FreeliteTM. Bulletin

des Syndicat National des Biologistes des Hopitaux 2005; 11

82. Pratt G, Mead G, Bradwell A. Changes in serum free light chain concentrations as a marker of chemosensitivity after high-dose melphalan and autologous stem cell transplant in myeloma patients. Haematologica 2005;90: PO413a

83. Rajkumar SV. MGUS and smoldering multiple myeloma: Update on pathogenesis, natural history and management. Hematology Am Soc Hematol Educ Program 2005: 340 – 5

84. Rajkumar SV, Kyle RA, Therneau TM, Melton JL, Bradwell AR, Clark RJ, et al. Abnormal serum free light chain ratio is an independent risk factor for progression in monoclonal gammopathy of undetermined significance (MGUS). Haematologica 2005; 90: PO1301a

85. Rajkumar SV, Kyle RA, Therneau TM, Melton LJ, 3rd, Bradwell AR, Clark RJ, et al. Serum free light chain ratio is an independent risk factor for progression in monoclonal gammopathy of undetermined significance. Blood 2005; 106: 812 – 7

86. Ramasamy I. Free immunoglobulin light chain serum levels in B cell dyscrasias. Clin Chem 2005; 51: B122a

87. Reid SD, Augustson BM, Katsavara H, Drayson MT, Mead GP, Shirfield M, et al. Monoclonal gammopathy screening: Improved sensitivity using serum free light chain assays. Clin Chem 2005; 51: B20a

88. Reid SD, Augustson BM, Katsavara H, Mead GP, Shirfield M, Drayson MT, et al. Screening for monoclonal gammopathy: Improved sensitivity using serum free light chain assays. Br J Haem 2005; 129: 81a

89. Reid SD, Leleu X, Moreau AS, Coiteux V, Guieze R, Hennache B, et al. Serum free light chain assays in solitary bone plasmacytoma. Clin Chem 2005; 51: D13a

90. Reid SD, Cockwell P, Hewins P, Millard JL, Mead GP, Bradwell AR. Haemodialysis removes free light chains from serum. Clin Chem 2005; 51: D12a

91. Reid SD, Cockwell P, Hewins P, Mead GP, Bradwell AR. Efficient removal of serum free light chains by haemodialysis. Haematologica 2005; 90: PO512a

92. Reid SD, Hewins P, Cockwell P, Millard JL, Mead GP, Bradwell AR. Haemodialysis removes free light chains from serum. Clin Chim Acta 2005; 355: MP10.5a

93. Romeril KR, White G, Ritchie D. Thalidomide therapy in relapsed/refractory light chain myeloma. Haematologica 2005; 90: PO708a

94. Sanchorawala V, Seldin DC, Magnani B, Skinner M, Wright DG. Serum free light-chain responses after high-dose intravenous melphalan and autologous stem cell transplantation for AL (primary) amyloidosis. Bone Marrow Transplant 2005; 36: 597 – 600

95. Sanchorawala V, Wright DG, Quillen K, Fisher C, Skinner M, Sedlin DC. Early serum free light chain responses following high-dose melphalan and stem cell transplantation for AL Amyloidosis predict treatment outcomes. Blood 2005; 106: 1160a

96. Shimojima Y, Matsuda M, Gono T, Ishii W, Fushimi T, Hoshii Y, et al. Correlation between serum levels of free light chain and phenotype of plasma cells in bone marrow in primary AL amyloidosis. Amyloid 2005; 12: 33 – 40

97. Showell PJ, Lynch EA, Overton J, Carr-Smith HD, Bradwell AR. Evaluation of latex-enhanced nephelometric reagents for measuring free immunoglobulin light chains on the Dade Behring ProSpec. Clin Chem 2005; 51: B38a

98. Smith DE, Abadie J, Bankson D, Mead G. Assessment of serum free light chain assay for screening for plasma cell disorders. Blood 2005; 106: 2563a

99. Szarka C. Serum free light chain ratio predicts outcome in MGUS. Clin Lab Invest 2005; December

100. Tate J, Mollee P, Gill D. Serum free light chains for monitoring multiple myeloma. Br J Haematol 2005; 128: 405 - 6; author reply 406 – 407

101. Tate JR, Mollee P, Dimeski G, Gill D. Falsely low serum free light chain concentrations in patients with monoclonal light chain diseases. Clin Chem 2005; 51: B31a

102. Tate JR, Mollee P, Gill D. Utility of serum free light chains for monitoring myeloma post autologous stem cell transplantation. Haematologica 2005; 90: PO412a

103. van Steijn J, Bijzet J, de Wit H, Hazenberg BPC, van Gameren II, van de Belt K, Vellenga E. Serum levels of free kappa and lambda light chains in patients with systemic AL, AA and ATTR Amyloidosis. In: Grateau G, Kyle RA, Skinner M, eds. Amyloid and Amyloidosis, CRC Press, 2005: 105 – 6

104. Walker R, Rasmussen E, Cavallo F, Jones-Jackson L, Anaissie E, Alpe T, et al. Correlation of suppression of FDG PET uptake with serum free light chain levels - both FDG PET-CT and serum clonal free light chain response precede and predict the likelihood of subsequent complete remission in newly diagnosed multiple myeloma. Blood 2005; 106: 3493a

105. Walker SA, Roddie PH, Ashby JP. Evaluation of serum free light chain (FLC) measurements as an alternative to urine Bence Jones (BJP) analysis in a routine laboratory setting. Clin Chim Acta 2005; 355: WP18.32a

106. Wechalekar AD, Goodman HJB, Gillmore JD, Lachmann HJ, Offer M, Bradwell AR, Hawkins PN. Clinical profile and treatment outcome in 92 Patients with AL amyloidosis associated with IgM paraproteinaemia. Blood 2005; 106: 3498a

107. Wechalekar AD, Goodman HJB, Gillmore JD, Lachmann HJ, Offer M, Bradwell AR, Hawkins PN. Efficacy of risk adapted cyclophosphamide, thalidomide and dexamethasone in systemic AL amyloidosis. Blood 2005; 106: 3496a

108. Wechalekar AD, Lachmann HJ, Goodman HJB, Bradwell AR, Hawkins PN. Role of serum free light chains in diagnosis and monitoring response to treatment in light chain deposition disease. Haematologica 2005; 90: PO1414a

109. Yagmur E, Mertens PR, Gressner AM, Kiefer P. Free light chain kappa/lambda ratios are also of diagnotic value for patients with renal insufficiency Nephro-News, Proceedings of Congress of Nephrology 2005; 102: P12.04a

2006

1. Abadie JM, Bankson DD. Assessment of Serum Free Light Chain Assays for Plasma Cell Disorder Screening in a Veterans Affairs Population. Ann Clin Lab Sci 2006; 36: 157 - 62

2. Blade J. Clinical practice. Monoclonal gammopathy of undetermined significance. N Engl J Med 2006;355: 2765 – 70

3. Bradwell AR, Harding SJ. Paraprotein management: Response to Smellie WSA, Spickett GP: BMJ 2006; 333: 185 - 187. BMJ, 7th August 2006

4. Campbell P, Murdock C. Cardiac amyloidosis-- sustained clinical and free light chain response to low dose thalidomide and corticosteroids. Intern Med J 2006; 36: 137 – 9

5. Catarini M, Pieroni S, Monachesi A. Serum free light chains: A potential marker for diagnosis, early assessment of response to treatment and relapse in plasma cell disorders. Haematologica 2006; 91: PO153a

6. Chang-Ki M, Ki-Seong E, Seok L, Jong-Wook L, Woo-Sung M, Chong-Won P, et al. Changes in serum free light chain and biochemical markers in patients with multiple myeloma receiving bortezomib treatment. Blood 2006; 108: 5028a

7. Comenzo RL. Amyloidosis. Curr Treat Options Oncol

2006; 7: 225 – 36

8. **Dingli D, Kyle RA, Rajkumar SV, Nowakowski GS, Larson DR, Bida JP, et al.** Immunoglobulin free light chains and solitary plasmacytoma of bone. Blood 2006; 108: 1979 – 83

9. **Dispenzieri A, Lacy MQ, Katzmann JA, Rajkumar SV, Abraham RS, Hayman SR, et al.** Absolute values of immunoglobulin free light chains are prognostic in patients with primary systemic amyloidosis undergoing peripheral blood stem cell transplantation. Blood 2006; 107: 3378 – 83

10. **Dispenzieri A, Rajkumar SV, Plevak MF, Katzmann JA, Kyle RA, Larson D, et al.** Early immunoglobulin free light chain (FLC) response post autologous peripheral blood stem cell transplant predicts for hematologic complete response in patients with multiple myeloma. Blood 2006; 108: 3097a

11. **Durie BG, Harousseau JL, Miguel JS, Blade J, Barlogie B, Anderson K, et al.** International uniform response criteria for multiple myeloma. Leukemia 2006; 20: 1467 – 73

12. **Franke DD, Gousha N, Levinson SS.** Analysis of immunoglobulin free light chains in Urine: A sensitive assay to characterize Bence Jones proteinuria. Clin Chem 2006; 52: E9a

13. **Friedman JF, Al-Zoubi A, Kaminski M, Kendall T, Jakubowiak A.** A new model predicting at least a very good partial response in patients with multiple myeloma (MM) after 2 cycles of bortezomib-based therapy. Haematologica 2006; 91: PO741a

14. **Gertz MA.** Monitoring organ response in amyloidosis. Blood 2006; 107: 3814a

15. **Gertz MA, Lacy MQ, Dispenzieri A, Hayman SR, Kumar SK, Gastineau DA.** Extent of hematologic response is important in determining outcome in transplanted patients with primary amyloidosis AL. Importance of achieving CR. Blood 2006; 108: 611a

16. **Giarin MM, Giaccone L, Bruno B, Omedè P, Battaglio S, Falco P, et al.** Serum free light chains: a potential useful marker for diagnosis and early assessment of response to treatment and relapse in plasma cell disorders. Haematologica Reports 2006; 2: 4a

17. **Giarin MM, Giaccone L, Caracciolo D, Bruno B, Falco P, Omedè P, et al.** Serum free light chains (SFLC) assay: a suggestive new criteria for evaluating disease response, progression and relapse in plasma-cell disorders (PD) and a prognostic factor in monoclonal gammopathy of undetermined significance (MGUS). Haematologica 2006; 91: PO151a

18. **Gillmore JD, Goodman HJ, Lachmann HJ, Offer M, Wechalekar AD, Joshi J, et al.** Sequential heart and autologous stem cell transplantation for systemic AL amyloidosis. Blood 2006; 107: 1227 – 9

19. **Goodman HJ, Gillmore JD, Lachmann HJ, Wechalekar AD, Bradwell AR, Hawkins PN.** Outcome of autologous stem cell transplantation for AL amyloidosis in the UK. Br J Haematol 2006; 134: 417 – 25

20. **Hajdu SI.** A note from history: the first biochemical test for detection of cancer. Ann Clin Lab Sci 2006; 36: 222 – 3

21. **Hassoun H, Reich L, Klimek VM, Dhodapkar M, Cohen A, Kewalramani T, et al.** Doxorubicin and dexamethasone followed by thalidomide and dexamethasone is an effective well tolerated initial therapy for multiple myeloma. Br J Haematol 2006; 132: 155 – 61

22. **Hill PG, Forsyth JM, Rai B, Mayne S.** Serum free light chains: an alternative to the urine Bence Jones proteins screening test for monoclonal gammopathies. Clin Chem 2006; 52: 1743 – 8

23. **Hutchison C, Bradwell AR, Mead G, Chandler K, Harper J, Cook M, et al.** Free light chain removal by extended hemodialysis in patients with cast nephropathy from multiple myeloma. J Am Soc Nephrol 2006; 17: TH-FC102a

24. **Hutchison C, Chandler K, Mead G, Cockwell P, Bradwell AR.** Free light chain removal from serum by haemofiltration and haemodialysis: a comparison of dialysis membranes in vitro. J Am Soc Nephrol 2006; 17: PUB237a

25. **Hutchison C, Cockwell P, Chandler K, Harper J, Mead G, Barnett AH, Bradwell AR.** Light chain abnormalities in diabetic patients with and without microalbuminuria. J Am Soc Nephrol 2006; 17: TH-PO231a

26. **Hutchison C, Cockwell P, Hatersley J, Evans N, Chappell MJ, Mead G, Bradwell A.** Simulations of free light chain removal by extended haemodialysis using a mathematical model. J Am Soc Nephrol 2006; 17: PUB238a

27. **Hutchison C, Cockwell P, Reid C, Chandler K, Mead GP, Harrison J, et al.** Efficient removal of immunoglobulin free light chains by hemodialysis in multiple myeloma: In-vitro and in-vivo studies. Blood 2006; 108: 5112a

28. **Hutchison CA, Cockwell P, Reid SD, Chandler KL, Millard JL, Mead GP, et al.** Removal of serum free light chains by hemodialysis in patients with multiple myeloma. Clin Chem 2006; 52: A81a

29. **Hutchison C, Cook M, Bradwell AR, Cockwell P.** Renal recovery following light chain removal by extended haemodialysis in a patient with cast nephropathy. Blood 2006; 108: 5115a

30. **Hutchison C, Cook M, Bradwell AR, Cockwell P.** Renal recovery following light chain removal by extended hemodialysis in a patient with cast nephropathy from multiple myeloma. J Am Soc Nephrol 2006; 17: TH-PO268a

31. **Hutchison CA, Mead G, Chandler K, Harper J, Bradwell AR, Cockwell P.** Free light chain abnormalities in patients with chronic kidney disease. J Am Soc Nephrol 2006; 17: PUB393a

32. **Hutchison CA, Mead G, Harper J, Chandler K, Bradwell AR, Cockwell P.** Free light chain abnormalities in patients with renal transplants. J Am Soc Nephrol 2006; 17: TH-PO060a

33. **Hutchison CA, Townend J, Mead G, Chandler K, Harper J, Bradwell AR, et al.** Serum free light chains as a determinant of adverse outcomes in patients with chronic kidney disease. J Am Soc Nephrol 2006; 17: PUB391a

34. **Hutchison C, Townend J, Mead GP, Chandler K, Harper J, Bradwell AR, et al.** Monoclonal gammopathy as a determinant of adverse outcomes in patients with chronic kidney disease. Blood 2006; 108: 5053a

35. **Itzykson R, Garff-Tavernier M, Katsahian S, Diemert M-C, Guis L, Choquet S, et al.** Serum free light chain (SFLC) elevation Is associated with high beta 2-microglobulin and with a shorter time to treatment in Waldenström's Macroglobulinemia (WM). Blood 2006; 108: 2419a

36. **Jaskowski TD, Litwin CM, Hill HR.** Detection of kappa and lambda light chain monoclonal proteins in human serum: automated immunoassay versus immunofixation electrophoresis. Clin Vaccine Immunol 2006; 13: 277 – 80

37. **Katzmann JA.** Serum free light chain specificity and sensitivity: a reality check. Clin Chem 2006; 52: 1638 – 9

38. **Katzmann JA.** Serum free light chains - quantitation and clinical utility in assessing monoclonal gammopathies. Clinical Lab News 2006; June: 12 – 14

39. **Katzmann JA, Dispenzieri A, Kyle RA, Snyder MR, Plevak MF, Larson DR, et al.** Elimination of the need for urine studies in the screening algorithm for monoclonal gammopathies by using serum immunofixation and free light chain assays. Mayo Clin Proc 2006; 81: 1575 – 8

40. **Katzmann JA, Dispenzieri A, Kyle RA, Snyder MR, Plevak MF, Larson DR, et al.** Elimination of the need for urine

studies during diagnostic studies of monoclonal gammopathies by the combined use of serum immunofixation and serum free light chain assays. Blood 2006; 108: 5011a

41.   Kumar S, Fonseca R, Dispenzieri A, Katzmann JA, Kyle RA, Clark R, Rajkumar SV. High incidence of IgH translocations in monoclonal gammopathies with abnormal free light chain levels. Blood 2006; 108: 3514a

42.   Kumar S, Rajkumar VS, Plevak M, Kyle RA, Katzmann JA, Dispenzieri A. Comparison of serum free light chain levels and 24 hour urinary monoclonal protein secretion in patients with myeloma: Concordant changes with response to therapy. Blood 2006; 108: 5064a

43.   Kyle RA, Rajkumar SV. Monoclonal gammopathy of undetermined significance. Br J Haematol 2006; 134: 573 – 89

44.   Kyrtsonis M-C, Vassilakopoulos TP, Kafasi N, Sachanas S, Tzenou T, Papadogiannis A, et al. Serum free light chain ratio (FLCR) at diagnosis constitute a powerful prognostic factor of survival in multiple myeloma (MM). Blood 2006; 108: 3522a

45.   Lopez J, Dauwalder O, Joly P, Dimet I, Bienvenu J, Bernon H. Interest and limit of a free light chain immunoassay in serum and urine for the diagnosis and the follow-up of monoclonal dysglobulinemia. Ann Biol Clin (Paris) 2006; 64: 287 – 97

46.   Lueck N, Agrawal YP. Lack of utility of free light chain-specific antibodies in the urine immunofixation test. Clin Chem 2006; 52: 906 – 7

47.   Mazumder A, Jagannath S. Use of free light chain measurements in clinical trials of novel agents in multiple myeloma. Blood 2006; 108: 3584a

48.   Mead G, Beardsmore C, Reid S, Hattersley J, Moss P, Pratt G, et al. Kinetics of tumour kill during induction chemotherapy for multiple myeloma using frequent free light chain measurements. Haematologica 2006; 92: PO1219a

49.   Mehta AB. Practice: Paraprotein management. Reply to Smellie WSA, Spickett GP. BMJ 2006; 333: 185 – 7

50.   Mehta J, Stein R, Vickrey E, Resseguie W, Singhal S. Significance of serum free light chain estimation with detectable serum monoclonal protein on immunofixation electrophoresis. Blood 2006; 108: 5048a

51.   Min C-K, Eom K-S, Lee S, Lee J-W, Min W-S, Park C-W, et al. Changes in serum free light chain and biochemical markers in patients with multiple myeloma receiving bortezomib treatment. Blood 2006; 108: 5028a

52.   Moreau A-S, Leleu X, Manning R, Coiteux V, Darre S, Hatjiharisi E, et al. Serum free light chains in Waldenstrom Macroglobulinemia. Blood 2006; 108: 2420a

53.   Nakano T, Miyazaki S, Shinoda Y, Inoue I, Katayama S, Komoda T, Nagata A. Proposed reference material for human free immunoglobulin light chain measurement. J Immunoassay Immunochem 2006; 27: 129 – 37

54.   Nakano T, Miyazaki S, Takahashi H, Matsumori A, Maruyama A, Komoda T, Nagata A. Immunochemical quantification of free immunoglobulin light chains from an analytical perspective. Clin Chem Lab Med 2006; 44: 522 – 32 .

55.   Offidani M, Polloni C, Corvatta L, Marconi M, Piersantelli MN, Scortechini I, et al. Usefulness of serum free light chain assay (sFLC) in predicting response to chemotherapy in patients affected by multiple myeloma. Haematologica Reports 2006; 2: 7a.

56.   Palladini G, Lavatelli F, Russo P, Perlini S, Perfetti V, Bosoni T, et al. Circulating amyloidogenic free light chains and serum N-terminal natriuretic peptide type B decrease simultaneously in association with improvement of survival in AL. Blood 2006; 107: 3854 – 8

57.   Palladini G, Russo P, Lavatelli F, Nuvolone M,

Bosoni T, D'Eril GM, et al. Diagnostic performance of the quantitative assay for free light chains (Freelite™) in AL amyloidosis. Clin Chem 2006; 52: D123a

58.   Palladini G, Perfetti V, Merlini G. Therapy and management of systemic AL (primary) amyloidosis. Swiss Med Wkly 2006; 136: 715 – 20

59.   Polloni C, Offidani M, Corvatta L, Marconi M, Piersantelli MN, Amoroso B, Leoni P. Role of serum free light chain (sFLC) level and ratio (FLCr) as early predictors of response in multiple myeloma. Haematologica 2006; 91: PO152a

60.   Pratt G, Mead GP, Godfrey KR, Hu Y, Evans ND, Chappell MJ, et al. The tumor kinetics of multiple myeloma following autologous stem cell transplantation as assessed by measuring serum-free light chains. Leuk Lymphoma 2006; 47: 21 – 8

61.   Presslauer S, Milosavljevic D, Hubl W, Brucke T, Bayer P. Elevated levels of kappa free light chains in CSF support the diagnosis of multiple sclerosis. Multiple Sclerosis 2006; 12: 783a

62.   Rajkumar SV, Dispenzieri A, Kyle RA. Monoclonal gammopathy of undetermined significance, Waldenstrom macroglobulinemia, AL amyloidosis, and related plasma cell disorders: diagnosis and treatment. Mayo Clin Proc 2006; 81: 693 – 703

63.   Rajkumar SV, Kyle RA, Plevak M, Clark RJ, Larson D, Therneau T, et al. Prevalence of light-chain monoclonal gammopathy of undetermined significance (LC-MGUS) among Olmsted Country, Minnesota residents aged 50 years of greater. Blood 2006; 108: 5060a

64.   Reid SD, Cockwell P, Millard JL, Chandler KL, Hutchison CA, Mead GP, Bradwell AR. MGUS incidence in a chronic kidney disease population. Clin Chem 2006; 52: E38a

65.   Reid SD, Katsavara H, Augustson BM, Hutchison CA, Mead GP, Shirfield M, et al. Screening for monoclonal gammopathy: inclusion of serum free light chain immunoassays produce an increased detection rate. Clin Chem 2006; 52: E37a

66.   Reid S, Chandler K, Hutchison C, Harrison J, Cockwell P, Mead G, Bradwell A. Free light chain removal from serum by haemofiltration and haemodialysis: a comparison of dialysis membranes in vitro. Nephrol Dial Transplant 2006; 21: SP676a

67.   Reid S, Cockwell P, Chandler K, Hutchison CA, Mead G, Bradwell AR. MGUS incidence in a chronic kidney disease population. Nephrol Dial Transplant 2006; 21: MP287a

68.   Rinker JR, 2nd, Trinkaus K, Cross AH. Elevated CSF free kappa light chains correlate with disability prognosis in multiple sclerosis. Neurology 2006; 67: 1288 – 90

69.   Rolinski B, Hammer F, Scherberich J. Clearance of kappa and lambda free light chains during hemodialysis. Clin Chem Lab Med 2006; 44: 117a

70.   Rolinski B, Hammer F, Scherberich JE. Serum concentrations of kappa and lambda free light chains in chronic renal failure. Clin Chem Lab Med 2006; 44: 118a

71.   Rutherford C, Wians FH. A large skull mass in a 67 year-old woman. Lab Med 2006; 37: 417 – 21

72.   Sanchorawala V. Light-chain (AL) amyloidosis: diagnosis and treatment. Clin J Am Soc Nephrol 2006;1:1331-41

73.   Scherberich J, Hammer F, Rolinski B. Impact of chronic renal failure and hemadialysis on serum free polyclonal immunoglobulin kappa/lambda light-chains. Nephrology Dialysis Transplantation 2006; 21: SP021a

74.   Shaw GR. Nonsecretory plasma cell myeloma--becoming even more rare with serum free light-chain assay: a brief review. Arch Pathol Lab Med 2006; 130: 1212 - 5

75.   Shimizu K, Itoh J, Sugiura I, Tsushita K, Kosugi H,

Nagura E. Evaluation of the clinical relevance of serum measurements of free-light chains in patients with multiple myeloma. Rinsho Ketsueki 2006; 47: 303 – 9

76. **Showell PJ, Scurvin M, Chnivimba A, Carr-Smith HC, Bradwell AR.** Evaluation of latex-enhanced turbidimetric reagents for measuring free immunoglobulin light-chains on The Binding Site automated analyser. Clin Chem 2006; 52: E41a

77. **Singhal S, Stein R, Vickrey E, Resseguie W, Mehta J.** The serum free light chain assay cannot replace 24-hour urine protein estimation in plasma cell dyscrasias. Blood 2006; 108: 3516a

78. **Smellie WSA.** Practice: Paraprotein management. Reply to Mehta AB. BMJ 2006; 333: 185 – 7

79. **Smellie WS, Spickett GP.** Paraprotein management. BMJ 2006; 333: 185 – 7

80. **Smellie WSA, Spickett GP.** Practice: Paraprotein management. Reply to Bradwell AR and Harding SJ. BMJ 2006; 333: 185 – 6

81. **Smith A, Wisloff F, Samson D.** Guidelines on the diagnosis and management of multiple myeloma 2005. Br J Haematol 2006; 132: 410 – 51

82. **Stein R, Mehta J, Vickrey E, Resseguie W, Singhal S.** Serum free light chain measurement in myeloma patient samples with oligoclonal protein bands. Blood 2006; 108: 5044a

83. **Stein R, Mehta J, Vickrey E, Resseguie W, Singhal S.** Correlation of serum free light chain levels with other parameters in myeloma. Blood 2006; 108: 5019a

84. **Tate JR, Mollee P, Carter A, Gill D.** Clinical utility and technical pitfalls of the serum free light (FLC) assay for monitoring of monoclonal light-chain diseases. Clin Chem 2006; 52: E44a

85. **Tucker KF.** Computer program for monitoring disease status in multiple myeloma with serum free kappa, and free lambda chains. Clin Chem 2006; 52: D137a

86. **Van der Heijden M, Kraneveld A, Redegeld F.** Free immunoglobulin light chains as target in the treatment of chronic inflammatory diseases. Eur J Pharmacol 2006; 533: 319 – 26

87. **Van Gysel M, Marien G, Verhoef G, Delforge M, Bossuyt X.** Free light chain testing in follow-up of multiple myeloma. Clin Chem Lab Med 2006; 44: 1044 – 6

88. **Wechalekar AD, Gillmore JD, Lachmann HJ, Offer M, Hawkins PN.** Efficacy and safety of bortezomib in systemic AL amyloidosis - a preliminary report. Blood 2006; 108: 129a

2007

1. **Abdalla SH.** Use of Freelite assay to monitor myeloma with renal failure. Haematologica 2007; 92: PO1016a

2. **Anderson KC, Alsina M, Bensinger W, Biermann JS, Chanan-Khan A, Comenzo RL, et al.** Multiple myeloma. Clinical practice guidelines in oncology. J Natl Compr Canc Netw 2007; 5: 118 – 47

3. **Ayliffe MJ, Davies FE, de Castro D, Morgan GJ.** Demonstration of changes in plasma cell subsets in multiple myeloma. Haematologica 2007; 92: 1135 – 8

4. **Barlogie B, Anaissie E, van Rhee F, Haessler J, Hollmig K, Pineda-Roman M, et al.** Incorporating bortezomib into upfront treatment for multiple myeloma: early results of total therapy 3. Br J Haematol 2007; 138: 176 – 85

5. **Basu S, Cook M, Hutchison C, Galvin GP, Harding S, Mead G, et al.** High rate of renal recovery in patients with cast nephropathy treated by removal of free light chains using extended hemodialysis. Haematologica 2007; 92: PO1109a

6. **Beetham R, Wassell J, Wallage MJ, Whiteway AJ, James JA.** Can serum free light chains replace urine electrophoresis in the detection of monoclonal gammopathies? Ann Clin Biochem 2007; 44: 516 – 22

7. **Behrens J.** Fighting fit. Med Lab World 2007;August:7

8. **Bergner R, Hoffmann M, Landmann T, Uppenkamp M.** Free light chains in urine - an additional diagnostic advantage? Haematologica 2007; 92: PO1012a

9. **Bradwell AR, Cockwell P, Hutchison CA.** Removal of nephrotoxic free light chains by haemodialysis in patients with myeloma kidney. NHS Nat Library for Health 2007

10. **Clark RJ, Lockington KS, Tostrud LJ, Katzmann JA.** Incidence of antigen excess in serum free light chain assays. Clin Chem 2007; 53: C145a

11. **Cohen AD, Zhou P, Chou J, Teruya-Feldstein J, Reich L, Hassoun H, et al.** Risk-adapted autologous stem cell transplantation with adjuvant dexamethasone +/- thalidomide for systemic light-chain amyloidosis: results of a phase II trial. Br J Haematol 2007; 139: 224 – 33

12. **Comenzo RL.** Current and emerging views and treatments of systemic immunoglobulin light-chain (Al) amyloidosis. Contrib Nephrol 2007; 153: 195 – 210

13. **Comenzo RL.** Managing systemic light-chain amyloidosis. J Natl Compr Canc Netw 2007; 5: 179 – 87

14. **Comenzo RL, Maurer M, Cohen A, Hassoun H, Zhou P, Lu P, et al.** Predictors of survival in de novo cardiac Amyloidosis. Blood 2007; 110: 2870a

15. **Cook M, Hutchison C, Basu S, Harding S, Basnayake K, Mead G, et al.** Renal recovery in 75% of myeloma patients with cast nephropathy following FLC removal by hemodialysis. Haematologica 2007; 92: PO1106a

16. **Corso A, Barbarano L, Mangiacavalli S, Montalbetti L, Brasca P, Zappasodi P, et al.** Bortezomib with HIG-dose dexamethasone as first line therapy in patients with multiple myeloma candidates to high-dose therapy. Blood 2007; 110: 3595a

17. **Cserti C, Haspel R, Stowell C, Dzik W.** Light-chain removal by plasmapheresis in myeloma-associated renal failure. Transfusion 2007; 47: 511 – 4

18. **Daval S, Tridon A, Mazeron N, Ristori JM, Evrard B.** Risk of antigen excess in serum free light chain measurements. Clin Chem 2007; 53: 1985 – 6

19. **Dawson MA, Patil S, Spencer A.** Extramedullary relapse of multiple myeloma associated with a shift in secretion from intact immunoglobulin to light chains. Haematologica 2007; 92: 143 – 4

20. **De Souza CA, Oliveira-Duarte G, Dias DF, Bottini PV, Vigorito AC, Aranha FJP, et al.** Serum free light-chain measurements for monitoring minimal residual disease after stem cell transplantation in multiple myeloma patients. Haematologica 2007; 92: PO1181a

21. **Dingli D, Pacheco JM, Dispenzieri A, Hayman SR, Kumar SK, Lacy MQ, et al.** Serum M-spike and transplant outcome in patients with multiple myeloma. Cancer Sci 2007; 98: 1035 – 40

22. **Dispenzieri A, Kyle RA, Katzmann JA, Larson D, Benson J, Clark RJ, et al.** Immunoglobulin free light chain ratio is an independent risk factor for progression of smoldering multiple myeloma. Blood 2007; 110: 1487a

23. **Dispenzieri A, Lacy MQ, Zeldenrust SR, Hayman SR, Kumar SK, Geyer SM, et al.** The activity of lenalidomide with or without dexamethasone in patients with primary systemic amyloidosis. Blood 2007; 109: 465 – 70

24. **Dispenzieri A, Zhang L, Snyder M, Blood E, Katzmann JA, DeGoey R, et al.** FLC as a marker of response in a mature dataset: EC0G E9486. Haematologica 2007; 92: PO1013a

25. **Émond JP, Harding S, Lemieux B.** Aggregation of serum free light chains (FLC) causes overestimation of FLC nephelometric results as compared to serum protein electrophoresis

(SPE) while preserving clinical usefulness. Blood 2007; 110: 4767a

26. **Émond JP, Lemieux B, Biron G.** Two algorithms for screening or follow up of monoclonal gammopathies: advantages and pitfalls of serum free light chains. Clin Chem Lab Med 2007; 45: M057a

27. **Gertz MA, Lacy MQ, Dispenzieri A, Hayman SR, Kumar S.** Transplantation for amyloidosis. Current Opinion in Oncology 2007; 19: 136 – 41

28. **Gertz MA, Lacy MQ, Dispenzieri A, Hayman SR, Kumar SK, Leung N, Gastineau DA.** Effect of hematologic response on outcome of patients undergoing transplantation for primary amyloidosis: importance of achieving a complete response. Haematologica 2007; 92: 1415 – 8

29. **Gironella M.** Free-light chain contribution to the evaluation of response after stem-cell transplantation in multiple myeloma. Haematologica 2007; 92: 1312a

30. **Gottenberg JE, Aucouturier F, Goetz J, Sordet C, Jahn I, Busson M, et al.** Serum immunoglobulin free light chain assessment in rheumatoid arthritis and primary Sjogren's syndrome. Ann Rheum Dis 2007; 66: 23 – 7

31. **Grosbois B.** New insights into monoclonal gammopathy and multiple myeloma. Rev Med Interne 2007; 28: 667 – 9

32. **Guenet L, Decaux O, Lechartier H, Ropert M, Grosbois B.** Usefulness of a free light chain immunoassay in serum for the diagnosis and the follow-up of monoclonal gammopathy. Rev Med Interne 2007; 28: 689 – 97

33. **Hajek R, Hajek R, Cermakova Z, Pour L, Novotna H, Maisnar V, et al.** Free light chain assays for early detection of resistance to bortezomib-based regimens. Haematologica 2007; 92: 257a

34. **Harding S, Drayson M, Ayliffe M, Mead G.** Flow cytometry detection of free light chains. Haematologica 2007; 92: PO205a

35. **Harding S, Mead G, McSkeane T, Beardsmore C, Hatterskey SJ, Moss P, et al.** Use of free light chain measurements as prediction of response to induction chemotherapy. Br J Haematol 2007; 137: 134a

36. **Harousseau JL.** Multiple myeloma: ESMO clinical recommendations for diagnosis, treatment and follow-up. Ann Oncol 2007;18 ii44 – 6

37. **Hassoun H, Rafferty BT, Flombaum C, D'Agati VD, Klimek VM, Cohen A, et al.** High dose chemotherapy and autologous stem cell transplantation with melphalan in patients with monoclonal immunoglobulin deposition disease associated with Multiple Myeloma. Blood 2007; 110: 5113a

38. **Herrera GA, Sanders PW.** Paraproteinemic Renal Diseases that Involve the Tubulo-Interstitium. Contrib Nephrol 2007;153: 105 – 15

39. **Herrera GA.** The kidney in plasma cell dyscrasias: a current view and a look at the future. Contrib Nephrol 2007; 153: 1 – 4

40. **Higgins C.** Science Review: Multiple myeloma and serum free light chain measurement. The Biomedical Scientist 2007; 51: 428 – 31

41. **Higgins T, Cho C, Dayton J.** Evaluation of a method to measure free light chains using Bayer Advia 1650 chemistry analyzer. Clin Chem Lab Med 2007; 45: M316a

42. **Hildebrandt B, Muller C, Pezzutto A, Daniel P, Dorken B, Scholz C.** Assessment of free light chains in the cerebrospinal fluid of patients with lymphatous meningitis - a pilot study. BMC Cancer 2007; 7: 185

43. **Holding S, Spradbery D, Hoole R, Wilmot R, Shields ML, Patmore R, Dore PC.** Prevalence of abnormal serum free light chain ratio in monoclonal gammopathies at presentation and sensitivity for Bence Jones proteinuria. Blood 2007; 110: 1495a

44. **Holding S, Spradbery D, Robson EJD, Dore PC, Wilmot R, Shields ML.** Combination of serum free light chain analysis with capillary zone electrophoresis improves screening for monoclonal gammopathies. Blood 2007; 110: 1497a

45. **Hutchison C, Cockwell P, Basnayake K, Cook M, Harding S, Basu S, et al.** Removal of free light chains by extended hemodialysis in patients with cast nephropathy: A phase 1/2 clinical trial. Nephrol Dial Transplant 2007; 22: FP122a

46. **Hutchison CA, Cockwell P, Harding S, Barnett A, Bradwell AR.** Quantitative asssessment of serum and urinary polyclonal free light chains in diabetics patients: An early marker of diabetic kidney disease? J Am Soc Nephrol 2007; 18: FPO1010a

47. **Hutchison CA, Cook M, Basu S, Harding S, Basnayake K, Mead G, et al.** Renal recovery in myeloma patients with cast nephropathy following light chain removal by hemodialysis. Proceedings of 29th Congress of Nordic Society of Nephrology, Gothenburg 2007: K28a

48. **Hutchison CA, Cockwell P, Reid S, Chandler K, Mead GP, Harrison J, et al.** Efficient removal of immunoglobulin free light chains by hemodialysis for multiple myeloma: in vitro and in vivo studies. J Am Soc Nephrol 2007; 18: 886 – 95

49. **Hutchison CA, Cook M, Basnayake K, Harding S, Bradwell AR, Cockwell P.** Free light chain removal hemodialysis increases rates of renal recovery from myeloma kidney. J Am Soc Nephrol 2007; 18: PUB582a

50. **Hutchison CA, Cook M, Basu S, Cockwell P, Basnayake K, R. BA.** Combined chemotherapy and high cut-off hemodialysis improve outcomes in multiple myeloma patients with severe renal failure. Blood 2007; 110: 3610a

51. **Hutchison CA, Cook M, Basu S, Harding S, Mead G, Cockwell P, Bradwell AR.** High rate of renal recovery in patients with cast nephropathy treated by removal of free light chains using extended hemodialysis: a phase 1/2 clinical trial. Br J Haematol 2007; 137: 137a

52. **Hutchison CA, Harding S, Basnayake K, Townsend J, Landray M, Mead GP, et al.** Increased MGUS prevalence in chronic kidney disease patients. Haematologica 2007; 92: PO905a

53. **Hutchison C, Harding S, Bradwell AR, Cockwell P.** Serum free light chain removal by high cut-off hemodialysis: Optimising removal and supportive care. J Am Soc Nephrol 2007; 18: SAPO511a

54. **Hutchison C, Harding S, Basnayake K, Mead G, Townsend J, Landray M, et al.** Chronic kidney disease patients and renal transplant recipients have a high prevalent rate of MGUS. Nephrol Dial Transplant 2007; 22: SaP072a

55. **Hutchison CA, Cook M, Harding S, Mead G, Cockwell P, Bradwell AR.** Renal function recovery following Velcade and extended hemodialysis in patients with refractory multiple myeloma and cast nephropathy: case studies. Am J Kidney Dis 2007; 49: 90a

56. **Hutchison CA, Cook M, Harding S, Mead G, Cockwell P, Bradwell AR.** Removal of free light chains by extended hemodialysis in patients with cast nephropathy: a phase 1/2 clinical trial. Am J Kidney Dis 2007; 49: 89a

57. **Hutchison CA, Cook M, Harding S, Mead G, Hattersley J, Evans N, et al.** Mathematical modelling of free light chain removal by plasma exchange and extended hemodialysis in patients with cast nephropathy. Am J Kidney Dis 2007; 49: 88a

58. **Hutchison CA, Harding S, Mead G, Townsend J, Landray M, Cockwell P, Bradwell AR.** Chronic kidney disease patients and renal transplant recipients have an increased incidence of monoclonal gammopathies. Am J Kidney Dis 2007; 49: 87a

59. **Jaccard A, Moreau P, Leblond V, Leleu X, Benboubker L, Hermine O, et al.** High-dose melphalan versus

melphalan plus dexamethasone for AL amyloidosis. N Engl J Med 2007; 357: 1083 – 93

60.   **Jagannath S.** Value of serum free light chain testing for the diagnosis and monitoring of monoclonal gammopathies in hematology. Clin Lymphoma Myeloma 2007; 7: 518 – 23

61.   **Kastritis E, Anagnostopoulos A, Roussou M, Toumanidis S, Pamboukas C, Migkou M, et al.** Treatment of light chain (AL) amyloidosis with the combination of bortezomib and dexamethasone. Haematologica 2007; 92: 1351 – 8

62.   **Kastritis E, Anagnostopoulos A, Roussou M, Toumanidis S, Pamboukas C, Tassidou A, et al.** Treatment of light chain (AL) amyloidosis and light chain deposition disease (LCDD) with the combination of bortezomib and dexamethasone. Blood 2007; 110: 191a

63.   **Katzel JA, Hari P, Vesole DH.** Multiple myeloma: charging toward a bright future. CA Cancer J Clin 2007;57:301-18

64.   **Khoriaty RN, Hussein MA, Lally J, Kelly M, Kalaycio M, Baz R.** Prediction of response and progression in multiple myeloma (MM) with serum-free light chains (sFLC): Corroboration of the International Myeloma Working Group (IMWG) criteria. J Clin Oncol 2007; 25: 8047a

65.   **Kilvington F.** Eliminating urine testing during initial investigations for monoclonal gammopathies. Clin Lab Int 2007; 3: 16 – 7

66.   **Kruger WH, Kiefer T, Schuler F, Lotze C, Busemann C, Dolken G.** Complete remission and early relapse of refractory plasma cell leukemia after bortezomib induction and consolidation by HLA-mismatched unrelated allogeneic stem cell transplantation. Onkologie 2007; 30: 193 – 5

67.   **Kumar S, Pérez WS, Zhang MJ, Bredeson CN, Lacy MQ, Milone G, et al.** Comparable outcomes in secretory (SM) versus non-secretory (NSM) multiple myeloma (MM) with autologous hematopoietic stem cell transplantation (AuHCT). Blood 2007; 110: 944a

68.   **Kuypers DR, Lerut E, Claes K, Evenepoel P, Vanrenterghem Y.** Recurrence of light chain deposit disease after renal allograft transplantation: potential role of rituximab? Transpl Int 2007; 20: 381 – 5

69.   **Kyle RA, Rajkumar SV.** Monoclonal gammopathy of undetermined significance and smouldering multiple myeloma: emphasis on risk factors for progression. Br J Haematol 2007; 139: 730 – 43

70.   **Kyle RA, Remstein ED, Therneau TM, Dispenzieri A, Kurtin PJ, Hodnefield JM, et al.** Clinical course and prognosis of smoldering (asymptomatic) multiple myeloma. N Engl J Med 2007; 356: 2582 - 90

71.   **Kyle RA.** Serum free light chain assays - their role in multiple myeloma. Touch Briefings 2007; US Hematology: 24 –27

72.   **Kyrtsonis MC, Vassilakopoulos TP, Kafasi N, Maltezas D, Anagnostopoulos A, Terpos E, et al.** The addition of sFLCR improves ISS prognostication in multiple myeloma (MM). Blood 2007; 110: 1490a

73.   **Kyrtsonis M-C, Vassilakopoulos TP, Kafasi N, Anagnostopoulos A, Delimpasi S, Terpos E, et al.** Prognostic implication of the combination SFLCR + ISS in MM. Haematologica 2007; 92: PO1023a

74.   **Kyrtsonis MC, Vassilakopoulos TP, Kafasi N, Sachanas S, Tzenou T, Papadogiannis A, et al.** Prognostic value of serum free light chain ratio at diagnosis in multiple myeloma. Br J Haematol 2007; 137: 240 – 243

75.   **Lakshminarayanan R, Li Y, Janatpour K, Beckett L, Jialal I.** Detection by immunofixation of M proteins in hypogammaglobulinemic patients with normal serum protein electrophoresis results. Am J Clin Pathol 2007; 127: 746 – 51

76.   **Lebovic D, Kendall T, Brozo C, McAllister A, Hari M, Alvi S, et al.** Serum free light chain analysis improves monitoring of multiple myeloma patients receiving first-line therapy with the combination of Velcade, Doxil, and Dexamethasone (VDD). Blood 2007; 110: 2736a

77.   **Leleu X, Hatjiharissi E, Roccaro AM, Moreau AS, Leduc R, Nelson M, et al.** Serum immunoglobulin free light chain (sFLC) is a sensitive marker of response in Waldenstrom Macroglobulinemia (WM). Blood 2007; 110: 1486a

78.   **Leleu X, Moreau AS, Manning R, Coiteux V, Darre S, Nelson M, et al.** Serum free light chain is a marker of tumor burden and of prognostic impact in Waldenstrom's macroglobulinemia. Haematologica 2007; 92: PO1230a

79.   **Leung N, Zeldenrust S, Dispenzieri A, Fervenza F, Gertz M, Kumar S, et al.** Serum free light chain and renal biopsy accurately predict renal response to plasma exchange in patients with light chain cast nephropathy. J Am Soc Nephrol 2007; 18: SUP074a

80.   **Leung N, Dispenzieri A, Lacy MQ, Kumar SK, Hayman SR, Fervenza FC, et al.** Severity of baseline proteinuria predicts renal response in immunoglobulin light chain–associated amyloidosis after autologous stem cell transplantation. Clin J Am Soc Nephrol 2007; 2: 440 – 4

81.   **Lewis E, Young C, Watson D.** Comparison of western blotting and free light chains in the laboratory diagnosis of multiple sclerosis. Annals Clin Biochem 2007; 44: T51a

82.   **Ludwig H, Drach J, Graf H, Lang A, Meran JG.** Reversal of acute renal failure by bortezomib-based chemotherapy in patients with multiple myeloma. Haematologica 2007; 92: 1411 – 4

83.   **Lynch EA, Showell PJ, Johnson-Brett B, Mead GP, Bradwell AR.** Evaluation of latex-enhanced turbidimetric reagents for measuring free immunoglobulin light chains on the Roche Cobas® Integra 400. Clin Chem Lab Med 2007; 45: W089a

84.   **Ma ES, Lee ET.** A case of IgM paraproteinemia in which serum free light chain values were within reference intervals. Clin Chem 2007; 53: 362 – 3

85.   **Martin W, Abraham R, Shanafelt T, Clark RJ, Bone N, Geyer SM, et al.** Serum-free light chain-a new biomarker for patients with B-cell non-Hodgkin lymphoma and chronic lymphocytic leukemia. Transl Res 2007; 149: 231 – 5

86.   **Mayo Medical Laboratories.** Laboratory analysis for monoclonal gammopathies. Mayo Communiqué 2007; 32: 1 – 4

87.   **Mayo MM, Johns GS.** Serum free light chains in the diagnosis and monitoring of patients with plasma cell dyscrasias. Contrib Nephrol 2007; 153: 44 – 65

88.   **Mazurkiewicz J, Teal T, Reddy R.** Evaluation of The Binding Site serum free light chain assay in a DGH: analytical aspects. Annals Clin Biochem 2007; 44: T46a

89.   **Mazurkiewicz J, Teal T, Reddy R, Grace R, Gover P.** Evaluation of The Binding Site serum free light chain assay in a DGH: clinical aspects. Annals Clin Biochem 2007; 44: T47a

90.   **Mehta J, Stein R, Vickrey E, Fellingham S, Singhal S.** The relationship between serum free light chain assay and serum immunofixation electrophoresis. Haematologica 2007; 92: PO1010a

91.   **Mitchell F, Showell PJ, Harding SJ, Lynch EA, Mead GP, Hutchison CA, Bradwell AR.** Correlation of Freelite™ and Cystatin C in chronic kidney disease on The Binding Site SPAplus™ bench-top analyser. Clin Chem Lab Med 2007; 45: T225a

92.   **Morgan GJ, Davies FE, Owen RG, Rawstron AC, Bell S, Cocks K, et al.** Thalidomide combinations improve response rates; results from the MRC IX study. Blood 2007; 110: 3593a

93.   **Morris KL, Tate JR, Gill D, Kennedy G, Wellwood**

J, Marlton P, et al. Diagnostic and prognostic utility of the serum free light chain assay in patients with AL amyloidosis. Intern Med J 2007; 37: 456 – 63

94. **Mösbauer U, Ayuk F, Schieder H, Lioznov M, Zander AR, Kroger N.** Monitoring serum free light chains in patients with multiple myeloma who achieved negative immunofixation after allogeneic stem cell transplantation. Haematologica 2007; 92: 275 – 276

95. **Nakano T, Komoda T, Nagata A.** To the Editor: Author reply; Nakano et al. Clin Chem Lab Med 2006; 44: 522 – 532. Clin Chem Lab Med 2007; 45: 266 – 7

96. **Nakorn TN, Watanaboonyongcharoen P, Suwannabutra S, Theerasaksilp S, Paritpokee N.** Early reduction of serum free light chain can predict therapeutic responses in multiple myeloma. Blood 2007; 110: 4742a

97. **Nguyen T, Bocquet J, Troncy J, Chapuis Cellier C.** Ability of capillary zone electrophoresis in detecting monoclonal light chains. Haematologica 2007; 92: PO246a

98. **Offidani M, Polloni C, Corvatta L, Busco F, Catarini M, Alesiani F, et al.** Usefulness of serum FLC assay in predicting outcome in MM. Haematologica 2007; 92: PO236a

99. **Orlowski R, Sutherland H, Blade J, Miguel JS, Hajek R, Nagler A, et al.** Early normalization of serum free light chains is associated with prolonged time to progression following bortezomib {+/-} pegylated liposomal doxorubicin treatment of relapsed/ refractory multiple myeloma. Blood 2007; 110: 2735a

100. **Owen RG, Child JA, Rawstron AC, Bell S, Cocks K, Davies FE, et al.** Defining complete response in multiple myeloma: Role of the serum free light chain assay and multiparameter flow cytometry. Blood 2007; 110: 1479a

101. **Palladini G, Nuvolone M, Russo P, Lavatelli F, Perfetti V, Obici L, Merlini G.** Treatment of AL amyloidosis guided by biomarkers. Haematologica 2007; 92: PO910a

102. **Pasquali S, De Sanctis LB, Mambelli E, Di Felice A, Wratten M, Santoro A.** Coupled plasmafiltration adsorption: A new technology for free light chain removal. J Am Soc Nephrol 2007; 18: FPO691a

103. **Pattenden RJ, Rogers SY, Wenham PR.** Serum free light chains; the need to establish local reference intervals. Ann Clin Biochem 2007; 44: 512 – 5

104. **Pika T, Minarik J, Zemanova M, Budikova M, Backovsky J, Scudla V.** Relationship of serum free light chain levels and selected biological parameters in patients with multiple myeloma. Haematologica 2007; 92: 1113a

105. **Pineda-Roman M, Tricot G.** High-dose therapy in patients with plasma cell dyscrasias and renal dysfunction. Contrib Nephrol 2007; 153: 182 – 94

106. **Polloni C, Offidani M, Corvatta L, Busco F, Catarini M, Alesiani F, et al.** Serum FLC as a measure of treatment outcome in MM. Haematologica 2007; 92: PO1014a

107. **Puig N, Chen C, Mikhael J, Reece D, Trudel S, Kukreti V.** Spontaneous tumoural regression in a patient with t(4;14) translocation multiple myeloma. A case report. Blood 2007; 110: 4777a

108. **Rajkumar SV, Lacy MQ, Kyle RA.** Monoclonal gammopathy of undetermined significance and smoldering multiple myeloma. Blood Rev 2007; 21: 255 – 265

109. **Ramasamy I.** Serum free light chain analysis in B-cell dyscrasias. Ann Clin Lab Sci 2007; 37: 291 – 294

110. **Rawstron AC, Davis B, D. DS, de Tute RM, Kerr MA, Owen RG, Ashcroft AJ.** Plasma cell phenotype and SFLC provide independent prognostic information in MGUS. Haematologica 2007; 92: PO907a

111. **Rawstron AC, Jones RA, Ferguson C, Hughes G, Selby P, Reid C, et al.** Outreach monitoring service for patients with indolent B-cell and plasma cell disorders: a UK experience. Br J Haematol 2007; 139: 845 – 8

112. **Rezazadeh A, Laber DA, Sharma VR, Kloecker GH.** Amyloidosis presenting with platelet dysfunction and recurrent retroperitoneal haemorrhage (RH). Blood 2007; 110: 3944a

113. **Richards T, Horowitz S, Temple S, Nguyen T, Thomas S, Wang M, et al.** Evaluation of MM response by free light chain assay. Haematologica 2007; 92: PO1015a

114. **Richter AG, Harding S, Rimmer S, Pratt G, Huissoon A, Drayson M.** Biclonal multiple myeloma with monoclonal free IgG3 heavy chain and kappa free light chains. Blood 2007; 110: 4768a

115. **Ricotta D, Radeghieri A, Amoroso B, Caimi L.** Serum free light chains in MM: comparison of 2 assays. Haematologica 2007; 92: PO1018a

116. **Robson E, Mead G, Bradwell A.** To the editor: in reply to Nakano et al. Clin Chem Lab Med 2006; 44: 522 – 532. Clin Chem Lab Med 2007; 45: 264 – 5

117. **Robson E, Mead G, Carr-Smith H, Bradwell A. In reply to Tate et al.** Clin Chim Acta 2007; 376: 30 – 6. Clin Chim Acta 2007; 380: 247

118. **Robson E, Mead G, Das M, Cavet J, Liakpoulou E.** Free Light Chain analysis in patients receiving Bortezomib. Haematologica 2007; 92: PO1019a

119. **Ruchlemer R, Reinus C, Paz E, Cohen A, Melnikov N, Ronson A, Rudensky B.** Free light chains, monoclonal proteins, and chronic lymphocytic leukemia. Blood 2007; 110: 4697a

120. **Ruckser R, Tatzreiter G, Kitzweger E, Strecker K, Hraby S, Buxhofer V, et al.** Combination therapy with lenalidomide, bortezomib, liposomal doxorubicin and dexamethasone (LBlipDD) may overcome resistance to prior treatment with doxorubicin, lenalidomide, and bortezomib in high-risk multiple myeloma. Blood 2007; 110: 4838a

121. **Sanchorawala V, Wright DG, Rosenzweig M, Finn KT, Fennessey S, Zeldis JB, et al.** Lenalidomide and dexamethasone in the treatment of AL amyloidosis: Results of a phase 2 trial. Blood 2007; 109: 492 – 496

122. **Sheldon J.** Free light chains. Ann Clin Biochem 2007; 44: 503 – 5

123. **Showell PJ, Hutchison CA, Cockwell P, Harding S, Mead GP, Mitchell F, Bradwell AR.** Correlation of FREELITE and Cystatin C in chronic kidney disease on The Binding Site SPAPLUS bench-top analyser. Clin Chem 2007; 53: C82a

124. **Showell PJ, Lynch EA, Johnson-Brett B, Mead G, Bradwell AR.** Evaluation of latex-enhanced turbidimetric reagents for measuring free immunoglobulin light-chains on the Roche COBAS Integra 400 automated analyser. Clin Chem 2007; 53: C18a

125. **Sinclair D, Wainwright L.** How lab staff and the estimation of free light chains can combine to aid the diagnosis of light chain disease. Clin Lab. 2007; 53: 267 – 271

126. **Singhal S, Stein R, Vickrey E, Mehta J.** The serum-free light chain assay cannot replace 24-hour urine protein estimation in patients with plasma cell dyscrasias. Blood 2007; 109: 3611 – 3612

127. **Singhal S, Stein R, Vickrey E, Fellingham S, Mehta J.** Serum free light chain levels in myeloma patients with oligoclonal bands. Haematologica 2007; 92: PO1011a

128. **Snozek CLH, Katzmann JA, Kyle RA, Dispenzieri A, Larson DR, Clark RJ, et al.** Prognostic value of the serum free light chain ratio in patients with newly diagnosed myeloma: proposed incorporation into the international staging system. Blood 2007; 110: 659a

129. **Spradberry D, Holding S.** Analytical variability of the

Binding Site Freelite assay in serum. Presented at IBMS Congress 2007

130. Storr M, Goehl H, Cockwell P, Hutchison C, Bradwell A. Efficient removal of immunoglobulin free light chains by hemodialysis using a novel high cut-off membrane. Blood Purif 2007; 25: O17a

131. Sui WW, Yao HJ, Wang YF, Wang L, Zhang K, Xu Y, Qiu LG. Assessment of serum free light chain of patients with multiple myeloma and clinical significance thereof. Zhonghua Yi Xue Za Zhi 2007; 87: 3158 – 3160

132. Sweeting R, Notkin A, Leung N, Ballard H, Pillon L. Approach to acute renal failure in biopsy proven myeloma cast nephropathy. Is there a role for plasmapheresis? J Am Soc Nephrol 2007; 18: PUB570a

133. Tan T-S, Dispenzieri A, Lacy MQ, Hayman SR, Buadi FK, Zeldenrust SR, et al. Melphalan and dexamethasone is an effective therapy for primary systemic amyloidosis. Blood 2007; 110: 3608a

134. Tate JR, Mollee P, Dimeski G, Carter AC, Gill D. Analytical performance of serum free light-chain assay during monitoring of patients with monoclonal light-chain diseases. Clin Chim Acta 2007; 376: 30 – 36

135. Trieu Y, Xu W, Chen C, Kukreti V, Mikhael J, Trudel S, Reece D. Evaluation of serum free light chains and outcome in multiple myeloma patients with an intact monoclonal immunoglobulin treatment with autologous stem cell transplantation. Haematologica 2007; 92: PO1021a

136. Trigo F, Guimaraes C, Bodas A, Teixeira-Pinto A, Guimaraes JE. Serum free light chain assessment in 311 patients with gammopathy. Blood 2007; 110: 4750a

137. Van Rhee F, Bolejack V, Hollmig K, Pineda-Roman M, Anaissie E, Epstein J, et al. High serum-free light chain levels and their rapid reduction in response to therapy define an aggressive multiple myeloma subtype with poor prognosis. Blood 2007; 110: 827 – 32

138. Wechalekar AD, Goodman HJB, Lachmann HJ, Offer M, Hawkins PN, Gillmore JD. Safety and efficacy of risk-adapted cyclophosphamide, thalidomide, and dexamethasone in systemic AL amyloidosis. Blood 2007; 109: 457 – 64

139. Wolff F, Thiry C, Willems D. Assessment of the analytical performance and the sensitivity of serum free light chains immunoassay in patients with monoclonal gammopathy. Clin Biochem 2007; 40: 351 – 354

140. Zarabpouri F, Popov J, Montana L, Ferruzza AS, Naides SJ. Role of nephelometry-based detection of serum free light chain kappa/ lambda in diagnosis of multiple myeloma. Clin Chem 2007; 53: C148a

141. Zibrat S, Gloria CT, Joseph L, W. MR. Analytical and clinical performance limitations of immunoturbidimetric vs. immunonephelometric methods for free light chain analysis in serum. Clin Chem 2007; 53: B79a

142. Zinke-Cerwenka W, Stojakovic T, Scharnagl H, Wasserthurer S, Linkesch W. Therapy response by decrease of free light chain concentration in multiple myeloma patients treated with bortezomib/ doxorubicin/ dexamethasone as first-line therapy. Blood 2007; 110: 4818a

2008

1. Aggarwal R, Mikolaitis R, Alvi S, Assi L, Block J, Jolly M, Sequeira W. Serum free light chains as a biomarker in systemic lupus erythematosus. Presented at American College of Rheumatology Annual Scientific Meeting 2008: 1089a

2. Allen S, Vickrey E, Mehta J, Singhal S. The relationship between serum free light chain levels and serum immunofixation electrophoresis: Implications for the definition of

"Stringent CR" in myeloma. Blood 2008; 112: 2724a

3. Alvi A, Harding S, Pietrantuono A, Berard A. Serum free light chain immunoassay (Freelite) as an adjunct to SPE and IFE in the detection of multiple myeloma and other B cell malignancies. Hematology meeting reports 2008; 2: C25a

4. Bachmann U, Schindler R, Storr M, Kahl A, Joerres A, Sturm I. Combination of bortezomib-based chemotherapy and extracorporeal free light chain removal for treating cast nephropathy in multiple myeloma. Nephrol Dial Transplant 2008; 1: 106 – 108

5. Barlogie B, van Rhee F, Shaughnessy JD, Jr., Epstein J, Yaccoby S, Pineda-Roman M, et al. Seven-year median time to progression with thalidomide for smoldering myeloma: partial response identifies subset requiring earlier salvage therapy for symptomatic disease. Blood 2008; 112: 3122 – 3125

6. Basnayake K, Hutchison C, Kamel D, Sheaff M, Ashman N, Oakervee H, et al. Resolution of cast nephropathy following free light chain removal by hemodialysis in a patient with multiple myeloma. Am J Kidney Dis 2008; 51: 24a

7. Basnayake K, Hutchison C, Kamel D, Sheaff M, Ashman N, Cook M, et al. Resolution of cast nephropathy following free light chain removal by haemodialysis in a patient with multiple myeloma: a case report. J Med Case Reports 2008; 2: 380

8. Basnayake K, Cheung CK, Hutchison C, Kamel D, Sheaff M, Fuggle W, et al. The differential evolution of renal scarring in myeloma kidney despite early reductions in free light chains. Hematology meeting reports 2008; 2: E35a

9. Basnayake K, Ghonemy T, Hutchison C, Bradwell AR, Cockwell P. Immunoglobulin free light chains in situ in patients with chronic kidney disease. Hematology meeting reports 2008; 2: E34a

10. Basnayake K, Hutchison C, Kamel D, Sheaff M, Fuggle W, Cook M, et al. Rapid reduction of serum free light chains by high cut-off haemodialysis in cast nephropathy: supporting histological evidence from two cases. Renal Association Annual Congress 2008: P60a

11. Bergner R, Vilardi S, Hoffmann M, Uppenkamp M. Free light chains in urine - an additional diagnostic advantage? Hematology meeting reports 2008; 2: B7a

12. Bernon H, Karame A, Dimet I, Villar E, Regad-Pellagru AL, Bienvenu J. About a case of lambda free light chain multiple myeloma treated by bortezomib chemotherapy and long haemodialysis on polymethylmethacrylate membrane. Hematology meeting reports 2008; 2: E40a

13. Bidet A, Marit G, Berard AM. Clinical value of the blood measurement of the free light chains of immunoglobulins. Ann Biol Clin (Paris) 2008; 66: 427 – 31

14. Blade J, Rosinol L. SMM: Towards pretter predictors of progression. Blood 2008; 111: 479 – 480

15. Blade J, Rosinol L, Cibeira MT, de Larrea CF. Pathogenesis and progression of monoclonal gammopathy of undetermined significance. Leukemia 2008; 22: 1651 – 1657

16. Bochtler T, Hegenbart U, Heiss C, Benner A, Cremer F, Volkmann M, et al. Evaluation of the serum-free light chain test in untreated patients with AL amyloidosis. Haematologica 2008; 93: 459 – 462

17. Brandhorst D, Bockling M, Sammet C, Daniels R, Schuett P, Buttkereit U, et al. Monoclonal free light chains and renal dysfunction in multiple myeloma. Hematology meeting reports 2008; 2: E42a

18. Bruck P, Atta J, Hildegund S-E, Bohme A, Oremek GM. Lambda and kappa free light chains in haematologic patients. Clin Exp Med Lett 2008; 49: 19 – 22

19. Burg M, Niederstadt C, Dotzeva E, Kliem V. Serum

free light chains in patients with renal failure stage II-V of native kidneys or renal transplants. Renal Association Annual Congress 2008: PUB249a

20. **Callis M, Garcia L, Gironella M, Rodrigo MJ, Garcia E, Castella D.** Serum free light chain reference intervals. A study of 133 blood donors. Haematologica 2008; 93: 1332a

21. **Callis M, Garcia L, Gironella M, Rodrigo MJ, Garcia E, Castella D.** Serum fee light chain reference intervals. A study of 133 blood donors. Hematology meeting reports 2008; 2: B6a

22. **Chan PC, Lem - Ragosnig B, Oczak M, Rozmanc M.** Follow-up investigations for 'nephrotic syndrome' pattern in serum protein electrophoresis. Clin Biochem 2008; 41: 108a

23. **Chattopadhyay A, Nath UK, De R, Singh A, Sanyal S, Chatterjee SK, Chaudhuri U.** Primary plasma cell leukemia with initial cutaneous involvement and IgA biclonal gammopathy. Ann Hematol 2008; 87: 249 – 251

24. **Chen HF, Hou J, Yuan ZG, Wang DX, Fu WJ, Chen YB.** Detection of serum free light chain and its clinical significance in nonsecretory multiple myeloma. Zhonghua Xue Ye Xue Za Zhi 2008; 29: 113 – 116

25. **Civini S, Radeghieri A, Amoroso B, Caimi L, Ricotta D.** Free light chain behaviour in sera from patients producing cryoglobulins. Hematology meeting reports 2008; 2: C16a

26. **Clark WF, Garg AX.** Plasma exchange for myeloma kidney: cast(s) away? Kidney Int 2008; 73: 1211 – 1213

27. **Coriu D, Talmaci R, Badelita S, Dobrea D, Dogaru M, Ostroveanu D, et al.** Clinical study of 35 cases of primary systemic amyloidosis. Haematologica 2008; 93: 1301a

28. **Davern S, Tang LX, Williams TK, Macy SD, Wall JS, Weiss DT, Solomon A.** Immunodiagnostic capabilities of anti-free immunoglobulin light chain monoclonal antibodies. Am J Clin Pathol 2008; 130: 702 – 711

29. **De Filippi R, Laccarino G, Frigeri F, Di Francia R, Amoroso B, Marchei A, Pinto A.** The presence of serum free-immunoglobulin light chains and abnormal k/l ratios is a frequent finding in patients with Hodgkin's and B-cell non-Hodgkin's lymphoma. Hematology meeting reports 2008; 2: C18a

30. **de Fost M, Out TA, de Wilde FA, Tjin EP, Pals ST, van Oers MH, et al.** Immunoglobulin and free light chain abnormalities in Gaucher disease type I: data from an adult cohort of 63 patients and review of the literature. Ann Hematol 2008; 87: 439 – 449

31. **Decaux O, Briand PY, Guenet L, Ropert M, Sebillot M, Grosbois B.** An abnormal serum free light chain ratio is associated with higher monoclonal concentration and bone marrow plasmacytosis in a French cohort of patients with monoclonal gammopathy of undetermined significance. Hematology meeting reports 2008; 2: C22a

32. **Dimopoulos MA, Kastritis E, Rosinol L, Blade J, Ludwig H.** Pathogenesis and treatment of renal failure in multiple myeloma. Leukemia 2008; 22: 1485 – 1493

33. **Dispenzieri A.** Impact of serum free light chain evaluation in monoclonal gammopathies. Hematology Education: the education program for the annual congress of the European Hematology Association 2008; 2: 279 – 286

34. **Dispenzieri A.** To the editor: Is early, deep free light chain response really an adverse prognostic factor? Blood 2008; 111: 2490 – 2491

35. **Dispenzieri A.** How and when I use serum free light chain tests. Hematology meeting reports 2008; 2: 4 – 6

36. **Dispenzieri A, Kyle RA, Katzmann JA, Therneau TM, Larson D, Benson J, et al.** Immunoglobulin free light chain ratio is an independent risk factor for progression of smoldering (asymptomatic) multiple myeloma. Blood 2008; 111: 785 – 789

37. **Dispenzieri A, Zhang L, Katzmann JA, Snyder M, Blood E, Degoey R, et al.** Appraisal of immunoglobulin free light chain as a marker of response. Blood 2008; 111: 4908 – 4915

38. **Dotzeva E, Niederstadt C, Burg M, Kliem V.** Serum free light chains in patients with renal failure stage II-V of native kidneys or renal transplants. Hematology meeting reports 2008; 2: E43a

39. **Doyle A, Soutar R, Geddes CC.** Multiple myeloma in chronic kidney disease. Utility of discretionary screening using serum electrophoresis. Nephron Clin Pract 2008; 111: c7 - c11

40. **Drengler T, Kumar S, Gertz M, Lacy M, Katzmann J, Hayman S, et al.** Serum immunoglobulin free light chain measurements provide insight into disease biology in patients with POEMS syndrome. Proceedings of Portuguese myeloma meeting 2008

41. **Encinas Madrazo A, González García ME, Cepeda Piorno J, González Huerta AJ, Fernández Rodríguez E.** Serum free light chains: Clinical utility. An Med Interna 2008; 25: 249 – 250

42. **Espinilla HA.** Determinaciande valories esperados de referencia Kappa libre en suero, Lambda libre in suero y Kappa/Lambda suero en nuestro departamento de salud. Rev Lab Clinic 2008; s1

43. **Exner I, Eigner M, Kraus R, Leithner C.** Acute renal failure because of light chain deposition disease and its successful combined treatment with chemotherapy and dialysis with high flux membrane. Hematology meeting reports 2008; 2: E36a

44. **Forsyth J, Hill P.** Serum free light chains. Reply to Beetham et al. Can serum free light chains replace urine electrophoresis? Ann Clin Biochem 2008; 45: 444 – 445

45. **Forsyth J, Mayne S, Lawson N.** A rare case of polyclonal Mu Heavy Chain Disease. Ann Clin Biochem 2008; 45: 4a

46. **Gertz MA.** Current therapy of myeloma induced renal failure. Leuk Lymphoma 2008; 49: 833 - 834

47. **Giroux M, Sheridan B.** The serum free light chain assay: an alternative to Bence Jones protein analysis in screening for monoclonal gammopathies. Hematology meeting reports 2008; 2: C21a

48. **Giroux M, Sheridan B.** The serum free light chain assay: An alternative to Bence Jones protein analysis in screening for monoclonal gammopathy. Clin Biochem 2008; 41: 113a

49. **Gottenberg JE, Dillon SR, Combe B, Harder B, Hahne M, Morel J, et al.** Serum April levels correlate with disease activity and severity of radiologic lesions in early rheumatoid arthritis patients: Results from the French multicenter prospective cohort study (espoir) Presented at American College of Rheumatology Annual Scientific Meeting 2008: 328a

50. **Granátová J, Bolková M, Fantová L, Hornová L, Mašková Z, Horák J, Lánská V.** Free light chains of immunoglobulins in patient with renal insufficiency. Klin Biochem Metab 2008; 16: 106 – 110

51. **Graubaum K, Heymann GA, Hunger T, Ostapowicz B.** Case report: Shift in secretion from intact immunoglobulin to free light chains in multiple myeloma at relapse: early detection by free light chain assay. Hematology meeting reports 2008; 2: D26a

52. **Gregorini GA, Mazzola G, Valerio F, Econimo I, Gasparotti I, Ricotta D, Cancarini G.** Extensive use of serum free light chain test in patients with nephropathy increases the diagnosis number of AL amyloidosis and light chain deposition disease. Hematology meeting reports 2008; 2: E41a

53. **Harding S, Pratt G, Holder R, Pepper C, Fegan C, Oscier D, Mead G.** Abnormal serum free light chain ratios are associated with poor survival in patients with chronic lymphocytic leukemia. Hematology meeting reports 2008; 2: C20a

54. **Harding S, Pratt G, Pepper C, Fegan C, Oscier D, Mead** G. Abnormal serum free light chain ratios are associated with poor survival in patients with chronic lymphocytic leukemia. Lymphoma & Myeloma congress 2008

55. **Harding SJ, Pratt G, Holder R, Fegan C, Pepper C, Oscier D, Mead GP.** Abnormal serum free light chain ratios in chronic lymphocytic leukaemia are significantly associated with earlier disease specific death. Haematologica 2008; 93: 89a

56. **Harousseau JL, Dreyling M.** Multiple myeloma: ESMO clinical recommendations for diagnosis, treatment and follow-up. Ann Oncol 2008; s2: ii55 – 57

57. **Hassoun H, Flombaum C, D'Agati VD, Rafferty BT, Cohen A, Klimek VM, et al.** High-dose melphalan and auto-SCT in patients with monoclonal Ig deposition disease. Bone Marrow Transplant 2008; 42: 405 – 412

58. **Hatjiharissi E, Ciccarelli BT, Ioakimidis L, Borok R, Soumerai JD, Manning RJ, et al.** Serum immunoglobulin free light chains as markers of disease burden and response to treatment in patients with Waldenstrom's macroglobulinemia. J Clin Oncol 2008; 26: 8617a

59. **Hobbs JA, Kilvington F, Sharp K, Harding S, Drayson M.** Incidence of light chain escape in UK MRC Myeloma VII trial. Hematology meeting reports 2008; 2: D31a

60. **Hobbs JA, Kilvington F, Sharp K, Harding S, Drayson M, Bradwell AR, Mead GP.** Incidence of light chain escape in UK MRC Myeloma VII Trial. Clin Chem 2008; 54: C109a

61. **Hobbs JAR, Sharp K, Harding SJ, Drayson M, Bradwell AR, Mead GP.** Incidence of light chain escape in UK MRC myeloma VII trial Haematologica 2008; 93: 672a

62. **Hodkinson F, Galligan L, Drain S, Catherwood MA, Morris TCM, Drake MB, et al.** Investigation of plasma cell dyscrasias with low level plasma cell aspirate involvement. Haematologica 2008; 93: 1205a

63. **Hoffman-Snyder C, Smith BE.** Neuromuscular disorders associated with paraproteinemia. Phys Med Rehabil Clin N Am 2008; 19: 61 – 79

64. **Hutchison C, Basnayake K, Bradwell A, Cockwell P.** Serum free light chain removal by high cut-off hemodialysis: Optimising removal and supportive care. Renal Association Annual Congress 2008: P129a

65. **Hutchison C, Basnayake K, Cockwell P.** Chronic kidney disease patients and renal transplant recipients have a high prevalence of MGUS. Hematology meeting reports 2008; 2: E38a

66. **Hutchison C, Basnayake K, Cook M, Bradwell A, Cockwell P.** High rates of renal recovery and patient survival in myeloma kidney using free light chain removal hemodialysis. Renal Association Annual Congress 2008: P245a

67. **Hutchison C, Basnayake K, Cook M, Bradwell A, Cockwell P.** Free light chain removal hemodialysis increases renal recovery rate and improves patient suvival in patients with cast nephropathy. Nephrol Dial Transplant 2008; 23: S0019a

68 .**Hutchison C, Basnayake K, Cook M, Cockwell P, Bradwell AR.** High renal recovery rate from cast nephropathy following free light chain removal hemodialysis. Am J Kidney Dis 2008; 49: B47a

69. **Hutchison C, Basnayake K, Plant T, Drayson M, Cockwell P.** Serum free light chain measurement aids the diagnosis of myeloma in patients with acute renal failure. Hematology meeting reports 2008; 2: E39a

70. **Hutchison C, Cook M, Basnayake K, Cockwell P, Bradwell A.** Free light chain removal hemodialysis increases renal recovery rate and improves patient survival in patients with cast nephropathy. Lymphoma & Myeloma congress 2008

71. **Hutchison C, Drayson M, Basnayake K, Mead G,**

Bradwell A, Plant T, Cockwell P. Serum free light chain measurement aids the diagnosis of myeloma in patients with acute renal failure. Renal Association Annual Congress 2008: SA-PO2498a

72. **Hutchison C, Grima DT.** Long-term clinical outcomes and cost-effectiveness of hemodialysis with the HCO 1100 dialyzer versus standard hemodialysis in patients with renal failure secondary to myeloma. Renal Association Annual Congress 2008: SA-PO3060a

73. **Hutchison C, Hewins P, Landray MJ, Townsend J, Cockwell P.** Quantitative assessment of serum and urinary polyclonal free light chains in patients with chronic kidney disease. Hematology meeting reports 2008; 2: E37a

74. **Hutchison C, Meryon I, Drayson M, Bradwell A, Cockwell P.** High cut-off hemodialysis lowers inflammatory status in chronic dialysis patients. Renal Association Annual Congress 2008: TH-PO707a

75. **Hutchison C, Plant T, Drayson M, Basnayake K, Cockwell P.** Serum free light chains - a new highly sensitive and specific immunoassay for detecting monoclonal gammopathies in patients with acute renal failure. Renal Association Annual Congress 2008: P67a

76. **Hutchison C, Plant T, Drayson M, Cockwell P, Basnayake K, Harding S, et al.** Serum immunoassay is highly sensitive and specific for the diagnosis of monoclonal free light chains in patients with severe renal failure. Nephrol Dial Transplant 2008; 23: SP113a

77. **Hutchison CA, Cook M, Heyne N, Weisel K, Billingham L, Bradwell AR, Cockwell P.** European trial of free light chain removal by extended haemodialysis in cast nephropathy (EuLITE): a randomised control trial. Trials 2008; 9: 55

78. **Hutchison CA, Harding S, Hewins P, Mead GP, Townsend J, Bradwell AR, Cockwell P.** Quantitative assessment of serum and urinary polyclonal free light chains in patients with chronic kidney disease. Clin J Am Soc Nephrol 2008; 3: 1684 – 1690

79. **Hutchison CA, Harding S, Mead G, Goehl H, Storr M, Bradwell A, Cockwell P.** Serum free-light chain removal by high cutoff hemodialysis: optimizing removal and supportive care. Artif Organs 2008; 32: 910 – 917

80. **Hutchison CA, Plant T, Drayson M, Cockwell P, Kountouri M, Basnayake K, et al.** Serum free light chain measurement aids the diagnosis of myeloma in patients with severe renal failure. BMC Nephrol 2008; 9: 11

81. **Hutchison CA, Cockwell P, Harding S, Mead GP, Bradwell AR, Barnett AH.** Quantitative assessment of serum and urinary polyclonal free light chains in patients with type II diabetes: an early marker of diabetic kidney disease? Expert Opin Ther Targets 2008;12: 667 – 676

82. **Itzykson R, Le Garff-Tavernier ML, Katsahian S, DiemertM-C, Musset L, Leblond V.** Serum Free Light Chain Elevation Is Associated with a Shorter Time to Treatment in Waldenstrom's Macroglobulinemia. Haematologica. 2008 May;93(5):793-794

83. **Jiminez J.** Utilidad de la determinacion de cedenas ligera libres ensuero para el diagnostico y monitorizacion de gammopatias monoclonales. Rev Lab Clinic 2008; s1

84. **Jung S, Kim M, Lim J, Kim Y, Han K, Min CK, Min WS.** Serum free light chains for diagnosis and follow-up of multiple myeloma. Korean J Lab Med 2008; 28: 169 – 173

85. **Kaplan B, Ramirez-Alvarado M, Dispenzieri A, Zeldenrust SR, Leung N, Livneh A, Gallo G.** Isolation and biochemical characterization of plasma monoclonal free light chains in amyloidosis and multiple myeloma: a pilot study of intact and truncated forms of light chains and their charge properties. Clin

Chem Lab Med 2008; 46: 335 – 341

86. **Kar R, Dutta S, Bhargava R, Kumar R, Pati HP.** Immunoglobulin free light chains: do they have a role in plasma cell leukemia? Hematology 2008;13:344 - 347

87. **Katzmann JA, Dispenzieri A.** Editorial: Screening algorithms for monoclonal gammopathies. Clin Chem 2008; 54: 1753 – 1755

88. **Katzmann JA.** Laboratory strategies for detecting Multiple Myeloma. Advance for Administrators of the Laboratory 2008; 17: 46

89. **Kimby E.** Treatment of Waldenstrom's macroglobulinaemia. Hematology Education: the education program for the annual congress of the European Hematology Association 2008; 2: 294 – 301

90. **Krizalkovicova V, Maisnar V, Pour L, Radocha J, Hajek R.** Monoclonal gammopathies of undetermined significance. Klin Onkol 2008; 21: 160 – 4

91. **Kumar K, Dispenzieri A, Lacy M, Hayman S, Buadi F, Zeldenrust S, et al.** A novel staging system for light chain amyloidosis incorporating free light chain levels. Hematology meeting reports 2008; 2: C19a

92. **Kumar K, Dispenzieri A, Lacy MQ, Hayman SR, Buadi FK, Zeldenrust SR, et al.** A novel staging system for light chain amyloidosis incorporating free light chain levels. Haematologica 2008; 93: 917a

93. **Kumar S, Dispenzieri A, Clark R, Larson D, Colby CL, Zeldenrust S, et al.** Serum immunoglobulin free light chain in primary amyloidosis: Prognostic value and correlations with clinical features. Blood 2008; 112: 2733a

94. **Kumar S, Dispenzieri A, Larson D, Colby CL, Kyle R, Gertz M, Rajkumar SV.** Normalization of the serum free light chain (FLC) ratio is associated with superior overall survival among myeloma patients achieving immunofixation negative state: Results support incorporation of serum FLC ratio in stringent CR definition. Blood 2008; 112: 1692a

95. **Kumar S, Perez WS, Zhang MJ, Ballen K, Bashey A, To LB, et al.** Comparable outcomes in nonsecretory and secretory multiple myeloma after autologous stem cell transplantation. Biol Blood Marrow Transplant 2008; 14: 1134 – 1140

96. **Kumar S, Zhang L, Dispenzieri A, Van Wier S, Katzmann J, Snyder M, et al.** Association of high levels of free light chain and the presence of IgH translocations in multiple myeloma and prognostic implications. Blood 2008; 112: 1677a

97. **Kyle RA.** The role of serum free light chains in monoclonal gammopathy of undetermined significance, solitary plasmacytoma of bone and smoldering multiple myeloma. Hematology meeting reports 2008; 2: 11 – 12

98. **Kyrtsonis MC, Maltezas D, Repousis P, Vassilakopoulos KT, Tzenou T, Papanikolaou X, et al.** Possible role of myeloma biologically important cytokines in the inappropriate light chain secretion of malignant plasma cells. Hematology meeting reports 2008; 2: B13a

99. **Kyrtsonis MC, Tzenou T, Vassilakopoulos N, Kafasi N, Sachanas S, Maltezas D, et al.** Free light chain ratio in the serum of patients with Waldenstroms macroglobulinemia at diagnosis. Hematology meeting reports 2008; 2: C24a

100. **Kyrtsonis MC, Vassilakopoulos D, Maltezas D, Kafasi N, Terpos E, Elefterakis-Papaiakovou E, et al.** Suggested risk-stratification models including serum free light chain ratio for improved prognostication in mulitple myeloma. Hematology meeting reports 2008; 2: C23a

101. **Kyrtsonis M-C, Vassilakopoulos TP, Tzenou T, Kafasi N, Sachanas S, Koulieris E, et al.** Serum free light chains in Waldenstrom's Macroglobulinaemia (WM) patients at diagnosis. Haematologica 2008; 93: 1057a

102. **Lacy MQ, Dispenzieri A, Hayman SR, Kumar S, Kyle RA, Rajkumar SV, et al.** Autologous stem cell transplant after heart transplant for light chain (AL) amyloid cardiomyopathy. J Heart Lung Transplant 2008; 27: 823 – 9

103. **Lebovic D, Hoffman J, Levine BM, Hassoun H, Landau H, Goldsmith Y, et al.** Predictors of survival in patients with systemic light-chain amyloidosis and cardiac involvement initially ineligible for stem cell transplantation and treated with oral melphalan and dexamethasone. Br J Haematol 2008; 143: 369 – 73

104. **Leleu X, Moreau AS, Hennache B, Dupire S, Faucompret JL, Facon T, et al.** Serum free light chain immunoassays measurements for monitoring solitary bone plasmacytoma. Haematologica 2008; 91: PO410a

105. **Leleu X, Moreau AS, Weller E, Roccaro AM, Coiteux V, Manning R, et al.** Serum immunoglobulin free light chain correlates with tumor burden markers in Waldenstrom macroglobulinemia. Leuk Lymphoma 2008; 49: 1104 – 1107

106. **Leleu X, Roccaro AM, Leduc R, Poulain S, Moreau A-S, Gay J, et al.** Early response using serum immunoglobulin free light chain (sFLC) measurement predicts response to therapy in Waldenstrom Macroglobulinemia (WM). Blood 2008; 112: 3755a

107. **Leung N, Gertz MA, Zeldenrust SR, Rajkumar SV, Dispenzieri A, Fervenza FC, et al.** Improvement of cast nephropathy with plasma exchange depends on the diagnosis and on reduction of serum free light chains. Kidney Int 2008; 73: 1282 – 1288

108. **Lonial S, Gertz MA.** To the editor: Eliminating the complete response penalty from myeloma response assessment. Blood 2008; 111: 3297 – 3298

109. **Lopez de la Guila A, Pollmar IR, Salvatierra MG, Rodrigo E, Martin-Salces M, De Paz R, et al.** IgD multiple myeloma: Case report. Hematology meeting reports 2008; 2: D28a

110. **Ludwig J, Aulmann M, Zorn M, Nawroth P.** Evaluation of the SPAPLUS for serum free light chain measurement. Hematology meeting reports 2008; 2: A1a

111. **Lynch EA, Showell PJ, Johnson-Brett B, Mead GP, Bradwell AR.** Evaluation of latex-enhanced turbidimetric reagents for measuring free immunoglobulin light chains on the Roche cobas c501. Clin Chem 2008; 54: C93a

112. **Lynch EA, Showell PJ, Johnson-Brett B, Mead GP, Bradwell AR.** Evalutation of latex-enhanced turbidimetric reagents for measuring free immunoglobulin light-chains on the Roche cobas Integra 800. Clin Chem 2008; 54: C94a

113. **Lynch EA, Showell PJ, Kaur A, Mead GP, Bradwell AR.** Evaluation of latex-enhanced turbidimetric reagents for measuring free immunoglobulin light-chains on the Beckman Synchron LX20 Pro and Beckman UniCel DxC Synchron. Clin Chem 2008; 54: C97a

114. **Maceira AM, Prasad SK, Hawkins PN, Roughton M, Pennell DJ.** Cardiovascular Magnetic Resonance and prognosis in cardiac amyloidosis. J Cardiovasc Magn Reson 2008; 10: 54

115. **Maisnar V, Hájek R.** Revision of criteria for the diagnosis and evaluation of response to therapy in multiple myeloma Klin Biochem Metab 2008; 16: 84 – 88

116. **Mao XB, Chen XQ, Zhai YP, Liang R, Gao GX, Ma GG, et al.** Measurement of serum free light chains and its clinical significance in 20 newly diagnosed patients of multiple myeloma. Zhongguo Shi Yan Xue Ye Xue Za Zhi 2008; 16: 829 – 832

117. **Marionneaux S, Zetlmeisl M, van Hoeven KH, Fagan D, Elkins B, Shulman S.** Analytical evaluation of kappa and lambda serum free light chains on The Binding Site SPAplus TM. Clin Chem 2008; 54: C79a

118. **Mark T, Jayabalan D, Coleman M, Pearse RN, Wang YL, Lent R, et al.** Atypical serum immunofixation patterns

frequently emerge in immunomodulatory therapy and are associated with a high degree of response in multiple myeloma. Br J Haematol 2008; 143: 654 – 660

119. **Matsuda M, Gono T, Katoh N, Yoshida T, Tazawa K, Shimojima Y, et al.** Nephrotic syndrome due to primary systemic AL amyloidosis, successfully treated with VAD (vincristine, doxorubicin and dexamethasone) alone. Intern Med 2008; 47: 543 – 549

120. **Matters DJ, Smith L.** Comparison of New Scientific Company and Binding Site kits for free light chain measurements in serum. Hematology meeting reports 2008; 2: B14a

121. **Matters DJ, Solanki M, Jones R, Rose S.** Statistical analysis of free light chain results between eight nephelometirc and turbidimetric platforms. Hematology meeting reports 2008; 2: B5a

122. **Matters DJ, Solanki M, Jones R, Rose S, Carr-Smith H, Bradwell AR.** Statistical analysis of free light chain results between seven nephelometric and turbidimetric platforms. Clin Chem 2008; 54: C107a

123. **McCudden CR, Hammett-Stabler CA.** A case of hook effect in the serum free light chain assay using the Olympus AU400e. Clin Biochem 2008; 41: 148a

124. **Mead G, Harding S, Pratt G, Basu S, Jacob A, Beardsmore C, Bradwell AR.** Serum immunoglobulin and free light chain abnormalities in non Hodgkin Lymphoma. Ann Oncol 2008; 19: 315a

125. **Mead GP, Carr-Smith HD, Bradwell AR.** Free light chains. Reply to Beetham et al., Can serum free light chains replace urine electrophoresis? Ann Clin Biochem 2008; 45: 444

126. **Mead GP, Hobbs JAR, Sharp K, Harding S, Drayson M.** Incidence of light chain escape in myeloma patients at relapse. Br J Haematol 2008; 141: 98a

127. **Mead GP, Stubbs PD, Goodman HJB, Hawkins PN.** Comparison of serum versus urine for the identification of monoclonal free light chain production in 219 patients with AL amyloidosis. Hematology meeting reports 2008; 2: B12a

128. **Merlini G, Palladini G.** Advances in AL amyloidosis. Hematology Education: the education program for the annual congress of the European Hematology Association 2008; 2: 287 – 293

129. **Mignot A, Bridoux F, Thierry A, Varnous S, Pujo M, Delcourt A, et al.** Successful heart transplantation following melphalan plus dexamethasone therapy in systemic AL amyloidosis. Haematologica 2008; 93: e32 – 5

130. **Mignot A, Varnous S, Redonnet M, Jaccard A, Epailly E, Vermes E, et al.** Heart transplantation in systemic (AL) amyloidosis: a retrospective study of eight French patients. Arch Cardiovasc Dis 2008; 101: 523 – 532

131. **Morandeira F, Soriano A, Quandt E, Castro P, Ruiz E. Oriol A. Pujol R, Herrero MJ.** Comparacion de 2 ensayos para la determinacion de cadenas ligeras libres en suero en pacientes con gammapatia monoclonal. Immunologia 2008; 27: 118a

132. **Morgera S, Kleeberg L, Jakob C, Schneider M, Muller C, Kaiser M, et al.** Elimination of serum free light chains by novel replacement strategies in combination with chemotherapy in patients with multiple myeloma and kappa light chain nephropathy. Hematology meeting reports 2008; 2: E34a

133. **Munshi NC.** Investigative tools for diagnosis and management. Hematology Am Soc Hematol Educ Program 2008: 298 – 305

134. **Nam M, McRoberts J, Moody M, Al Swanamy T, Poynton C.** Light chain escape from paraproteinemia - myth or reality? Hematology meeting reports 2008; 2: D32a

135. **Niesvizky R, Jayabalan DS, Christos PJ, Furst JR, Naib T, Ely S, et al.** BiRD (Biaxin [clarithromycin]/Revlimid [lenalidomide]/dexamethasone) combination therapy results in high complete- and overall-response rates in treatment-naive symptomatic multiple myeloma. Blood 2008; 111: 1101 – 1109

136. **Offidani M, Polloni C, Piersantelli M-N, Corvatta L, Gentili S, Catarini M, et al.** Impact of free light chains ratio normalization on response duration in multiple myeloma (MM) patients treated with new drugs. Haematologica 2008; 93: 1476a

137. **Park M, Song W, Kim J, Hang H, Lee K.** Diagnostic performance of serum free light chain assay in monoclonal gammopathies. Clin Chem 2008; 54: C108a

138. **Pasquali S, Mancini E, Santoro A.** Removal of free circulating light chains (LC) by a high cut-off membrane: Different dialysis strategies. Renal Association Annual Congress 2008: F-PO1560a

139. **Piehler AP, Gulbrandsen N, Kierulf P, Urdal P.** Quantitation of serum free light chains in combination with protein electrophoresis and clinical information for diagnosing multiple myeloma in a general hospital population. Clin Chem 2008; 54: 1823 – 1830

140. **Pika T, Minarik J, Schneiderka P, Budikova M, Langova K, Lochman P, et al.** The correlation of serum immunoglobulin free light chain levels and selected biological markers in multiple myeloma. Biomed Pap Med Fac Univ Palacky Olomouc Czech Repub 2008; 152: 61 – 64

141. **Pika T, Minarik J, Zemanová M, Schneiderka P, Bacovský J, Šlézar J, Ščudla V.** Serum free light chain immunoglobulin levels in multiple myeloma and monoclonal gammopathy of undetermined significance. Klin Biochem Metab 2008; 16: 102 – 105

142. **Pillon L, Sweeting RS, Arora A, Notkin A, Ballard HS, Wieczorek RL, Leung N.** Approach to acute renal failure in biopsy proven myeloma cast nephropathy: is there still a role for plasmapheresis? Kidney Int 2008; 74: 956 – 961

143. **Pinto A, De Filippi R, Iaccarino G, Di Francia R, Distinto M, Frigeri F, et al.** Abnormalities in serum free-immunoglobulin light chains show a high and differential frequency among WHO subtypes of B-cell non-Hodgkin's lymphoma (NHL) and may turn of value for therapeutic monitoring: A study of 354 newly diagnosed patients. Blood 2008; 112: 2813a

144. **Pisani F, Cordone I, Dessanti ML, Cigliana G, Frollano B, Masi S, et al.** Bortezomib, liposomal doxorubicin and dexamethasone regimen for relapsed or refractory multiple myeloma and introduction of serum free light chains (sFLC) in response monitoring: Preliminary data. Haematologica 2008; 93: 1106a

145. **Popat R, Oakervee H, Williams C, Cook M, Craddock C, Basu S, et al.** Bortezomib, low dose intravenous melphalan and dexamethasone for patients with relapsed multiple myeloma. Haematologica 2008; 93: 918a

146. **Pratt G, Harding S, Holder R, Fegan C, Pepper C, Oscier D, Mead G.** Abnormal serum free light chain ratios are associated with poor survival in patients with chronic lymphocytic leukaemia. Br J Haematol 2008; 141: 80a

147. **Pratt G, Holding S.** More studies are needed to assess the performance of serum free light chain measurement for the diagnosis of B-cell disorders in routine clinical practice – response to Vermeersch et al. Br J Haematol 2008; 143: 145 – 146

148. **Pratt G.** The evolving use of serum free light chain assays in haematology. Br J Haematol 2008; 141: 413 – 422

149. **Pratt G, Mead G, Basu S, Jacobs A, Holder R, Fegan C, et al.** Abnormal serum free light chain ratios are associated with earlier time to first treatment and poor survival in patients with chronic lymphocytic leukaemia. Blood 2008; 112: 3141a

150. **Presslauer S, Milosavljevic D, Brucke T, Bayer P, Hubl W.** Elevated levels of kappa free light chains in CSF support

the diagnosis of multiple sclerosis. J Neurol 2008; 255: 1508 – 1514

151. Recasens V, Rubio-Escuin R, Rubio-Martinez A, Pascual T, Lopez N, Sanjuan M, Giraldo P. Serum free light chains profile as early biomarker with predictive value after autologous transplant in multiple myeloma. Blood 2008; 112: 5142a

152. Recasens V, Rubio-Escuin R, Rubio-Martinez A, Rodriguez T, Lopez N, Giraldo P. Early predictive value of serum free light chains profile in relapsed of multiple myeloma after autologous transplant. Hematology meeting reports 2008; 2: D33a

153. Robin H, Gaertner R, Parton A, Zetlmeisl M, Garcia A, Kuus-Reichel K. Utility of serum free light chain analysis in the diagnosis of monoclonal gammopathies in a community hospital. Hematology meeting reports 2008; 2: C17a

154. Rodriguez GP, Jimenez JJ, Rodriguez MJR, de Carvalho NB, Manzanal RM, de Larramendi CH, Godoy PS. Patient with multiple myeloma, in extramedullary relapse associated to light chain excretion and disappearing of intact immunoglobulin 'light chain escape'. Hematology meeting reports 2008; 2: D29a

155. Roussou M, Kastritis E, Migkou M, Psimenou E, Grapsa I, Matsouka C, et al. Treatment of patients with multiple myeloma complicated by renal failure with bortezomib-based regimens. Leuk Lymphoma 2008; 49: 890 – 895

156. Sackmann F, Pavlovsky AP, Corrado C, Pizzolato M, Alejandre M, Pavlovsky S. Prognostic factors in monoclonal gammopathy of undetermined significance. Haematologica 2008; 93: 153 – 154

157. Schonland SO, Hansmann J, Mechtersheimer G, Goldschmidt H, Ho AD, Hegenbart U. Bone involvement in patients with systemic AL amyloidosis mimics lytic myeloma bone disease. Haematologica 2008; 93: 955 –956

158. Šcudla V, Schneiderka P, Pika T, Minarík J, Bacovský J, Farbiaková V. Clinical importance of evaluating serum levels of free light chains of immunoglobulins in monoclonal gammopathies Klin Biochem Metab 2008; 16: 76 – 83

159. Seldin DC, Ward JE. Serum free light chain analysis, 4th Edition. Amyloid 2008; 15: 142 – 143

160. Shaheen SP, Talwalkar SS, Medeiros LJ. Multiple myeloma and immunosecretory disorders: an update. Adv Anat Pathol 2008; 15: 196 – 210

161. Sharp K, Emond JP, Giroux M, Steiner A, Russ G, Harding S. Serum free light chain measurements by electrophoretic and nephelometric methods may differ due to the presence of aggregated light chains. Hematology meeting reports 2008; 2: B11a

162. Sheldon J. Free light chains: author's response. Ann Clin Biochem 2008; 45: 445 – 446

163. Showell PJ, Lynch EA, Johnson-Brett B. Evaluation of latex-enhanced turbidimetric reagents for measuring free immunoglobulin light chains on the Roche COBAS C501. Hematology meeting reports 2008; 2: A2a

164. Showell PJ, Lynch EA, Johnson-Brett B. Evaluation of latex-enhanced turbidimetric reagents for measuring free immunoglobulin light-chains on the Roche COBAS INTEGRA 800. Hematology meeting reports 2008; 2: A3a

165. Showell PJ, Lynch EA, Mitchell F. Comparison of serum free immunoglobulin light chain assays on eight nephelometric/ turbidimetric analysers. Hematology meeting reports 2008; 2: A4a

166. Showell PJ, Lynch EA, Mitchell F, Mead GP, Bradwell AR. Comparison of serum free immunoglobulin light-chain assays on eight Nephelometric/Turbidimetric analysers. Clin Chem 2008; 54: C92a

167. Siegel DS, McBride L, Lendvai N, Gonsky J, Berges

T, Schillen D, et al. Pitfalls in the calculation of 24-urine test results in patients with monoclonal gammopathies: serum free light chain assays provide a more accurate assessment in this setting. Hematology meeting reports 2008; 2: B8a

168. Snozek CL, Katzmann JA, Kyle RA, Dispenzieri A, Larson DR, Therneau TM, et al. Prognostic value of the serum free light chain ratio in newly diagnosed myeloma: proposed incorporation into the international staging system. Leukemia 2008; 22: 1933 – 1937

169. Snyder MR, Clark R, Bryant SC, Katzmann JA. Quantification of urinary light chains. Clin Chem 2008; 54: 1744 – 1746

170. Sobas M, Perez L, de Carvalho N, Encinas MP. Serum immunoglobulin free light chains for monitoring patients with oligosecretory myeloma. Hematology meeting reports 2008; 2: D30a

171. Špicka I, Mecl J, Benáková H, Nohejlová A, Straub J, Novotová E, Zima T. Comparing the detection of monoclonal protein by means of presently available biochemical methods Klin Biochem Metab 2008; 16: 89 – 92

172. Tang G, Snyder M, Rao LV. Assessment of serum free light chain (FLC) assays with immunofixation electrophoresis (IFE) and bone marrow (BM) immunophenotyping in the diagnosis of plasma cell disorders. Clin Chem 2008; 54: A96a

173. Tate J, Bazeley S, Sykes S, Mollee P. Serum free light chain assay for diagnosis and monitoring of monoclonal light-chain diseases: analytical and clinical correlations. Annals Clin Biochem 2008; 45: Th40a

174. Tate JR, Bazeley S, Sykes S, Mollee P. Serum free light chain assay for diagnosis and monitoring of monoclonal light-chain diseases - analytical and clinical correlations. Clin Chem 2008; 54: C75a

175. Thio M, Blokhuis BR, Nijkamp FP, Redegeld FA. Free immunoglobulin light chains: a novel target in the therapy of inflammatory diseases. Trends Pharmacol Sci 2008; 29: 170 – 174

176. Tichý M, Vávrová J, Friedecký B, Maisnar V. Methods for detection of free immunoglobulin light chains Klin Biochem Metab 2008; 16: 93 – 96

177. Trieu Y, Xu W, Anglin P, Chen C, Kukreti V, Trudel S, Reece DE. Evaluation of serum free light chains and outcome in multiple myeloma patients with an intact monoclonal immunoglobulin treated with autologous stem cell transplantation. Blood 2008; 112: 3323a

178. van Gameren II, Bijzet J, Limburg PC, Vellenga E, Van Rijswijk MH, Hazenberg BP. Diagnostic performance of measuring free light chains in fat tissue of patients with AL amyloidosis Presented at American College of Rheumatology Annual Scientific Meeting 2008: 831a

179. Van Hoeven KH, McBride L, Lendvai N, Gonsky J, Berges T, Siegel DS. Serum free light chain testing is a more sensitive baseline marker than urine protein electrophoresis among patients undergoing primary autologous stem cell transplantation in multiple myeloma. Hematology meeting reports 2008; 2: B9a

180. Van Hoeven KH, Wells J, Abadie A. Serum protein electrophoresis (SPEP) and serum free light chain assays: A more sensitive approach that SPEP and urine protein electrophoresis (UPEP) for the diagnosis of monoclonal gammopathy. Renal Association Annual Congress 2008: SA-PO2501a

181. Van Rhee F, Crowley J, Barlogie B. Response: Top tertile SFLC reduction indeed is an independent feature of myeloma aggressiveness. Blood 2008; 111: 2491

182. Vavrova J, Friedecky B, Tichy M, Holeckova M, Maisnar V, Hajek R. The evaluation of light chains using ELISA method. Blood 2008; 112: 5147a

183. Vávrová J, Friedecký B, Tichý M, Holecková M,

Maisnar V, Hájek R. The determination of free light chains by immunoturbidimetry and ELISA Klin Biochem Metab 2008; 16: 97 – 101

184. Vermeersch P, Marien G, Bossuyt X. More studies are needed to assess the performance of serum free light chain measurement for the diagnosis of B-cell disorders in routine clinical practice. Br J Haematol 2008; 143: 143 – 145

185. Vermeersch P, Van Hoovels L, Delforge M, Marien G, Bossuyt X. Diagnostic performance of serum free light chain measurement in patients suspected of a monoclonal B-cell disorder. Br J Haematol 2008; 143: 496 – 502

186. Vescio R. Advances in the diagnosis of multiple myeloma. Clin Adv Hematol Oncol 2008; 6: 299 – 300

187. Vickrey E. Multiple myeloma: vague symptoms can challenge diagnostic skills. JAAPA 2008; 21: 19 – 22

188. Vickrey E, Allen S, Krishnamurthy J, Singh V, Mehta J, Singhal S. The utility of serum free light chain measurement in myeloma patients with oligoclonal bands on serum or urine Immunofixation. Blood 2008; 112: 5121a

189. Wechalekar AD, Hawkins PN, Gillmore JD. Perspectives in treatment of AL amyloidosis. Br J Haematol 2008; 140: 365 – 77

190. Wechalekar AD, Lachmann HJ, Goodman HJ, Bradwell A, Hawkins PN, Gillmore JD. AL amyloidosis associated with IgM paraproteinemia: clinical profile and treatment outcome. Blood 2008; 112: 4009 – 16

191. Wechalekar AD, Lachmann HJ, Offer M, Hawkins PN, Gillmore JD. Efficacy of bortezomib in systemic AL amyloidosis with relapsed/refractory clonal disease. Haematologica 2008; 93: 295 – 298

192. Wells JM, van Hoeven KH, Abadie JM. Serum free light chains assays in clinical practice: Impact of impaired renal function on interpretation. Clin Chem 2008; 54: C91a

193. Went R, Chan-Lam D, Thornhill M. Isolated tongue amyloid in a patient with multiple myeloma. Br J Haematol 2008; 143: 606

194. White DA, Rose S, Galsinh S, Perry A, Parker A, Smith A. Comparison of normal and multiple myeloma samples measured by an ELISA free light chain assay (Biovendor) and the Binding Site Ltd nephelometric Freelite assay. Hematology meeting reports 2008; 2: B15a

195. Willenbacher E, Gastl G, Nachbaur D, Clausen J, Willenbacher W. Free light chain analysis: Is it useful to monitor patients undergoing hematological stem cell therapy for multiple myeloma? An analysis of 'real life' data. Hematology meeting reports 2008; 2: D27a

196. Willenbacher E, Gastl G, Nachbaur D, Willenbacher W. Free light chain analysis: Is it useful to monitor patients undergoing haematological stem cell therapy for multiple myeloma? An analysis of "real life" data. Bone Marrow Transplant 2008; 41: P674a

197. Wood P, McElroy Y, Stone MJ. Comparison of serum immunofixation electrophoresis and free light chain assays in detection of monoclonal gammopathies. Hematology meeting reports 2008; 2: B10a

198. Wurster U. Cerebrospinal fluid analysis in multiple sclerosis and neuroborreliosis: free light chains compared with oligoclonal bands and IgG, IgA, IgM synthesis. Ann Clin Biochem 2008; 45: T9a

199. Yoshida T, Matsuda M, Katoh N, Tazawa K, Shimojima Y, Gono T, et al. Long-term follow-up of plasma cells in bone marrow and serum free light chains in primary systemic AL amyloidosis. Intern Med 2008; 47: 1783 - 90

2009

1. Abadie JM, van Hoeven KH, Wells JM. Are renal reference intervals required when screening for plasma cell disorders with serum free light chains and serum protein electrophoresis? Am J Clin Pathol 2009; 131: 166 - 71

2. Aggarwal R, Sequeira W, Mikolaitis R, Kokebie R, Block JA, Jolly M. Measurement of serum free light chains performs better than known immunological biomarkers for systemic lupus erythematosus disease activity Presented at ACR/AHRP 2009: 916a

3. Airoldi A, Quaglia M, Fenoglio R, Lazzarich E, Crespi I, Bellomo G, et al. Pitfalls in nephrology: Renal threshold for serum free light chains required to produce bence-jones proteinuria. Presented at World Congress Nephrology, Milan 2009: M182a

4. Almeshhedani M, Salamat A, Al-Ismail S. Evaluating the potential use of risk categorisation in the management of MGUS. Haematologica 2009; 94: 969a

5. Arneth B. Author's response to professor reiber's second letter concerning our article: High sensitivity of free lambda and free kappa light chains for the detection of intrathecal immunoglobulin synthesis in cerebrospinal fluid. Acta Neurol Scand 2009

6. Arneth B, Birklein F. High sensitivity of free lambda and free kappa light chains for detection of intrathecal immunoglobulin synthesis in cerebrospinal fluid. Acta Neurol Scand 2009; 119: 39 - 44

7. Arrobas Velilla T, Bermudo Guitarte C, Duro Millan R, Santos Rey K, Sanchez Margalet V, Arjona Rueda I, Goberna Ortiz R. Use of serum free light chain concentrations as an early response marker in multiple myeloma (MM). Clin Chem 2009: C69a

8. Asenova S, Bacher U, Gerritzen A, Zander AR, Kröger N. Role of free light chain assay ratio to distinguish between CR according EBMT criteria or stringent complete remission (SCR) according IMWG in multiple myeloma patients Presented at ASH 2009 2009: 1787a

9. Audard V, Matignon M, Weiss L, Remy P, Pardon A, Haioun C, et al. Successful long-term outcome of the first combined heart and kidney transplant in a patient with systemic AL amyloidosis. Am J Transplant 2009; 9: 236 - 40

10. Bakker AJ, Bierma-Ram A, der Werf CE, Strijdhaftig ML, van Zanden JJ. Quantitation of serum free light chains. Clin Chem 2009; 55: 1585 - 1587

11. Bakker AJ, Bierma-Ram A, Elderman-van der Werf C, Strijdhaftig ML, van Zanden JJ. Screening for m-proteinemia: Serum protein electrophoresis and free light chains compared. Clin Chem Lab Med. 2009; 47: 1507 - 11

12. Basnayake K, Cheung CK, Hutchison C, Cook M, Rylance P, Stoves J, et al. Differential evolution of renal scarring in cast nephropathy despite early reductions in serum free light chains. Presented at NKF Spring Clinical Meeting 2009

13. Basnayake K, Ghonemy T, Hutchison C, Bradwell AR, Cockwell P. Immunoglobulin free light chains in situ in patients with chronic kidney disease. Presented at NKF Spring Clinical Meeting 2009

14. Bensinger WI, Jagannath S, Vescio R, Camacho E, Wolf J, Irwin D, et al. Phase 2 study of two sequential three-drug combinations containing bortezomib, cyclophosphamide and dexamethasone, followed by bortezomib, thalidomide and dexamethasone as frontline therapy for multiple myeloma. Br J Haematol 2009; 148: 562 - 568

15. Bird J, Behrens J, Westin J, Turesson I, Drayson M, Beetham R, et al. Uk myeloma forum (UKMF) and nordic myeloma study group (NMSG): Guidelines for the investigation of newly detected m-proteins and the management of monoclonal gammopathy of undetermined significance (MGUS). Online publication 2009; http://www.bcshguidelines.com/pdf/MGUS_Guidelines_Final_090209.pdf

16. **Blade J, Rosinol L, Cibeira MT.** Are all myelomas preceded by MGUS? Blood 2009; 113: 5370

17. **Boer K, Deufel T.** Quantitation of serum free light chains does not compensate for serum immunofixation only when screening for monoclonal gammopathies. Clin Chem Lab Med 2009; 47: 1109 - 1115

18. **Bradwell AR, Harding SJ, Fourrier NJ, Wallis GL, Drayson MT, Carr-Smith HD, Mead GP.** Assessment of monoclonal gammopathies by nephelometric measurement of individual immunoglobulin kappa/lambda ratios. Clin Chem 2009; 55: 1646 - 1655

19. **Bryce AH, Ketterling RP, Gertz MA, Lacy M, Knudson RA, Zeldenrust S, et al.** Translocation t(11;14) and survival of patients with light chain (AL) amyloidosis. Haematologica 2009; 94: 380 - 6

20. **Chanan-Khan AA, Jagannath S, Heffner LT, Avigan D, Lee KP, Lutz RJ, et al.** Phase I study of bt062 given as repeated single dose once every 3 weeks in patients with relapsed or relapsed/refractory multiple myeloma Presented at ASH 2009 2009: 1862a

21. **Chee CE, Kumar S, Larson DR, Kyle RA, Dispenzieri A, Gertz MA, et al.** The importance of bone marrow examination in determining complete response to therapy in patients with multiple myeloma. Blood 2009; 114: 2617 - 2618

22. **Cohen G, Horl WH.** Free immunoglobulin light chains as a risk factor in renal and extrarenal complications. Semin Dial 2009; 22: 369 - 372

23. **Comenzo RL.** How i treat amyloidosis. Blood 2009; 114: 3147 - 3157

24. **Cook M, Hutchison C, Basu S, Cockwell P, Bradwell AR.** Flc removal hd improves outcome in myeloma kidney. Clin Lymphoma Myeloma 2009; 9: A069a

25. **De Filippi R, Iaccarino G, Frigeri F, Di Francia R, Crisci S, Capobianco G, et al.** Elevation of clonal serum free light chains in patients with HIV-negative primary effusion lymphoma (PEL) and pel-like lymphoma. Br J Haematol 2009; 147: 405 - 408

26. **De Filippi R, Laccarino G, Frigeri F, Russo F, Di Francia R, Distinyo M, et al.** Serum free-immunoglobulin light chains testing is frequently abnormal in patients with B-cell Non-Hodgkin and Hodgkin Lymphoma and may change in value for prognosis and therapeutic monitoring. Clin Lymphoma Myeloma 2009; 9: A412a

27. **De Filippi R, Russo F, Iaccarino G, Crisci S, Frigeri F, Riemma C, et al.** Abnormally elevated levels of serum free-immunoglobulin light chains are frequently found in Classic Hodgkin Lymphoma (CHL) and predict outcome of patients with early stage disease Presented at ASH 2009 2009: 267a

28. **de Larrea CF, Cibeira MT, Elena M, Arostegui JI, Rosinol L, Rovira M, et al.** Abnormal serum free light chain ratio in patients with multiple myeloma in complete remission has strong association with the presence of oligoclonal bands: Implications for stringent complete remission definition. Blood 2009; 114: 4954 - 4956

29. **Decaux O, Briand PY, Guenet L, Sebillot M, Jego P, Grosbois B.** Frequency of abnormal serum free light chain ratio in a french cohort of patients with MGUS. Clin Lymphoma Myeloma 2009; 9: B373a

30. **Decaux O, Karras A.** Update in multiple myeloma: International response criteria and renal complications. Rev Med Interne 2009

31. **Delbes S, Bridoux F, Tennent G, Valleix S, Touchard G, Goujon JM.** Predominant renal hereditary amyloidosis caused by a novel apolipoproteine (APO) a1 variant. Presented at World Congress Nephrology, Milan 2009: Su046a

32. **Diagnostika a lebca mnohocetneho myelomu** [czech hematological society and the slovak myeloma society: Diagnosis and treatment of multiple myeloma]. Transfuze Hematol 2009

33. **Diamantidis MD, Ioannidou-Papagiannaki E, Ntaios G.** Novel extended reference range for serum kappa/lambda free light chain ratio in diagnosing monoclonal gammopathies in renal insufficient patients. Clin Biochem 2009;42:1202 - 1203

34. **Dimopoulos MA, Kyle RA, Jagannath S.** Guidelines for standard investigative workup: Report of the international myeloma workshop consensus panel 3. Online publication 2009;http://www.mw-delhi09.com/spargoDocs/Consensuspanelthree.pdf

35. **Dimopoulos MA, Roussou M, Gavriatopoulou M, Zagouri F, Migkou M, Matsouka C, et al.** Reversibility of renal impairment in patients with multiple myeloma treated with bortezomib-based regimens: Identification of predictive factors. Clin Lymphoma Myeloma 2009; 9: 302 - 306

36. **Dispenzieri A, Gertz MA, Hayman SR, Buadi F, Kumar S, Reeder CB, et al.** A pilot study of pomalidomide and dexamethasone in previously treated light chain amyloidosis patients Presented at ASH 2009: 3854a

37. **Dispenzieri A, Kyle R, Merlini G, Miguel JS, Ludwig H, Hajek R, et al.** International myeloma working group guidelines for serum-free light chain analysis in multiple myeloma and related disorders. Leukemia 2009; 23: 215 - 224

38. **Drayson M, Cobbold M, Goodall M, Wang Y, Birtwhistle J, Jefferis R.** Monoclonal antibodies that provide a reliable new assay for serum immunoglobulin FLC in myeloma. Clin Lymphoma Myeloma 2009; 9: A355a

39. **Drayson MT, Behrens J, Cohen DR, Iggo N, Gregory WM, Bell SE, et al.** Serum FLC levels can be reduced rapidly; lower levels are associated with renal recovery. Clin Lymphoma Myeloma 2009; 9: A351a

40. **Drayson MT, Morgan GJ, Jackson GH, Davies FE, Owen RG, Ross FM, et al.** Prospective study of serum FLC and other M-protein assays: When and how to measure response? Clin Lymphoma Myeloma 2009; 9: A346a

41. **Eom HS, Min CK, Cho BS, Lee S, Lee JW, Min WS, et al.** Retrospective comparison of bortezomib-containing regimens with vincristine-doxorubicin-dexamethasone (VAD) as induction treatment prior to autologous stem cell transplantation for multiple myeloma. Jpn J Clin Oncol 2009; 39: 449 - 455

42. **Esperanza E, Sanchez-Navarro MR, Caffarena C, Mora-Vallellano J, Zafra-Ceres M, Samaniego-Sanchez C.** Algorithm for monoclonal gammopathies diagnosis. To prospective study of incidence from our population. Haematologica 2009; 94: 1611a

43. **Fernández de Larrea C, Cibeira MT, Elena M, Rosiñol L, Rovira M, Arostegui JJ, et al.** Abnormal serum free-light chain ratio in patients with multiple myeloma in long-lasting complete remission: Strong association with oligoclonal bands. Haematologica 2009; 94: 375a

44. **Fulton RB, Fernando SL.** Serum free light chain assay reduces the need for serum and urine immunofixation electrophoresis in the evaluation of monoclonal gammopathy. Ann Clin Biochem 2009; 46: 407 - 12

45. **Gera M, Jeevan R, Dhar S, Shaheen M, Phillips CL.** Idiopathic light chain deposition disease (LCDD) with light chain cast nephropathy (LCCN) treated with bortezomib. A case report and review of treatment/prognostic factors. Presented at NKF Spring Clinical Meeting 2009

46. **Giarin MM, Giaccone L, Sorasio R, Sfiligoi C,**

Amoroso B, Cavallo F, et al. Serum free light chain ratio, total kappa/lambda ratio, and immunofixation results are not prognostic factors after stem cell transplantation for newly diagnosed multiple myeloma. Clin Chem 2009; 55: 1510 - 1516

47. **Gibbs SDJ, De Cruz M, Sattianayagam PT, Lachmann HJ, Gillmore JD, Hawkins PN, Wechalekar AD.** Transient post chemotherapy rise in nt Pro-BNP in AL amyloidosis: Implications for organ response assessment Presented at ASH 2009 2009: 1791a

48. **Gibbs SDJ, Gillmore JD, Sattianayagam PT, Offer M, Lachmann HJ, Hawkins PN, Wechalekar AD.** In AL amyloidosis, both oral melphalan and dexamethasone (Mel-Dex) and risk-adapted cyclophosphamide, thalidomide and dexamethasone (CTD) have similar efficacy as upfront treatment Presented at ASH 2009: 745a

49. **Gibbs SDJ, Sattianayagam PT, Gillmore JD, Offer M, Lachmann HJ, Hawkins PN, Wechalekar AD.** Is there a role for thalidomide maintenance in the treatment of al amyloidosis? Presented at ASH 2009: 1863a

50. **Gillmore J, Cocks K, Gibbs SDJ, Sattianayagam PT, Lane T, Lachmann H, et al.** Cyclophosphamide, thalidomide and dexamethasone (CTD) versus melphalan plus dexamethasone (MD) for newly-diagnosed systemic AL amyloidosis – results from the UK amyloidosis treatment trial Presented at ASH 2009: 2869a

51. **Giraldo P, Recasens V, Rubio-Escuin R, Rubio-Martinez A, Pascual T, Giraldo P.** Predictive value of serum free light chains ratio in relapsed of multiple myeloma after autologous transplant Haematologica 2009; 94: 1615a

52. **Girnius S, Tsai F, Seldin DC, Quillen K, Yanarella L, Finn KT, et al.** Early serum free light chain responses following high-dose melphalan and stem cell transplantation for AL amyloidosis. Presented at ASH 2009: 4352a

53. **Gottenberg JE, Miceli-Richard C, Ducot B, Goupille P, Combe B, Mariette X.** Markers of B-lymphocyte activation are elevated in patients with early rheumatoid arthritis and correlated with disease activity in the ESPOIR cohort. Arthritis research & therapy 2009; 11: R114

54. **Harding S, Koulieris E, Kyrtsonis MC, Bradley C, Drayson M, Morgan G, et al.** Serum free light chain and serum heavy/light chain ratios are independently prognostic in multiple myeloma patients. Presented at ASH 2009

55. **Harding SJ, Mead GP, Bradwell AR, Berard AM.** Serum free light chain immunoassay as an adjunct to serum protein electrophoresis and immunofixation electrophoresis in the detection of multiple myeloma and other B-cell malignancies. Clin Chem Lab Med 2009; 47: 302 - 304

56. **Harding SJ, Sharp K, Steiner A, Russ G, Blackmore S, Mead G, et al.** Quantification of polymerising serum free light chains. Clin Lymphoma Myeloma 2009; 9: B101a

57. **Harousseau JL, Attal M, Avet-Loiseau H.** The role of complete response in multiple myeloma. Blood 2009; 114: 3139 - 3146

58. **Harousseau JL, Dreyling M.** Multiple myeloma: ESMO clinical recommendations for diagnosis, treatment and follow-up. Ann Oncol 2009; 20: 97 - 99

59. **Hata H, Nishi K, Oshihara W, Arai J, Shimizu K, Kawakita T, et al.** Letter to the editor: Adsorption of Bence–Jones protein to polymethylmethacrylate membrane in primary amyloidosis. Amyloid 2009; 16: 108 - 110

60. **Haynes R, Hutchison CA, Dasgupta T, Emberson J, Wheeler DC, Townend J, et al.** Associations of serum free light chains (FLCs) with the risks of end-stage renal disease (ESRD) and of death in chronic kidney disease. Presented at ASN Renal Week 2009: TH-PO552a

61. **Heher EC, Spitzer TR, Goes NB.** Light chains: Heavy burden in kidney transplantation. Transplantation 2009; 87: 947 - 952.

62. **Herrera GA.** Renal lesions associated with plasma cell dyscrasias: Practical approach to diagnosis, new concepts, and challenges. Arch Pathol Lab Med 2009; 133: 249 - 267

63. **Higgins C.** Myeloma and monoclonal gammopathy of undetermined significance. Biomed Sci 2009; September

64. **Hillengass J, Neben K, Goldschmidt H.** Current status and developments in diagnosis and therapy of multiple myeloma. J Cancer Res Clin Oncol 2009. Published online Sept 2009

65. **Hobbs J, Legg A, Giroux M.** Clinical applications of serum free light chain analysis. Biomed Sci 2009; April: 285 - 286

66. **Hobbs JA, Drayson MT, Sharp K, Harding S, Bradwell AR, Mead GP.** Frequency of altered monoclonal protein production at relapse of multiple myeloma. Br J Haematol 2009; 148: 659 - 661

67. **Hobbs JA, Sharp K, Harding S, Drayson M, Bradwell AR, Mead GP.** Changes in M-protein at multiple myeloma relapse. Clin Lymphoma Myeloma 2009; 9: A162a

68. **Holding S.** Serum free light chain used with electrophoresis increases sensitivity for monoclonal gammopathies. Clin Lymphoma Myeloma 2009; 9: A189a

69. **Hughes M, Soutar R, Lucraft H, Owen R, Bird J.** Guidelines on the diagnosis and management of solitary plasmacytoma of bone, extramedullary plasmacytoma and multiple solitary plasmacytomas: Online publication 2009; http://www.bcshguidelines.com/pdf/SBP_guideline_update_FINAL_190109.pdf

70. **Hutchison C, Airia P, Cook M, Grima D.** Improving outcomes in patients with renal failure secondary to multiple myeloma: Results from CHAR2M2 Presented at ASH 2009: 1875a

71. **Hutchison C, Meryon I, Hewins P, Drayson M, Cockwell P.** High cut-off haemodialysis improves pro-inflammatory status when used in a chronic haemodialysis schedule: An open-label crossover study. Renal Association Annual Congress 2009: P104a

72. **Hutchison C, Meryon I, Hewins P, Drayson M, Cockwell P.** High cut-off haemodialysis improves pro-inflammatory status when used in a chronic haemodialysis schedule: An open label crossover study. Presented at World Congress Nephrology, Milan 2009: Su384a

73. **Hutchison C, Pinney J, Jain P, Ghonemy T, Cockwell P.** Incidence of significant haemorrhagic complications of renal biopsies in patients with and without monoclonal gammopathies. Renal Association Annual Congress 2009: P123a

74. **Hutchison CA.** Reduction of serum free light chains predict renal recovery. Ann Hematol 2009. Online publication September 2009

75. **Hutchison CA, Airia P, Cockwell P, Grima D.** Free light chain removal by high cut-off hemodialysis is associated with high rates of renal recovery in patients with renal failure secondary to multiple myeloma. Presented at ASN Renal Week 2009: SA-PO2448a

76. **Hutchison CA, Basnayake K, Cockwell P.** Serum free light chain assessment in monoclonal gammopathy and kidney disease. Nat Rev Nephrol 2009

77. **Hutchison CA, Bradwell AR, Cook M, Basnayake K, Basu S, Harding S, et al.** Treatment of acute renal failure secondary to multiple myeloma with chemotherapy and extended high cut-off hemodialysis. Clin J Am Soc Nephrol 2009; 4: 745 - 754

78. **Iancu DM.** Correlation of serum free light chains (Freelite) with urine immunofixation (IFE) free light chains (FLC) in the diagnostic and follow-up of patients with intact immunoglobulin multiple myeloma (IIMM) in a comprehensive

cancer center. Clin Chem 2009: C37a

79. **Inman Z, Martin H, Chubb SP.** Reporting of quantitative protein electrophoresis in Australia and New Zealand: A call for standardisation. Clin Biochem Rev 2009; 30: 141 - 51

80. **Jimenez-Zepeda VH, Reeder CB, Mikhael JR, Dispenzieri A, Gertz M, Mayo A, et al.** Cyclophosphamide, bortezomib and dexamethasone (CYBORD) induces rapid and complete responses in patients with amyloidosis not eligible for peripheral blood stem cell transplant Presented at ASH 2009: 1857a

81. **Kaplan B, Ramirez-Alvarado M, Sikkink L, Golderman S, Dispenzieri A, Livneh A, Gallo G.** Free light chains in plasma of patients with light chain amyloidosis and non-amyloid light chain deposition disease. High proportion and heterogeneity of disulfide-linked monoclonal free light chains as pathogenic features of amyloid disease. Br J Haematol 2009; 144: 705 - 715

82. **Kaposztas Z, Kahan BD, Katz SM, Van Buren CT, Cherem L.** Bortezomib successfully reverses early recurrence of light-chain deposition disease in a renal allograft: A case report. Transplant Proc 2009; 41: 4407 - 4410

83. **Kastritis E, Migkou M, Gavriatopoulou M, Zirogiannis P, Hadjikonstantinou V, Dimopoulos MA.** Treatment of light chain deposition disease with bortezomib and dexamethasone. Haematologica 2009; 94: 300 - 302

84. **Katodritou E, Gastari V, Verrou E, Hadjiaggelidou C, Varthaliti M, Georgiadou S, et al.** Extramedullary (EMP) relapse in unusual locations in multiple myeloma: Is there an association with precedent thalidomide administration and a correlation of special biological features with treatment and outcome? Leuk Res 2009; 33: 1137 - 40

85. **Katzmann J, Clark R, Dispenzieri A, Kyle R, Landgren O, Bradwell AR, Rajkumar SV.** Isotype-specific heavy/light chain (HLC) suppression as a predictor of myeloma development in monoclonal gammopathy of undetermined significance (MGUS) Presented at ASH 2009: 1788a

86. **Katzmann JA.** Screening panels for monoclonal gammopathies: Time to change. Clin Biochem Rev 2009; 30: 105 - 111

87. **Katzmann JA, Kyle RA, Benson J, Larson DR, Snyder MR, Lust JA, et al.** Screening panels for detection of monoclonal gammopathies. Clin Chem 2009; 55: 1517 - 1522

88. **Keren DF.** Heavy/light-chain analysis of monoclonal gammopathies. Clin Chem 2009; 55: 1606 - 1608

89. **Kim JS, Kim K, Cheong JW, Min YH, Suh C, Kim H, et al.** Complete remission status before autologous stem cell transplantation is an important prognostic factor in patients with multiple myeloma undergoing upfront single autologous transplantation. Biol Blood Marrow Transplant 2009; 15: 463 - 470

90. **Kleeberg L, Morgera S, Jakob C, Hocher B, Schneider M, Peters H, et al.** Novel renal replacement strategies for the elimination of serum free light chains in patients with kappa light chain nephropathy. Eur J Med Res 2009; 14: 47 - 54

91. **Kristinsson SY, Bjorkholm M, Andersson TML, Eloranta S, Dickman PW, Goldin LR, et al.** Patterns of survival and causes of death following a diagnosis of monoclonal gammopathy of undetermined significance: A population-based study. Haematologica 2009; 94: 1714 - 1720

92. **Kuhnemund A, Liebisch P, Bauchmuller K, Zur Hausen A, Veelken H, Wasch R, Engelhardt M.** 'Light-chain escape-multiple myeloma'-an escape phenomenon from plateau phase: Report of the largest patient series using LC-monitoring. J Cancer Res Clin Oncol 2009; 135: 477 - 484

93. **Kumar R, Gupta D, Raina V, Sharma A, Kumar L, Irshad M, et al.** sFLC in healthy indian subjects and myeloma patients. Clin Lymphoma Myeloma 2009; 9: A576a

94. **Kumar S.** Treatment of newly diagnosed multiple myeloma: Advances in current therapy. Med Oncol 2009; s1: S 14 - 24.

95. **Kumar S, Dispenzieri A, Lacy MQ, Hayman S, Buadi F, Zeldenrust SR, et al.** A novel prognostic staging system for light chain amyloidosis (AL) incorporating markers of plasma cell burden and organ involvement Presented at ASH 2009: 2797a

96. **Kumar S, Gertz MA, Lacy MQ, Hayman SR, Buadi F, Zeldenrust SR, et al.** Changes in serum free light chain rather than intact monoclonal immunoglobulin levels predict outcome with therapy in patients with light chain amyloidosis Presented at ASH 2009: 747a

97. **Kumar S, Hayman SR, Buadi F, Allred J, Laumann K, Roy V, et al.** A phase II trial of lenalidomide, cyclophosphamide and dexamethasone (RCD) in patients with light chain amyloidosis Presented at ASH 2009: 3853a

98. **Kumar SK, Mikhael JR, Buadi FK, Dingli D, Dispenzieri A, Fonseca R, et al.** Management of newly diagnosed symptomatic multiple myeloma: Updated mayo stratification of myeloma and risk-adapted therapy (MSMART) consensus guidelines. Mayo Clin Proc 2009; 84: 1095 - 1110

99. **Kyle RA, Benson J, Larson D, Therneau T, Dispenzieri A, Melton III LJ, Rajkumar SV.** IgM monoclonal gammopathy of undetermined significance and smoldering waldenstrom's macroglobulinemia. Clin Lymphoma Myeloma 2009; 9: 17 - 18

100. **Kyle RA, Kumar S.** The significance of monoclonal gammopathy of undetermined significance. Haematologica 2009; 94: 1641 - 1644

101. **Kyle RA, Rajkumar SV.** Criteria for diagnosis, staging, risk stratification and response assessment of multiple myeloma. Leukemia 2009; 23: 3 - 9

102. **Kyle RA, Rajkumar SV.** Treatment of multiple myeloma: A comprehensive review. Clin Lymphoma Myeloma 2009; 9: 278 - 288

103. **Kyrtsonis MC, Maltezas D, Tzenou T, Koulieris E, Bradwell AR.** Staging systems and prognostic factors as a guide to therapeutic decisions in multiple myeloma. Semin Hematol 2009; 46: 110 - 117

104. **Landgren O, Goedert J, Rabkin C, Wilson W, Dunleavy K, Kyle R, et al.** Risk of AIDS Non-Hodgkin's Lymphoma is strongly predicted by elevated levels of circulating immunoglobulin-free light chains. Sixteenth Conference on Retroviruses and Opportunistic Infections, Montreal 2009: 29a

105. **Landgren O, Kyle RA, Pfeiffer RM, Katzmann JA, Caporaso NE, Hayes RB, et al.** Monoclonal gammopathy of undetermined significance (MGUS) consistently precedes multiple myeloma: A prospective study. Blood 2009; 113: 5412 - 5417

106. **Landgren O, Kyle RA, Rajkumar V.** Response: Multiple myeloma is universally preceded by a prolonged premalignant stage: Novel clinical insights and future directions. Blood 2009; 114: 2356 - 2357

107. **Landgren O, Weiss BM.** Patterns of monoclonal gammopathy of undetermined significance and multiple myeloma in various ethnic/racial groups: Support for genetic factors in pathogenesis. Leukemia 2009; 23: 1691 - 1697

108. **Lee CK, Berg M, Ryder J, Myint H, Kolhouse JF, Robinson W.** Presence of dysregulated immune recovery (IMD) following autotransplant may change the type of predominant serum light chain and elevate the level in patients with multiple myeloma (MM). Presented at ASH 2009: 4870a

109. **Legg A, Hobbs JA, Mead GP, Bradwell AR, Davern S, Solomon A.** Monoclonal vs polyclonal free light chain assays. Am J Clin Pathol 2009; 131: 901 - 902

110. **Leleu X, Leduc2 R, Rourke M, Chuma S, Sam A,**

Harris B, et al. Serum immunoglobulin free light chain (sFLC) measurement as a new marker of response to therapy and survival in Waldenstrom macroglobulinemia (WM). Presented at ASH 2009: 3952a

111. LeSourd S, Sweat K, Johnston M, Marshall J, Bornhorst J. Method comparison of quantitative serum free light chain immunoassays performed on nephelometric versus turbidimetric platforms. Clin Chem 2009: D160a

112. Lonial S, Francis D, Karanes C, Trudel S, Dollard AM, Aceo F, et al. A phase I clinical trial testing the combination of bortezomib and tipifarnib in relapsed/refractory multiple myeloma. Presented at ASH 2009: 3851a

113. Madan S, Greipp PR. The incidental monoclonal protein: Current approach to management of monoclonal gammopathy of undetermined significance (MGUS). Blood Rev 2009; 23: 257 - 265

114. Magarotto V, Londhe A, Lantz KC, Lowery C, Sonneveld P, Blade J, et al. Impact of baseline free light chain ratio (rFLC) on clinical outcomes and change in rFLC during treatment in patients with relapsed/refractory multiple myeloma treated with pegylated liposomal doxorubicin plus bortezomib or bortezomib alone. ASH Annual Meeting Abstracts 2009; 114: 4778

115. Maisnar V, Vavrova J, Friedecky B, Tichy M, Radocha J, Hajek R. The evaluation of free light chains using the ELISA method. Clin Lymphoma Myeloma 2009; 9: B391a

116. Mark T, Coleman M, Jayabalan DS, Pearse R, Zafar F, Ely S, et al. T-bird (thalidomide, clarithromycin/[biaxin®], lenalidomide/[revlimid®], dexamethasone) therapy for upfront use in symptomatic multiple myeloma Presented at ASH 2009: 1861a

117. Marshall G, Tate J, Mollee P. Borderline high serum free light chain kappa/lambda ratios are seen not only in dialysis patients but also in non-dialysis-dependent renal impairment and inflammatory states. Am J Clin Pathol 2009; 132: 309

118. Matschke J, Eisele L, Sellmann L, Duehrsen U, Duerig J, Nückel H. Abnormal free light chain ratios in chronic lymphocytic leukemia: A new prognostic factor? Presented at ASH 2009: 1237a

119. Matsue K, Fujiwara H, Iwama KI, Kimura SI, Yamakura M, Takeuchi M. Reversal of dialysis-dependent renal failure in patients with advanced multiple myeloma: Single institutional experiences over 8 years. Ann Hematol 2009

120. Maurer MJ, Micallef IN, Katzmann JA, Nikcevich D, Witzig TE. Elevated pre-treatment serum immunoglobulin free light chains (FLC) are associated with poor event-free and overall survival in diffuse large B-cell lymphoma (DLBCL) Presented at ASH 2009: 136a

121. McCudden CR, Voorhees PM, Hammett-Stabler CA. A case of hook effect in the serum free light chain assay using the Olympus AU400e. Clin Biochem 2009; 42: 121 - 124

122. Mead GP, Drayson MT. Sensitivity of serum free light chain measurement of residual disease in multiple myeloma patients. Blood 2009; 114: 1717

123. Mead GP, Stubbs PD, Goodman HJB, Hawkins PN. Serum and urine light chains in al amyloidosis. Clin Lymphoma Myeloma 2009; 9: B153a

124. Mehta J, Vickrey E, Ramadurai D, Voigt E, Fitzpatrick C, Singhal S. The relationship between serum free light chain levels and urine protein in patients with IgG or IgA myeloma Presented at ASH 2009: 1810a

125. Meng QH, Chibbar R, Pearson D, Kappel J, Krahn J. Heat-insoluble cryoglobulin in a patient with essential type II cryoglobulinemia and cryoglobulin-occlusive membranoproliferative glomerulonephritis: Case report and literature review. Clin Chim Acta 2009; 406: 170 - 173

126. Merlini G. Serum-free light chain analysis: Works in progress. Clin Chem Lab Med 2009; 47: 1021 - 1022

127. Middela S, Kanse P. Nonsecretory multiple myeloma. Indian J Orthop 2009; 43: 408 - 411

128. Mollee P. Current trends in the diagnosis, therapy and monitoring of the monoclonal gammopathies. Clin Biochem Rev 2009; 30: 93 - 103

129. Morgan GJ, Davies FE, Gregory WM, Bell SE, Szubert AJ, Cocks K, et al. The addition of thalidomide to the induction treatment of newly presenting myeloma patients increases the CR rate which is likely to translate into improved PFS and OS Presented at ASH 2009: 352a

130. Nair B, Waheed S, Szymonifka J, Shaughnessy JD, Jr., Crowley J, Barlogie B. Immunoglobulin isotypes in multiple myeloma: Laboratory correlates and prognostic implications in total therapy protocols. Br J Haematol 2009; 145: 134 - 137

131. Palladini G, Merlini G. Current treatment of AL amyloidosis. Haematologica 2009; 94: 1044 - 1048

132. Palladini G, Russo P, Bosoni T, Sarais G, Lavatelli F, Foli A, et al. Al amyloidosis associated with IgM monoclonal protein: A distinct clinical entity. Clin Lymphoma Myeloma 2009; 9: 80 - 83

133. Palladini G, Russo P, Bosoni T, Verga L, Sarais G, Lavatelli F, et al. Identification of amyloidogenic light chains requires the combination of serum-free light chain assay with immunofixation of serum and urine. Clin Chem 2009; 55: 499 - 504

134. Palumbo A, Gay F. How to treat elderly patients with multiple myeloma: Combination of therapy or sequencing. Hematology Am Soc Hematol Educ Program 2009: 566 - 577

135. Palumbo A, Rajkumar SV. Treatment of newly diagnosed myeloma. Leukemia 2009; 23: 449 - 456

136. Palumbo A, Sezer O, Kyle R, Miguel JS, Orlowski RZ, Moreau P, et al. International myeloma working group guidelines for the management of multiple myeloma patients ineligible for standard high-dose chemotherapy with autologous stem cell transplantation. Leukemia 2009; 23: 1716 - 1730

137. Pasquali S, Mancini E, Morabito S, Perego A, Besso L, Frascà G, et al. A kinetic study on free light chains (FLC) removal by means of a high cut-off (HCO) membrane: Comparison between hemodialysis (HD) and hemodiafiltration (HDF). Presented at World Congress Nephrology, Milan 2009: M449a

138. Patriarca F, Petrucci MT, Bringhen S, Baldini L, Caravita T, Corradini P, et al. Considerations in the treatment of multiple myeloma: A consensus statement from Italian experts. Eur J Haematol 2009; 82: 93 - 105

139. Pattenden RJ, Davidson KL, Wenham PR. The value of serum free light chains in a case of waldenstrom's macroglobulinaemia that produces a type I cryoglobulinaemia. Ann Clin Biochem 2009;46: 531 - 532

140. Petchey M, Fletcher M, Hobbs J. Screening for monoclonal gammopathies in patients with unexplained renal failure. Renal Association Annual Congress 2009: P126a

141. Piehler AP, Urdal P. In reply. Clin Chem 2009; 55: 1587 - 1588

142. Poshusta TL, Sikkink LA, Leung N, Clark RJ, Dispenzieri A, Ramirez-Alvarado M. Mutations in specific structural regions of immunoglobulin light chains are associated with free light chain levels in patients with AL amyloidosis. PLoS ONE 2009; 4: e5169

143. Pratt G, Harding S, Fegan C, Pepper CJ, Oscier D, Gardiner A, et al. Serum FLC levels at presentation have independent prognostic significance in CLL and levels above 50mg/L identify patients with progressive disease Presented at ASH 2009: 2355a

144. Pratt G, Harding S, Holder R, Fegan C, Pepper C,

Oscier D, et al. Abnormal serum free light chain ratios are associated with poor survival and may reflect biological subgroups in patients with chronic lymphocytic leukaemia. Br J Haematol 2009; 144: 217 - 222

145. Pratt G, Richter A, Huisson A, Drayson M, Harding S. Multiple myeloma with monoclonal free IgG3 heavy chains and free kappa light chains. Clin Lymphoma Myeloma 2009 ; 9:A177a.

146. Radovic V. Recommendations for use of FLC tests in monoclonal gammopathies. Haematologica 2009; 94: 1628a

147. Radovic VR. Advantage of serum measurments free light chain in plasma cells disorders. Haematologica 2009; 94: 1635a

148. Rajkumar SV. Multiple myeloma. Curr Probl Cancer 2009; 33: 7 - 64

149. Rajkumar SV. Prevention of progression in monoclonal gammopathy of undetermined significance. Clin Cancer Res. 2009; 15: 5606 - 8

150. Rajkumar SV, San Miguel J, Harousseau JL. Guidelines for the uniform reporting of clinical trials: Report of the international myeloma workshop consensus panel I. Online publication 2009:http://www.mw-delhi09.com/spargoDocs/Consensuspanelone.pdf

151. Reece DE, Sanchorawala V, Hegenbart U, Merlini G, Palladini G, Fermand JP, et al. Weekly and twice-weekly bortezomib in patients with systemic AL amyloidosis: Results of a phase 1 dose-escalation study. Blood 2009; 114: 1489 - 1497

152. Reiber H. Free light chains in CSF - pushing a method with biased interpretations. Acta Neurol Scand 2009; 120: 445 - 446

153. Robson EJD, Taylor J, Beardsmore C, Basu S, Mead G, Lovatt T. Utility of serum free light chain analysis when screening for lymphoproliferative disorders. Lab Med 2009; 40: 325 - 329

154. Rogoski RR. Serum free light chain assays: Detecting plasma cell disorders. MLO Med Lab Obs 2009; 41: 2 - 6

155. Sanchez-Jimenez F, Bermudo-Guitarte C, Cisneros-Barrera G, Goberna R. Demonstration of serum monoclonal immunoglobulin light chain in a case of non-secretory multiple myeloma. Presented at 2nd European Congress of Immunology 2009

156. San-Miguel JF, Mateos MV. How to treat a newly diagnosed young patient with multiple myeloma. Hematology Am Soc Hematol Educ Program 2009: 555 - 565

157. Sauter A, Weisel K, Horger M. Imaging findings in relapsing multiple myeloma. Br J Haematol 2009. Mar 26 Advance online publication

158. Schaar CG, le Cessie S, Snijder S, Franck PF, Wijermans PW, Ong C, Kluin-Nelemans H. Long-term follow-up of a population based cohort with monoclonal proteinaemia. Br J Haematol 2009; 144: 176 - 184

159. Scudla V, Pika T, Bacovsky J, Lochman P, Faber E, Minarik J. The high-dose chemotherapy with autologous stem-cell transplantation sweeps away the prognostic importance of serum levels of free light chains and kappa/lambda ratio in multiple myeloma. Lymphoma & Myeloma congress 2009: P-00011a

160. Sengul S, Li M, Batuman V. Myeloma kidney: Toward its prevention--with new insights from in vitro and in vivo models of renal injury. J Nephrol 2009; 22: 17 - 28

161. Shaheen SP, Levinson SS. Serum free light chain analysis may miss monoclonal light chains that urine immunofixation electrophoreses would detect. Clin Chim Acta 2009; 406: 162 - 166

162. Shustik C, Harding S, Ding K, Zhu L, Rassenti LZ, Kipps TJ, et al. Analysis of the serum free light chain ratio and its prognostic value in a cohort of patients with chronic lymphocytic leukemia Presented at ASH 2009: 2631a

163. Siegel DS, Bilotti E, Van Hoeven K. Serum free light chain analysis for diagnosis, monitoring, and prognosis of monoclonal gammopathies. Lab Med 2009; 40: 363 - 366

164. Siegel DS, McBride L, Bilotti E, Lendvai N, Gonsky J, Berges T, et al. Inaccuracies in 24-hour urine testing for monoclonal gammopathies. Lab Med 2009; 40: 341 - 344

165. Singhal S, Vickrey E, Krishnamurthy J, Singh V, Allen S, Mehta J. The relationship between the serum free light chain assay and serum immunofixation electrophoresis, and the definition of concordant and discordant free light chain ratios. Blood 2009; 114: 38 - 39

166. Snyder M, Dispenzieri A, Rajkumar SV, Kyle R, Benson J, Katzmann JA. The biologic and analytic variability of serum protein electrophoresis M-spike, nephelometric Ig quantitation, serum FLC quantitation, and urine M-spike in monoclonal gammopathies. Presented at ASH 2009: 1803a

167. Stankowski-Drengler TJ, Lockington KS, Snyder MR, Dispenzieri A, Lust JA, Kyle RA, Katzmann JA. Reflex testing for the diagnosis of monoclonal gammopathies. Clin Chem 2009: C85a

168. Stewart AK, Richardson PG, San-Miguel JF. How I treat multiple myeloma in younger patients. Blood 2009; 114: 5436 - 5443

169. Sthaneshwar P, Nadarajan V, Maniam JA, Nordin N, Gin Gin G. Serum free light chains: Diagnostic and prognostic value in multiple myeloma. Clin Chem Lab Med 2009; 47: 1101 - 1107

170. Stringer S, Harding S, Koulieris E, Kyrtsonis MC, Drayson M, Morgan G, et al. Serum free light chain and serum heavy/light chain ratios are independently prognostic in multiple myeloma patients. Lymphoma & Myeloma congress 2009: P00028a

171. Tate J, Bazeley S, Sykes S, Mollee P. Quantitative serum free light chain assay - analytical issues. Clin Biochem Rev 2009; 30:131-140

172. Tate J, Mollee P, Johnson R. Monoclonal gammopathies - clinical and laboratory issues. Clin Biochem Rev 2009; 30: 89 - 91

173. Tate JR, Bazeley S, Sykes S, Mollee P. Dilution anomalies in serum free light chain measurement. Clin Chem 2009: C11a

174. Telio D, Bailey D, Chen C, Crump M, Reece DE, Kukreti V. Two distinct syndromes of lymphoma-associated amyloidosis: Findings from a case series Presented at ASH 2009: 2938a

175. Terrier B, Sene D, Saadoun D, Ghillani-Dalbin P, Thibault V, Delluc A, et al. Serum-free light chain assessment in hepatitis c virus-related lymphoproliferative disorders. Ann Rheum Dis 2009; 68: 89 - 93

176. Tong W, Wienholt L, Adelstein S. Cerebrospinal fluid free kappa light chain quantification by nephelometry compared to oligoclonal bands by isoelectric focusing in the diagnosis of multiple sclerosis. Internal Medicine Journal 2009; 39: 53a

177. Tsai H-T, Caporaso NE, Kyle RA, Katzmann JA, Dispenzieri A, Hayes RB, et al. Evidence of serum immunoglobulin abnormalities up to 9.8 years before diagnosis of chronic lymphocytic leukemia: A prospective study. Blood 2009; 114: 4928 - 4932

178. Tsai WL, Lin PH, Chang WW, Shiesh SC. Assessment of serum free light chains in patients with monoclonal gammopathies. Clin Chem Lab Med 2009; 47: M-C030a

179. Tuck HH, James GW. Spontaneous tumor lysis syndrome caused by high grade plasma cell dyscrasia. Presented at ASN Renal Week 2009: PUB652a

180. **Tzenou T, Vassilakopoulos TP, Kafasi N, Sachanas S, Koulieris E, Maltezas D, et al.** sFLCr in LPL/WM patients at diagnosis. Clin Lymphoma Myeloma 2009; 9: A424a

181. **Uljon SN, Richardson PG, Grudzien C, Schur PH, Tanasijevic MJ, Lindeman NI.** Changes in serum free light chain values do not precede changes in M-spike in a series of patients with relapsed/refractory intact immunoglobulin multiple myeloma. Clin Chem 2009: D176a

182. **Vachon CM, Kyle RA, Therneau TM, Foreman BJ, Larson DR, Colby CL, et al.** Increased risk of monoclonal gammopathy in first-degree relatives of patients with multiple myeloma or monoclonal gammopathy of undetermined significance. Blood 2009; 114: 785 - 790

183. **van Gameren I, van Rijswijk MH, Bijzet J, Vellenga E, Hazenberg BP.** Histological regression of amyloid in AL amyloidosis is exclusively seen after normalization of serum free light chain. Haematologica 2009; 94: 1094 - 1100

184. **van Hoeven K, Wells J, Abadie J.** Serum protein electrophoresis (SPEP) plus serum free light chain (sFLC) testing: A sensitive panel for diagnosis of monoclonal gammopathy (MG). Presented at NKF Spring Clinical Meeting 2009

185. **van Hoeven KH, Bilotti E, McBride L, Berges T, McNeill A, Schillen D, Siegel D.** Serum free light chain assays are more sensitive than urinary tests for light chain monoclonal proteins. Clin Chem 2009: C30a

186. **Vavrova J, Friedecky B, Tichy M, Holeckova M, Maisnar V, Spacilova J.** Measurement of free light chains using elisa method. Clin Chem Lab Med 2009; 47: W-B019a

187. **Vermeersch P, Vercammen M, Holvoet A, Broeck IV, Delforge M, Bossuyt X.** Use of interval-specific likelihood ratios improves clinical interpretation of serum FLC results for the diagnosis of malignant plasma cell disorders. Clin Chim Acta 2009; 410: 54 - 58

188. **Watanaboonyongcharoen P, Suwannabutra S, Theerasaksilp S, Paripokee N, Na Nakorn T.** The roles of serum free light chain assay on the prognosis of multiple myeloma. Clin Lymphoma Myeloma 2009; 9: A420a

189. **Wechalekar A, Hawkins PN, Wassef NJ, Gillmore ND, Lachmann HJ, Morris J, et al.** Monoclonal protein detection in AL amyloidosis revisited: Analysis of 642 patients. Clin Lymphoma Myeloma 2009; 9: A300a

190. **Wechalekar AD, Wassef N, Lachmann H, Sattianayagam P, Gibbs SDJ, Gillmore J, Hawkins PN.** High early mortality and poor outcomes for patients with AL amyloidosis presenting with high serum free light chains - a new risk stratification model. Haematologica 2009; 94: 544a

191. **Wechalekar AD, Wassef NL, Gibbs SDJ, Gillmore J, Dike F, Lachmann H, et al.** A new staging system for AL amyloidosis incorporating serum free light chains, cardiac troponin-T and NT-proBNP Presented at ASH 2009: 2796a

192. **Weisel KC, Hänel M, Niederwieser D, Naumann R, Kröger N, Engelhardt M, et al.** Speed of response with lenalidomide and dexamethasone in patients with relapsed or refractory multiple myeloma: First results of the MM-019 german compassionate use protocol. Haematologica 2009; 94: 397a

193. **Weiss B, Minter A, Laquer M, Howard R, Abadie J, Ascencao J, et al.** Patterns of monoclonal immunoglobulins and serum free light chains are significantly different in African-American compared to Caucasian MGUS patients Presented at ASH 2009: 2838a

194. **Weiss BM, Abadie J, Verma P, Howard RS, Kuehl WM.** A monoclonal gammopathy precedes multiple myeloma in most patients. Blood 2009; 113: 5418 - 5422

195. **Wians FH.** Use of excel spreadsheets to create interpretive reports for laboratory tests requiring complex calculations. Lab Med 2009; 40: 5 - 12

196. **Willenbacher E, Gasser S, Gastl G, Willenbacher W.** Serum & urine free light chain analysis compared to conventional paraprotein measurements: Usefullness for clinical decision making in real life haematology. Presented at ASH 2009: 2889a

197. **Willenbacher E, Nachbaur D, Gastl G, Gasser S.** Free light chain (by rFLC) monitoring reveals early relapse after HSCT. Clin Lymphoma Myeloma 2009; 9: A326a

198. **Yegin ZA, Yagci M, Özkurt ZN, Kursunoglu N, Aki SZ.** Abnormal serum free light chain ratio as a predictor of adverse outcome in chronic lymphocytic leukemia: Has the time come for a simple prognostic marker? Haematologica 2009; 94: 929a

199. **Zabek-Adamska A, Dyrka A, Drozdz R.** Investigation of factors affecting immunoglobulin free light chain level. Clin Chem Lab Med 2009; 47: M-C075a

2010

1. **Avet-Loiseau H, Mirbahai L, Harousseau J-L, Moreau M, Mathiot C, Facon T, et al.** Serum immunoglobulin heavy/light chain ratios are independent risk factors for predicting progression free survival in multiple myeloma. Haematologica 2010; 95: 0953a

2. **Azais I, Brault R, Debiais F.** New treatments for myeloma. Joint Bone Spine. 2010; 77: 20 - 6

3. **Basnayake K, Ying WZ, Wang PX, Sanders PW.** Immunoglobulin light chains activate tubular epithelial cells through redox signaling. J Am Soc Nephrol 2010; Online advance publication

4. **Berenson JR, Anderson KC, Audell RA, Boccia RV, Coleman M, Dimopoulos MA, et al.** Monoclonal gammopathy of undetermined significance: A consensus statement. Br J Haematol 2010; Online advance publication

5. **Bianchi G, Kyle RA, Colby CL, Larson DR, Kumar S, Katzmann JA, et al.** Impact of optimal follow-up of monoclonal gammopathy of undetermined significance (MGUS) on early diagnosis and prevention of myeloma-related complications. Blood 2010; Online advance publication

6. **Blade J, Dimopoulos M, Rosinol L, Rajkumar SV, Kyle RA.** Smoldering (asymptomatic) multiple myeloma: Current diagnostic criteria, new predictors of outcome, and follow-up recommendations. J Clin Oncol 2010; Feb 1; 28: 690 - 7

7. **Briand PY, Decaux O, Caillon H, Grosbois B, Le Treut A, Guenet L.** Analytical performance of the serum free light chain assay. Clin Chem Lab Med 2010; 48: 73 - 9

8. **Brunvand MW, Bitter M.** Amyloidosis relapsing after autologous stem cell transplantation treated with bortezomib: Normalization of detectable serum-free light chains and reversal of tissue damage with improved suitability for transplant. Haematologica 2010; 5: 519 - 21

9. **Chanan-Khan AA, Giralt S.** Importance of achieving a complete response in multiple myeloma, and the impact of novel agents. J Clin Oncol 2010; Online advance publication

10. **Corazzelli G, De Filippi R, Capobianco G, Frigeri F, De Rosa V, Iaccarino G, et al.** Tumor flare reactions and response to lenalidomide in patients with refractory classic hodgkin lymphoma. Am J Hematol 2010; 85: 87 - 90

11. **de Kat Angelino CM, Raymakers R, Teunesen MA, Jacobs JFM, Klasen IS.** Overestimation of serum kappa free light chain concentration by immunonephelometry. Clinical Chemistry 2010; 56: 1188 - 90

12. **de Larrea CF, Cibeira MT, Elena M, Arostegui JI, Rosinol L, Rovira M, et al.** Free light chain assay and stringent complete remission in multiple myeloma: More questions than answers. Blood 2010; 115: 2414 - 2415

13. **Dispenzieri A, Katzmann JA, Kyle RA, Larson DR,**

Melton LJ, 3rd, Colby CL, et al. Prevalence and risk of progression of light-chain monoclonal gammopathy of undetermined significance: A retrospective population-based cohort study. Lancet 2010; 375: 1721 - 1728

14. **Dobson R, Miller RF, Palmer HE, Feldmann M, Thompson EJ, Thompson AJ, et al.** Increased urinary free immunoglobulin light chain excretion in patients with multiple sclerosis. J Neuroimmunol 2010; 220: 99 - 103

15. **Elliott BM, Peti S, Lee D, Osman K, Isola LM, Scigliano E, Kostakoglu L.** Use of combined FDG-PET and laboratory data for predicting relapse in multiple myeloma. J Clin Oncol 2010; 28: P8148a

16. **Emile C.** Freelite: Chronologie et dernieres mises a jour. Option Bio 2010; 430: 15 - 9

17. **Faiman B, Licata AA.** New tools for detecting occult monoclonal gammopathy, a cause of secondary osteoporosis. Cleve Clin J Med 2010; 77: 273 - 278

18. **Garcia de Veas Silva J, Bermudo Guitarte C, Arrobas Velilla T, Sanchez Margalet V, Martin Ruiz J, Merchan Iglesias M, Goberna Ortiz R.** Utility of free light chains ratio in the monitoring of the treatment of a patient with light chain lambda multiple myeloma. AACC 2010: P113a

19. **Go RS.** Monoclonal gammopathy of undetermined significance: To screen or not to screen for multiple myeloma? Br J Haematol 2010; 149: 620 - 621

20. **Guastafierro S, Falcone U, Ferrara MG, Parascandola RR, Fiorillo P, Cuomo C, et al.** Biclonal gammopathies: Review of 19 cases. AACC 2010: PD59a

21. **Haynes RJ, Read S, Collins GP, Darby SC, Winearls CG.** Presentation and survival of patients with severe acute kidney injury and multiple myeloma: 20-year experience from a single centre. Nephrol Dial Transplant 2010; 25: 419 - 26

22. **Heher EC, Goes NB, Spitzer TR, Raje NS, Humphreys BD, Anderson KC, Richardson PG.** Kidney disease associated with plasma cell dyscrasias. Blood 2010. Online advance publication

23. **Hoffmann-Luecke E, Hojskov C, Solling H, Nielsen JL, Moller HJ.** Local experience from measurements of serum free light chains - a necessity to establish local reference ranges. XXXII Nordic Congress in Medical Biochemistry 2010: P-040a

24. **Honma R, Fukase S, Suzuki M, Omoto E.** Serum free light-chain assay for nonsecretory multiple myeloma with light chain cast nephropathy and light chain deposition disease. Rinsho Ketsueki 2010; 51: 270 - 274

25. **Isik M.** Renal disorder preceding multiple myeloma. Med Oncol 2010. Online advance publication

26. **Kastritis E, Wechalekar AD, Dimopoulos MA, Merlini G, Hawkins PN, Perfetti V, et al.** Bortezomib with or without dexamethasone in primary systemic (light chain) amyloidosis. J Clin Oncol 2010; 28: 1031 - 1037

27. **Katzmann JA, Kyle RA, Benson J, Larson DR, Snyder MR, Lust J, et al.** In reply. Clin Chem 2010; 56 :679-680

28. **Kayserova J, Capkova S, Skalicka A, Vernerova E, Polouckova A, Malinova V, et al.** Serum immunoglobulin free light chains in severe forms of atopic dermatitis. Scand J Immunol 2010; 71: 312 - 316

29. **Khoriaty R, Hussein MA, Faiman B, Kelly M, Kalaycio M, Baz R.** Prediction of response and progression in multiple myeloma with serum free light chain assays: Corroboration of the serum free light chain response definitions Clin Lymphoma, Myeloma and Leukemia 2010: E10 - E4.

30. **Kroger N, Asenova S, Gerritzen A, Bacher U, Zander A.** Questionable role of free light chain assay ratio to determine stringent complete remission in Multiple myeloma patients. Blood 2010; 115: 3413 - 3414

31. **Kumar S, Zhang L, Dispenzieri A, Van Wier S, Katzmann JA, Snyder M, et al.** Relationship between elevated immunoglobulin free light chain and the presence of igh translocations in multiple myeloma. Leukemia 2010; Online advance publication

32. **Kyle RA, Durie BG, Rajkumar SV, Landgren O, Blade J, Merlini G, et al.** Monoclonal gammopathy of undetermined significance (MGUS) and smoldering (asymptomatic) multiple myeloma: IMWG consensus perspectives risk factors for progression and guidelines for monitoring and management. Leukemia 2010; 24: 1121 - 1127

33. **Kyle RA, Rajkumar SV.** Monoclonal gammopathy of undetermined significance and smoldering multiple myeloma. Curr Hematol Malig Rep 2010; 5: 62 - 69

34. **Landgren O, Goedert JJ, Rabkin CS, Wilson WH, Dunleavy K, Kyle RA, et al.** Circulating serum free light chains as predictive markers of AIDS-related lymphoma. J Clin Oncol 2010; 28: 773 - 779

35. **Levoguer AM, Legg A, Hughes RG.** Serum free lightchains in chronic lymphocytic leukemia: Light at the end of the tunnel? Clinical Lab International 2010: 16 - 19

36. **Lorenz EC, Sethi S, Poshusta TL, Ramirez-Alvarado M, Kumar S, Lager DJ, et al.** Renal failure due to combined cast nephropathy, amyloidosis and light-chain deposition disease. Nephrol Dial Transplant 2010; 25: 1340 - 1343

37. **Madan S, Dispenzieri A, Lacy MQ, Buadi F, Hayman SR, Zeldenrust SR, et al.** Clinical features and treatment response of light chain (AL) amyloidosis diagnosed in patients with previous diagnosis of multiple myeloma. Mayo Clin Proc 2010; 85: 232 - 238

38. **Mark TM, Koirala A, Pearse RN, Zafar F, Jayabalan D, Leonard JP, et al.** An evaluation of the role of bone marrow biopsy in patients with multiple myeloma who achieve an unconfirmed stringent complete remission. J Clin Oncol 2010; 28: 8144a

39. **Mead GP, Carr-Smith HD.** Overestimation of serum {kappa} free light chain concentration by immunonephelometry. Clin Chem 2010. Online advance publication

40. **Miles OA, Miller E.** Serum free light chain assay monitoring in plasma cell disorders - a single centre audit. Br J Haematol 2010; 149: P144a

41. **Mollee P, Tate J, Weiss D, Solomon A.** Assessment of a monoclonal antibody-based free light chain ELISA and the polyclonal antibody-based freelite assay in AL amyloidosis. Amyloid 2010; 17: P196

42. **Morabito F, De Filippi RD, Gentile M, Iaccarino GI, Attolico I, Cutrona G, et al.** Abnormal immunoglobulin serum free light chains (sFLC) ratio is associated with unfavourable prognostic markers and is a predictor of treatment free survival (TFS) in chronic lymphocytic leukemia (CLL). Haematologica 2010

43. **Morabito F, Mazzone C, Gigliotti V, Caruso N, Madeo A, Recchia A, et al.** Flow cytometry evaluation of bone marrow plasma cells in monoclonal gammopathy of undetermined significance (MGUS): Correlation with monoclonal component, free light chain and magnetic resonance imaging. EHA 2010

44. **Moscetti A, Saltarelli F, Bianchi MP, Monarca B, Bandiera G, De Biase L, La Verde G.** Two cases of localised amyloidosis. Amyloid 2010; 17: P200a

45. **Moscetti A, Saltarelli F, Bianchi MP, Monarca B, De Biase L, Porrini R, et al.** Quick response to bortezomib plus dexamethasone in a patient with al amyloidosis in first relapse. Amyloid 2010; 17: P219a

46. **Munshi NC, Lee S, Roodman GD, Behler C, Kambhampati S, Rose MG, et al.** Phase II multisite study to evaluate efficacy and safety of single weekly administration of bortezomib (BZ) and dexmethasone (DEX) in newly diagnosed multiple myeloma (NDMM) patients (pts). J Clin Oncol 2010; 28: TPS307a

47.   Murata K, Clark RJ, Lockington KS, Tostrud LJ, Greipp PR, Katzmann JA. Sharply increased serum free light-chain concentrations after treatment for multiple myeloma. Clin Chem 2010; 56: 16 - 18

48.   Piehler AP. Serum, free light chains in multiple myeloma and related disorders. XXXII Nordic Congress in Medical Biochemistry 2010: P039a

49.   Pretorius CJ, Ungerer JPJ, Wilgen U, Klingberg S. Screening panels for detection of monoclonal gammopathies: Confidence intervals Clin Chem 2010; 56: 677 - 679

50.   Richter AG, Harding S, Huissoon A, Drayson M, Pratt G. Multiple myeloma with monoclonal free IgG3 heavy chains and free kappa light chains. Acta Haematol 2010; 123: 158 - 161

51.   Saltarelli F, Bianchi MP, Moscetti A, Ferranti G, Capogreco F, Corsetti MT, et al. Correlation between serum levels of free light chain and circulating cytokines in a patient with AL cardiac amyloidosis. Amyloid 2010; 17: P196a

52.   Schroers R, Baraniskin A, Heute C, Kuhnhenn J, Alekseyev A, Schmiegel W, et al. Detection of free immunoglobulin light chains in cerebrospinal fluids of patients with central nervous system lymphomas. Eur J Haematol 2010; Online advance publication

53.   Shimazaki C, Murakami H, Sawamura M, Matsuda M, Kinoshita T, Hata H, et al. Clinical usefulness of serum free light chain measurement in monoclonal gammopathy. Rinsho Ketsueki 2010; 51: 245 - 252

54.   Spencer A, Walker P, Asvadi P, Wong M, Campbell D, Reed K, et al. A phase I study of the anti-kappa monoclonal antibody, mdx-1097, in previously treated multiple myeloma patients. J Clin Oncol 2010; 28: 8143a

55.   Stankowski-Drengler T, Gertz MA, Katzmann JA, Lacy MQ, Kumar S, Leung N, et al. Serum immunoglobulin free light chain measurements and heavy chain isotype usage provide insight into disease biology in patients with POEMS syndrome. Am J Hematol 2010; 85: 431 - 434

56.   Tate JR, Bazeley S, Sykes S, Hawley C, Mollee P. Serum free light chain anomalies in a case of light-chain cast nephropathy. AACC 2010: P112a

57.   Testa A, Dejoie T, Lecarrer D, Wratten M, Sereni L, Renaux JL. Reduction of free immunoglobulin light chains using adsorption properties of hemodiafiltration with endogenous reinfusion. Blood purification 2010; 30: 34 - 36

58.   Turesson I. Guidelines on diagnosis and follow-up of multiple myeloma and MGUS. XXXII Nordic Congress in Medical Biochemistry 2010: P-038a

59.   van Rhee F. Light-chain MGUS: Implications for clinical practice. Lancet 2010; 375: 1670 - 1671

60.   Varsi K, Aakre KM, Sandberg S. Targeted requesting followed by a specific laboratory test algorithm; a simple but expensive method for diagnosing monoclonal gammopathies. XXXII Nordic Congress in Medical Biochemistry 2010: P-006a

61.   Vavrova J, Tichy M, Friedecky B, Maisnar V, Hajek R, Cermakova Z, et al. Mezilaboratorni studie stanoveni volnych monoklonalnich lehkych retezcu imunoglobulinu (inter - laboratory study for the measurement of free monoclonal light chain of immunoglobulins.). Klin Biochem Metab 2010; 18: 73 - 76.

62.   Weiss BM, Kuehl WM. Advances in understanding monoclonal gammopathy of undetermined significance as a precursor of multiple myeloma. Expert Rev Hematol 2010; 3: 165 - 174

63.   Wiwanitkit V. Nonsecretory multiple myeloma, how to make a diagnosis? Indian J Orthop 2010; 44: 112

64.   Yegin ZA, Ozkurt ZN, Yagci M. Free light chain : A novel predictor of adverse outcome in chronic lymphocytic leukemia. Eur J Haematol 2010; 84: 406 - 411

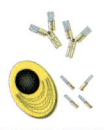

Chapter

32

Analysis of immunoglobulin heavy chain/light chain pairs (Hevylite™)

32.1. *Introduction: limitations of immunoglobulin measurements* 301
32.2. *Concept: immunoglobulin heavy chain/light chain assays* 302
32.3. *Antibody specificity* 302
32.4. *Normal ranges of Hevylite (HLC) assays* 304
32.5. *Clinical sensitivity of HLC assays for monoclonal gammopathies* 306
32.6. *HLC assays for monitoring monoclonal gammopathies* 309
32.7. *Prognostic value of HLC assays in monoclonal gammopathies* 312
32.8. *HLC assays in non-Hodgkin's lymphoma (NHL)* 317
32.9. *HLC assays for immunohistochemistry* 317
32.10. *Publications on FcRn receptors and related issues* 318
32.11. *Publications on HLC assays.* 319

Summary: Immunoglobulin HLC immunoassays:

1. Are of clinical value when monitoring patients with monoclonal gammopathies.
2. Are more sensitive than serum protein electrophoresis (SPE) for quantifying monoclonal immunoglobulins.
3. Provide quantitative information compared with immunofixation electrophoresis (IFE) and can be more sensitive.
4. Are prognostic in multiple myeloma (MM) and monoclonal gammopathy of undetermined significance (MGUS).

32.1. Introduction: limitations of immunoglobulin measurements

Typical analytical tests for monoclonal gammopathies are SPE with scanning densitometry together with serum free light chain (sFLC) immunoassays. While SPE is a simple, cheap test, it is not particularly sensitive so that quantification of proteins at low concentrations (1-3g/L) is inaccurate. This is particularly apparent for monoclonal IgA since its anodal electrophoretic migration positions it over other bands such as transferrin. Improved sensitivity is achieved with IFE but it is a non-quantitative assay. Nephelometry is also used for immunoglobulin measurements and is analytically accurate to low concentrations. However, patients' samples also contain non-tumour polyclonal immunoglobulins of both κ and λ types that are included in the analysis so that results are clinically inaccurate at normal serum concentrations. Furthermore, assessments of monoclonal IgG are unreliable because of variable catabolism as FcRn recycling receptors become saturated or reduced by chemotherapy *(Chapter 10)*.

In contrast, one of the great diagnostic benefits of sFLC analysis is the κ/λ ratio. This is because: 1) it provides a quantitative assessment of FLC clonality, 2) it has typical high immunoassay sensitivity, 3) clinical ranges are wide due to immunosuppression of the non-tumour FLCs and 4) there is automatic compensation for variable renal metabolism and changes in blood volume *(Chapter 10)*.

Immunoglobulin heavy chain/light chain immunoassays - *"Hevylite" (HLC)*, have similar analytical advantages. This chapter provides early data on these new immunoglobulin reagents.

32.2. Concept: immunoglobulin heavy chain/light chain assays

Intact immunoglobulin molecules contain unique junctional epitopes between the heavy chain (C_H1) and light chain (C_L) constant regions *(Figure 32.1)*. These are the target of HLC antibodies. Hence, they can separately identify the different light chain types of each immunoglobulin class, i.e. IgGκ, IgGλ, IgAκ, IgAλ, IgMκ and IgMλ *(Figure 32.2)*. These molecules are then measured in pairs, e.g., IgGκ/IgGλ, to produce ratios of monoclonal immunoglobulin/background polyclonal immunoglobulin concentrations, in the same manner as sFLC κ/λ ratios.

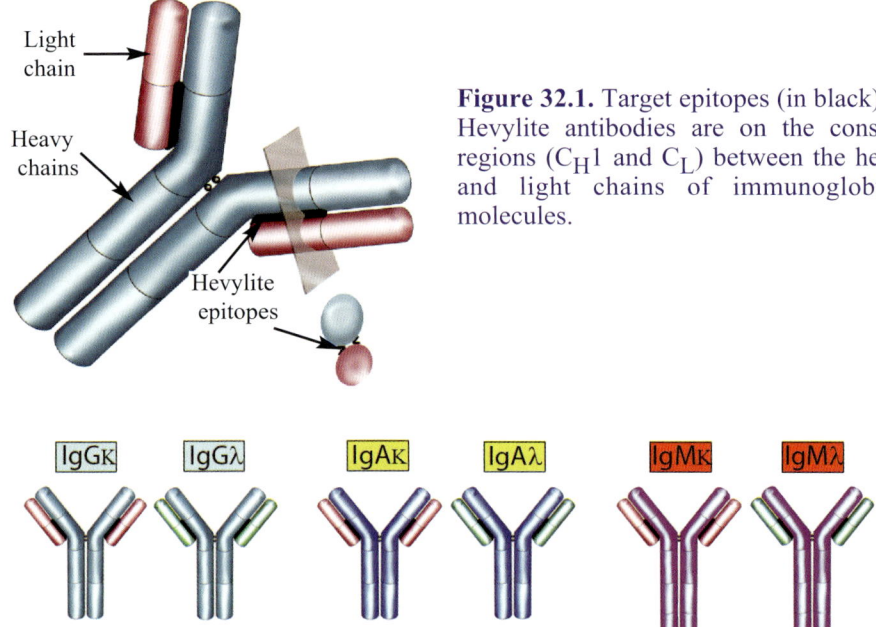

Figure 32.1. Target epitopes (in black) for Hevylite antibodies are on the constant regions (C_H1 and C_L) between the heavy and light chains of immunoglobulin molecules.

Figure 32.2. Heavy chain/light chain pairs of IgG, IgA and IgM molecules showing the target epitopes for Hevylite immunoassays in black.

32.3. Antibody specificity

Inevitably, one of the most demanding aspects of HLC assay production is ensuring good specificity. As for FLC immunoassays, the reagents are polyclonal antibodies produced in sheep. Immunisation and subsequent purification techniques are designed to ensure no cross reactivity. For example, IgGκ reagents do not cross react with kappa, either free or bound to other heavy chain classes, or IgGλ. Cross reactivity with non target immunoglobulins and interference is tested as part of the validation process

(Figure 32.3). There are 4 HLC epitope regions per molecule - one on each side of the heavy chain / light chain contact regions. Because there are 4 per molecule, immune complexes readily form to produce good homogeneous immunoassays that are suitable for nephelometry and turbidimetry. Latex enhancement is not necessary for IgG and IgA HLC assays, but is required for IgM and IgD assays and may be required for analysis of cerebrospinal fluid samples.

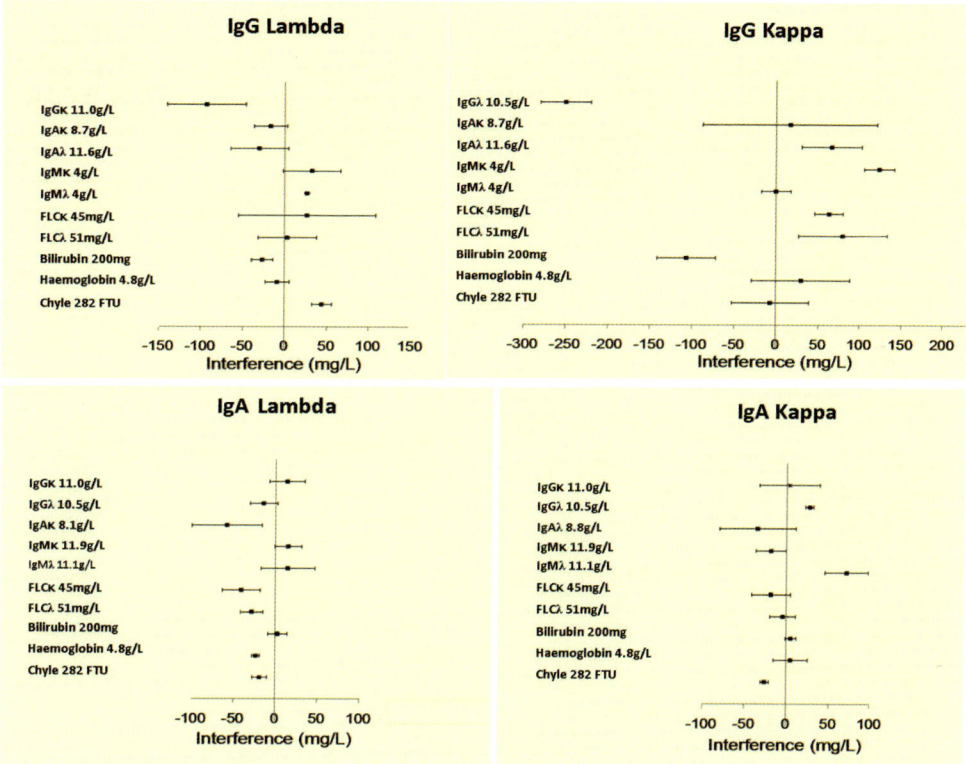

Figure 32.3. Interference data for IgG and IgA HLC assays *(Bradwell 7)*. Data points are the mean values of 3 analyses with 95% CIs shown as bars. Changes in the measured Ig concentrations upon the addition of potentially interfering substances are shown. Concentration of added substances (× normal median serum values) were as follows: IgGκ 11.0 g/L (×1.5); IgGλ 10.5 g/L (×2.5); IgAκ 8.7 g/L (×7); IgAλ 11.6 g/L (×13), IgMκ 4 g/L (×6); IgMλ 4 g/L (×8), FLCκ 45 mg/L (×6); FLCλ 51 mg/L (×4); Chyle 282 formazin turbidity units (FTU) (×25); haemoglobin 4.8 g/L (×100); bilirubin 200 mg/L (×30).

32.4. Normal ranges of Hevylite (HLC) assays

Intact immunoglobulin concentrations are normally controlled within narrow limits, as are their HLC κ and λ subsets *(Bradwell 7; Katzmann 10)*. The results from testing blood donor panels are shown in Table 32.1 and Figures 32.4 and 32.5. Pearson rank correlations for summation of IgGκ+IgGλ Hevylite samples to total IgG was ~ 0.9 (p<0.01), IgAκ+IgAλ to total IgA was 0.9 (p<0.01) and IgMκ+IgMλ to total IgM was 0.9 (p<0.001) *(Figure 32.6)*.

Ranges that include older individuals, hospital populations and patients with chronic infections and autoimmune diseases are required. Initial studies have indicated that HLC κ/λ ratios in diseases with raised polyclonal immunoglobulins are maintained within the narrow normal limits observed for blood donors (as with sFLC κ/λ ratios).

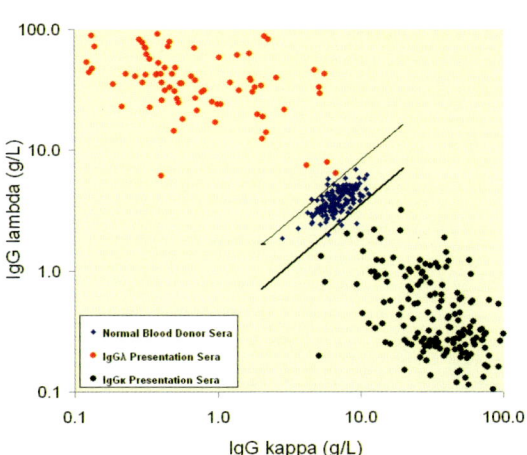

Figure 32.4. IgG HLC tests on 146 blood donor sera (95% range) and 245 IgG presentation sera (166 IgGκ, black circles and 79 IgGλ, red circles) from the IFM 2005-01 MM trial *(Avet-Loiseau 1)*. Immunosuppression of the non-tumour HLC from hypercatabolism is apparent in high concentration samples. (Courtesy of H. Avet-Loiseau).

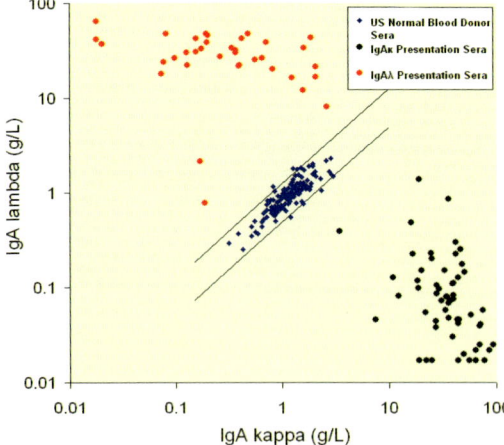

Figure 32.5. IgA HLC tests on 146 blood donor sera (95% range) and 94 IgA presentation sera (60 IgAκ, black circles and 34 IgAλ, red circles) from the IFM 2005-01 MM trial *(Avet-Loiseau 1)*. All MM samples were abnormal by Hevylite while 31 of the samples could not be quantified by SPE because of overlying protein bands or low concentration. (Courtesy of H. Avet-Loiseau).

Immunoglobulin	Mean	Median	95 % range
IgGκ (mg/L)	6900	6850	4030-9780
IgGλ (mg/L)	3840	3810	1970-5710
IgGκ/IgGλ ratio	1.86	1.87	0.98-2.75
IgAκ (mg/L)	1250	1190	480-2820
IgAλ (mg/L)	1000	980	360-1980
IgAκ/IgAλ ratio	1.28	1.27	0.80-2.04
IgMκ (mg/L)	717	634	289-1820
IgMλ (mg/L)	451	417	168-940
IgMκ/IgMλ ratio	1.59	1.59	0.96-2.30

Table 32.1. Normal concentration ranges of HLC immunoglobulins and HLC ratios in blood donors (IgG 130 samples; IgA 138 samples; IgM 120 samples).

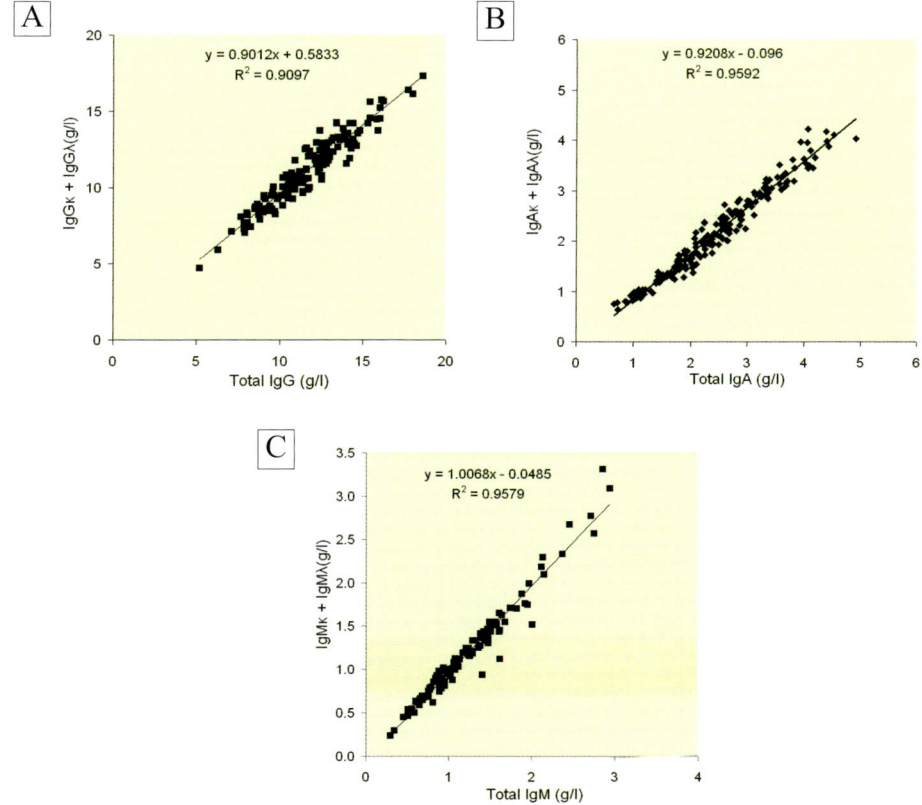

Figure 32.6. Pearson rank correlations of IgG, IgA and IgM to summation of their respective Hevylite pairs; Figures A, B and C respectively.

32.5. Clinical sensitivity of HLC assays for monoclonal gammopathies

Serum samples from patients with multiple myeloma (MM) enrolled onto the IFM 2005 trial *(courtesy of H Avet-Loiseau and the IFM, France; Figures 32.4, 32.5 and Table 32.2)* and AL amyloidosis *(courtesy of P Hawkins and the UK, National Amyloidosis Centre; Figures 32.7-32.10)* were tested using the IgG, IgA and IgM HLC assays. The results are shown as HLC κ/λ plots (as for FLC assays) alongside normal (blood donor) samples to show normal ranges.

As with FLC assays, in the majority of MM patients the isotype specific monoclonal protein production level was greater than the upper limit of the normal range *(Table 32.2)*. In addition all of the patients tested had the appropriate abnormal HLC κ/λ ratio and matched IFE sensitivity. The suppression of the non-tumour isotype matched immunoglobulins (eg. suppression of IgGκ by an IgGλ tumour) and so abnormal ratios were present even when immunoglobulin concentrations were within the normal range. The degree of suppression varied greatly in individual patients although suppression was generally greater with IgA producing tumours compared with IgG tumours *(Table 32.2)*. However, the correlation between suppression and production was greater in IgG patients (IgGκ $r = -0.456$; $p=8.7 \times 10^{-10}$, IgGλ patients $r = -0.310$; $p=0.005$) than in IgA patients (IgAκ $r = -0.28$; $p=0.031$, IgAλ $r = -0.33$; $p=0.05$). The correlation was greater for IgG patients than IgA patients due to saturation of the FcRn receptor: thus greater concentrations of monoclonal IgG directly result in more rapid catabolism of any polyclonal IgG. 33% (31/94) of IgA patients could not be accurately quantified by SPE because of their co-migration with other serum proteins or low concentration, but all had abnormal HLC ratios.

Serum samples were also tested in patients with AL amyloidosis to assess sensitivity compared with IFE *(Wechalekar 5)*. Patients were categorized as having a detectable M-component in serum and/or urine, or no detectable M-component and normal free light chain ratio. Initially 8 IgM, 14 IgA and 58 IgG IFE positive sera were tested. 8/8 patients had IgM proteins quantifiable by SPE, all of whom had an abnormal IgM HLC ratio *(Figure 32.7)*. 7/14 IgA patients had monoclonal proteins quantifiable by SPE and 7/14 patients had monoclonal proteins which were hidden / non quantifiable; the IgA HLC ratio was abnormal in all (14/14) cases *(Figure 32.8)*. 54 IgG patients had monoclonal proteins quantifiable by SPE, and in 4 patients the concentration was below the sensitivity of SPE. 54/58 patients had abnormal IgG HLC ratios; whilst in 4/58 patients the HLC ratio was normal, in all cases the monoclonal protein level was not quantifiable by SPE densitometry *(Figure 32.9)*. The normal ratio here is likely to be because of very low monoclonal protein production (below the sensitivity of SPE) against a normal polyclonal background. Among 46 AL amyloidosis patients with no detectable serum *(Figure 32.10)* or urinary bands and a normal serum free light chain ratio, the HLC ratio was abnormal in 9 cases (19%) identifying 2 IgAκ, 3 IgAλ, and 4 IgGκ clones *(Figures 32.8 &32.9)*.

These results show that HLC assays can quantify monoclonal proteins and can identify monoclonality (by abnormal HLC ratios) even when SPE and in some instances IFE are negative. This high sensitivity should help clarify disease status in many patients with subtle monoclonal gammopathies.

	IgGκ	IgGλ	IgAκ	IgAλ
Ig by SPE densitometry	40.5 8.02-90: (4.9-109)	46.0 10.86-85.02: (5.7-86.0)	34.0 11.35-88.55: (10.0-98.0)	28.0 3.46-62.1: (2.2-67.0)
Involved HLC (g/L)	33 5.6-91.81: (4.07-107)	35.5 4.77-91.12: (4.58-97.3)	36.7 5.36-110.2: (3.49-125)	30.1 1.95-49.9: (0.78-64.1)
Un-involved HLC (g/L)	0.33 0.072-2.03: (0.04-3.19)	0.41 0.12-5.84: (0.09-6.76)	0.02 0.017-1.13: (0.017-1.38)	0.02 0.018-2.17: (0.018-2.65)
HLC ratio	93.52 5.13-864 (3.94-1334.88)	0.018 0.001-0.878 (0.001-1.05)	462 11.2-6020 (8.8-7352)	0.01 0.018-0.255 (0.001-0.32)

Table 32.2. Concentrations of serum Igs and other proteins in 339 patients with multiple myeloma showing median values, 95% limits and total ranges.

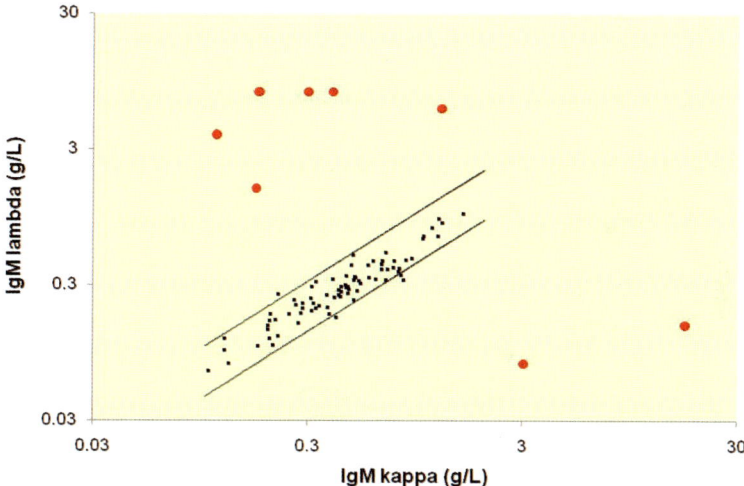

Figure 32.7. Plot of IgMκ v IgMλ for blood donor sera (solid black squares), 8 IgM immunofixation positive amyloid patients (solid red circles SPE positive) *(Wechalekar 5)*. The parallel lines indicate the 95% range for IgMκ / IgMλ ratios. (Courtesy of A. Wechalekar).

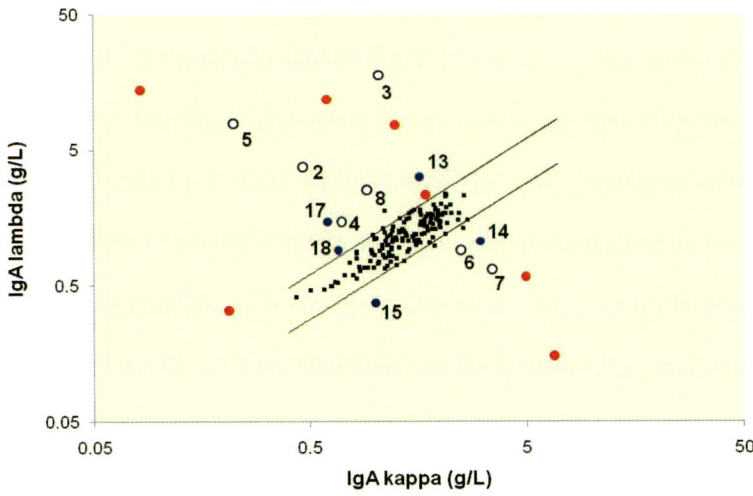

Figure 32.8. Plot of IgAκ v IgAλ for 146 blood donor sera (solid black squares), 14 IgA immunofixation positive amyloid patients (solid red circles SPE positive, hollow black circles SPE negative) and 5 IgA amyloid patients where clonality was only detectable using IgAκ / IgAλ ratios (solid blue circles) *(Wechalekar 5)*. The parallel lines indicate the 95% range for IgAκ / IgAλ ratios. Numbers correspond to positions on the SPE gel in Figure 32.10. (Courtesy of A. Wechalekar).

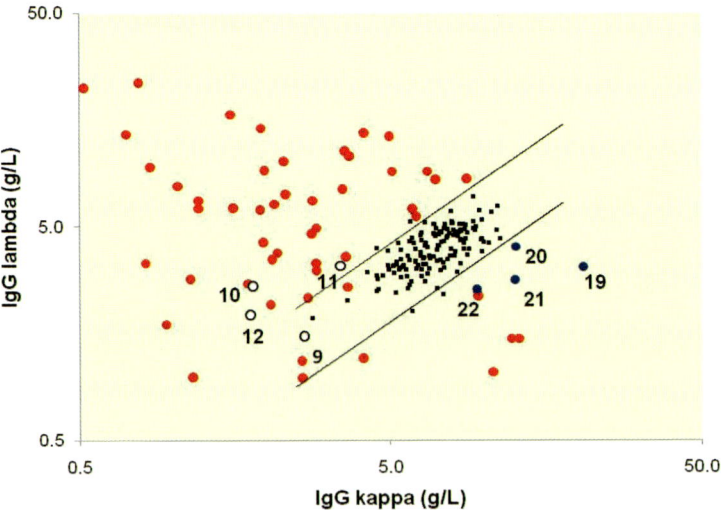

Figure 32.9. Plot of IgGκ v IgGλ for 146 blood donor sera (solid black squares), 58 IgG immunofixation positive amyloid patients (solid red circles SPE positive, hollow black circles SPE negative) and 4 IgG amyloid patients where clonality was only detectable using IgGκ / IgGλ ratios (solid blue circles) *(Wechalekar 5)*. The parallel lines indicate the 95% range for IgGκ / IgGλ ratios. Numbers correspond to positions on the SPE gel in Figure 32.10. (Courtesy of A. Wechalekar).

Figure 32.10. SPE of National Amyloid Centre samples *(Wechalekar 5).* Normal human serum controls (lanes 1 and 16), 7 IgA (lanes 2-8) and 4 IgG (lanes 9-12) IFE positive patients whose monoclonal band was not accurately quantifiable by SPE densitometry and 9 IFE negative patients (lanes 13–22). (Courtesy of A. Wechalekar).

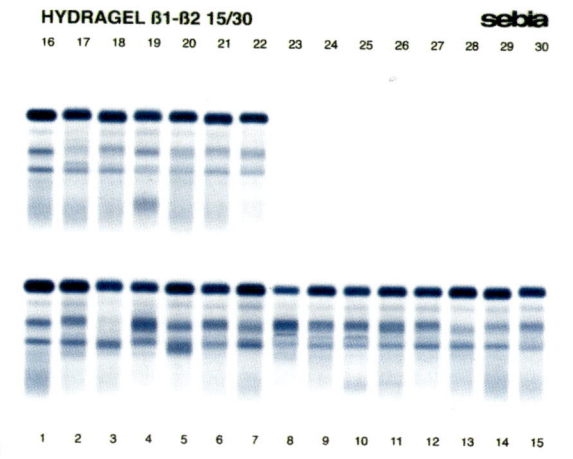

32.6. HLC assays for monitoring monoclonal gammopathies

There are several reasons why HLC assays might be useful for monitoring patients with monoclonal gammopathies.

1. They are more sensitive than SPE and IFE in many samples, so they can help in the assessment of residual disease.
2. They provide numerical results for patients who are only IFE positive.
3. HLC κ/λ ratios have a greater range of changes than monoclonal immunoglobulin measurements because the non-tumour immunoglobulin allows assessment of immunosuppression.
4. HLC κ/λ ratios are not affected by changes in blood volume, haematocrit and variable metabolism (via FcRn receptors for IgG) that affect current assays for serum immunoglobulins *(Chapter 10).*
5 HLC κ/λ ratios provide information about the tumour selective killing rates versus non-tumour plasma cell kill rates. This assessment of selective tumour killing rates may help with making decisions on effective chemotherapies.

In order to determine these features of HLC assays we assessed 9 patients with IgG MM and 5 with IgA MM who were undergoing treatment in the UK, MRC VII trial. For the IgG patients (4 IgGκ and 5 IgGλ), 25 samples were available. In 4/4 who did not achieve complete response, the ratio remained abnormal throughout. In 3/5 patients achieving complete remission, IgGκ/λ ratios were abnormal at relapse earlier than IFE measurements.

For the IgA patients (4 IgAκ and 1 IgAλ), all of the 26 samples that were positive by IFE had abnormal IgAκ/λ ratios. In 2/5 patients, abnormal ratios indicated residual disease when IFE was negative. For one patient the IgA monoclonal protein was obscured by another protein in the SPE gel but could be monitored by HLC ratios. In a second patient, abnormal HLC ratios indicated a slow relapse more than a year before IFE became positive.

One IgG MM patient was studied in detail during 2 remissions and relapses, and illustrates the main features of HLC assays *(Figures 32.11 - 32.13) (Bradwell 7)*. HLC ratios had a greater range of values than IgG quantitation by scanning densitometry or nephelometry and were more sensitive during remissions and indicated relapse earlier. Of particular interest were the HLC results during the second course of chemotherapy. IgG measurements indicated a tumour response but HLC ratios did not change, indicating no selective tumour cell killing. This was in contrast to the first course of chemotherapy that had huge selective tumour cell killing. Indeed, the patient terminally relapsed after the second round of chemotherapy. This suggested that the HLC ratio provided the correct interpretation of the lack of response to the chemotherapy. The discrepancy between total IgG measurements and IgG HLC ratios may, in part, be due to inhibition of the FcRn receptor by the chemotherapy. This would cause a fall in total IgG (because of faster turnover) but IgG HLC ratios would be unaffected.

Measurement of HLC IgGλ *(Figure 32.12)* did not provide more information than total IgG. IgGκ quantitation showed the functional activity of the bone marrow plasma cells, the response to chemotherapy and the subsequent tumour relapse *(Figure 32.13)*. However, it was the HLC IgG κ/λ ratio that provided the most interesting results.

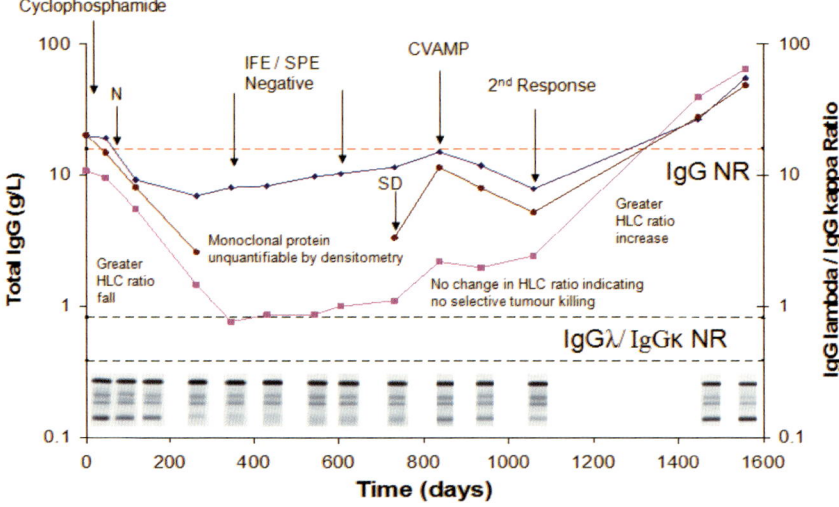

Figure 32.11. Monitoring of a patient with IgGλ MM using HLC ratios (IgGλ/ IgGκ), SPE, scanning densitometry (SD) of IgG and nephelometry (N). NR: Normal Range, CVAMP: cyclophosphamide and vincristine, adriamycin, melphalan, methyl-prednisolone.

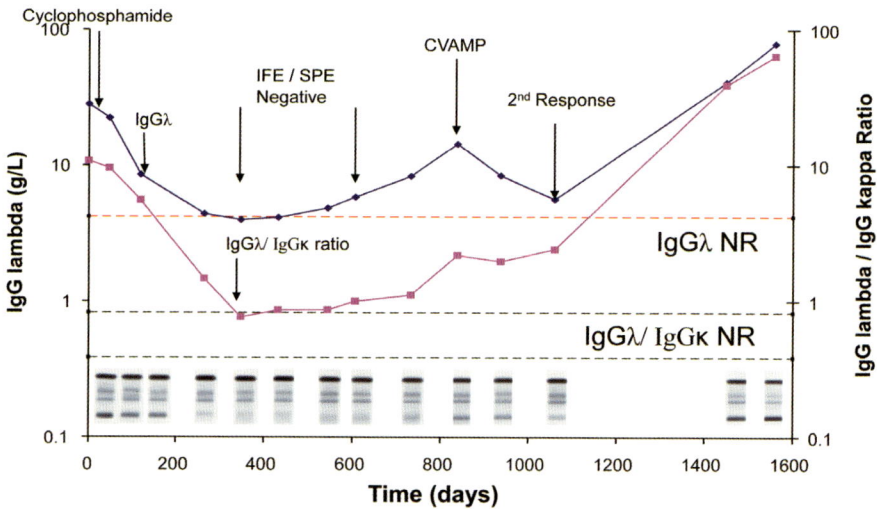

Figure 32.12. Monitoring of a patient with IgGλ MM using IgGλ HLC and IgG HLC ratios. Comparison with SPE is shown and see Figure 32.10 for other details. NR: Normal Range.

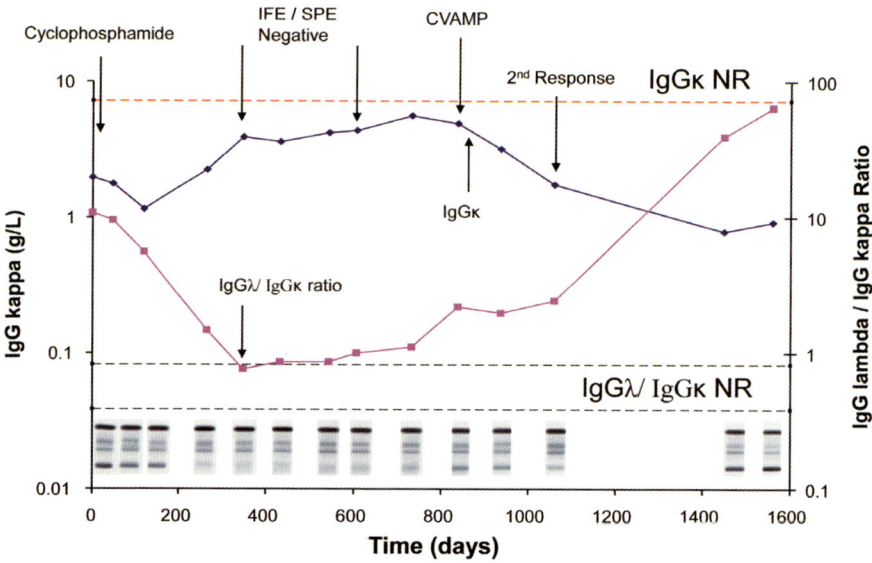

Figure 32.13. Monitoring of a patient with IgGλ MM using IgGκ HLC and IgG HLC ratios. Comparison with SPE is shown and see Figure 32.10 for other details. NR: Normal Range.

32.7. Prognostic value of Hevylite assays in monoclonal gammopathies

Several recent studies have assessed the relationship between HLC ratios and outcome in MM both at presentation and at maximum response. Avet Loiseau *(1, 2)* et al. investigated the prognostic value of baseline HLC ratios in 339 patients (166 IgGκ/79 IgGλ and 60 IgAκ/34 IgAλ). The median observed ratios for the individual immunoglobulin isotypes are summarised in Table 32.2. During the study period 125 patients (37%) progressed and 46 patients (14%) died. Progression free survival (PFS) and overall survival (OS) were used as the time dependent variables. Figure 32.14a shows the relationship between PFS and involved monoclonal protein production (measured by Hevylite) for both IgG and IgA patients. There was a weak correlation (p=0.039) in PFS when comparing patients with involved monoclonal protein concentrations in the upper tertile (red line) compared to those in the lower two tertiles (blue line). However, there was a stronger correlation between HLC ratios and PFS. Figure 32.14b compares patients within (blue line) or outside (red line) of the ratio range <0.01 to >200 (p=0.0002); approximately 2/3 of the patients had ratios within this range. IgG and IgA HLC ratios were also analysed individually to predict the relative risk of progression. Increasingly abnormal IgG HLC ratios were associated with an increased risk of progression *(Figure 32.14c*, p<0.0001), but this did not apply to IgA HLC ratios (p=0.32). It is likely that the IgA HLC ratios would have been similarly associated with poorer survival if more patients had been included. HLC ratios are determined by the concentration of both the tumour-derived immunoglobulin and that of the non-tumour immunoglobulin of the same heavy chain class. Whilst the monoclonal, tumour immunoglobulin concentration was significant (p=0.039, *Figure 32.14a*) suppression of the uninvolved polyclonal immunoglobulins (p=0.002, *Figure 32.14d*) accounted for most of the association between HLC ratios and PFS. As there is no international guideline for describing systemic immunoparesis in this study it was defined as a reduction in the immunoglobulin concentration 33% below the normal range. Polyclonal levels of IgG and IgM in IgA MM patients and IgA and IgM in IgG MM were measured. Figure 32.15 shows the relationship between patients with no immunoparesis (blue lines) and those with systemic immunoparesis (red lines). Decreased concentrations of polyclonal immunoglobulins were not associated with shorter PFS in either IgA (IgG p=0.169, IgM p=0.477) or IgG (IgA p=0.952, IgM p=0.977) MM patents.

The current international staging system (ISS) for MM relies upon serum beta-2-microglobulin (β_2M) and albumin measurements (Stage I β_2M<3.5mg/L, albumin>35g/L, Stage II not stage I or III, Stage III β_2M>5.5mg/L). The correlation between these measurements and PFS is shown in Figure 32.16 (p=0.023). In a Cox multivariant regression model β_2M and extreme HLC ratios were the only independent risk factors identified. Figure 32.17 shows an alternative staging system where albumin is replaced by HLC ratios <0.01 or >200. Patients with 0, 1 or 2 risk factors had significantly different PFS (p=0.000013) *(Avet-Loiseau 1)*.

IgG but not IgA HLC ratios have been shown to predict malignant transformation in MGUS patients *(Katzmann 10)*. Table 32.3 shows comparison of 105 IgG MGUS samples: 36 with stable disease, 30 initial samples from patients who subsequently progressed and 39 samples collected shortly before malignant transformation was diagnosed. HLC suppression was present in 22%, 53% and 90% respectively, whereas non isotype suppression was present in 6%, 7% and 46%. Thus, IgG HLC pair suppression was more frequent than suppression of Igs from other heavy chain classes, and MGUS patients who eventually progressed had a 2-fold higher rate of isotype specific suppression than stable MGUS patients. In IgA MGUS patients there was no difference between the degree of non isotype suppression and isotype specific suppression *(Table 32.3)*.

Systemic (or non isotype matched) immunoparesis is not associated with shortened PFS in MM. The prognostic value of HLC ratios, therefore, supports the existence of separate "niches" within the bone marrow for plasma cells producing IgG or IgA.

IgG MGUS	N	Abnormal IgGκ/ IgGλ	A or M heavy chain isotype suppression*	Heavy-light pair suppression*
MGUS, no progression, initial sample**	36	64%	6%	22%
MGUS, progression, initial sample**	30	83%	7%	53%
MGUS, progression, pre-MM sample***	39	87%	46%	90%

IgA MGUS	N	Abnormal IgAκ/ IgAλ	G or M heavy chain isotype suppression*	Heavy-light pair suppression*
MGUS, no progression, initial sample**	4	100%	25%	25%
MGUS, progression, initial sample**	10	100%	50%	40%
MGUS, progression, pre-MM sample***	14	93%	71%	71%

Table 32.3. Prognostic value of Hevylite ratios in IgG and IgA MGUS patients *(Katzmann 10)*. Clonal IgG plasma cells appear to suppress other IgG producing non-clonal plasma cells more effectively than IgA or IgM secreting cells and this occurs more frequently with cells that will eventually undergo malignant transformation. * Suppression is defined as below the lower limit of the normal reference range. ** MGUS, no progression (or progression) are from Olmsted County, long-term study of MGUS (mean follow-up = 8 years). *** MGUS, last sample, pre-MM sample are from NIH PLCO cohort. (Courtesy of J. Katzmann).

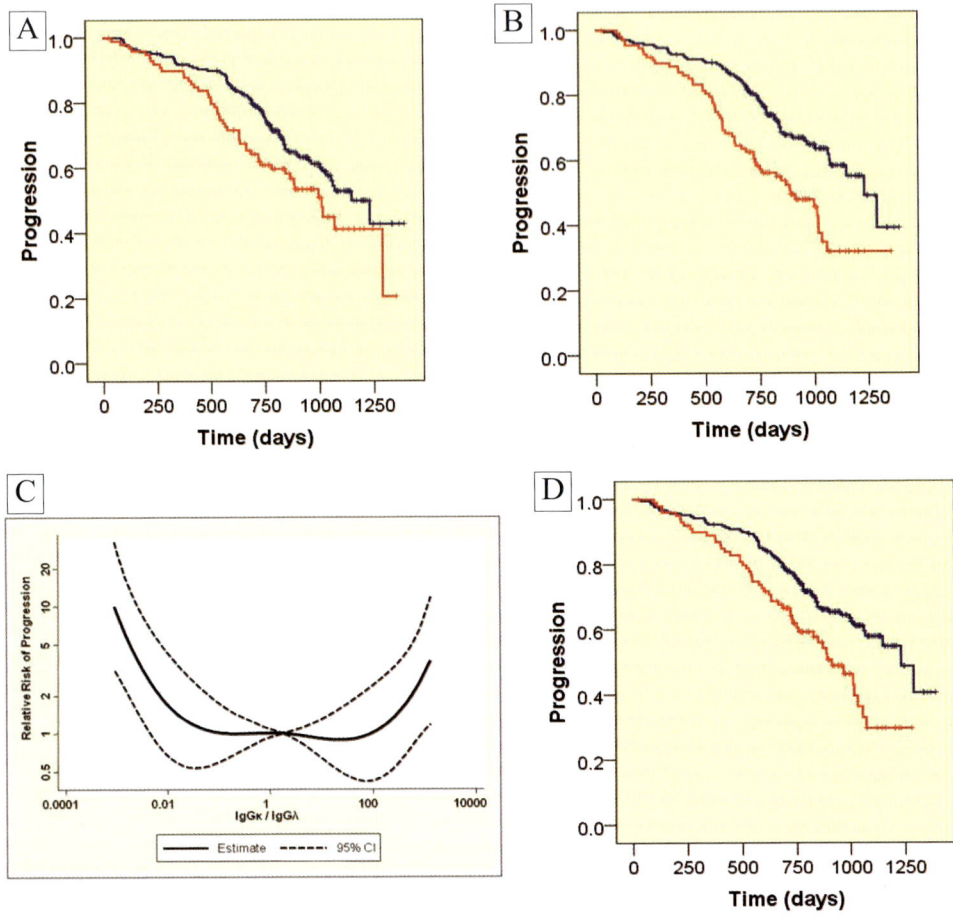

Figure 32.14. Kaplan Meier survival curves for 308 multiple myeloma patients. Heavy/light chain values were measured in presentation sera and progression free survival (PFS) calculated *(Avet-Loiseau 2)*. (Courtesy of H. Avet-Loiseau).

A) Patients with involved monoclonal protein values in upper tertile (red, n = 101) were compared with those in the lower two tertiles (blue, n = 207). Increased concentrations of involved monoclonal protein were weakly associated with shorter PFS (p=0.039).

B) Patients with heavy/light chain ratios in the upper tertile (red, n = 101) were compared to those in the lower two tertiles (blue, n = 207). Increased heavy/light chain ratios were significantly associated with shorter PFS (p=0.0002).

C) Effect of increasingly abnormal IgG ratios on the relative risk of progression free survival. Increasingly abnormal Hevylite ratios were associated with decreased PFS (p=0.0001).

D) Patients with uninvolved polyclonal protein values in the lower tertile (red, n = 101) were compared to those in the upper two tertiles (blue, n = 207). Decreased concentrations of polyclonal isotype specific protein were associated with shorter PFS (p=0.002).

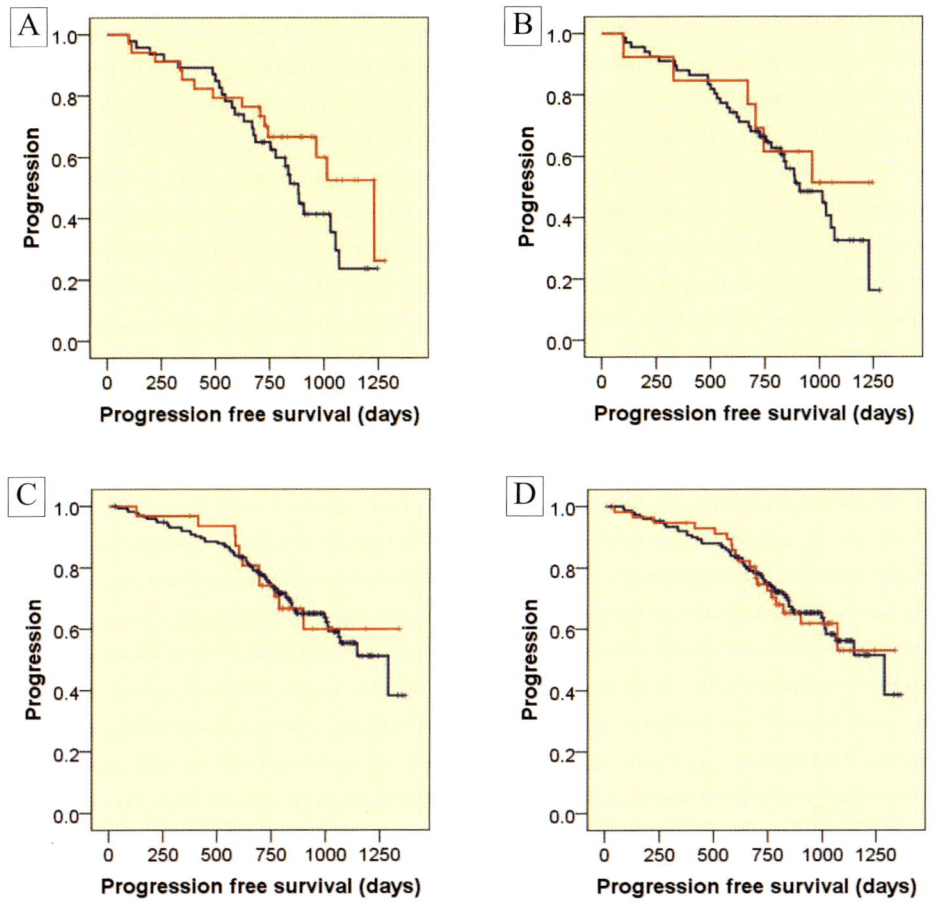

Figure 32.15. Kaplan Meier survival curves for 79 IgA and 208 IgG MM patients *(Avet-Loiseau 2)*. Patients with no immunoparesis (blue) and with systemic immunoparesis (red) were compared. Systemic immunoparesis was defined as a reduction of the immunoglobulin measurement 33% below the normal range. A and B: decreased IgG or IgM concentrations in IgA MM patients were not associated with shorter progression free survival (PFS) (p=0.169 and p=0.477 respectively). C and D: decreased IgA or IgM concentrations in IgG MM patients were not associated with shorter PFS (p=0.952 and 0.977 respectively). (Courtesy of H. Avet-Loiseau).

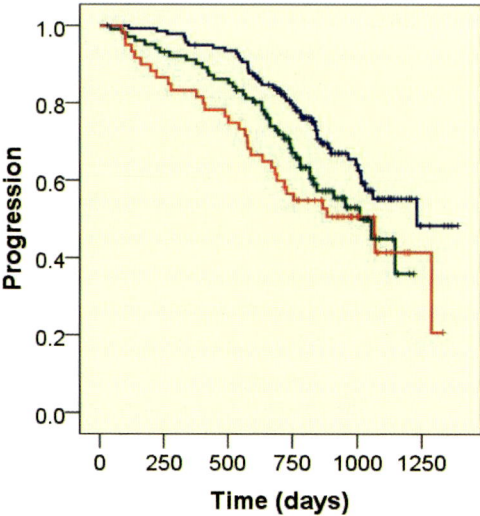

Figure 32.16. Kaplan Meier survival curves for 310 MM patients *(Avet-Loiseau 1)*. International Staging System criteria (Stage I β_2M <3.5mg, albumin >35g/L, blue; Stage II not Stage I or Stage III, green; Stage III β_2M >5.5mg/L, red) were used. Patients with Stage I, II or III disease had significantly different progression free survival times (p=0.023). (Courtesy of H. Avet-Loiseau).

Figure 32.17. Kaplan Meier survival curves for 310 MM patients *(Avet-Loiseau 1)*. Extreme heavy/ light chain ratios (>200 or <0.01) were combined with β_2M >3.5mg/L to produce a three tiered risk stratification model. Patients with 0 (β_2M <3.5mg/L + low HLCr, blue), 1 (β_2M >3.5mg/L or extreme HLCr, green), or 2 (β_2M >3.5mg/L and extreme HLCr, red) risk factors had significantly different progression free survival times (p=0.000013). (Courtesy of H. Avet-Loiseau).

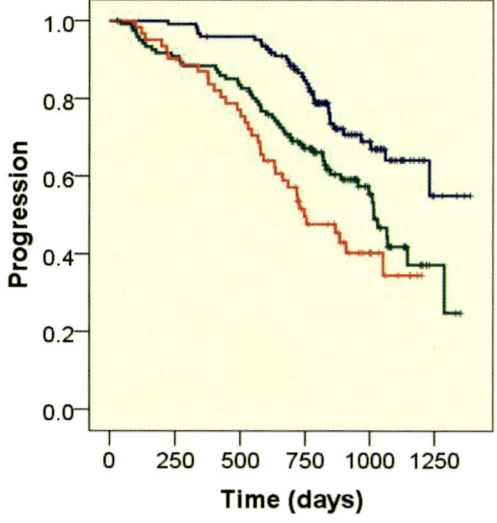

32.8. HLC assays in non-Hodgkin's lymphoma (NHL)

Quantitative abnormalities of sFLC and/or HLC were identified in 45/93 (48%) patients with NHL compared with 17/93 (18%) by SPE alone. The frequency of abnormalities varied markedly between disease i.e. 8/8 with Waldenström's Macroglobulinaemia, 13/20 (65%) with diffuse B cell lymphomas, 17/27 (63%) with marginal zone lymphomas and 5/17 (29%) with follicular lymphomas.

The most frequent abnormalities were in sFLC ratios (22/93) and IgDκ/IgDλ ratios (18/93). For both, the abnormalities predominantly indicated an excess production of κ clones (20/22 for κFLC and 17/18 for IgDκ) while only 1 patient had both sFLC and IgD abnormalities. HLC abnormalities were usually present (37/46 positive patients) in only one immunoglobulin class, as would be expected in monoclonal diseases. Abnormally low concentrations of IgM (with normal IgMκ/IgMλ ratios) were found in 28% (26/93) of the sera. This degree of immunoparesis was not seen with the other immunoglobulins or sFLCs. Further studies will determine whether sFLC ratios and HLC ratios have utility for prognosis or disease monitoring in NHL.

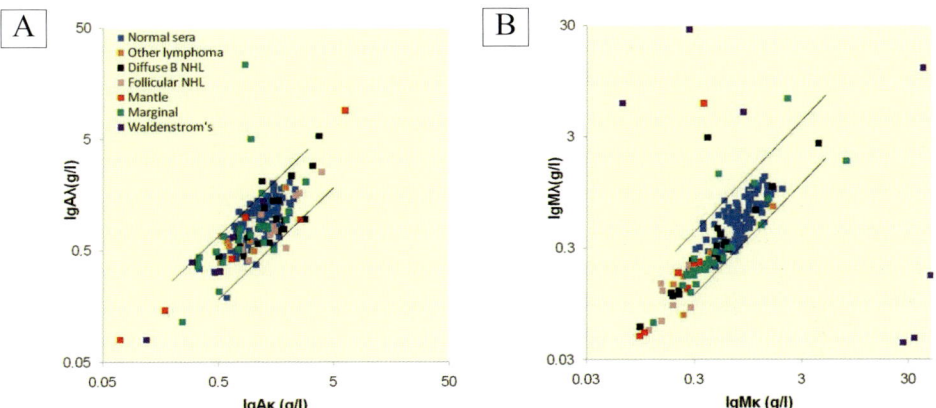

Figure 32.18. HLC assays for (A) IgA and (B) IgM in NHL (Patient identification for A + B are the same).(Courtesy of G. Pratt).

32.9. HLC assays for immunohistochemistry

HLC antibodies may also find use in immunohistochemistry *(Figures 32.19 and 32.20)* for assessing immunoglobulin light chain subsets. Commonly, lymph node, bone marrow and other tissue biopsies contain B-cells and plasma cells that need testing for clonality. Measurements of total light chain ratios are not always reliable and in any case, are not specific for each of the IgG, A, M, or D producing cells. These subsets can be evaluated using double staining techniques but such methods are cumbersome. HLC reagents labelled with peroxidase or fluorochromes could be used to provide sensitive analysis of clonality in a variety of immune derived tumours.

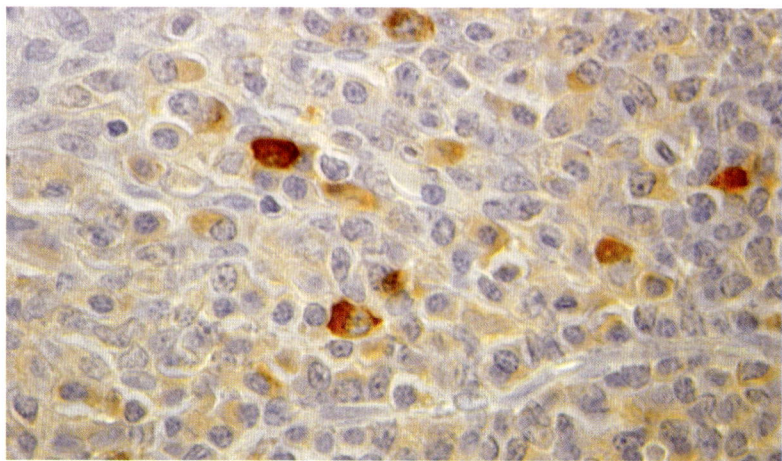

Figure 32.19. Immunohistochemistry of a lymph node stained with peroxidase-labelled IgGλ HLC reagent.

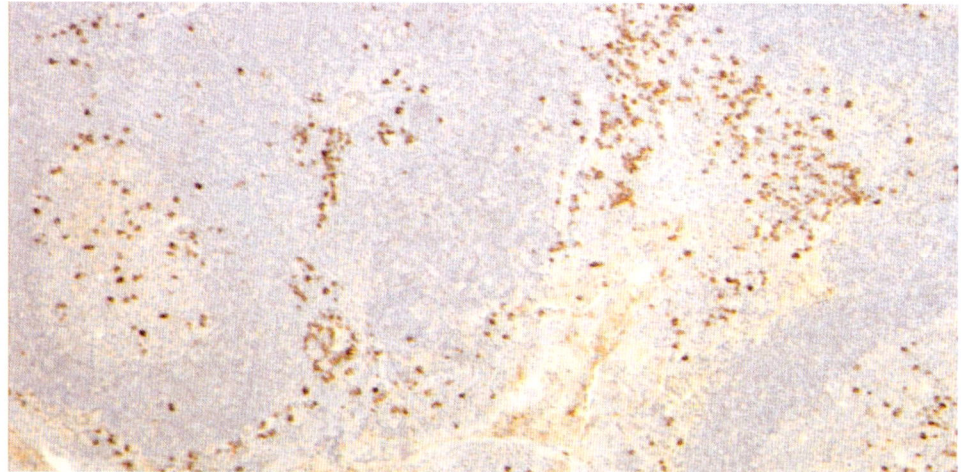

Figure 32.20. Immunohistochemistry of a lymph node stained with peroxidase labelled IgMκ HLC reagent.

32.10. Publications on FcRn receptors and related issues.

Akilesh S, Christianson GJ, Roopenian DC, Shaw AS. Neonatal FcR expression in bone marrow-derived cells functions to protect serum IgG from catabolism. J Immunol 2007; 179: 4580 – 8

Alexanian R. Blood volume in monoclonal gammopathy. Blood 1977; 49: 301 – 7

Anderson CL, Chaudhury C, Kim J, Bronson CL, Wani MA, Mohanty S. Perspective - FcRn transports albumin: relevance to immunology and medicine. Trends Immunol 2006; 27: 343 – 8 Chaudhury C,

Brooks CL, Carter DC, Robinson JM, Anderson CL. Albumin binding to FcRn: distinct from the FcRn-IgG interaction. Biochemistry 2006;45: 4983 – 90

Dingli D, Pacheco JM, Nowakowski GS, Kumar SK, Dispenzieri A, Hayman SR, et al. Relationship between depth of response and outcome in multiple myeloma. J Clin Oncol 2007; 25: 4933 – 7

Dingli D, Pacheco JM, Dispenzieri A, Hayman SR, Kumar SK, Lacy MQ, et al. Serum M-spike and transplant outcome in patients with multiple myeloma. Cancer Sci 2007; 98: 1035 – 40

Durie BG, Jacobson J, Barlogie B, Crowley J. Magnitude of response with myeloma frontline therapy does not predict outcome: importance of time to progression in southwest oncology group chemotherapy trials. J Clin Oncol 2004; 22: 1857 – 63

Gurbaxani B. Mathematical modeling as accounting: predicting the fate of serum proteins and therapeutic monoclonal antibodies. Clin Immunol 2007; 122: 121 – 4

Kim J, Hayton WL, Robinson JM, Anderson CL. Kinetics of FcRn-mediated recycling of IgG and albumin in human: pathophysiology and therapcutic implications using a simplified mechanism-based model. Clin Immunol 2007; 122: 146 – 55

Lencer WI, Blumberg RS. A passionate kiss, then run: exocytosis and recycling of IgG by FcRn. Trends Cell Biol 2005; 15: 5 – 9

MG, Wu SV, Walsh JH. Hormonal control of intestinal Fc receptor gene expression and immunoglobulin transport in suckling rats. J Clin Invest 1993; 91: 2844 – 9

Sachs UJ, Socher I, Braeunlich CG, Kroll H, Bein G, Santoso S. A variable number of tandem repeats polymorphism influences the transcriptional activity of the neonatal Fc receptor alpha-chain promoter. Immunology 2006; 119: 83 – 9

Salmon SE, Smith BA. Immunoglobulin synthesis and total body tumor cell number in IgG multiple myeloma. J Clin Invest 1970; 49: 1114 – 21

Sullivan PW, Salmon SE. Kinetics of tumor growth and regression in IgG multiple myeloma. J Clin Invest 1972; 51: 1697 – 708

Telleman P, Junghans RP. The role of the Brambell receptor (FcRB) in liver: protection of endocytosed immunoglobulin G (IgG) from catabolism in hepatocytes rather than transport of IgG to bile. Immunology 2000; 100: 245 – 51

Wang L, Young DC. Suppression of polyclonal immunoglobulin production by M-proteins shows isotype specificity. Ann Clin Lab Sci 2001; 31: 274 – 8

32.11. Publications on HLC assays.

1. Avet-Loiseau H, Harousseau JL, Moreau P, Mathiot C, Facon T, Attal M, et al. Heavy/light chain specific immunoglobulin ratios at presentation are prognostic for progression free survival in the IFM 2005-01 myeloma trial. Blood 2009;114:1818a.

2. Avet-Loiseau H, Mirbahai L, Harousseau J-L, Moreau M, Mathiot C, Facon T, et al. Serum immunoglobulin heavy/light chain ratios are independent risk factors for predicting progression free survival in multiple myeloma. EHA 2010; 953a

3. Bradwell AR, Harding S, Drayson M, Mead G. Novel nephelometric assays give a sensitive marker of residual disease in multiple myeloma (MM). Br J Haematol 2008;141:107a.

4. Bradwell AR, Harding S, Fourrier N, Harris J, Sharp K, Hobbs J, et al. Separate nephelometric immunoassays for IgA kappa and IgA lambda for the assessment of patients with multiple myeloma (MM). Clin Chem 2008;54:C116a.

5. Bradwell AR, Harding SJ, Fourrier NJ, Wallis GL, Drayson MT, Carr-Smith HD, Mead GP. Assessment of monoclonal gammopathies by nephelometric measurement of individual immunoglobulin kappa/lambda ratios. Clin Chem 2009;55:1646-55.

6. Fourrier NJB, Sharp K, Walsh P, Wallis GLF, Drayson MT, Bradwell AR, Harding S. Nephelometric immunoassay measurements of IgMκ and IgMλ for the assessment of patients with IgM monoclonal gammopathies. Clin Chem 2010;56:C120a.

7. Harding S, Drayson M, Hobbs J, Mead G, Bradwell AR. Analysis of the involved IgG kappa/IgG lambda ratios may give a more sensitive measure of response to treatment in multiple myeloma. Haematologica 2008;93:662a.

8. Harding S, Drayson M, Lachmann H, Hawkins P, Hobbs J, Mead G, Bradwell AR. Novel nephelometric immunoassays for the sensitive detection of IgA monoclonal gammopathies in multiple myeloma and AL amyloidosis. Haematologica 2008;93:668a.

9. Harding S, Harris J, Fourrier N, Drayson M, Mead G, Bradwell AR. Quantification of IgA kappa and IgA lambda in human serum using nephelometric assays. Clin Chem Lab Med 2009;47:M-B077a.

10. Harding S, Margetts C, Fourrier N, Drayson M, Mead G, Bradwell AR. Quantification of IgAκ/IgAλ in monoclonal gammopathies. Clin Lymphoma Myeloma 2009;February:073a.

11. Harding S, Mead G, Drayson M, Bradwell AR. Monitoring of residual disease in multiple myeloma (MM) patients using novel immunoglobulin assays. Ann Oncol 2008;19:538a.

12. Harding SJ, Alvi A, Margetts C, Plant T, Drayson M, Mead G, Bradwell AR. Immunoglobulin ratios: an alternative to immunofixation. Clin Lymphoma Myeloma 2009:596a.

13. Harding SJ, Drayson M, Hobbs J, Mead G, Bradwell AR. Response to treatment in multiple myeloma may be monitored more sensitively using novel IgGκ and IgGλ nephelometric assays. Hematology meeting reports 2008;2:F45.

14. Harding SJ, Drayson MT, Mead GP, Bradwell AR. Prognostic value of free and heavy/light chain analysis. Clin Lymphoma Myeloma 2009;February:555a.

15. Harding SJ, Margetts C, Bradwell AR, Mead G, Hawkins P, Lachmann H. Hevylite detects residual IgGk in IgG Heavy-chain disease. Clin Lymphoma Myeloma 2009;February:081a.

16. Harding SJ, Mead GP, Hobbs JAR, Drayson MT, Bradwell AR. Free light and Heavy/light chain monitoring in IgGL sera. Clin Lymphoma Myeloma 2009;February:225a.

17. Harris JC, Drayson M, Wood CR, Mitchell F, Harding SJ. Nephelometric assays for the quantification of IgA kappa and IgA lambda in human serum. Hematology meeting reports 2008;2:F44.

18. Katzmann J, Clark R, Dispenzieri A, Kyle R, Landgren O, Bradwell AR, Rajkumar SV. Isotype-specific heavy/light chain (HLC) suppression as a predictor of myeloma development in monoclonal gammopathy of undetermined significance (MGUS). Blood 2009;114:1788a.

19. Keren DF. Heavy/Light-chain analysis of monoclonal gammopathies. Clin Chem 2009;55:1606-8.

20. Margetts C, Drayson M, Sharp K, Harper J, Fourrier N, Harding S. Serial sample analysis of 3 IgA multiple myeloma patients using a novel immunoassay measuring IgA kappa and IgA lambda. Hematology meeting reports 2008;2:F46.

21. Margetts CD, Harding SJ, Drayson MT, Sharp K, Fourrier N, Mead GP, Bradwell AR. Serum IgGk/IgGL measurements in monoclonal gammopathies. Clin Lymphoma Myeloma 2009;February:260a.

22. Mead G, Harding S, Pratt G, Basu S, Jacob A, Beardsmore C, Bradwell AR. Serum immunoglobulin and free light chain abnormalities in non Hodgkin Lymphoma. Ann Oncol 2008;19:315a.

23. Wallis GF, Walsh P, White E, Fourrier N, Harding S, Mead G, Bradwell AR. Preparation of polyclonal immunoglobulin G and A reference material for calibration of nephelometric HevyLiteTM assays. Clin Chem 2008;54:C111a.

24. Wechalekar A, Harding S, Lachmann H, Gillmore J, Wassef NJ, Thomas M, et al. Serum immunoglobulin heavy/light chain ratios (Hevylite) in patients with systemic AL amyloidosis. Amyloid 2010;17:P186a.

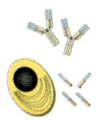

1) Abadie JM, Bankson DD. Assessment of serum free light chain assays for plasma cell disorder screening in a Veterans Affairs population. Ann Clin Lab Sci 2006;36:157-62.

2) Abadie JM, van Hoeven KH, Wells JM. Are renal reference intervals required when screening for plasma cell disorders with serum free light chains and serum protein electrophoresis? Am J Clin Pathol 2009;131:166-71.

Abdalla IA, Tabbara IA. Nonsecretory multiple myeloma. South Med J 2002;95:761-4.

Abdalla SH. Use of Freelite assay to monitor myeloma with renal failure. Haematologica 2007;92:PO1016a.

Abe M, Goto T, Kosaka M, Wolfenbarger D, Weiss DT, Solomon A. Differences in kappa to lambda (kappa:lambda) ratios of serum and urinary free light chains. Clin Exp Immunol 1998;111:457-62.

Abraham GN, Waterhouse C. Evidence for defective immunoglobulin metabolism in severe renal insufficiency. Am J Med Sci 1974;268:227-33.

1) Abraham RS, Bergen RH, Naylor S, Katzmann JA, Bradwell AR, Kyle RA, Fonseca R. Characterisation of free immunoglobulin light chains (LC) by mass spectroscopy in light chain-associated (AL) amyloidosis. Blood 2001;98:3772a.

2) Abraham RS, Charlesworth MC, Owen BA, Benson LM, Katzmann JA, Reeder CB, Kyle RA. Trimolecular complexes of lambda light chain dimers in serum of a patient with multiple myeloma. Clin Chem 2002;48:1805-11.

3) Abraham RS, Clark RJ, Bryant SC, Lymp JF, Larson T, Kyle RA, Katzmann JA. Correlation of serum immunoglobulin free light chain quantification with urinary Bence Jones protein in light chain myeloma. Clin Chem 2002;48:655-7.

4) Abraham RS, Katzmann JA, Clark RJ, Bradwell AR, Kyle RA, Gertz MA. Quantitative analysis of serum free light chains. A new marker for the diagnostic evaluation of primary systemic amyloidosis. Am J Clin Pathol 2003;119:274-8.

Aggarwal R, Sequeira W, Mikolaitis R, Kokebie R, Block JA, Jolly M. Measurement of serum free light chains performs better than known immunological biomarkers for systemic lupus erythematosus disease activity Presented at ACR/AHRP 2009:916a.

1) Akar H, Seldin DC, Magnani B, O'Hara C, Berk JL, Schoonmaker C, et al. Quantitative serum free light chain assay in the diagnostic evaluation of AL amyloidosis. Amyloid 2005;12:210-5.

2) Akar H, Seldin DC, Magnani B, O'Hara C, Berk JL, Schoonmaker C, et al. Quantitative serum free light-chain assay in the diagnostic evaluation of AL amyloidosis. In: Grateau G, Kyle RA, Skinner M, eds. Amyloid and Amyloidosis, CRC Press, 2005.

1) Akilesh S, Christianson GJ, Roopenian DC, Shaw AS. Neonatal FcR expression in bone marrow-derived cells functions to protect serum IgG from catabolism. J Immunol 2007;179:4580-8.

2) Akilesh S, Huber TB, Wu H, Wang G, Hartleben B, Kopp JB, et al. Podocytes use FcRn to clear IgG from the glomerular basement membrane. Proc Natl Acad Sci U S A 2008;105:967-72.

Ala-Houhala I, Parviainen MT, Pasternack A. A comparison of three different methods of concentration of urinary proteins. Clin Chim Acta 1984;142:339-42.

Alexanian R. Blood volume in monoclonal gammapathy. Blood 1977;49:301-7.

Alyanakian MA, Abbas A, Delarue R, Arnulf B, Aucouturier P. Free immunoglobulin light-chain serum levels in the follow-up of patients with monoclonal gammopathies: correlation with 24-hr urinary light-chain excretion. Am J Hematol 2004;75:246-8.

Anderson CL, Chaudhury C, Kim J, Bronson CL, Wani MA, Mohanty S. Perspective-- FcRn transports albumin: relevance to immunology and medicine. Trends Immunol 2006;27:343-8.

Anderson KC, Alsina M, Bensinger W, Biermann JS, Chanan-Khan A, Comenzo RL, et al. Multiple myeloma. Clinical practice guidelines in oncology. J Natl Compr Canc Netw 2007;5:118-47.

Arfors KE, Rutili G, Svensjo E. Microvascular transport of macromolecules in normal and inflammatory conditions. Acta Physiol Scand Suppl 1979;463:93-103.

Arneth B, Birklein F. High sensitivity of free lambda and free kappa light chains for detection of intrathecal immunoglobulin synthesis in cerebrospinal fluid. Acta Neurol Scand 2009;119:39-44.

Attaelmannan M, Levinson SS. Understanding and identifying monoclonal gammopathies. Clin Chem 2000;46:1230-8.

1) Augustson BM, Katsavara H, Reid SD, Mead GP, Shirfield M, Bradwell AR. Monoclonal gammopathy screening: Improved sensitivity using the serum free light chain assay. Haematologica 2005;90 PO1302a.

2) Augustson BM, Reid SD, Mead GP, Drayson MT, Child JA, Bradwell AR. Serum free light chain levels in asymptomatic myeloma. Blood 2004;104:4880a.

1) Avet-Loiseau H, Harousseau JL, Moreau P, Mathiot C, Facon T, Attal M, et al. Heavy/light chain specific immunoglobulin ratios at presentation are prognostic for progression free survival in the IFM 2005-01 myeloma trial. Blood 2009;114:1818a.

2) Avet-Loiseau H, Mirbahai L, Harousseau J-L, Moreau M, Mathiot C, Facon T, et al. Serum immunoglobulin heavy/light chain ratios are independent risk factors for predicting progression free survival in multiple myeloma. EHA 2010; 953a.

Ayliffe MJ, Davies FE, de Castro D, Morgan GJ. Demonstration of changes in plasma cell subsets in multiple myeloma. Haematologica 2007;92:1135-8.

Bachmann U, Schindler R, Storr M, Kahl A, Joerres A, Sturm I. Combination of bortezomib-based chemotherapy and extracorporeal free light chain removal for treating cast nephropathy in multiple myeloma. Nephrol Dial Transplant 2008;1:106-8.

Bailey EM, McDermott TJ, Bloch KJ. The urinary light-chain ladder pattern. A product of improved methodology that may complicate the recognition of Bence Jones proteinuria. Arch Pathol Lab Med 1993;117:707-10.

Bakshi NA, Gulbranson R, Garstka D, Bradwell AR, Keren DF. Serum free light chain (FLC) measurement can aid capillary zone electrophoresis in detecting subtle FLC-producing M proteins. Am J Clin Pathol 2005;124:214-8.

Baldini L, Guffanti A, Cesana BM, Colombi M, Chiorboli O, Damilano I, Maiolo AT. Role of different hematologic variables in defining the risk of malignant transformation in monoclonal gammopathy. Blood 1996;87:912-8.

Basnayake K, Ying WZ, Wang PX, Sanders PW. Immunoglobulin light chains activate tubular epithelial cells through redox signaling. J Am Soc Nephrol 2010;21:1165-73

Bayne-Jones S, Wilson DW. Immunological reactions of Bence-Jones proteins. II. Differences between Bence-Jones proteins from various sources. Bulletin of the John Hopkins Hospital 1922;33:119-25.

Beetham R, Wassell J, Wallage MJ, Whiteway AJ, James JA. Can serum free light chains replace urine electrophoresis in the detection of monoclonal gammopathies? Ann Clin Biochem 2007;44:516-22.

Berenson JR, Anderson KC, Audell RA, Boccia RV, Coleman M, Dimopoulos MA, et al. Monoclonal gammopathy of undetermined significance: a consensus statement. Br J Haematol 2010; 150:28-38.

Berggard I, Peterson PA. Polymeric forms of free normal kappa and lambda chains of human immunoglobulin. J Biol Chem 1969;244:4299-307.

Bergner R, Hoffmann M, Uppenkamp M. Reply. Kastritis E. et al. Reversibility of renal failure in newly diagnosed multiple myeloma patients treated with high dose dexamethasone containing regimes and the impact of novel agents. Haematologica 2008;93:e18-9.

Bidart JM, Thuillier F, Augereau C, Chalas J, Daver A, Jacob N, et al. Kinetics of serum tumor marker concentrations and usefulness in clinical monitoring. Clin Chem 1999;45:1695-707.

1) Bird JM, Cavenagh J, Samson D, Mehta A, Hawkins P, Lachmann H. Guidelines on the diagnosis and management of AL amyloidosis. Br J Haematol 2004;125:681-700.

2) Bird J, Behrens J, Westin J, Turesson I, Drayson M, Beetham R, et al. UK Myeloma Forum (UKMF) and Nordic Myeloma Study Group (NMSG): guidelines for the investigation of newly detected M-proteins and the management of monoclonal gammopathy of undetermined significance (MGUS). Br J Haematol 2009;147:22-42.

1) Bladé J. Management of renal, hematologic, and infectious complications. In: Malpas JS, Bergsagel DE, Kyle RA, Anderson KC, eds. Myeloma: Biology and management, Saunders, 2004.

2) Bladé J. Clinical practice. Monoclonal gammopathy of undetermined significance. N Engl J Med 2006;355:2765-70.

3) Bladé J, Kyle RA. Nonsecretory myeloma, immunoglobulin D myeloma and plasma cell leukemia. Hematol/Oncology Clinics of North America 1999;13:1259-72.

4) Bladé J, Montoto S, Rosinol L, Montserrat E. Appropriateness of applying the response criteria for multiple myeloma to Waldenstrom's macroglobulinemia? Semin Oncol 2003;30:329-31.

5) Bladé J, Rosinol L. SMM: Towards pretter predictors of progression. Blood 2008;111:479-80.

6) Bladé J, Samson D, Reece D, Apperley J, Bjorkstrand B, Gahrton G, et al. Criteria for evaluating disease response and progression in patients with multiple myeloma treated by high-dose therapy and haemopoietic stem cell transplantation. Myeloma Subcommittee of the EBMT. European Group for Blood and Marrow Transplant. Br J Haematol 1998;102:1115-23.

7) Bladé J, Dimopoulos M, Rosinol L, Rajkumar SV, Kyle RA. Smoldering (asymptomatic) multiple myeloma: current diagnostic criteria, new predictors of outcome, and follow-up recommendations. J Clin Oncol 2010;28:690-7

Bochtler T, Hegenbart U, Heiss C, Benner A, Cremer F, Volkmann M, et al. Evaluation of the serum-free light chain test in untreated patients with AL amyloidosis. Haematologica 2008;93:459-62.

Boege F. Measuring Bence Jones proteins with antibodies against bound immunoglobulin light-chains: how reliable are the results? Eur J Clin Chem Clin Biochem 1993;31:403-5.

1) Bossuyt X, Bogaerts A, Schiettekatte G, Blanckaert N. Detection and classification of paraproteins by capillary immunofixation/subtraction. Clin Chem 1998;44:760-4.

2) Bossuyt X, Marien G. False-negative results in detection of monoclonal proteins by capillary zone electrophoresis: a prospective study. Clin Chem 2001;47:1477-9.

3) Bossuyt X, Schiettekatte G, Bogaerts A, Blanckaert N. Serum protein electrophoresis by CZE 2000 clinical capillary electrophoresis system. Clin Chem 1998;44:749-59.

1) Bradwell AR, Carr-Smith HD, Mead GP, Harvey TC, Drayson MT. Serum test for assessment of patients with Bence Jones myeloma. Lancet 2003;361:489-91.

2) Bradwell AR, Carr-Smith HD, Mead GP, Tang LX, Showell PJ, Drayson MT, Drew R. Highly sensitive, automated immunoassay for immunoglobulin free light chains in serum and urine. Clin Chem 2001;47:673-80.

3) Bradwell AR, Drayson MT, Mead GP. Measurement of free light chains in urine (letter - reply). Clin Chem 2001;47:2069-70.

4) Bradwell AR, Galvin GP, Mead GP, Carr-Smith HD. Practical use of serial measurements of serum free light chains to monitor response to treatment of multiple myeloma. Blood 2003;102:5234a.

5) Bradwell AR, Mead GP, Drayson MT, Carr-Smith HD. Serum immunoglobulin free light chain measurement in intact immunoglobulin myeloma. Blood 2002;100:5054a.

6) Bradwell AR. Editorial: Clinical importance of serum free light chain analysis. Personalized Medicine 2010;7:229-31.

7) Bradwell AR, Harding SJ, Fourrier NJ, Wallis GL, Drayson MT, Carr-Smith HD, Mead GP. Assessment of monoclonal gammopathies by nephelometric measurement of individual immunoglobulin kappa/lambda ratios. Clin Chem 2009;55:1646-55.

1) **Brigden ML**, Neal ED, McNeely MD, Hoag GN. The optimum urine collections for the detection and monitoring of Bence Jones proteinuria. Am J Clin Pathol 1990;93:689-93.

2) **Brigden ML**, Webber D. Clinical Pathology Rounds: The case of anaplastic carcinoma that was not - potential problems in the interpretation of monoclonal proteins. Lab Med 2000;31:661-5.

Brockhurst I, Harris KP, Chapman CS. Diagnosis and monitoring a case of light-chain deposition disease in the kidney using a new, sensitive immunoassay. Nephrol Dial Transplant 2005;20:1251-3.

Brouwer J, Otting-van de Ruit M, Busking-van der Lely H. Estimation of free light chains of immunoglobulins by enzyme immunoassay. Clin Chim Acta 1985;150:267-74.

Brown LM, Gridley G, Check D, Landgren O. Risk of multiple myeloma and monoclonal gammopathy of undetermined significance among white and black male United States veterans with prior autoimmune, infectious, inflammatory, and allergic disorders. Blood 2008;111:3388-94.

1) **Buxbaum JN**, Chuba JV, Hellman GC, Solomon A, Gallo GR. Monoclonal immunoglobulin deposition disease: light chain and light and heavy chain deposition diseases and their relation to light chain amyloidosis. Clinical features, immunopathology, and molecular analysis. Ann Intern Med 1990;112:455-64.

2) **Buxbaum J**, Gallo G. Nonamyloidotic monoclonal immunoglobulin deposition disease. Light-chain, heavy-chain, and light- and heavy-chain deposition diseases. Hematol Oncol Clin North Am 1999;13:1235-48.

1) **Carr Smith HD**, Edwards J, Showell P, Drew R, Tang LX, Bradwell AR. Preparation of an immunoglobulin free light-chain reference material. Clin Chem 2000;46:699a.

2) **Carr-Smith HD**, Harland B, Anderson J, Overton J, Wieringa G, Bradwell AR. The effect on laboratory organisation of introducing serum free light chain assays. Clin Chem 2004;50:A76a.

3) **Carr-Smith HD**, Harland B, Anderson J, Overton J, Wieringa G, Bradwell AR. Evaluation of latex-enhanced turbidimetric reagents for measuring free immunoglobulin light chains on the Bayer Advia 1650. Clin Chem 2004;50:C44a.

4) **Carr-Smith HD**, Mead GP, Bradwell AR. Serum free light chain assays as a replacement for urine electrophoresis. Haematologica 2005;90:PO404a.

5) **Carr-Smith HD**, Showell P, Bradwell AR. Antigen excess assessment of free light chain assays on the Dade-Behring BNII nephelometer. Clin Chem 2002;48:23a.

6) **Carr-Smith HD**. Production Director, Binding Site Group Ltd. Personal communication 2008.

Cavallo F, Rasmussen E, Zangari M, Tricot G, Fender B, Fox M, et al. Serum free-lite chain (sFLC) assay in Multiple Myeloma (MM): Clinical correlates and prognostic implications in newly diagnosed MM patients treated with Total Therapy 2 or 3 (TT2/3). Blood 2005;106:3490a.

Cesana C, Klersy C, Barbarano L, Nosari AM, Crugnola M, Pungolino E, et al. Prognostic factors for malignant transformation in monoclonal gammopathy of undetermined significance and smoldering multiple myeloma. J Clin Oncol 2002;20:1625-34.

Chan DW, Schwartz MK. Tumor markers: Introduction and general principles. In: Diamandis EP, Fritsche HA, Lilja H, Chan DW, Schwartz MK, eds. Tumor markers: Physiology, pathology, technology and clinical applications: AACC Press, 2002.

Chanan-Khan AA, Kaufman JL, Mehta J, Richardson PG, Miller KC, Lonial S, et al. Activity and safety of bortezomib in multiple myeloma patients with advanced renal failure: a multicenter retrospective study. Blood 2007;109:2604-6.

Chaudhury C, Mehnaz S, Robinson JM, Hayton WL, Pearl DK, Roopenian DC, Anderson CL. The major histocompatibility complex-related Fc receptor for IgG (FcRn) binds albumin and prolongs its lifespan. J Exp Med 2003;197:315-22.

Clamp JR. Some aspects of the first recorded case of multiple myeloma. Lancet 1967;2:1354-6.

Clark RJ, Lockington KS, Tostrud LJ, Katzmann JA. Incidence of antigen excess in serum free light chain assays. Clin Chem 2007;53:C145a.

Clark WF, Stewart AK, Rock GA, Sternbach M, Sutton DM, Barrett BJ, et al. Plasma exchange when myeloma presents as acute renal failure: a randomized, controlled trial. Ann Intern Med 2005;143:777-84.

Cockcroft DW, Gault MH. Prediction of creatinine clearance from serum creatinine. Nephron 1976;16:31-41.

1) **Cohen AD**, Zhou P, Chou J, Teruya-Feldstein J, Reich L, Hassoun H, et al. Risk-adapted autologous stem cell transplantation with adjuvant dexamethasone +/- thalidomide for systemic light-chain amyloidosis: results of a phase II trial. Br J Haematol 2007;139:224-33.

2) **Cohen AD**, Zhou P, Xiao Q, Fleisher M, Kalakonda N, Akhurst T, et al. Systemic AL amyloidosis due to non-Hodgkin's lymphoma: an unusual clinicopathologic association. Br J Haematol 2004;124:309-14.

1) **Cohen G**, Rudnicki M, Horl WH. Uraemic toxins modulate the spontaneous apoptotic cell death and essential function of neutrophils. Kidney Int 2001;59:S48-S52.

2) **Cohen G**, Rudnicki M, Schmaldienst S, Horl WH. Effect of dialysis on serum/plasma levels of free immunoglobulin light chains in end-stage renal disease patients. Nephrol Dial Transplant 2002;17:879-83.

Cole PW, Durie BG, Salmon SE. Immunoquantitation of free light chain immunoglobulins: applications in multiple myeloma. J Immunol Methods 1978;19:341-9.

Comenzo RL, Gertz MA. Autologous stem cell transplantation for primary systemic amyloidosis. Blood 2002;99:4276-82.

Corazzelli G, De Filippi R, Capobianco G, Frigeri F, De Rosa V, Iaccarino G, et al. Tumor flare reactions and response to lenalidomide in patients with refractory classic Hodgkin lymphoma. Am J Hematol 2010;85:87-90.

Corlu D, Weaver K, Schell M, Eulitz M, Murphy CL, Weiss DT, Solomon A. A molecular basis for nonsecretory myeloma. Blood 2004;104:829-31.

Cserti C, Haspel R, Stowell C, Dzik W. Light-chain removal by plasmapheresis in myeloma-associated renal failure. Transfusion 2007;47:511-6.

Das M, Mead GP, Sreekanth V, Anderson J, Blair S, Howe T, et al. Serum free light chain (sFLC) concentration kinetics in patients receiving Bortezomib: Temporary inhibition of protein synthesis and early biomarker for disease response. Blood 2005;106:5094a.

Daval S, Tridon A, Mazeron N, Ristori JM, Evrard B. Risk of antigen excess in serum free light chain measurements. Clin Chem 2007;53:1985-6.

Dawson MA, Patil S, Spencer A. Extramedullary relapse of multiple myeloma associated with a shift in secretion from intact immunoglobulin to light chains. Haematologica 2007;92:143-4.

1) De Filippi R, Iaccarino G, Frigeri F, Di Francia R, Crisci S, Capobianco G, et al. Elevation of clonal serum free light chains in patients with HIV-negative primary effusion lymphoma (PEL) and PEL-like lymphoma. Br J Haematol 2009.

2) De Filippi R, Russo F, Iaccarino G, Crisci S, Frigeri F, Riemma C, et al. Abnormally elevated levels of serum free-immunoglobulin light chains are frequently found in classic Hodgkin Lymphoma (cHL) and predict outcome of patients with early stage disease Presented at ASH 2009 2009:267a.

DeCarli C, Menegus MA, Rudick RA. Free light chains in multiple sclerosis and infections of the CNS. Neurology 1987;37:1334-8.

Deegan MJ, Abraham JP, Sawdyk M, Van Slyck EJ. High incidence of monoclonal proteins in the serum and urine of chronic lymphocytic leukemia patients. Blood 1984;64:1207-11.

Deinum J, Derkx FH. Cystatin for estimation of glomerular filtration rate? Lancet 2000;356:1624-5.

Desplat-Jego S, Feuillet L, Pelletier J, Bernard D, Cherif AA, Boucraut J. Quantification of immunoglobulin free light chains in cerebrospinal fluid by nephelometry. J Clin Immunol 2005;25:338-45.

Diemert MC, Musset L, Gaillard O, Escolano S, Baumelou A, Rousselet F, Galli J. Electrophoretic study of the physico-chemical characteristics of Bence-Jones proteinuria and its association with kidney damage. J Clin Pathol 1994;47:1090-7.

Dillon JJ, Sedmak DD, Cosio FG. Rapid-onset diabetic nephropathy in type II diabetes mellitus. Ren Fail 1997;19:819-22.

1) Dimopoulos MA, Moulopoulos A, Smith T, Delasalle KB, Alexanian R. Risk of disease progression in asymptomatic multiple myeloma. Am J Med 1993;94:57-61.

2) Dimopoulos MA, Roussou M, Gavriatopoulou M, Zagouri F, Migkou M, Matsouka C, et al. Reversibility of renal impairment in patients with multiple myeloma treated with bortezomib-based regimens: identification of predictive factors. Clin Lymphoma Myeloma 2009;9:302-6.

1) Dingli D, Kyle RA, Rajkumar SV, Nowakowski GS, Larson DR, Bida JP, et al. Immunoglobulin free light chains and solitary plasmacytoma of bone. Blood 2006;108:1979-83.

2) Dingli D, Larson DR, Plevak MF, Grande JP, Kyle RA. Focal and segmental glomerulosclerosis and plasma cell proliferative disorders. Am J Kidney Dis 2005;46:278-82.

3) Dingli D, Pacheco JM, Dispenzieri A, Hayman SR, Kumar SK, Lacy MQ, et al. Serum M-spike and transplant outcome in patients with multiple myeloma. Cancer Sci 2007;98:1035-40.

1) Dispenzieri A, Gertz MA, Kyle RA. Response: Determining appropriate treatment options for patients with primary systemic amyloidosis. Blood 2004;104:2992 - 3.

2) Dispenzieri A, Gertz MA, Therneau TM, Kyle RA. Retrospective cohort study of 148 patients with polyclonal gammopathy. Mayo Clin Proc 2001;76:476-87.

3) Dispenzieri A, Kyle RA, Katzmann JA, Therneau TM, Larson D, Benson J, et al. Immunoglobulin free light chain ratio is an independent risk factor for progression of smoldering (asymptomatic) multiple myeloma. Blood 2008;111:785-9.

4) Dispenzieri A, Lacy MQ, Katzmann JA, Rajkumar SV, Abraham RS, Hayman SR, et al. Absolute values of immunoglobulin free light chains are prognostic in patients with primary systemic amyloidosis undergoing peripheral blood stem cell transplantation. Blood 2006;107:3378-83.

5) Dispenzieri A, Rajkumar SV, Plevak MF, Katzmann JA, Kyle RA, Larson D, et al. Early immunoglobulin free light chain (FLC) response post autologous peripheral blood stem cell transplant predicts for hematologic complete response in patients with multiple myeloma. Blood 2006;108:3097a.

6) Dispenzieri A, Zhang L, Katzmann JA, Snyder M, Blood E, Degoey R, et al. Appraisal of immunoglobulin free light chain as a marker of response. Blood 2008;111:4908-15.

7) Dispenzieri A, Kyle R, Merlini G, Miguel JS, Ludwig H, Hajek R, et al. International Myeloma Working Group guidelines for serum-free light chain analysis in multiple myeloma and related disorders. Leukemia 2009;23:215-24.

8) Dispenzieri A, Katzmann JA, Kyle RA, Larson DR, Melton LJ, 3rd, Colby CL, et al. Prevalence and risk of progression of light-chain monoclonal gammopathy of undetermined significance: a retrospective population-based cohort study. Lancet 2010;375:1721-8.

1) Drayson M, Begum G, Basu S, Makkuni S, Dunn J, Barth N, Child JA. Effects of paraprotein heavy and light chain types and free light chain load on survival in myeloma: an analysis of patients receiving conventional-dose chemotherapy in Medical Research Council UK multiple myeloma trials. Blood 2006;108:2013-9.

2) Drayson M, Tang LX, Drew R, Mead GP, Carr-Smith H, Bradwell AR. Serum free light-chain measurements for identifying and monitoring patients with nonsecretory multiple myeloma. Blood 2001;97:2900-2.

3) Drayson MT, Morgan GJ, Jackson GH, Davies FE, Owen RG, Ross FM, et al. Prospective study of serum FLC and other M-protein assays: when and how to measure response? Clin Lymphoma Myeloma 2009;9:A346a.

Dreicer R, Alexanian R. Nonsecretory multiple myeloma. Am J Hematol 1982;13:313-8.

Drengler T, Kumar S, Gertz M, Lacy M, Katzmann J, Hayman S, et al. Serum immunoglobulin free light chain measurements provide insight into disease biology in patients with POEMS syndrome. Proceedings of Portuguese myeloma meeting 2008.

1) Durie BG, Harousseau JL, Miguel JS, Blade J, Barlogie B, Anderson K, et al. International uniform response criteria for multiple myeloma. Leukemia 2006;20:1467-73.

2) Durie BG, Jacobson J, Barlogie B, Crowley J. Magnitude of response with myeloma frontline therapy does not predict outcome: importance of time to progression in southwest oncology group chemotherapy trials. J Clin Oncol 2004;22:1857-63.

Edelman GM, Gally JA. The nature of Bence-Jones proteins. Chemical similarities to polypetide chains of myeloma globulins and normal gamma-globulins. J Exp Med 1962;116:207-27.

Émond JP, Harding S, Lemieux B. Aggregation of serum free light chains (FLC) causes overestimation of FLC nephelometric results as compared to serum protein electrophoresis (SPE) while preserving clinical usefulness. Blood 2007;110:4767a.

Epstein WV, Tan M. Increase of L-chain proteins in the sera of patients with systemic lupus erythematosus and the synovial fluids of patients with peripheral rheumatoid arthritis. Arthritis Rheum 1966;9:713-9.

Evans LS, Hancock BW. Non-Hodgkin lymphoma. Lancet 2003;362:139-46.

Fine JD, Rees ED. Letter: Bence-Jones protein: detection and implications. N Engl J Med 1974;290:106-7.

Fischer C, Arneth B, Koehler J, Lotz J, Lackner KJ. Kappa free light chains in cerebrospinal fluid as markers of intrathecal immunoglobulin synthesis. Clin Chem 2004;50:1809-13.

Foray V, Chapuis-Cellier C. Contribution of serum free light chain immunoassays in diagnosis and monitoring of free light chain monoclonal gammapathies. Immuno-analyse et Biologie Specialisee 2005;20:385-93.

Fulton RB, Fernando SL. Serum free light chain assay reduces the need for serum and urine immunofixation electrophoresis in the evaluation of monoclonal gammopathy. Ann Clin Biochem 2009;46:407-12.

1) Gertz MA, Comenzo R, Falk RH, Fermand JP, Hazenberg BP, Hawkins PN, et al. Definition of organ involvement and treatment response in immunoglobulin light chain amyloidosis (AL): a consensus opinion from the 10th International Symposium on Amyloid and Amyloidosis, Tours, France, 18-22 April 2004. Am J Hematol 2005;79:319-28.

2) Gertz MA, Lacy MQ, Dispenzieri A, Hayman SR. Amyloidosis: diagnosis and management. Clin Lymphoma Myeloma 2005;6:208-19.

Giarin MM, Giaccone L, Caracciolo D, Bruno B, Falco P, Omedè P, et al. Serum free light chains (SFLC) assay: a suggestive new criteria for evaluating disease response, progression and relapse in plasma-cell disorders (PD) and a prognostic factor in monoclonal gammopathy of undetermined significance (MGUS). Haematologica 2006;91:PO151a.

Gillmore JD, Goodman HJ, Lachmann HJ, Offer M, Wechalekar AD, Joshi J, et al. Sequential heart and autologous stem cell transplantation for systemic AL amyloidosis. Blood 2006;107:1227-9.

1) Goodman HJ, Bridoux F, Lachmann HJ, Gilbertson JA, Gallimore R, Joshi J, et al. Localised amyloidosis: Clinical features and outcomes in 235 cases. Haematologica 2005;90:PO1413a.

2) Goodman HJ, Gillmore JD, Lachmann HJ, Wechalekar AD, Bradwell AR, Hawkins PN. Outcome of autologous stem cell transplantation for AL amyloidosis in the UK. Br J Haematol 2006;134:417-25.

3) Goodman HJB, Wechalekar AD, Lachmann HJ, Bradwell AR, Hawkins PN. Clonal disease response and clinical outcome in 229 patients with AL amyloidosis treated with VAD-like chemotherapy. Haematologica 2005;90:PO1408a.

1) Gottenberg JE, Aucouturier F, Goetz J, Sordet C, Jahn I, Busson M, et al. Serum immunoglobulin free light chain assessment in rheumatoid arthritis and primary Sjogren's syndrome. Ann Rheum Dis 2007;66:23-7.

2) Gottenberg JE, Miceli-Richard C, Ducot B, Goupille P, Combe B, Mariette X. Markers of B-lymphocyte activation are elevated in patients with early rheumatoid arthritis and correlated with disease activity in the ESPOIR cohort. Arthritis Res Ther 2009;11:R114.

Grabar P, Williams CA. [Method permitting the combined study of the electrophoretic and the immunochemical properties of protein mixtures; application to blood serum.]. Biochim Biophys Acta 1953;10:193-4.

Graziani MS, Merlini G. Measurement of free light chains in urine. Clin Chem 2001;47:2069-70.

Greipp PR, San Miguel J, Durie BG, Crowley JJ, Barlogie B, Blade J, et al. International staging system for multiple myeloma. J Clin Oncol 2005;23:3412-20.

Groop L, Makipernaa A, Stenman S, DeFronzo RA, Teppo AM. Urinary excretion of kappa light chains in patients with diabetes mellitus. Kidney Int 1990;37:1120-5.

Guinan JE, Kenny DF, Gatenby PA. Detection and typing of paraproteins: comparison of different methods in a routine diagnostic laboratory. Pathology 1989;21:35-41.

Guis L, Diemert MC, Ghillani P, Choquet S, Leblond V, Vernant JP, Musset L. The quantitation of serum free light chains: Three case reports. Clin Chem 2004;50:F38a.

Gupta S, Comenzo RL, Hoffman BR, Fleisher M. National Academy of Clinical Biochemistry guidelines for the use of tumor markers in monoclonal gammopathies. http://www.aacc.org/sitecollectiondocuments/nacb/lmpg/tumor/chp3k_gammopathies.pdf

Hajdu SI. A note from history: the first biochemical test for detection of cancer. Ann Clin Lab Sci 2006;36:222-3.

Harding SJ, Mead GP, Bradwell AR, Berard AM. Serum free light chain immunoassay as an adjunct to serum protein electrophoresis and immunofixation electrophoresis in the detection of multiple myeloma and other B-cell malignancies. Clin Chem Lab Med 2009;47:302-4.

Harris J, Tang LX, Showell PJ, Carr-Smith HD, Drew R, Bradwell AR. Assays for immunoglobulin free light chains in serum on the Beckman IMMAGE™. Clin Chem 2000;46:A180a.

Harousseau JL, Dreyling M. Multiple myeloma: ESMO clinical recommendations for diagnosis, treatment and follow-up. Ann Oncol 2009;20:97-9.

Harrison HH. The "ladder light chain" or "pseudo-oligoclonal" pattern in urinary immunofixation electrophoresis (IFE) studies: a distinctive IFE pattern and an explanatory hypothesis relating it to free polyclonal light chains. Clin Chem 1991;37:1559-64.

1) Hassoun H, Rafferty BT, Flombaum C, D'Agati VD, Klimek VM, Cohen A, et al. High dose chemotherapy and autologous stem cell transplantation with melphalan in patients with monoclonal immunoglobulin deposition disease associated with Multiple Myeloma. Blood 2007;110:5113a.

2) **Hassoun H**, Reich L, Klimek VM, Dhodapkar M, Cohen A, Kewalramani T, et al. The serum free light chain ratio after one or two cycles of treatment is highly predictive of the magnitude of final response in patients undergoing initial treatment for Multiple Myeloma. Blood 2005;106:972a.

3) **Hassoun H**, Reich L, Klimek VM, Dhodapkar M, Cohen A, Kewalramani T, et al. Doxorubicin and dexamethasone followed by thalidomide and dexamethasone is an effective well tolerated initial therapy for multiple myeloma. Br J Haematol 2006;132:155-61.

Haymann JP, Levraud JP, Bouet S, Kappes V, Hagege J, Nguyen G, et al. Characterization and localization of the neonatal Fc receptor in adult human kidney. J Am Soc Nephrol 2000;11:632-9.

Heino J, Rajamaki A, Irjala K. Turbidimetric measurement of Bence-Jones proteins using antibodies against free light chains of immunoglobulins. An artifact caused by different polymeric forms of light chains. Scand J Clin Lab Invest 1984;44:173-6.

1) **Hemmingsen L**, Skaarup P. The 24-hour excretion of plasma proteins in the urine of apparently healthy subjects. Scand J Clin Lab Invest 1975;35:347-53.

2) **Hemmingsen L**, Skaarup P. Urinary excretion of ten plasma proteins in patients with febrile diseases. Acta Med Scand 1977;201:359-64.

1) **Herrera GA**, Joseph L, Gu X, Hough A, Barlogie B. Renal pathologic spectrum in an autopsy series of patients with plasma cell dyscrasia. Arch Pathol Lab Med 2004;128:875-9.

2) **Herrera GA**, Sanders PW. Paraproteinemic Renal Diseases that Involve the Tubulo-Interstitium. Contrib Nephrol 2007;153:105-15.

Herzog W, Hofmann W. Detection of free kappa and lambda light chains in serum and urine in patients with monoclonal gammopathy. Blood 2003;102:5190a.

Herzum I, Heinz R, Bruder-Burzlaff B, Renz H, Wahl HG. Reliability of the new freelite assay for quantification of free light chains in urine. Clin Chem 2004;50:C36a.

Hess PP, Mastropaolo W, Thompson GD, Levinson SS. Interference of polyclonal free light chains with identification of Bence Jones proteins. Clin Chem 1993;39:1734-8.

Higgins T, Cho C, Dayton J. Evaluation of a method to measure free light chains using Bayer Advia 1650 chemistry analyzer. Clin Chem Lab Med 2007;45:M316a.

Hildebrandt B, Muller C, Pezzutto A, Daniel P, Dorken B, Scholz C. Assessment of free light chains in the cerebrospinal fluid of patients with lymphomatous meningitis - a pilot study. BMC Cancer 2007;7:185.

Hill PG, Forsyth JM, Rai B, Mayne S. Serum free light chains: an alternative to the urine Bence Jones proteins screening test for monoclonal gammopathies. Clin Chem 2006;52:1743-8.

1) **Hobbs JAR**, Sharp K, Harding SJ, Drayson M, Bradwell AR, Mead GP. Incidence of light chain escape in UK MRC myeloma VII trial Haematologica 2008;93:672a.

2) **Hobbs JA**, Drayson MT, Sharp K, Harding S, Bradwell AR, Mead GP. Frequency of altered monoclonal protein production at relapse of multiple myeloma. Br J Haematol 2009;148:659 - 61.

Hobbs JR. Modes of escape from therapeutic control in myelomatosis. Br Med J 1971;2:325.

Hoffman U, Opperman M, Kuchler S, Ventur Y, Teuber W, Michels H, et al. Free immunoglobulin light chains in patients with rheumatic diseases. J Rheumatology 2003;62:Fr40a.

Holding S, Spradbery D, Robson EJD, Dore PC, Wilmot R, Shields ML. Combination of serum free light chain analysis with capillary zone electrophoresis improves screening for monoclonal gammopathies. Blood 2007;110:1497a.

1) **Hopper JE**, Golbus J, Meyer C, Ferrer GA. Urine free light chains in SLE: clonal markers of B-cell activity and potential link to in vivo secreted Ig. J Clin Immunol 2000;20:123-37.

2) **Hopper JE**, Sequeira W, Martellotto J, Papagiannes E, Perna L, Skosey JL. Clinical relapse in systemic lupus erythematosus: correlation with antecedent elevation of urinary free light-chain immunoglobulin. J Clin Immunol 1989;9:338-50.

1) **Hutchison CA**, Cockwell P, Harding S, Mead GP, Bradwell AR, Barnett AH. Quantitative assessment of serum and urinary polyclonal free light chains in patients with type II diabetes: an early marker of diabetic kidney disease? Expert Opin Ther Targets 2008;12:667-76.

2) **Hutchison CA**, Cockwell P, Reid S, Chandler K, Mead GP, Harrison J, et al. Efficient removal of immunoglobulin free light chains by hemodialysis for multiple myeloma: in vitro and in vivo studies. J Am Soc Nephrol 2007;18:886-95.

3) **Hutchison CA**, Harding S, Basnayake K, Townsend J, Landray M, Mead GP, et al. Increased MGUS prevalance in chronic kidney disease patients. Haematologica 2007;92:PO905a.

4) **Hutchison CA**, Mead G, Chandler K, Harper J, Bradwell AR, Cockwell P. Free light chain abnormalities in patients with chronic kidney disease. J Am Soc Nephrol 2006;17:899a.

5) **Hutchison CA**, Plant T, Drayson M, Cockwell P, Kountouri M, Basnayake K, et al. Serum free light chain measurement aids the diagnosis of myeloma in patients with severe renal failure. BMC Nephrol 2008;9:11.

6) **Hutchison CA**, Bradwell AR, Cook M, Basnayake K, Basu S, Harding S, et al. Treatment of acute renal failure secondary to multiple myeloma with chemotherapy and extended high cut-off hemodialysis. Clin J Am Soc Nephrol 2009;4:745-54.

7) **Hutchison CA**, Harding S, Hewins P, Mead GP, Townsend J, Bradwell AR, Cockwell P. Quantitative assessment of serum and urinary polyclonal free light chains in patients with chronic kidney disease. Clin J Am Soc Nephrol 2008;3:1684-90.

8) **Hutchison CA**, Bradwell AR, Cook M, Basnayake K, Basu S, Harding S, et al. Treatment of acute renal failure secondary to multiple myeloma with chemotherapy and extended high cut-off hemodialysis. Clin J Am Soc Nephrol 2009;4:745-54.

Itzykson R, Le Garff-Tavernier M, Katsahian S, Diemert MC, Musset L, Leblond V. Serum-free light chain elevation is associated with a shorter time to treatment in Waldenstrom's macroglobulinemia. Haematologica 2008;93:793-4.

Jaccard A, Moreau P, Aucouturier P, Ronco P, Fermand J-P, Hermine O. Amylose immunoglobulinique. Hematologie 2003;9:485-95.

Jagannath S. Value of serum free light chain testing for the diagnosis and monitoring of monoclonal gammopathies in hematology. Clin Lymphoma Myeloma 2007;7:518-23.

Jaskowski TD, Litwin CM, Hill HR. Detection of kappa and lambda light chain monoclonal proteins in human serum: automated immunoassay versus immunofixation electrophoresis. Clin Vaccine Immunol 2006;13:277-80.

Jemal A, Siegel R, Ward E, Hao Y, Xu J, Thun MJ. Cancer statistics, 2009. CA ancer J Clin 2009;59:225-49.

Jenkins MA, Cheng L, Ratnaike S. Multiple sclerosis: use of light-chain typing to assist diagnosis. Ann Clin Biochem 2001;38:235-41.

Johnson WJ, Kyle RA, Pineda AA, O'Brien PC, Holley KE. Treatment of renal failure associated with multiple myeloma. Plasmapheresis, hemodialysis, and chemotherapy. Arch Intern Med 1990;150:863-9.

1) Jones HB. Papers on Chemical Pathology, Lecture III. Lancet 1847;2:88-92

2) Jones HB. On the new substance occurring in the urine of a patient with mollities ossium. Phil Trans R Soc B 1848;138:55-62.

Kang SY, Suh JT, Lee HJ, Yoon HJ, Lee WI. Clinical usefulness of free light chain concentration as a tumor marker in multiple myeloma. Ann Hematol 2005;84:588-93.

Kanoh T, Niwa Y. Nonsecretory IgD (kappa) multiple myeloma. Report of a case and review of the literature. Am J Clin Pathol 1987;88:516-9.

Kastritis E, Anagnostopoulos A, Roussou M, Gika D, Matsouka C, Barmparousi D, et al. Reversibility of renal failure in newly diagnosed multiple myeloma patients treated with high dose dexamethasone-containing regimens and the impact of novel agents. Haematologica 2007;92:546-9.

Katzel JA, Hari P, Vesole DH. Multiple myeloma: charging toward a bright future. CA Cancer J Clin 2007;57:301-18.

1) Katzmann JA. Serum free light chain specificity and sensitivity: a reality check. Clin Chem 2006;52:1638-9.

2) Katzmann JA. Laboratory analysis for monoclonal gammopathies. Mayo Communiqué 2007;32:1-4.

3) Katzmann JA, Abraham RS, Dispenzieri A, Lust JA, Kyle RA. Diagnostic performance of quantitative kappa and lambda free light chain assays in clinical practice. Clin Chem 2005;51:878-81.

4) Katzmann JA, Dispenzieri A, Kyle RA, Snyder MR, Plevak MF, Larson DR, et al. Elimination of the need for urine studies in the screening algorithm for monoclonal gammopathies by using serum immunofixation and free light chain assays. Mayo Clin Proc 2006;81:1575-8.

5) Katzmann JA, Clark R, Sanders E, Landers JP, Kyle RA. Prospective study of serum protein capillary zone electrophoresis and immunotyping of monoclonal proteins by immunosubtraction. Am J Clin Pathol 1998;110:503-9.

6) Katzmann JA, Clark RJ, Abraham RS, Bryant S, Lymp JF, Bradwell AR, Kyle RA. Serum reference intervals and diagnostic ranges for free kappa and free lambda immunoglobulin light chains: relative sensitivity for detection of monoclonal light chains. Clin Chem 2002;48:1437-44.

7) Katzmann JA, Clark RJ, Rajkumar VS, Kyle RA. Monoclonal free light chains in sera from healthy individuals: FLC MGUS. Clin Chem 2003;49:A74a.

8) Katzmann JA, Kyle RA, Benson J, Larson DR, Snyder MR,

Lust JA, et al. Screening panels for detection of monoclonal gammopathies. Clin Chem 2009;55:1517-22.

9) Katzmann JA. Screening panels for monoclonal gammopathies: time to change. Clin Biochem Rev 2009;30:105-11.

10) Katzmann J, Clark R, Dispenzieri A, Kyle R, Landgren O, Bradwell AR, Rajkumar SV. Isotype-specific heavy/light chain (HLC) suppression as a predictor of myeloma development in monoclonal gammopathy of undetermined significance (MGUS). Blood 2009;114:1788a.

Kayserova J, Capkova S, Skalicka A, Vernerova E, Polouckova A, Malinova V, et al. Serum immunoglobulin free light chains in severe forms of atopic dermatitis. Scand J Immunol 2010;71:312-6.

Khoury SJ, Weiner HL. Kappa light chains in spinal fluid for diagnosing multiple sclerosis. JAMA 1994;272:242-3.

Kim J, Hayton WL, Robinson JM, Anderson CL. Kinetics of FcRn-mediated recycling of IgG and albumin in human: pathophysiology and therapeutic implications using a simplified mechanism-based model. Clin Immunol 2007;122:146-55.

Kleeberg L, Morgera S, Jakob C, Hocher B, Schneider M, Peters H, et al. Novel renal replacement strategies for the elimination of serum free light chains in patients with kappa light chain nephropathy. Eur J Med Res 2009;14:47-54.

Kobayashi N, Suzuki Y, Tsuge T, Okumura K, Ra C, Tomino Y. FcRn-mediated transcytosis of immunoglobulin G in human renal proximal tubular epithelial cells. Am J Physiol Renal Physiol 2002;282:F358-65.

Korngold L, Lipari R. Multiple-myeloma proteins. III. The antigenic relationship of Bence Jones proteins to normal gammaglobulin and multiple-myeloma serum proteins. Cancer 1956;9:262-72.

Krakauer M, Schaldemose Nielsen H, Jensen J, Sellebjerg F. Intrathecal synthesis of free immunoglobulin light chains in multiple sclerosis. Acta Neurol Scand 1998;98:161-5.

Kriz W, LeHir M. Pathways to nephron loss starting from glomerular diseases-insights from animal models. Kidney Int 2005;67:404-19.

Kruger WH, Kiefer T, Schuler F, Lotze C, Busemann C, Dolken G. Complete remission and early relapse of refractory plasma cell leukemia after bortezomib induction and consolidation by HLA-mismatched unrelated allogeneic stem cell transplantation. Onkologie 2007;30:193-5.

1) Kühnemund A, Liebisch P, Bauchmüller K, Haas P, Kleber M, Bisse E, et al. Secondary light chain multiple myeloma with decreasing IgA paraprotein levels correlating with renal insufficiency and progressive disease: Clinical course of two patients and review of the literature. Onkologie 2005;28

2) Kühnemund A, Liebisch P, Bauchmuller K, Zur Hausen A, Veelken H, Wasch R, Engelhardt M. 'Light-chain escape-multiple myeloma'-an escape phenomenon from plateau phase: report of the largest patient series using LC-monitoring. J Cancer Res Clin Oncol 2009;135:477-84.

1) Kumar S, Fonseca R, Dispenzieri A, Katzmann JA, Kyle RA, Clark R, Rajkumar SV. High incidence of IgH translocations in monoclonal gammopathies with abnormal free light chain levels. Blood 2006;108:3514a.

2) Kumar S, Gertz MA, Hayman SR, Lacy MQ, Dispenzieri A,

Zeldenrust SR, et al. Use of the serum free light chain assay in assessment of response to therapy in multiple myeloma: Validation of recently proposed response criteria in a prospective clinical trial of lenalidomide plus dexamethasone for newly diagnosed multiple myeloma. Blood 2005;106 3479a.

3) **Kumar S**, Rajkumar VS, Dispenzieri A, Lacy MQ, Hayman SR, Buadi FK, et al. Improving survival in multiple myeloma: Impact of novel therapies. Blood 2007;110:3594a.

4) **Kumar S**, Perez WS, Zhang MJ, Ballen K, Bashey A, To LB, et al. Comparable outcomes in nonsecretory and secretory multiple myeloma after autologous stem cell transplantation. Biol Blood Marrow Transplant 2008;14:1134-40.

5) **Kumar K**, Dispenzieri A, Lacy MQ, Hayman SR, Buadi FK, Zeldenrust SR, et al. A novel staging system for light chain amyloidosis incorporating free light chain levels. Haematologica 2008;93:917a.

Kuypers DR, Lerut E, Claes K, Evenepoel P, Vanrenterghem Y. Recurrence of light chain deposit disease after renal allograft transplantation: potential role of rituximab? Transpl Int 2007;20:381-5.

1) **Kyle RA**. Multiple myeloma and other plasma cell disorders. Haematology: Basic principles & practice, 2nd ed: Churchill Livingstone, 1995:1354-74.

2) **Kyle RA**. Sequence of testing for monoclonal gammopathies. Arch Pathol Lab Med 1999;123:114-8.

3) **Kyle RA**. Multiple myeloma: an odyssey of discovery. Br J Haematol 2000;111:1035-44.

4) **Kyle RA**. Henry Bence Jones - physician, chemist, scientist and biographer: a man for all seasons. Br J Haematol 2001;115:13-8.

5) **Kyle R**, Child JA, Anderson K, Barlogie B, Bataille R, Bensinger W, et al. Criteria for the classification of monoclonal gammopathies, multiple myeloma and related disorders: a report of the International Myeloma Working Group. Br J Haematol 2003;121:749-57.

6) **Kyle RA**, Gertz MA. Primary systemic amyloidosis: clinical and laboratory features in 474 cases. Semin Hematol 1995;32:45-59.

7) **Kyle RA**, Gertz MA, Witzig TE, Lust JA, Lacy MQ, Dispenzieri A, et al. Review of 1027 patients with newly diagnosed multiple myeloma. Mayo Clin Proc 2003;78:21-33.

8) **Kyle RA**, Greipp PR. "Idiopathic" Bence Jones proteinuria: long-term follow-up in seven patients. N Engl J Med 1982;306:564-7.

9) **Kyle RA**, Rajkumar SV. Monoclonal gammopathies of undetermined significance. Hematol Oncol Clin North Am 1999;13:1181-202.

10) **Kyle R**, Child JA, Anderson K, Barlogie B, Bataille R, Bensinger W, et al. Criteria for the classification of monoclonal gammopathies, multiple myeloma and related disorders: a report of the International Myeloma Working Group. Br J Haematol 2003;121:749-57.

11) **Kyle RA**, Rajkumar SV. Monoclonal gammopathy of undetermined significance. Br J Haematol 2006;134:573-89.

12) **Kyle RA**, Remstein ED, Therneau TM, Dispenzieri A, Kurtin PJ, Hodnefield JM, et al. Clinical course and prognosis of smoldering (asymptomatic) multiple myeloma. N Engl J Med 2007;356:2582-90.

13) **Kyle RA**, Therneau TM, Rajkumar SV, Offord JR, Larson DR, Plevak MF, Melton LJ, 3rd. A long-term study of prognosis in monoclonal gammopathy of undetermined significance. N Engl J Med 2002;346:564-9.

14) **Kyle RA**, Therneau TM, Rajkumar SV, Remstein ED, Offord JR, Larson DR, et al. Long-term follow-up of IgM monoclonal gammopathy of undetermined significance. Blood 2003;102:3759-64.

15) **Kyle RA**, Rajkumar SV. Criteria for diagnosis, staging, risk stratification and response assessment of multiple myeloma. Leukemia 2009;23:3-9.

16) **Kyle RA**, Durie BG, Rajkumar SV, Landgren O, Blade J, Merlini G, et al. Monoclonal gammopathy of undetermined significance (MGUS) and smoldering (asymptomatic) multiple myeloma: IMWG consensus perspectives risk factors for progression and guidelines for monitoring and management. Leukemia 2010.

1) **Kyrtsonis MC**, Sachanas S, Vassilakopoulos TP, Kafassi N, Tzenou T, Papadogiannis A, et al. Bortezomib in patients with relapsed-refractory multiple myeloma (MM). Clinical observations. Blood 2005;106:5193a.

2) **Kyrtsonis MC**, Vassilakopoulos TP, Kafasi N, Maltezas D, Anagnostopoulos A, Terpos E, et al. The addition of sFLCR improves ISS prognostication in multiple myeloma (MM). Blood 2007;110:1490a.

3) **Kyrtsonis MC**, Vassilakopoulos TP, Kafasi N, Sachanas S, Tzenou T, Papadogiannis A, et al. Prognostic value of serum free light chain ratio at diagnosis in multiple myeloma. Br J Haematol 2007;137:240-3.

Lachmann HJ, Gallimore R, Gillmore JD, Carr-Smith HD, Bradwell AR, Pepys MB, Hawkins PN. Outcome in systemic AL amyloidosis in relation to changes in concentration of circulating free immunoglobulin light chains following chemotherapy. Br J Haematol 2003;122:78-84.

Lamers KJ, de Jong JG, Jongen PJ, Kock-Jansen MJ, Teunesen MA, Prudon-Rosmulder EM. Cerebrospinal fluid free kappa light chains versus IgG findings in neurological disorders: qualitative and quantitative measurements. J Neuroimmunol 1995;62:19-25.

1) **Landgren O**, Gridley G, Turesson I, Caporaso NE, Goldin LR, Baris D, et al. Risk of monoclonal gammopathy of undetermined significance (MGUS) and subsequent multiple myeloma among African American and white veterans in the United States. Blood 2006;107:904-6.

2) **Landgren O**, Goedert J, Rabkin C, Wilson W, Dunleavy K, Kyle R, et al. Risk of AIDS non-Hodgkin's lymphoma is strongly predicted by elevated levels of circulating immunoglobulin-free light chains. Sixteenth Conference on Retroviruses and Opportunistic Infections, Montreal 2009:29a.

3) **Landgren O**, Kyle RA, Pfeiffer RM, Katzmann JA, Caporaso NE, Hayes RB, et al. Monoclonal gammopathy of undetermined significance (MGUS) consistently precedes multiple myeloma: a prospective study. Blood 2009;113:5412-7.

4) **Landgren O**, Goedert JJ, Rabkin CS, Wilson WH, Dunleavy K, Kyle RA, et al. Circulating serum free light chains as predictive markers of AIDS-related lymphoma. J Clin Oncol 2010;28:773-9.

Le Bricon T, Bengoufa D, Benlakehal M, Bousquet B, Erlich D. Urinary free light chain analysis by the Freelite immunoassay: a preliminary study in multiple myeloma. Clin Biochem 2002;35:565-7.

Lebovic D, Kendall T, Brozo C, McAllister A, Hari M, Alvi S, et al. Serum free light chain analysis improves monitoring of multiple myeloma patients receiving first-line therapy with the combination of Velcade, Doxil, and Dexamethasone (VDD). Blood 2007;110:2736a.

1) Leleu X, Hatjiharissi E, Roccaro AM, Moreau AS, Leduc R, Nelson M, et al. Serum immunoglobulin free light chain (sFLC) is a sensitive marker of response in Waldenstrom Macroglobulinemia (WM). Blood 2007;110:1486a.

2) Leleu X, Moreau AS, Hennache B, Dupire S, Faucompret JL, Facon T, et al. Serum free light chain immunoassays measurement for monitoring solitary bone plasmacytoma. Haematologica 2005;90:PO410a.

3) Leleu X, Moreau AS, Weller E, Roccaro AM, Coiteux V, Manning R, et al. Serum immunoglobulin free light chain correlates with tumor burden markers in Waldenstrom macroglobulinemia. Leuk Lymphoma 2008;49:1104-7.

1) Leung N, Lager DJ, Gertz MA, Wilson K, Kanakiriya S, Fervenza FC. Long-term outcome of renal transplantation in light-chain deposition disease. Am J Kidney Dis 2004;43:147-53.

2) Leung N, Gertz MA, Zeldenrust SR, Rajkumar SV, Dispenzieri A, Fervenza FC, et al. Improvement of cast nephropathy with plasma exchange depends on the diagnosis and on reduction of serum free light chains. Kidney Int 2008;73:1282-8.

Levey AS, Bosch JP, Lewis JB, Greene T, Rogers N, Roth D. A more accurate method to estimate glomerular filtration rate from serum creatinine: a new prediction equation. Modification of Diet in Renal Disease Study Group. Ann Intern Med 1999;130:461-70.

Levinson SA, MacFate RP. Bence-Jones proteins. In : Clinical laboratory diagnosis, 3rd edition, Lea and Febiger, Philadelphia, 1946.

Levinson SS, Keren DF. Free light chains of immunoglobulins: clinical laboratory analysis. Clin Chem 1994;40:1869-78.

Lewis E, Young C, Watson D. Comparison of western blotting and free light chains in the laboratory diagnosis of multiple sclerosis. Annals Clin Biochem 2007;44:T51a.

Li J, Zhou DB, Jiao L, Duan MH, Zhang W, Zhao YQ, et al. Bortezomib and dexamethasone therapy for newly diagnosed patients with multiple myeloma complicated by renal impairment. Clin Lymphoma Myeloma. 2009;9:394-8.

Lindstedt G, Lundberg PA. Loss of tubular proteinuria pattern during urine concentration with a commercial membrane filter cell (Minicon B-15 system). Clin Chim Acta 1974;56:125-6.

Ling NR, Lowe J, Hardie D, Evans S, Jefferis R. Detection of free kappa chains in human serum and urine using pairs of monoclonal antibodies reacting with C kappa epitopes not available on whole immunoglobulins. Clin Exp Immunol 1983;52:234-40.

Lokhorst H. Myeloma: Clinical features and diagnostic criteria. In: Mehta J, Singhal S, eds. Myeloma, M Dunitz, London, 2002.

Longsworth LG, Shedlovsky T, Macinnes DA. Electrophoretic Patterns of Normal and Pathological Human Blood Serum and Plasma. J Exp Med 1939;70:399-413.

Ludwig H, Drach J, Graf H, Lang A, Meran JG. Reversal of acute renal failure by bortezomib-based chemotherapy in patients with multiple myeloma. Haematologica 2007;92:1411-4.

Lueck N, Agrawal YP. Lack of utility of free light chain-specific antibodies in the urine immunofixation test. Clin Chem 2006;52:906-7.

Luxton RW, McLean BN, Thompson EJ. Isoelectric focusing versus quantitative measurements in the detection of intrathecal local synthesis of IgG. Clin Chim Acta 1990;187:297-308.

1) Lynch EA, Showell PJ, Kaur A, Mead GP, Bradwell AR. Evaluation of latex-enhanced turbidimetric reagents for measuring free immunoglobulin light-chains on the Beckman Synchron LX20 Pro and Beckman UniCel DxC Synchron. Clin Chem 2008;54:C97a.

3) Lynch EA, Showell PJ, Johnson-Brett B, Mead GP, Bradwell AR. Evaluation of latex-enhanced turbidimetric reagents for measuring free immunoglobulin light-chains on the Roche cobas Integra 800. Clin Chem 2008;54:C94a.

4) Lynch EA, Showell PJ, Johnson-Brett B, Mead GP, Bradwell AR. Evaluation of latex-enhanced turbidimetric reagents for measuring free immunoglobulin light chains on the Roche cobas c501. Clin Chem 2008;54:C93a.

Maack T, Johnson V, Kau ST, Figueiredo J, Sigulem D. Renal filtration, transport, and metabolism of low-molecular-weight proteins: a review. Kidney Int 1979;16:251-70.

MacIntyre W. Case of Mollities and Fragilitas Ossium, accompanied with urine strongly charged with animal matter. Med Chir Tran 1850;33:211-32.

MacNamara EM, Aguzzi F, Petrini C, Higginson J, Gasparro C, Bergami MR, et al. Restricted electrophoretic heterogeneity of immunoglobulin light chains in urine: a cause for confusion with Bence Jones protein. Clin Chem 1991;37:1570-4.

Malpas JS, Bergsagel DE, Kyle RA, Anderson KC. Myeloma biology and management: Saunders, 2004.

Marien G, Oris E, Bradwell AR, Blanckaert N, Bossuyt X. Detection of monoclonal proteins in sera by capillary zone electrophoresis and free light chain measurements. Clin Chem 2002;48:1600-1.

Marionneaux S, Zetlmeisl M, van Hoeven KH, Fagan D, Elkins B, Shulman S. Analytical evaluation of kappa and lambda serum free light chains on The Binding Site SPAPLUS. Clin Chem 2008;54:C79a.

Marshall G, Tate J, Mollee P. Borderline high serum free light chain kappa/lambda ratios are seen not only in dialysis patients but also in non-dialysis-dependent renal impairment and inflammatory states. Am J Clin Pathol 2009;132:309.

Martin MG, Wu SV, Walsh JH. Hormonal control of intestinal Fc receptor gene expression and immunoglobulin transport in suckling rats. J Clin Invest 1993;91:2844-9.

Martin W, Abraham R, Shanafelt T, Clark RJ, Bone N, Geyer SM, et al. Serum-free light chain-a new biomarker for patients with B-cell non-Hodgkin lymphoma and chronic lymphocytic leukemia. Transl Res 2007;149:231-5.

Matschke J, Eisele L, Sellmann L, Duehrsen U, Duerig J, Nückel

H. Abnormal free light chain ratios in chronic lymphocytic leukemia: A new prognostic factor? Presented at ASH 2009: 1237a

Matters DJ, Solanki M, Jones R, Rose S, Carr-Smith H, Bradwell AR. Statistical analysis of free light chain results between seven nephelometric and turbidimetric platforms. Clin Chem 2008;54:C107a.

Maurer MJ, Micallef IN, Katzmann JA, Nikcevich D, Witzig TE. Elevated pre-treatment serum immunoglobulin free light chains (FLC) are associated with poor event-free and overall survival in diffuse large B-Cell lymphoma (DLBCL) Presented at ASH 2009 2009:136a.

Mayo MM, Johns GS. Serum free light chains in the diagnosis and monitoring of patients with plasma cell dyscrasias. Contrib Nephrol 2007;153:44-65.

Mazurkiewicz J, Teal T, Reddy R, Grace R, Gover P. Evaluation of The Binding Site serum free light chain assay in a DGH: clinical aspects. Annals Clin Biochem 2007;44:T47a.

McIntyre OR. Laboratory investigation of myeloma. In: Malpas JS, Bergsagel DE, Kyle RA, eds. Myeloma biology and management, Oxford Medical Publications, 1995.

1) Mead GP, Carr-Smith HD, Drayson MT, Morgan GJ, Child JA, Bradwell AR. Serum free light chains for monitoring multiple myeloma. Br J Haematol 2004;126:348-54.

2) Mead GP, Carr-Smith HD, Drayson MT, Morgan GJ, Child AJ. Response to: Serum free light chains for monitoring multiple myeloma. Br J Haematol 2005;128:406-7.

3) Mead GP, Reid SD, Augustson BM, Drayson MT, Bradwell AR, Child JA. Correlation of serum free light chains and bone marrow plasma cell infiltration in multiple myeloma. Blood 2004;104:4865a.

4) Mead GP, Stubbs PD, Carr-Smith HD, Drew R, Drayson MT, Bradwell AR. Nephelometric measurement of serum free light chains in nonsecretory myeloma. Clin Chem 2002;48:70a.

5) Mead G, Harding S, Pratt G, Basu S, Jacob A, Beardsmore C, Bradwell AR. Serum immunoglobulin and free light chain abnormalities in non Hodgkin Lymphoma. Ann Oncol 2008;19:315a.

Mehta J, Stein R, Vickrey E, Resseguie W, Singhal S. Significance of serum free light chain estimation with detectable serum monoclonal protein on immunofixation electrophoresis. Blood 2006;108:5048a.

Merlini G. Editorial: Sharpening therapeutic strategy in AL amyloidosis. Blood 2004;104:1593-4.

Miettinen TA, Kekki M. Effect of impaired hepatic and renal function on [131I] Bence Jones protein catabolism in human subjects. Clin Chim Acta 1967;18:395-407.

Mignot A, Bridoux F, Thierry A, Varnous S, Pujo M, Delcourt A, et al. Successful heart transplantation following melphalan plus dexamethasone therapy in systemic AL amyloidosis. Haematologica 2008;93:e32-5.

Montseny JJ, Kleinknecht D, Meyrier A, Vanhille P, Simon P, Pruna A, Eladari D. Long-term outcome according to renal histological lesions in 118 patients with monoclonal gammopathies. Nephrol Dial Transplant 1998;13:1438-45.

Morandeira F, Soriano A, Quandt E, Castro P, Ruiz E, Oriol A, et al. Comparacion de 2 ensayos para la determination de cadenas ligeras libres en suero en pacientes con gammapatia monoclonal. Immunologia 2008;27:118a.

Morris KL, Tate JR, Gill D, Kennedy G, Wellwood J, Marlton P, et al. Diagnostic and prognostic utility of the serum free light chain assay in patients with AL amyloidosis. Intern Med J 2007;37:456-63.

Mösbauer U, Ayuk F, Schieder H, Lioznov M, Zander AR, Kroger N. Monitoring serum free light chains in patients with multiple myeloma who achieved negative immunofixation after allogeneic stem cell transplantation. Haematologica 2007;92:275-6.

Munshi NC. Editorial: Determining the undetermined. Blood 2005;106:767-8.

Murray DL, Ryu E, Snyder MR, Katzmann JA. Quantitation of serum monoclonal proteins: relationship between agarose gel electrophoresis and immunonephelometry. Clin Chem 2009;55:1523-9.

1) Nakano T, Miyazaki S, Shinoda Y, Inoue I, Katayama S, Komoda T, Nagata A. Proposed reference material for human free immunoglobulin light chain measurement. J Immunoassay Immunochem 2006;27:129-37.

2) Nakano T, Miyazaki S, Takahashi H, Matsumori A, Maruyama T, Komoda T, Nagata A. Immunochemical quantification of free immunoglobulin light chains from an analytical perspective. Clin Chem Lab Med 2006;44:522-32.

3) Nakano T, Nagata A. ELISAs for free human immunoglobulin light chains in serum: improvement of assay specificity by using two specific antibodies in a sandwich detection method. J Immunol Methods 2004;293:183-9.

Nakorn TN, Watanaboonyongcharoen P, Suwannabutra S, Theerasaksilp S, Paritpokee N. Early reduction of serum free light chain can predict therapeutic responses in multiple myeloma. Blood 2007;110:4742a.

Nelson M, Brown RD, Gibson J, Joshua DE. Measurement of free kappa and lambda chains in serum and the significance of their ratio in patients with multiple myeloma. Br J Haematol 1992;81:223-30.

Niewold TA, Murphy CL, Weiss DT, Solomon A. Characterization of a light chain product of the human JC lambda 7 gene complex. J Immunol 1996;157:4474-7.

Novak L, Cook WJ, Herrera GA, Sanders PW. AL-amyloidosis is underdiagnosed in renal biopsies. Nephrol Dial Transplant 2004;19:3050-3.

Nowrousian MR, Brandhorst D, Sammet C, Kellert M, Daniels R, Schuett P, et al. Serum free light chain analysis and urine immunofixation electrophoresis in patients with multiple myeloma. Clin Cancer Res 2005;11:8706-14.

Ohtani S, Ohtani H. Clinical significance of free light chain in urine by latex agglutination immunoassay. Kitsato Med 1998;28:435-45.

Ong S, Sethi S. Assessment of free light chain assay in serum on the Beckman ImageTM. Clin Chem 2003;49:D59a.

Orlowski R, Sutherland H, Blade J, Miguel JS, Hajek R, Nagler A, et al. Early normalization of serum free light chains is associated with prolonged time to progression following bortezomib {+/-} pegylated liposomal doxorubicin treatment of relapsed/ refractory multiple myeloma. Blood 2007;110:2735a.

Ouchterlony O. Antigen-antibody reactions in gels. IV. Types of reactions in coordinated systems of diffusion. Acta Pathol Microbiol Scand 1953;32:230-40.

Overton J, Goodier D, Carr-Smith HD, Bradwell A. Evaluation of latex-enhanced turbidimetric reagents for measuring free immunoglobulin light-chains on the Roche Modular P. Clin Chem 2003;49:D60a.

1) **Owen RG**, Child JA, Rawstron AC, Bell S, Cocks K, Davies FE, et al. Defining complete response in multiple myeloma: Role of the serum free light chain assay and multiparameter flow cytometry. Blood 2007;110:1479a.

2) **Owen RG**, Treon SP, Al-Katib A, Fonseca R, Greipp PR, McMaster ML, et al. Clinicopathological definition of Waldenstrom's macroglobulinemia: consensus panel recommendations from the Second International Workshop on Waldenstrom's Macroglobulinemia. Semin Oncol 2003;30:110-5.

1) **Palladini G**, Lavatelli F, Russo P, Perlini S, Perfetti V, Bosoni T, et al. Circulating amyloidogenic free light chains and serum N-terminal natriuretic peptide type B decrease simultaneously in association with improvement of survival in AL. Blood 2006;107:3854-8.

2) **Palladini G**, Nuvolone M, Russo P, Lavatelli F, Perfetti V, Obici L, Merlini G. Treatment of AL amyloidosis guided by biomarkers. Haematologica 2007;92:PO910a.

3) **Palladini G**, Russo P, Lavatelli F, Nuvolone M, Bosoni T, D'Eril GM, et al. Diagnostic performance of the quantitative assay for free light chains (FreeliteTM) in AL amyloidosis. Clin Chem 2006;52:D123a.

5) **Palladini G**, Russo P, Bosoni T, Verga L, Sarais G, Lavatelli F, et al. Identification of amyloidogenic light chains requires the combination of serum-free light chain assay with immunofixation of serum and urine. Clin Chem 2009;55:499-504.

1) **Pascali E**, Pezzoli A. The clinical spectrum of pure Bence Jones proteinuria. A study of 66 patients. Cancer 1988;62:2408-15.

2) **Pascali E**. Bence Jones proteinuria in chronic lymphocytic leukemia. Am J Clin Pathol 1995;103:665-7.

Patten PE, Ahsan G, Kazmi M, Fields PA, Chick GW, Jones RR, et al. The early use of the serum free light chain assay in patients with relapsed refractory myeloma receiving treatment with thalidomide analogue (CC-4047). Blood 2003;102:1640a.

Pattenden RJ, Rogers SY, Wenham PR. Serum free light chains; the need to establish local reference intervals. Ann Clin Biochem 2007;44:512-5.

Penders J, Fiers T, Delanghe JR. Quantitative evaluation of urinalysis test strips. Clin Chem 2002;48:2236-41.

Peterson PA, Berggard I. Urinary immunoglobulin components in normal, tubular, and glomerular proteinuria: quantities and characteristics of free light chains, IgG, igA, and Fc-gamma fragment. Eur J Clin Invest 1971;1:255-64.

Pezzoli A, Pascali E. Monoclonal Bence Jones proteinuria in chronic lymphocytic leukaemia. Scand J Haematol 1986;36:18-24.

Piehler AP, Gulbrandsen N, Kierulf P, Urdal P. Quantitation of serum free light chains in combination with protein electrophoresis and clinical information for diagnosing multiple myeloma in a general hospital population. Clin Chem 2008;54:1823-30.

Pillon L, Sweeting RS, Arora A, Notkin A, Ballard HS, Wieczorek RL, Leung N. Approach to acute renal failure in biopsy proven myeloma cast nephropathy: is there still a role for plasmapheresis? Kidney Int 2008;74:956-61.

Pinto A, De Filippi R, Iaccarino G, Di Francia R, Distinto M, Frigeri F, et al. Abnormalities in serum free-immunoglobulin light chains show a high and differential frequency among WHO subtypes of B-cell non-Hodgkin's lymphoma (NHL) and may turn of value for therapeutic monitoring: A study of 354 newly diagnosed patients. Blood 2008;112:2813a.

Pozzi C, D'Amico M, Fogazzi GB, Curioni S, Ferrario F, Pasquali S, et al. Light chain deposition disease with renal involvement: clinical characteristics and prognostic factors. Am J Kidney Dis 2003;42:1154-63.

1) **Pratt G**, Harding S, Holder R, Fegan C, Pepper C, Oscier D, et al. Abnormal serum free light chain ratios are associated with poor survival and may reflect biological subgroups in patients with chronic lymphocytic leukaemia. Br J Haematol 2009;144:217-22.

2) **Pratt G**, Mead GP, Godfrey KR, Hu Y, Evans ND, Chappell MJ, et al. The tumor kinetics of multiple myeloma following autologous stem cell transplantation as assessed by measuring serum-free light chains. Leuk Lymphoma 2006;47:21-8.

3) **Pratt G**. The evolving use of serum free light chain assays in haematology. Br J Haematol 2008;141:413-22.

Presslauer S, Milosavljevic D, Hubl W, Brucke T, Bayer P. Elevated levels of kappa free light chains in CSF support the diagnosis of multiple sclerosis. Multiple Sclerosis 2006;12:783a.

Prevoo ML, van 't Hof MA, Kuper HH, van Leeuwen MA, van de Putte LB, van Riel PL. Modified disease activity scores that include twenty-eight-joint counts. Development and validation in a prospective longitudinal study of patients with rheumatoid arthritis. Arthritis Rheum 1995;38:44-8.

1) **Rajkumar SV**. MGUS and smoldering multiple myeloma: Update on pathogenesis, natural history and management. Hematology Am Soc Hematol Educ Program 2005;340-5.

2) **Rajkumar SV**, Griepp PR. Prognostic factors in multiple myeloma. In: Kyle RA, Gertz MA, eds. Hematology/Oncology clinics of North America: Monoclonal gammopathies and related disorders, W B Saunders Co. Philadelphia, 1999.

3) **Rajkumar SV**, Kyle RA. Conventional therapy and approach to management. Best Pract Res Clin Haematol 2005;18:585-601.

4) **Rajkumar SV**, Kyle RA, Plevak M, Clark RJ, Larson D, Therneau T, et al. Prevalance of light-chain monoclonal gammopathy of undetermined significance (LC-MGUS) among Olmsted Country, Minnesota residents aged 50 years of greater. Blood 2006;108:5060a.

5) **Rajkumar SV**, Kyle RA, Therneau TM, Melton LJ, 3rd, Bradwell AR, Clark RJ, et al. Serum free light chain ratio is an independent risk factor for progression in monoclonal gammopathy of undetermined significance. Blood 2005;106:812-7.

6) **Rajkumar SV**, Lacy MQ, Kyle RA. Monoclonal gammopathy of undetermined significance and smoldering multiple myeloma. Blood Rev 2007;21:255-65.

Ramasamy I. Serum free light chain analysis in B-cell dyscrasias. Ann Clin Lab Sci 2007;37:291-4.

Raubenheimer EJ, Dauth J, Senekal JC. Non-secretory IgA kappa myeloma with distended endoplasmic reticulum: a case report. Histopathology 1991;19:380-2.

Rawstron AC, Davis B, D. DS, de Tute RM, Kerr MA, Owen RG, Ashcroft AJ. Plasma cell phenotype and SFLC provide independent prognostic information in MGUS. Haematologica 2007;92:PO907a.

2) Reid SD, Drayson MT, Mead GP, Augustson B, Bradwell AR. Serum free light chains are a more sensitive marker of serological remission in multiple myeloma patients. Clin Chem 2004;50:C34a.

3) Reid SD, Katsavara H, Augustson BM, Hutchison CA, Mead GP, Shirfield M, et al. Screening for monoclonal gammopathy: inclusion of serum free light chain immunoassays produce an increased detection rate. Clin Chem 2006;52:E37a.

Reilly BM, Clarke P, Nikolinakos P. Clinical problem-solving. Easy to see but hard to find. N Engl J Med 2003;348:59-64.

Ricotta D, Radeghieri A, Amoroso B, Caimi L. Serum free light chains in MM: comparison of 2 assays. Haematologica 2007;92:PO1018a.

Rinker JR, 2nd, Trinkaus K, Cross AH. Elevated CSF free kappa light chains correlate with disability prognosis in multiple sclerosis. Neurology 2006;67:1288-90.

Ritz E. Nephrology beyond JASN: Plasma exchange for acute renal failure of myeloma - logical, yet ineffective. J Am Soc Nephrol 2006;17:914 - 6.

Robinson EL, Gowland E, Ward ID, Scarffe JH. Radioimmunoassay of free light chains of immunoglobulins in urine. Clin Chem 1982;28:2254-8.

1) Robson E, Mead G, Bradwell A. To the editor: in reply to Nakano et al. Clin Chem Lab Med 2006;44(5):522-532. Clin Chem Lab Med 2007;45:264-5.

2) Robson E, Mead G, Carr-Smith H, Bradwell A. In reply to Tate et al Clin Chim Acta 2007; 376: 30 - 6. Clin Chim Acta 2007;380:247.

3) Robson E, Mead G, Das M, Cavet J, Liakpoulou E. Free Light Chain analysis in patients receiving Bortezomib. Haematologica 2007;92:PO1019a.

4) Robson EJD, Taylor J, Beardsmore C, Basu S, Mead G, Lovatt T. Utility of serum free light chain analysis when screening for lymphoproliferative disorders. Lab Med 2009;40:325-9.

Rolinski B, Hammer F, Scherberich J. Clearance of kappa and lambda free light chains during hemodialysis. Clin Chem Lab Med 2006;44:117a.

Roopenian DC, Christianson GJ, Sproule TJ, Brown AC, Akilesh S, Jung N, et al. The MHC class I-like IgG receptor controls perinatal IgG transport, IgG homeostasis, and fate of IgG-Fc-coupled drugs. J Immunol 2003;170:3528-33.

Rosenfeld L. Henry Bence Jones (1813-1873): the best "chemical doctor" in London. Clin Chem 1987;33:1687-92.

Rosinol L, Blade J, Esteve J, Aymerich M, Rozman M, Montoto S, et al. Smoldering multiple myeloma: natural history and recognition of an evolving type. Br J Haematol 2003;123:631-6.

Roussou M, Kastritis E, Christoulas D, Migkou M, Gavriatopoulou M, Grapsa I, et al. Reversibility of renal failure in newly diagnosed patients with multiple myeloma and the role of novel agents. Leuk Res. *In press*

Ruchlemer R, Reinus C, Paz E, Cohen A, Melnikov N, Ronson A, Rudensky B. Free light chains, monoclonal proteins, and chronic lymphocytic leukemia. Blood 2007;110:4697a.

Russo LM, Bakris GL, Comper WD. Renal handling of albumin: a critical review of basic concepts and perspective. Am J Kidney Dis 2002;39:899-919.

Rutherford C, Wians FH. A large skull mass in a 67 year-old woman. Lab Med 2006;37:417-21.

Salmon SE, Smith BA. Immunoglobulin synthesis and total body tumor cell number in IgG multiple myeloma. J Clin Invest 1970;49:1114-21.

1) Sanders PW, Booker BB, Bishop JB, Cheung HC. Mechanisms of intranephronal proteinaceous cast formation by low molecular weight proteins. J Clin Invest 1990;85:570-6.

2) Sanders PW, Booker BB. Pathobiology of cast nephropathy from human Bence Jones proteins. J Clin Invest 1992;89:630-9.

3) Sanders PW, Herrera GA, Kirk KA, Old CW, Galla JH. Spectrum of glomerular and tubulointerstitial renal lesions associated with monotypical immunoglobulin light chain deposition. Lab Invest 1991;64:527-37.

Scherberich J, Hammer F, Rolinski B. Impact of chronic renal failure and hameodialysis on serum free polyclonal immunoglobulin kapp/lambda light-chains. Nephrol Dial Transplant 2006;21:SP021a.

Shaw GR. Nonsecretory plasma cell myeloma--becoming even more rare with serum free light-chain assay: a brief review. Arch Pathol Lab Med 2006;130:1212-5.

Sheehan T, Sinclair D, Tansey P, O'Donnell JR. Demonstration of serum monoclonal immunoglobulin in a case of non-secretory myeloma by immunoisoelectric focusing. J Clin Pathol 1985;38:806-9.

1) Showell PJ, Hutchison CA, Cockwell P, Harding S, Mead GP, Mitchell F, Bradwell AR. Correlation of FREELITE and Cystatin C in Chronic Kidney Disease on The Binding Site SPAPLUS bench-top analyser. Clin Chem 2007;53:C82a.

2) Showell PJ, Long JM, Carr-Smith HC, Bradwell AR. Evaluation of latex-enhanced turbidimetric reagents for measuring free immunoglobulin light-chains on the Hitachi 911/912. Clin Chem 2002;48:A66a.

3) Showell PJ, Lynch EA, Carr-Smith H, Bradwell AR. Evaluation of latex-enhanced nephelometric reagents for measuring free immunoglobulin light-chains on the Radim Delta. Clin Chem 2004;50:C40a.

5) Showell PJ, Lynch EA, Overton J, Carr-Smith HD, Bradwell AR. Evaluation of latex-enhanced nephelometric reagents for measuring free immunoglobulin light chains on the Dade Behring ProSpec. Clin Chem 2005;51:B38a.

6) Showell PJ, Matters DJ, Long JM, Carr-Smith HD, Bradwell AR. Evaluation of latex-enhanced turbidimetric reagents for measuring free immunoglobulin light-chains on the Olympus AU400. Clin Chem 2003;49:D55a.

7) Showell PJ, Matters DJ, Long JM, Carr-Smith HD, Bradwell AR. Evaluation of latex-enhanced nephelometric reagents for measuring free immunoglobulin light-chains on a modified Minineph™. Clin Chem 2002;48:A67a.

8) Showell PJ, Scurvin M, Chnivimba A, Carr-Smith HC, Bradwell AR. Evaluation of latex-enhanced turbidimetric reagents for measuring free immunoglobulin light-chains on The Binding Site automated analyser. Clin Chem 2006;52:E41a.

9) Showell PJ, Lynch EA, Mitchell F, Mead GP, Bradwell AR. Comparison of serum free immunoglobulin light-chain assays on eight Nephelometric/Turbidimetric analysers. Clin Chem 2008;54:C92a.

10) Showell PJ, Lynch EA, Johnson-Brett B, Mead G, Bradwell AR. Evaluation of latex-enhanced turbidimetric reagents for measuring free immunoglobulin light-chains on the Roche COBAS Integra 400 automated analyser. Clin Chem 2007;53:C18a.

Shustik C, Harding S, Ding K, Zhu L, Rassenti LZ, Kipps TJ, et al. Analysis of the serum free light chain ratio and its prognostic value in a cohort of patients with chronic lymphocytic leukemia Presented at ASH 2009 2009:2631a.

1) Siegel DS, Bilotti E, Van Hoeven K. Serum free light chain analysis for diagnosis, monitoring, and prognosis of monoclonal gammopathies. Lab Med 2009;40:363-6.

2) Siegel DS, McBride L, Bilotti E, Lendvai N, Gonsky J, Berges T, et al. Inaccuracies in 24-hour urine testing for monoclonal gammopathies. Lab Med 2009;40:341-4.

1) Sinclair D, Dagg JH, Smith JG, Stott DI. The incidence and possible relevance of Bence-Jones protein in the sera of patients with multiple myeloma. Br J Haematol 1986;62:689-94.

2) Sinclair D, Wainwright L. How lab staff and the estimation of free light chains can combine to aid the diagnosis of light chain disease. Clin Lab 2007;53:267-71.

Singer M, Berg P. Part III, The Molecular Anatomy, Expression and Regulation of Eukaryotic Genes. Genes and Genomes: Blackwell Scientific Publications Ltd, 1991.

Singhal S, Stein R, Vickrey E, Mehta J. The serum-free light chain assay cannot replace 24-hour urine protein estimation in patients with plasma cell dyscrasias. Blood 2007;109:3611-2.

Sirohi B, Powles R, Kulkarni S, Carr-Smith HD, Patel G, Das M, et al. Serum free light chain assessment in myeloma patients who are in complete remission (CR) by immunofixation predicts early relapse. Blood 2003;102:5195a.

Smedby KE, Vajdic CM, Falster M, Engels EA, Martínez-Maza O, Turner J, et al. Autoimmune disorders and risk of non-Hodgkin lymphoma subtypes: a pooled analysis within the InterLymph Consortium. Blood 2008;111:4029-38

Smith A, Wisloff F, Samson D. Guidelines on the diagnosis and management of multiple myeloma 2005. Br J Haematol 2006;132:410-51.

Smith LJ, Long J, Carr-Smith HD, Bradwell AR. Measurement of immunoglobulin free light chains by automated homogeneous immunoassay in serum and plasma samples. Clin Chem 2003;49:D58a.

1) Snozek CLH, Katzmann JA, Kyle RA, Dispenzieri A, Larson DR, Clark RJ, et al. Prognostic value of the serum free light chain ratio in patients with newly diagnosed myeloma: proposed incorporation into the international staging system. Blood 2007;110:659a.

2) Snozek CL, Katzmann JA, Kyle RA, Dispenzieri A, Larson DR, Therneau TM, et al. Prognostic value of the serum free light chain ratio in newly diagnosed myeloma: proposed incorporation

into the international staging system. Leukemia 2008;22:1933-7.

Snyder MR, Clark R, Bryant SC, Katzmann JA. Quantification of urinary light chains. Clin Chem 2008;54:1744-6.

1) Sölling K. Free light chains of immunoglobulins in normal serum and urine determined by radioimmunoassay. Scand J Clin Lab Invest 1975;35:407-12.

2) Sölling K. Polymeric forms of free light chains in serum from normal individuals and from patients with renal diseases. Scand J Clin Lab Invest 1976;36:447-52.

3) Sölling K. Normal values for free light chains in serum different age groups. Scand J Clin Lab Invest 1977;37:21-5.

4) Sölling K. Free light chains of immunoglobulins. Scand J Clin Lab Invest Suppl 1981;157:1-83.

5) Sölling K, Nielsen JL, Solling J, Ellegaard J. Free light chains of immunoglobulins in scrum from patients with leukaemias and multiple myeloma. Scand J Haematol 1982;28:309-18.

6) Sölling K, Solling J, Lanng Nielsen J. Polymeric Bence Jones proteins in serum in myeloma patients with renal insufficiency. Acta Med Scand 1984;216:495-502.

7) Sölling K, Solling J, Romer FK. Free light chains of immunoglobulins in serum from patients with rheumatoid arthritis, sarcoidosis, chronic infections and pulmonary cancer. Acta Med Scand 1981;209:473-7.

1) Solomon A. Light chains of human immunoglobulins. Methods Enzymol 1985;116:101-21.

2) Solomon A, Weiss DT, Herrera GA. Light-chain deposition disease. In: Mehta J, Singhal S, eds. Myeloma: Informa Healthcare, 2002:507-18.

Stubbs P. The Binding Site Group Ltd. Personal communication.

Sullivan PW, Salmon SE. Kinetics of tumor growth and regression in IgG multiple myeloma. J Clin Invest 1972;51:1697-708.

Takagi K, Kin K, Itoh Y, Enomoto H, Kawai T. Human alpha 1-microglobulin levels in various body fluids. J Clin Pathol 1980;33:786-91.

Tan T-S, Dispenzieri A, Lacy MQ, Hayman SR, Buadi FK, Zeldenrust SR, et al. Melphalan and dexamethasone is an effective therapy for primary systemic amyloidosis. Blood 2007;110:3608a.

Tanenbaum ND, Howell DN, Middleton JP, Spurney RF. Lambda light chain deposition disease in a renal allograft. Transplant Proc 2005;37:4289-92.

Tang G, Snyder M, Rao LV. Assessment of serum free light chain (FLC) assays with immunofixation electrohphoresis (IFE) and bone marrow (BM) immunophenotyping in the diagnosis of plasma cell disorders. Clin Chem 2008;54:A96a.

1) Tate JR, Gill D, Cobcroft R, Hickman PE. Practical considerations for the measurement of free light chains in serum. Clin Chem 2003;49:1252-7.

2) Tate JR, Grimmett K, Mead GP, Cobcroft R, Gill D. Free light chain ratios in the serum of myeloma patients in complete remission following autologous peripheral blood stem cell transplantation. Clin Chem 2002;48:E50a.

3) Tate JR, Mollee P, Dimeski G, Carter AC, Gill D. Analytical performance of serum free light-chain assay during monitoring of patients with monoclonal light-chain diseases. Clin Chim Acta

2007;376:30-6.

4) Tate J, Mollee P, Gill D. Serum free light chains for monitoring multiple myeloma. Br J Haematol 2005;128:405-6; author reply 6-7.

1) Tencer J, Thysell H, Andersson K, Grubb A. Long-term stability of albumin, protein HC, immunoglobulin G, kappa- and lambda-chain-immunoreactivity, orosomucoid and alpha 1-antitrypsin in urine stored at -20 degrees C. Scand J Urol Nephrol 1997;31:67-71.

2) Tencer J, Thysell H, Andersson K, Grubb A. Stability of albumin, protein HC, immunoglobulin G, kappa- and lambda-chain immunoreactivity, orosomucoid and alpha 1-antitrypsin in urine stored at various conditions. Scand J Clin Lab Invest 1994;54:199-206.

Teppo AM, Groop L. Urinary excretion of plasma proteins in diabetic subjects. Increased excretion of kappa light chains in diabetic patients with and without proliferative retinopathy. Diabetes 1985;34:589-94.

Terrier B, Sene D, Saadoun D, Ghillani-Dalbin P, Thibault V, Delluc A, et al. Serum-free light chain assessment in hepatitis C virus-related lymphoproliferative disorders. Ann Rheum Dis 2009;68:89-93.

Testa A, Dejoie T, Lecarrer D, Wratten M, Sereni L, Renaux JL. Reduction of Free Immunoglobulin Light Chains Using Adsorption Properties of Hemodiafiltration with Endogenous Reinfusion. Blood purification 2010;30:34-6.

Thio M, Blokhuis BR, Nijkamp FP, Redegeld FA. Free immunoglobulin light chains: a novel target in the therapy of inflammatory diseases. Trends Pharmacol Sci 2008;29:170-4.

Tillyer CR, Iqbal J, Raymond J, Gore M, McIlwain TJ. Immunoturbidimetric assay for estimating free light chains of immunoglobulins in urine and serum. J Clin Pathol 1991;44:466-71.

Tryggvason K, Wartiovaara J. How does the kidney filter plasma? Physiology (Bethesda) 2005;20:96-101.

Tsai H-T, Caporaso NE, Kyle RA, Katzmann JA, Dispenzieri A, Hayes RB, et al. Evidence of serum immunoglobulin abnormalities up to 9.8 years before diagnosis of chronic lymphocytic leukemia: a prospective study. Blood 2009;114:4928-32.

Udey JA, Blomberg B. Human lambda light chain locus: organization and DNA sequences of three genomic J regions. Immunogenetics 1987;25:63-70.

Urban S, Oppermann M, Reucher SW, Schmolke M, Hoffmann U, Hiefinger-Schindlbeck R, Helmke KH. Free light chains (FLC) of immunoglobulins as parameter resembling disease activity in autoimmune rheumatic diseases. Ann Rheum Dis 2004;63:141a.

van der Heijden M, Kraneveld A, Redegeld F. Free immunoglobulin light chains as target in the treatment of chronic inflammatory diseases. Eur J Pharmacol 2006;533:319-26.

van Hoeven KH, Bilotti E, McBride L, Berges T, McNeill A, Schillen D, Siegel D. Serum free light chain assays are more sensitive than urinary tests for light chain monoclonal proteins. Clin Chem 2009;55:C30a.

van Rhee F, Bolejack V, Hollmig K, Pineda-Roman M,

Anaissie E, Epstein J, et al. High serum-free light chain levels and their rapid reduction in response to therapy define an aggressive multiple myeloma subtype with poor prognosis. Blood 2007;110:827-32.

van Zaanen HC, Diderich PP, Pegels JG, Ruizeveld de Winter JA. [Renal insufficiency due to light chain multiple myeloma]. Ned Tijdschr Geneeskd 2000;144:2133-7.

1) Vermeersch P, Vercammen M, Holvoet A, Broeck IV, Delforge M, Bossuyt X. Use of interval-specific likelihood ratios improves clinical interpretation of serum FLC results for the diagnosis of malignant plasma cell disorders. Clin Chim Acta 2009;410:54-8.

2) Vermeersch P, Van Hoovels L, Delforge M, Marien G, Bossuyt X. Diagnostic performance of serum free light chain measurement in patients suspected of a monoclonal B-cell disorder. Br J Haematol 2008;143:496-502.

Viedma JA, Garrigos N, Morales S. Comparison of the sensitivity of 2 automated immunoassays with immunofixation electrophoresis for detecting urine Bence Jones proteins. Clin Chem 2005;51:1505-7.

Vitali C, Bencivelli W, Isenberg DA, Smolen JS, Snaith ML, Sciuto M, et al. Disease activity in systemic lupus erythematosus: report of the Consensus Study Group of the European Workshop for Rheumatology Research. II. Identification of the variables indicative of disease activity and their use in the development of an activity score. The European Consensus Study Group for Disease Activity in SLE. Clin Exp Rheumatol 1992;10:541-7.

1) Wakasugi K, Sasaki M, Suzuki M, Azuma N, Nobuto T. Increased concentrations of free light chain lambda in sera from chronic hemodialysis patients. Biomater Artif Cells Immobilization Biotechnol 1991;19:97-109.

2) Wakasugi K, Suzuki H, Imai A, Konishi S, Kishioka H. Immunoglobulin free light chain assay using latex agglutinin. Int J Clin Lab Res 1995;25:211-5.

Waldeyer W. Ueber Bindegewebszellen. Archiv fur Microbiologie und Anatomie 1875;11:176-94.

Waldmann TA, Strober W, Mogielnicki RP. The renal handling of low molecular weight proteins. II. Disorders of serum protein catabolism in patients with tubular proteinuria, the nephrotic syndrome, or uremia. J Clin Invest 1972;51:2162-74.

Walker R, Rasmussen E, Cavallo F, Jones-Jackson L, Anaissie E, Alpe T, et al. Correlation of suppression of FDG PET uptake with serum free light chain levels - both FDG PET-CT and serum clonal free light chain response precede and predict the likelihood of subsequent complete remission in newly diagnosed multiple myeloma. Blood 2005;106:3493a.

Wands C, Powell M, Jupp R. The development of serum free light chain immunoassay on Bayer Advia. Proceedings of ACB National Meeting 2003;2:3a.

Wang L, Young DC. Suppression of polyclonal immunoglobulin production by M-proteins shows isotype specificity. Ann Clin Lab Sci 2001;31:274-8.

Ward AM, White PAE, Beetham, R. Monoclonal protein identification distribution 986. UK NEQAS Sheffield, 1998.

Weber D, Treon SP, Emmanouilides C, Branagan AR, Byrd JC, Blade J, Kimby E. Uniform response criteria in Waldenstrom's macroglobulinemia: consensus panel recommendations from the

Second International Workshop on Waldenstrom's Macroglobulinemia. Semin Oncol 2003;30:127-31.

Weber DM, Dimopoulos MA, Moulopoulos LA, Delasalle KB, Smith T, Alexanian R. Prognostic features of asymptomatic multiple myeloma. Br J Haematol 1997;97:810-4.

Weber FP, Ledingham JC. A Note on the Histology of a Case of Myelomatosis (Multiple Myeloma) with Bence-Jones Protein in the Urine (Myelopathic Albumosuria). Proc R Soc Med 1909;2:193-206.

1) Wechalekar AD, Goodman HJB, Lachmann HJ, Offer M, Hawkins PN, Gillmore JD. Safety and efficacy of risk-adapted cyclophosphamide, thalidomide, and dexamethasone in systemic AL amyloidosis. Blood 2007;109:457-64.

2) Wechalekar AD, Hawkins PN, Gillmore JD. Perspectives in treatment of AL amyloidosis. Br J Haematol 2008;140:365-77.

3) Wechalekar AD, Lachmann HJ, Goodman HJB, Bradwell AR, Hawkins PN. Role of serum free light chains in diagnosis and monitoring response to treatment in light chain deposition disease. Haematologica 2005;90:PO1414a.

4) Wechalekar AD, Lachmann HJ, Gillmore JD, Goodman HJB, Offer M, Hawkins PN. The role of renal transplantation in systemic AL amyloidosis. Blood 2005;106:3497a.

5) Wechalekar A, Harding S, Lachmann H, Gillmore J, Wassef NJ, Thomas M, et al. Serum immunoglobulin heavy/light chain ratios (Hevylite) in patients with systemic AL amyloidosis. Amyloid 2010;17:P186a.

Weiss BM, Abadie J, Verma P, Howard RS, Kuehl WM. A monoclonal gammopathy precedes multiple myeloma in most patients. Blood 2009;113:5418-22.

Wells JM, van Hoeven KH, Abadie JM. Serum free light chains assays in clinical practice: Impact of impaired renal function on interpretation. Clin Chem 2008;54:C91a.

Whicher JT, Hawkins L, Higginson J. Clinical applications of immunofixation: a more sensitive technique for the detection of Bence Jones protein. J Clin Pathol 1980;33:779-80.

Winearls CG. Acute renal failure: Myeloma kidney. In: Johnson RJ, Feehally J, eds. Comprehensive clinical nephrology, Mosby, 2003.

Wochner RD, Strober W, Waldmann TA. The role of the kidney in the catabolism of Bence Jones proteins and immunoglobulin fragments. J Exp Med 1967;126:207-21.

Wolff F, Thiry C, Willems D. Assessment of the analytical performance and the sensitivity of serum free light chains immunoassay in patients with monoclonal gammopathy. Clin Biochem 2007;40:351-4.

Yegin ZA, Ozkurt ZN, Yagci M. Free light chain : A novel predictor of adverse outcome in chronic lymphocytic leukemia. Eur J Haematol 2010;84:406-11.

Ying WZ, Sanders PW. Mapping the binding domain of immunoglobulin light chains for Tamm-Horsfall protein. Am J Pathol 2001;158:1859-66.

Zucchelli P, Pasquali S, Cagnoli L, Ferrari G. Controlled plasma exchange trial in acute renal failure due to multiple myeloma. Kidney Int 1988;33:1175-80.

Index

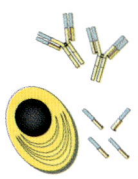

Important pages are marked in bold
Figures and clinical case histories are high-lighted in blue
Questions and answers are high-lighted in red - p263-71
Chapter summaries are high-lighted in red

A

Abbreviations	xvi
Acknowledgements	xv, 350
Age related changes	**38-40**, *Figs 5.1-5.3, Table 5.3,* 175, 178
Algorithms for monoclonal protein measurements	**198-9**, *Table 23.1*
Amyloidosis and free light chains	**130-47**, Q18
amyloid P scans	138-9, *Figs 15.12 -15.13, 15.18*
brain naturetic peptide	144-5
cardiac dysfunction	144-5
cardiac transplantation	144, *Figs 15.18, 15.20*
clinical case history No. 5: diagnosis	137-8, *Fig 15.11*
clinical case history No. 6: monitoring	142-3, *Fig 15.18*
clinical case history No. 7: renal failure	145-6, *Fig 15.21*
clinical features	130-1, *Figs 15.1-15.2*
comparison with other tests	132-7, *Figs 15.5-15.9*
diagnosis	**132-7**, 213-4, *Figs 15.3-15.11, Tables 15.1, 24.1-24.3*
discordant free light chain results	53, **136-7**, 217-8, 236-8, *Table 24.4*
guidelines	140, 222, **228-9**
importance	138
incidence	131, *Fig 7.0*, **Table 7.0**,
localised amyloid disease	**148**
monitoring	138-46, *Figs 15.12-15.21*
non-Hodgkin lymphomas	162
prognosis	137, *Fig 15.10*
renal failure	145-6, *Fig 15.21*, Q22
renal transplantation	144
sensitivity of tests	*Table 15.1*
stem cell transplantation	143-4, *Fig 15.19*
summary	130
treatment responses	143-6, *Figs 15.12-15.16*
Antibody structure	10-2, *Figs 1.0, 3.1-3.3*
Antigen excess	25-7, *Figs 4.8-4.9*, Q40

Antisera batch variation | 32-3, 237, *Fig 4.16*, Q38

Antisera for free light chains | **20-3**, 32-3, *Figs 4.1-4.4*

batch comparison | 32-3, 237, *Fig 4.16*

production scale-up | 34, *Fig 4.18-4.19*

Assays for free light chains | 18-20, **45-55**, *Figs 6.1, 6.3-6.9, Table 4.1*

accuracy | **23-4**, 31, 42, 52-4, 70-2, 198 *Tables 5.4, 23.1*, Q42

antigen excess | **25-7**, *Figs 4.8-4.9*, Q40

antisera | **20-23**, 32-3, *Figs 4.1-4.4*

borderline results | **41**, 176-9, 233-5, *Figs 20.1-20.2, 20.5-20.6, 26.2, Table 26.1*, Q33

calibration curves | 25, *Figs 4.6-4.7*

clonality assessment | **38**, 233-6, *Fig 26.2, Table 26.1*, Q31

comparison of assays | *(see Comparison of)*

development | 20-34

different batches | **237**

false positives & negatives | 41-2, **52-4**, 136-7, 211-2, *Tables 5.4, 24.2*, Q34

ideal test characteristics | 4, **18-20**

implementation | **231-6**

inaccuracy | 23-4, 52-4, **236-7**

instrumentation | 31, **239-51**, *Figs 27.1-27.11, Tables 4.3-4.4, 27.1-27.12*, Q41

interpretation | **233-4** *Fig 26.2, Table 26.1*, Q20

limitations | **236-7**

linearity | **27-8**, 237, *Fig 4.12*

monoclonal antibodies | 20, 36, *Tables 5.1, 5.5*

normal ranges | *(see Free light chains)*

plasma versus serum | 30-1, *Figure 4.14, Table 4.3*

polyclonal antisera | 20-3, *Figs 4.1-4.4*

polymerisation | 29, 71-2, *Figs 9.2-9.3*

precision | 27, *Figs 4.10-4.11, Tables 4.4, 27.1-27.12*

production | 20, 34, *Fig 4.19*

quality assurance | **252-7**, *Figs 28.1-28.7*

quality control | 32-4, *Fig 4.16*, Q38

ranges of assays | 25, *Figs 4.6-4.7, Tables 27.1-27.12*

scale-up | 34, *Fig 4.19*

sensitivity: analytical | 24, *Fig 6.1, Tables 4.2*, **27.1-27.12**, Q37

sensitivity: clinical | 41-2, **45-55**, 198-206, *Figs 6.1-6.9, 23.5, 24.6 Tables 5.4, 23.1, 23.3, 24.2*

specificity | **20-3**, 41-2, *Figs 4.1-4.4, 23.5, Table 5.4*

stability | 30, *Fig 4.13*

standardisation | **23-4**, 237, *Fig 4.5*

urine | *(see Urine free light chains)*

utility | 41-42, *Fig 23.5, Table 5.4*

Asymptomatic (smouldering) multiple myeloma | **126-9**, *Figs 14.1-14.4*

summary | 126

Audit of serum free light chain usage   207, *Table 23.4*

B

β_2 microglobulin   79, 97, 224, 312 *Figs 10.2, 12.3, Table 18.4, 25.1*
B-cell chronic lymphocytic leukaemia   162-4, *Figs 7.0*, **18.9-18.12**,
                                                                       Tables 18.1-18.3, 23.2
B-cell development   12-4, *Figs 3.3-3.4, 18.7-18.8*
B-cell non-Hodgkin lymphomas   159-62, *Figs 18.7-18.10, Tables 18.1-18.2*
       complicated by AL amyloidosis   162
       summary   164
Beckman instruments   *(see Instrumentation)*
Bence Jones, Henry   6-9 *Fig 2.1*
Bence Jones protein test   7-8
Biclonal gammopathies   238, *Fig 26.4*
Binding Site instruments   *(see Instrumentation)*
Biology of immunoglobulin light chains   10-17
       summary   10
Bone marrow biopsies   86-8, 97, *Figs 11.1-11.2, 12.3 Table 12.1*
Bone marrow suppression   233-4, *Fig 26.1-26.2*, Q23
Borderline results   **41**, 176-79, 233-5, *Figs 20.1-20.2, 20.5-20.6, 26.2,*
                                           Table 26.1, Q33
Brambell receptors (FcRn) and IgG recycling   77, **79**, 301, 309-10 *Fig 10.2*

C

Capillary zone electrophoresis (CZE)   **49**, 200, *Figs 6.1, 6.7, Tables 4.1,*
                                                                       23.2, Q37
Cast nephropathy   **14-6**, 113-4, 123-4, *Figs 3.7-3.9, 13.1-13.3, 13.20*
Cerebrospinal fluid and free light chains   194-6, *Figs 22.1-22.2*
       summary   194
Chronic kidney disease *(see Renal impairment)*   175-183
Chronic lymphocytic leukaemia   162-4, *Figs 7.0*, **18.9-18.12**,
                                                                       Tables 18.1-18.3, 23.2

Classification of monoclonal gammopathies   259-60
       associated with non-malignant disorders   260-2
       incidence and frequency   57-8, *Figs 7.0-7.1, Tables 7.0-7.1*
       malignant disorders - new classification   259, Q10
                                           - old classification   259-60
Classification of polyclonal gammopathies   260-2
Clinical case histories:
       No. 1: Light chain multiple myeloma   66-7, *Fig 8.8*
       No. 2: Light chain multiple myeloma   68, *Fig 8.9*
       No. 3: Nonsecretory multiple myeloma   73-5, *Figs 9.5-9.6*
       No. 4: Nonsecretory multiple myeloma   75-6, *Fig 9.7*
       No. 5: AL amyloidosis   137-8, *Fig 15.11*

No. 6: AL amyloidosis 142-3, *Fig 15.18*
No. 7: AL amyloidosis 145-6, *Fig 15.21*
No. 8: Light chain deposition disease 150-1, *Fig 17.2*
No. 9: Light chain deposition disease 152-3, *Fig 17.3*
No. 10: Effect of renal function on light chains 182-3, *Fig 20.11*
No. 11: Is urine examination a mandatory procedure? 219-20, *Table 24.7*
Clonality assessment **38**, 233-6, *Fig 26.2, Table 26.1* Q31
Comparison of immunoassays for FLCs 18-20 31-2, **45-55**, *Figs 6.1-6.9,
Tables 4.1, 4.4, 6.1-6.2,*

      summary 45
Contents v-xii
Cost benefits 218-9, *Table 24.5*
CRAB criteria 222
Creatinine *(see Renal impairment)*
Cryoglobulinaemia 165
Cystatin C 37, 177 *Fig 5.1, Table 20.1*

D

Dade-Behring BNII (see Instrumentation)
Diabetes mellitus 189-90, *Figs 21.7-21.8*
Discordant free light chain results 52-4, **136-7**, 217-8, 236-7, *Table 24.4*
Drug selection using FLC measurements 99-100, **108-9**, *Fig 12.19, Table 12.2*

E

Electrophoresis
    Capillary zone *(see Capillary zone electrophoresis)*
    Immunoelectrophoresis 20 *Fig.4.1*
    Immunofixation electrophoresis
        AL amyloidosis 132-5, *Figs 15.3-15.9 Table 15.1*
        chronic lymphocytic leukaemia 162-4, *Fig 18.10, Table 18.3*
        comparison with other tests *Figs 6.1, 6.9, Table 4.1*
        intact immunoglobulin myeloma **88-90**, 95-104, *Figs 11.3-11.6,
        12.5-12.10, Table 12.1*

        light chain deposition disease 149-53, *Figs 17.1-17.2, Table 17.1*
        light chain multiple myeloma 61-8, *Figs 8.1-8.2, 8.9*
        lymphoma 159-64, *Figs 18.5, 18.9-18.10, Tables 18.2-
        18.3*

        MGUS 167-8, *Figs 19.1, 19.5*
        nonsecretory myeloma 70-2, *Fig 9.1, Table 9.1*
        quality assurance 252-7, *Figs 28.1-28.5*
        questions and answers Q2, 6-7, Q13-14, Q28, Q30
        screening studies 198-206, *Figs 23.2-23.3, Table 23.2, 23.3*
        sensitivity for diagnosis 198-9, *Fig 24.6, Tables 23.1, 23.3,*
        urine 50-5, 208, *Fig 23.3*, **203-7**, *Tables 6.1, **24.2**,
        24.4*

Serum protein tests (SPE)   *45-50, Figs 6.1-6.5, 6.9, Tables 4.1, 6.3*
Urine protein tests (UPE)   *50-4, 208-20, Figs 6.1, 6.5, 24.1-*
                                  24.4, Tables 4.1, 6.2-6.3, 24.2- 24.6 Q1-9, 11
Extramedullary plasmacytoma   156

F

FcRn (neonatal) receptors and IgG recycling   77, **79**, 98, 301, 309-10, *Fig 10.2*
Free light chains
    as bioactive molecules   192-3, *Fig 21.12*
    clearance and metabolism   **14-17**, 77-85, *Figs 3.5-3.10, 10.1-10.10*
    comparison of assays   *(see Comparison of......)*
        of immunoassays   32
    epitopes   10-12, *Fig 3.1*
    escape (breakthrough)   104-5, *Figs 12.12-12.13*, Clinical case history
                                  No. 2
    kinetics/half-life   14-16, 78, **108-9**, *Figs 10.1, 10.4-10.5*, Q24
    normal ranges for serum   36-9, *Figs 5.1-5.2, Tables 5.1-5.3*, Q19
        in children   38-9, *Fig 5.3*
        urine   42-3, *Figs 5.5-5.6, Tables 5.4-5.5*
    polyclonal   184-93, *Figs 21.1-21.12*
    polymerisation   29, **71-2**, *Figs 9.2-9.3*
    production   12-4, 77, *Figs 3.5-3.6*
    prognosis   177, *Fig 20.3*
    renal tubular damage   **15-7**, 111-4, *Figs 3.7-3.9, 13.1-13.3*
    structure   10-2, *Figs 1.0, 3.1-3.3,*
    subgroups   **11-2**, 131, 149
    synthesis   12, *Figs 3.3-3.4*

G

Gel filtration   20, 34, 71-2, *Figs 4.19, 9.2-9.3*
Glomerular filtration rate   14-17, 37, 107, 177, *Chapters 13, 20, Figs 3.7,*
                                  12.15, 20.1-20.2, 20.4, Tables 20.1, 20.3
Guidelines for assessing monoclonal gammopathies   221-30
    AL amyloidosis   140, 228-9
    Assessing outcome in MGUS   168-70, 229-30
    Consensus statement for the screening, evaluation and
    management of MGUS   230
    European Society of Medical Oncology (ESMO)   226
    International guidelines for the classification of MM and MGUS   221-2
    International Myeloma Working Group guidelines for MGUS
    and smoldering (asymptomatic) multiple myeloma   229-30
    International Myeloma Working Group guidelines for serum
    free light chain (sFLC) analysis in multiple myeloma and
    related disorders (2009)   222-3

International staging system for multiple myeloma (2005)   224, *Table 25.1*
National Comprehensive Cancer Network   226
UK Myeloma Forum and Nordic Myeloma Study Group:
Guidelines for the investigation of newly detected M-proteins
and management of MGUS (2009)   223-4
UK, Nordic and British Committee for Standards in Haematology (2006) 223
"Uniform response criteria for MM" incorporating FLCs
(2009)   224-6, *Tables 25.2-25.4*
USA National Academy of Clinical Biochemistry guidelines   226

H

Haemodialysis and free light chains   **117-25**, 179-81, *Figs 13.8-19, 20.2. 20.5-20.6*
Haemodialysis membranes   **119-20**, 179-81, *Figure 13.10*
Half-life of serum free light chains   11-17, **78**, **108**, *Figs 10.1, 10.4-10.5*, Q24
    serum immunoglobulins *(see FcRn, and free light chains: clearance/kinetics)*
Hevylite - assays for heavy chain/light chain immunoglobulin pairs
    AL amyloidosis   307-9, *Figs 32.7-32.10*
    antibody specificity   302-3, *Fig 32.3*
    concept   302, *Figs 32.1-32.2*
    immunohistochemistry   317-8, *Figs 32.19-20*
    monitoring multiple myeloma   309-11, *Figs 32.11-32.13*
    non-Hodgkin's lymphoma   317, *Fig 32.18*
    normal ranges   304, *Figs 32.4-6, Table 32.1*
    prognostic value   312-16, *Figs 32.14-32.17, Table 32.3*
    publications on FcRn receptors   318-19
    publications on Hevylite assays   319-20
    sensitivity   306-9, *Figs 32.4-5, 32.7-9, Table 32.2*
    summary   301
History   6-9
Hitachi   *(see Instrumentation: Roche)*
Hodgkin's lymphoma   192

I

Immunoassays for FLC measurements   18-35
    summary   18
Immunoelectrophoresis   20, *Fig 4.1*
Immunofixation electrophoresis   *(see Electrophoresis)*
Immune stimulation and polyclonal FLCs   184-93
    summary   184
Implementation of free light chain assays   231-8, Q39
    getting started   232-3
    guidelines for FLC usage   221-30
    interpretation of results   233-6, *Fig 26.2, Table 26.1*, Q20
    limitations of assays   52-4, **136-7**, 217-8, 236-8, *Table 24.4*
    monitoring patients   235-6, *Figs 26.1, 26.3*

    quality assurance   252-7, *Figs 28.1-28.7*
    summary   231
Incidence of monoclonal gammopathies   57-8, *Figs 7.0-7.1*
Infectious diseases and free light chains   190-2
Instrumentation   31, **239-51**, *Figs 27-1-27.11, Tables 4.3-4.4, 27.1-27.12,* Q41
    Beckman Coulter AU® (AU400, 640, 2700 and
    5400)   241, *Tables 27.1,* ***27.5***
    Beckman Coulter IMMAGE and IMMAGE 800   242-3, *Figs 4.14, Tables
                                                       4.3-4.4,* ***27.1-27.2***
    Binding Site Minineph~PLUS~™   243, *Table* ***27.3***
    Binding Site SPA~PLUS~ bench-top analyser   244-5, *Tables 27.1,* ***27.4***
    Radim Delta   245, *Table* ***27.6***
    Roche COBAS c501 automated analyser   246, *Table* ***27.7***
    Roche COBAS Integra 400, 400 plus, 800
    automated analyser   246-7, *Tables 27.1,* ***27.8***
    Roche Hitachi 911/912/917 and Modular P   247-8, *Figs 4.6, 4.7, 4.10-4.12,
                                                       4.15, Tables 4.3,4.4, 27.1,* ***27.9***
    Siemens (Bayer) Advia 1650, 1800, 2400   248-9, *Tables 27.1,* ***27.10***
    Siemens BNII   249-50, *Fig 4.15, Tables 4.3-4.4, 27.1,* ***27.11***
    Siemens BN ProSpec   250-1, *Tables 27.1,* ***27.12***
Instrument comparisons   239-41, *Fig 4.15, Table 4.4,* ***27.1***
Intact immunoglobulin multiple myeloma   57-60, **77-110**, 301-20, *Fig 7.1, Table 7.0*
    blood volume and immunoglobulins   82-3, *Fig 10.7*
    bone marrow biopsies   86-8, 97, *Figs 11.1-11.2, 12.3, Table 12.1*
    correlation of serum light chains and IgG   90-1, *Figs 11.7, 11.8*
    disease stage and serum free light chains   91-3, *Figs 11.9-11.11, Table 11.1*
    drug selection and free light chain responses   108-9, *Fig 12.19*
    FcRn receptors and IgG recycling   *(see FcRn)*
    free light chain escape (breakthrough)   104-5, *Figs 12.12-12.13*
    free light chain levels at presentation   88-91, *Figs 11.3-11.8,*
    frequency of free light chains   88-90, *Fig 11.5*
    half-life of intact immunoglobulins   79-80, *Figs 10.2-10.6*
    half-life of serum free light chains   78, *Figs 10.1,10.4-10.5*
    incidence   58, *Table 7.0*
    international staging system and free light chains   91-3, *Fig 11.10, Table 11.1*
    kinetics   77-85, Q24
    monitoring   95-109
    relapse   102-4, *Figs 12.10-12.13*
    renal damage and free light chains   111-25, 175-83
    residual disease and free light chains   100-2, *Figs 12.5-12.9*
    response rates and free light chains   100-2, *Figs 12.6-12.8*
    speed of responses using free light chains   98-102, *Fig 12.4-12.5, Table 12.2*
    summaries   77, 86, 95
    theoretical considerations   77-85
    transplantation and free light chains   105-7, *Figs 12.14-18*

tumour killing rates	80-5, *Figs 10.4-10.5, 10.8-10.10*
utility of sFLCs in IIMM	109
Interpretation	233-6, *Fig 26.2, Table 26.1,* Q20
Intrathecal inflammation	194-6, *Figs 22.1-22.2*

K

Kappa/lambda ratios (normal)	36-9, *Tables 5.1-5.2*
effect of renal impairment	178-9, *Figs 20.2, 20.4, 20.5 Table 20.1*
Kidney and monoclonal free light chains	111-25
summary	111
Kinetics/half-life	14-17, 89-92, **108-9**, *Figs 10.1, 10.4-10.5,* Q24
Korngold L	7

L

Lapiri R	7
Leukaemia	*(see B cell chronic lymphocytic)*
plasma cell	157
Light chain casts	**15-6**, 111-4, *Figs 3.7-3.9, 13.1-13.3*
Light chain deposition disease	149-53, *Table 7.0*
clinical case histories 8 and 9	150-3, *Figs 17.2-17.3*
diagnosis using serum free light chains	149-51, *Fig 17.1, Table 17.1*
monitoring using serum free light chains	151-3
summary	149
Light chain multiple myeloma	61-8, *Fig 7.0, Table 7.0*
clinical case histories 1 and 2	66-8
diagnosis	61-4, *Figs 8.1-8.3, Table 8.1*
haemodialysis	*(see Haemodialysis)*
incidence	*Table 7.0*
monitoring	64-8, *Figs 8.4-8.9,* Q16
myeloma kidney	*(see Cast nephropathy)*
nephrotoxicity	111-25
plasma exchange	117-9, *Figs 13.8-13.9*
residual disease	65
screening trials	199-206, *Figs 23.1-23.5, Table 23.2*
serum versus urine tests	3, 65, **208-20**, *Figs 8.5, 24.1-24.6, Tables 24.1-24.7*
summary	61
Light chain structure	10-12, *Figs 1.0, 3.1-3.3*
Limitations of free light chain assays	236-8
Localised amyloid disease	148, *Table 16.1*
Lymphomas	*(see B-cell)*

M

Methods for free light chain measurement	*(see Assays for free light chains)*

MGUS   57-8, **166-74**
    as the precursor condition for multiple myeloma   173
    definition and frequency   166, *Fig 7.0, Table 7.0*
    guidelines for management   168-70, 223-4, 229-30, *Table 19.1*
    MGUS and monoclonal FLCs   168-70, *Fig 19.1*, Q15
    MGUS and heavy/light chain
    analysis   312-3, *Table 32.3*
    risk stratification using sFLCs   168-70, *Figs 19.2-19.4, Tables 19.1, 25.5*
    sFLC MGUS   170-2, *Fig 19.5*
    summary   166
Monoclonal gammopathies
    classification   259-60, Q10
    incidence and frequency   57-8, *Fig 7.0, Table 7.0*
Multiple myeloma   57-129, *Fig 7.0, Table 7.0*
    asymptomatic (smouldering) myeloma *(see Asymptomatic multiple myeloma)*
    classification   60, 259-60, *Fig 7.3*
    diversity of immunoglobulin and FLC production   86-8, *Figs 11.1-11.2*
    guidelines   *(see Guidelines)*
    incidence   57-8, *Tables 7.0-7.1*
    intact immunoglobulin multiple myeloma *(see Intact immunoglobulin MM)*
    introduction   59-60
    light chain multiple myeloma   *(see Light chain multiple myeloma)*
    light chains and early response   *(see Kinetics)*
    nonsecretory multiple myeloma   *(see Nonsecretory multiple myeloma)*
Multiple sclerosis   194-6, *Figs 22.1-22.2*
Multiple solitary plasmacytomas   156-7
Myeloma kidney   *(see Cast nephropathy)*

N

Neonatal (FcRn) receptors and IgG recycling   77, **79**, 98, 301, 309-10, *Fig 10.2*
Nephron   14-17, **77-85**, *Figs 3.5-3.10, 10.1-10.10*
Nephrotoxicity and free light chains   *(see Cast nephropathy and Haemodialysis)*
New Scientific Company FLC immunoassays   32
Nonsecretory multiple myeloma   69-76, *Figs 7.3, 9.1-9.7, Table 7.0*, Q16
    clinical case histories 3 and 4   73-6, *Figs 9.5-9.7*
    definition   69, 221
    diagnosis   70-72, *Figs 9.1-9.3, Table 9.1*
    monitoring   72-6, *Figs 9.4-9.7*
    summary   69
Normal/reference ranges   36-44, Q19
    borderline results   41
    children   39-40, *Fig 5.3*
    hospital ranges   39-40, *Fig 5.4*
    kappa/lambda ratios   37, *Table 5.2*
        effect of renal failure   117, 178-9, *Figs 5.2, 13.7, 20.1-20.2, 20.5,*

	Table 20.1
serum	36-9, *Figs 5.1-5.2, Tables 5.1-5.4*
historical studies	36-7, *Table 5.1*
summary	36
urine	42-3, *Figs 5.5-5.6, Table 5.5*
utility for disease diagnosis	41-2, *Table 5.4*
variations/hospital ranges	39-40, *Fig 5.4*

O

Olympus analysers	*(see Instrumentation)*
Overview	xviii-xxi

P

Peritoneal dialysis	179-81, *Figs 20.1, 20.5*
Plasma cell leukaemia	157
Plasma cells	12-14, 86-7, *Figs 3.4-3.6, 11.1-11.2, 18.7-18.8*
Plasma exchange	117-9, *Figs 13.8-13.9*
Plasmacytoma	154-7, *Fig 7.0, Table 7.0*
extramedullary	156
multiple solitary	156-7
solitary	154-6, *Figs 18.1-18.4*
Pneumonia and polyclonal free light chains	190-1, *Fig 21.9*
POEMS syndrome	164-5
Polyclonal free light chains and all cause mortality	177, *Fig 20.3*
Polyclonal free light chains in renal disease	(see Renal impairment)
Polyclonal hypergammaglobulinaemia	184-93, *Figs 21.1-21.12*
classification	260-2
lymphoma prognosis	192
Prefaces	*xiii-xv*
Protocols for monoclonal protein diagnosis	198-9, *Table 23.1, 23.3*
Publications on serum free light chains	272-300

Q

Quality control of free light chain assays	32-4, *Figs 4.15-4.16*, Q38
Quality assurance schemes	252-7
Binding Site QA scheme	252-3, *Figs 28.1-28.2*
practical aspects	255-7, *Figs 28.5-28.7*
College of American Pathologists (CAP)	254, *Fig 28.3*
French QA scheme	254-5, *Fig 28.4*
Instand e.V. Laboratories GmbH	255
Randox Laboratories	255
UK NEQAS	255
Questions and answers about free light chains:-	

clinical questions about serum tests 264-7
laboratory questions about serum tests 267-71
urine tests 263-4

R

Radim Delta (see Instrumentation)
Reference intervals (see Normal ranges)
References used in the book 321-35
Reimbursement 218-9, *Tables 24.5-24.6*
Renal biopsy 14-17, 113-5, 124, 181, *Figs 3.8-3.9, 13.3, 13.20*
Renal diseases
    cast nephropathy **14-17**, 111-4, 124, *Figs 3.7-3.9, 13.1-13.3, 13.20*
        biopsies after haemodialysis 124, *Fig 13.20*
    diagnosis of acute myeloma kidney using sFLCs 116, *Fig 13.7*
    FLC changes in renal impairment 176-9, *Figs 20.1-20.2, 20.4-20.5*
    haemodialysis: *"high cut-off"* dialysers 119-25, *Figs 13.10-13.18, Table 13.1*
    κ/λ ratio changes in renal impairment 117, **178-9**, *Figs 5.1, 13.7, 20.1-20.2, 20.5, Tables 5.3, 20.1*, Q43
    mechanisms of FLC toxicity 113-4
    model of FLC removal methods 119-21, *Figs 13.8, 13.11, Table 13.1*
        removal efficiency *Table 13.1*
    monitoring patients **145-6**, 181-3, *Figs 15.21, 20.9*, Q22
    monoclonal light chains 181-3, *Figs 20.4, 20.7-20.8, Table 20.3*
    normal FLC clearance and metabolism 14-7, **77-85**, *Figs 3.5-3.10, 10.1-10.10*
    peritoneal dialysis 179-80, *Fig 20.5*
    removal of FLCs by dialysis **117-9**, 179-81, *Figs 13.8-13.9, 20.5-20.8*
    removal of FLCs by plasma exchange 117-9, *Figs 13.8-13.9*
    renal impairment 15-7, 63, **175-83**, 185-6, *Figs 3.7-3.9, 8.2-8.3, 21.3*, Q22, 26, 31
    renal tubular damage **14-7**, 111-4, *Figs 3.7-3.9, 13.1-13.3*
    summary - monoclonal FLCs and acute renal failure 111
    summary - polyclonal FLCs and chronic renal failure 175
Renal threshold 208-10, *Figs 24.1-24.2*
Residual disease and free light chains 100-2, *Figs 12.5-12.9*
Rheumatic diseases 184-8, 261 *Figs 21.2-21.6*
Rheumatoid arthritis 187-8, *Figs 21.2, 21.4, 21.6*
Roche instruments *(see Instrumentation)*

S

Screening studies using serum free light chains 197-207
    accuracy of diagnostic protocols 198-9, *Table 23.1, 23.3*
    audit of serum free light chain usage 53, 207, *Tables 6.3, 23.4*
    comparison with capillary zone electrophoresis 200, *Fig 23.1, Table 23.2*
    comparison with SPE/IFE 200-206, *Figs 23.2-23.5*

comparison with urine tests	200-206, **208-220**, *Fig 23.3, 24.6*
cost analysis	202, 218-9, *Table 24.5*
summary	197

Sensitivity of serum free light chain immunoassays:-
analytical sensitivity	24, *Fig 6.1, Tables 4.2,* ***27.1-27.12***, Q37
clinical sensitivity	41, **45-55**, 198-206, *Figs 6.1-6.9, 23.5, Tables 5.4, 23.1, 24.2*
summary	45

| Serum amyloid P scans | 138-9, *Figs 15.12 -15.15, 15.18* |

Serum free light chains
clinical questions about serum tests	264-7
laboratory questions about serum tests	267-71
normal ranges	(*see Normal reference ranges*)
Serum protein electrophoresis	(*see Electrophoresis*)
Serum versus urine free light chain tests	xviii-xxi, 50-54, **208-20**, Q8-9, 11
clinical benefits of serum tests	3-4, **212-3**, 231
cost/benefit analysis	202, 218-9, *Table 24.5*
elimination of urine studies	213-7, *Tables 24.2-24.3*
problems collecting urine samples	210-11, *Fig 24.3, Table 24.3*
problems measuring urine samples	211-2, *Fig 24.4*
renal threshold	14-7, 208-10, *Figs 3.7-3.10, 24.1-24.4,*
sensitivity of serum vs urine tests	(*see clinical sections and Tables 24.6-24.6*)
summary	208, *Table 24.6*
Siemens instruments	(*see Instrumentation*)
Sjögren's syndrome	187-8, *Figs 21.2, 21.4-21.5*
Smouldering (asymptomatic) myeloma	126-9, *Figs 14.1-14.4*
summary	126
Solitary plasmacytoma of bone	154-6, *Fig 18.1-18.4*
Staging of disease and serum free light chains	91-4, *Figs 11.9-11.11, Table 11.1*
Structure of free light chains and immunoglobulins	10-12, *Figs 3.1-3.2*
Summary of serum free light chain benefits	3
Systemic lupus erythematosus	185-6, *Figs 21.2-21.5*
Systemic sclerosis	185, *Fig 21.2*

T

Total kappa and lambda assays	50, *Figs 6.1, 6.8, Tables 4.1, 6.1*, Q35-36
Transplantation & free light chains	
bone marrow	73-5, 83-85, 91-94, **105-8**, 138-44
	Figs 9.6, 10.8-10.9, 11.10-11.11, 12.14-12.16, 15.19, Q25
heart	144, *Figs 15.18, 15.20*
renal	144
Tumour markers	1-4, *Table 1*

U

Urine free light chains (*of little consequence, but for those remaining enthusiasts......*)

amyloidosis   *(see Amyloidosis)*
compared with serum tests   *(see Serum versus urine)*
diabetes mellitus   *(see Diabetes)*
discordant with serum   *(see Discordant free light chain results)*
guidelines for use   *(see Guidelines)*
immunofixation electrophoresis   *(see Electrophoresis)*
intact immunoglobulin myeloma   *(see Intact immunoglobulin multiple myeloma)*
light chain deposition disease   *(see Light chain deposition disease)*
light chain myeloma   *(see Light chain multiple myeloma)*
MGUS   *(see MGUS)*
non-Hodgkin lymphoma   *(see B-cell, non Hodgkin lymphoma)*
nonsecretory myeloma   *(see Nonsecretory myeloma)*
normal ranges   *(see Normal ranges)*
protein electrophoresis   *(see Electrophoresis)*
protein tests   18-20, 50-2, *Tables 4.1, 6.1-6.2,* Q1
quality assurance schemes   *(see Quality assurance schemes)*
questions and answers   263-71
renal clearance and metabolism   **14-17**, 112-3, *Figs 3.7-3.10, 13.1-13.3*
renal threshold   14-17, 208-10, *Figs 3.7-3.10, 24.1-21.4*
sensitivity in diagnosis   198-9, *Fig 23.3, Table 23.1,* **23.3**
versus serum FLCs   *(see Serum versus urine free light chain tests)*
Utility of sFLCs in IIMM - comments on utility   109

V

Velcade and free light chain responses   **84-5**, 121-2, *Fig 10.10, 13.14*
Vasculitis and elevated free light chains   184-6, *Figs 21.2-21.3*
Virus hepatitis elevated free light chains   190-2, *Figs 21.10-21.11*

W

Waldenström's macroglobulinaemia   157-9, *Figs 7.0, 18.5-18.6, Table 7.0*
    summary   159
Why measure free light chain in serum?   231

Acknowledgements

* Acknowledgements of permission according to the figure numbers and the journals in which they were originally published.

Figure 2.1A. Albumen print by Maull c.1850s, reproduced with permission from The Royal Institution, London and The Bridgeman Art Library.

Figure 2.1B. Charcoal and chalk on paper by Richmond 1865, reproduced with permission from The Royal Institution, London and The Bridgeman Art Library.

Figures 3.7 & 13.1. Acute renal failure: Myeloma kidney. Winearls CG. In: Johnson RJ, Feehally J, eds. Comprehensive clinical nephrology, Mosby: Page 238, figure 17.5 © Elsevier (2003).

Figures 3.8 & 13.2. Investigation of renal disease: Urinalysis. Fogazzi GB. In: Johnson RJ, Feehally J, eds. Comprehensive clinical nephrology, Mosby: Page 41, figure 4.3B © Elsevier (2003).

Figure 5.3. Solling K. Polymeric forms of free light chains in serum from normal individuals and from patients with renal diseases. Scand J Clin Lab Invest 1976;36:447-52.

Figure 7.1. Kumar S, Rajkumar VS, Dispenzieri A, Lacy MQ, Hayman SR, Buadi FK, et al. Improving survival in multiple myeloma: Impact of novel therapies. Blood 2007;110:3594a. © the American Society of Hematology.

Figure 10.6. Anderson CL, Chaudhury C, Kim J, Bronson CL, Wani MA, Mohanty S. Perspective-- FcRn transports albumin: relevance to immunology and medicine. Trends Immunol 2006;27:343-8.

Figure 10.7. Alexanian R. Blood volume in monoclonal gammopathy. Blood 1977;49:301-7. © the American Society of Hematology.

Figures 11.1 & 11.2. Ayliffe MJ, Davies FE, de Castro D, Morgan GJ. Demonstration of changes in plasma cell subsets in multiple myeloma. Haematologica 2007;92:1135-8.

Figure 11.10. van Rhee F, Bolejack V, Hollmig K, Pineda-Roman M, Anaissie E, Epstein J, et al. High serum-free light chain levels and their rapid reduction in response to therapy define an aggressive multiple myeloma subtype with poor prognosis. Blood 2007;110:827-32. © the American Society of Hematology.

Figure 11.11. Dispenzieri A, Zhang L, Katzmann JA, Snyder M, Blood E, Degoey R, et al. Appraisal of immunoglobulin free light chain as a marker of response. Blood 2008;111:4908-15. © the American Society of Hematology.

Figure 13.9. Cserti C, Haspel R, Stowell C, Dzik W. Light-chain removal by plasmapheresis in myeloma-associated renal failure. Transfusion 2007;47:511-4.

Figures 14.2, 14.3 & 14.4. Dispenzieri A, Kyle RA, Katzmann JA, Therneau TM, Larson D, Benson J, et al. Immunoglobulin free light chain ratio is an independent risk factor for progression of smoldering (asymptomatic) multiple myeloma. Blood 2008;111:785-9. © the American Society of Hematology.

Figure 15.10 & 15.19. Dispenzieri A, Lacy MQ, Katzmann JA, Rajkumar SV, Abraham RS, Hayman SR, et al. Absolute values of immunoglobulin free light chains are prognostic in patients with primary systemic amyloidosis undergoing peripheral blood stem cell transplantation. Blood 2006;107:3378-83. © the American Society of Hematology.

Figure 15.15. Wechalekar AD, Goodman HJB, Lachmann HJ, Offer M, Hawkins PN, Gillmore JD. Safety and efficacy of risk-adapted cyclophosphamide, thalidomide, and dexamethasone in systemic AL amyloidosis. Blood 2007;109:457-64. © the American Society of Hematology.

Figure 15.16. Tan T-S, Dispenzieri A, Lacy MQ, Hayman SR, Buadi FK, Zeldenrust SR, et al. Melphalan and dexamethasone is an effective therapy for primary systemic amyloidosis. Blood 2007;110:3608a. © the American Society of Hematology.

Figure 15.20. Mignot A, Bridoux F, Thierry A, Varnous S, Pujo M, Delcourt A, et al. Successful heart transplantation following melphalan plus dexamethasone therapy in systemic AL amyloidosis. Haematologica 2008;93:e32-5.

Figures 18.2, 18.3 & 18.4. Dingli D, Kyle RA, Rajkumar SV, Nowakowski GS, Larson DR, Bida JP, et al. Immunoglobulin free light chains and solitary plasmacytoma of bone. Blood 2006;108:1979-83. © the American Society of Hematology.

Figure 18.6. Itzykson R, Le Garff-Tavernier M, Katsahian S, Diemert MC, Musset L, Leblond V. Serum-free light chain elevation is associated with a shorter time to treatment in Waldenstrom's macroglobulinemia. Haematologica 2008;93:793-4.

Figure 19.2, 19.3 & 19.4. Rajkumar SV, Kyle RA, Therneau TM, Melton LJ, 3rd, Bradwell AR, Clark RJ, et al. Serum free light chain ratio is an independent risk factor for progression in monoclonal gammopathy of undetermined significance. Blood 2005;106:812-7. © the American Society of Hematology.

Figure 19.6. Dispenzieri A, Katzmann JA, Kyle RA, Larson DR, Melton LJ, 3rd, Colby CL, et al. Prevalence and risk of progression of light-chain monoclonal gammopathy of undetermined significance: a retrospective population-based cohort study. Lancet 2010;375:1721-8.

Figures 21.4 & 21.6. Gottenberg JE, Aucouturier F, Goetz J, Sordet C, Jahn I, Busson M, et al. Serum immunoglobulin free light chain assessment in rheumatoid arthritis and primary Sjogren's syndrome. Ann Rheum Dis 2007;66:23-7.

Figures 21.10 & 21.11. Terrier B, Sene D, Saadoun D, Ghillani-Dalbin P, Thibault V, Delluc A, et al. Serum-free light chain assessment in hepatitis C virus-related lymphoproliferative disorders. Ann Rheum Dis 2009;68:89-93.

Figure 21.12. Thio M, Blokhuis BR, Nijkamp FP, Redegeld FA. Free immunoglobulin light chains: a novel target in the therapy of inflammatory diseases. Trends Pharmacol Sci 2008;29:170-4.

Figure 24.7. Sinclair D, Wainwright L. How lab staff and the estimation of free light chains can combine to aid the diagnosis of light chain disease. Clin Lab 2007;53:267-71.